Ergebnisse der Mathematik und ihrer Grenzgebiete

Volume 4

3. Folge

A Series of Modern Surveys in Mathematics

Springer-Verlag Berlin Heidelberg GmbH

Wolf P. Barth
Klaus Hulek
Chris A. M. Peters
Antonius Van de Ven

Compact Complex Surfaces

Second Enlarged Edition

Springer

Mathematics Subject Classification (2000):
14B05, 14C30, 14E05, 14E30, 14F17, 14J15, 14J17, 14J25, 14J26, 14J27, 14J28, 14J29, 32G05, 32G13, 32J15, 32Q60, 53D45, 57R57

ISBN 978-3-642-57738-3 **ISBN 978-3-642-57739-0 (eBook)**
DOI 10.1007/978-3-642-57739-0

Springer-Verlag is a part of Springer Science+Business Media
springeronline.com

Softcover reprint of the hardcover 2nd edition 2004

Typeset by the authors using a Springer TEX macro package.
Printed on acid-free paper 44/3142LK - 5 4 3 2 1 0

Preface to the Second Edition

In the 19 years which passed since the first edition was published, several important developments have taken place in the theory of surfaces. The most sensational one concerns the differentiable structure of surfaces. Twenty years ago very little was known about differentiable structures on 4-manifolds, but in the meantime Donaldson on the one hand and Seiberg and Witten on the other hand, have found, inspired by gauge theory, totally new invariants. Strikingly, together with the theory explained in this book these invariants yield a wealth of new results about the differentiable structure of algebraic surfaces.

Other developments include the systematic use of nef-divisors (in accordance with the progress made in the classification of higher dimensional algebraic varieties), a better understanding of Kähler structures on surfaces, and Reider's new approach to adjoint mappings.

All these developments have been incorporated in the present edition, though the Donaldson and Seiberg-Witten theory only by way of examples. Of course we use the opportunity to correct some minor mistakes, which we ether have discovered ourselves or which were communicated to us by careful readers to whom we are much obliged.

We gratefully acknowledge the support of various bodies which helped us prepare this new edition; in particular the following grants and institutions: EAGER (European Algebraic Geometry Research Network) and the DFG (Deutsche Forschungsgemeinschaft) as well as the universities of Essen, Grenoble, Hannover and Leiden for the hospitality we were offered at various occasions. Our thanks go to those who have read and commented on parts of the manuscript: R. Eckert, C. Erdenberger, M. Friedland, A. Gathmann, M. Lönne, K. Ludwig, John D. McCarthy, M. Schütt, J. Spandaw and H. Verrill. We are in particular grateful to J.-P. Demailly, L. Bonavero and A. Teleman for all the advice they offered which helped us to understand some of the hard analysis needed in various new parts of the book.

Special thanks also to Mme. A. Guttin-Lombard, who efficiently prepared a major part of the book, and to Mrs. S. Guttner for the careful typing of several chapters.

Grenoble/Erlangen/Hannover/Leiden, July 2003

W. Barth
K. Hulek
C. Peters
A. Van de Ven

Preface to the First Edition

> Par une belle matinée du mois de mai, une élégante amazone parcourait, sur une superbe jument alezane, les allées fleuries de Bois de Boulogne.
>
> (A. Camus, La Peste)

Early versions of parts of this work date back to the mid-sixties, when the third author started to write a book on surfaces. But for several reasons, in particular the appearance of Šafarevič's book, he postponed the projects. It was revived about ten years later, when all three authors were in Leiden. It is impossible to cover in one book the vast and rapidly developing theory of surfaces. Choices have to be made, with respect to content as well as to presentation. We have chosen for a complex-analytic point of view; this distinguishes our text from most of the existing treatments. Relations with the case of characteristic p are not discussed.

We hope to have succeeded in writing a readable book; a book that can be used by non-specialists. The specialist will find very little that is new to him anyhow.

As to acknowledgements, the authors certainly have to thank the Koninklijke Shellprijs, awarded to the third author in 1964. The numerous contacts with colleagues from other countries made possible by that award have had a very favourable influence on this book. Our thanks are furthermore due to G. Angermüller, G. Barthel, G. Fischer, G. van der Geer, N. Hitchin, D. Husemoller, M. Reid, T. A. Springer, D. Zagier and S. Zucker. Each of them has read some part of the manuscript and has made valuable suggestions.

Editor and printer have done an excellent job, and the Springer-Verlag has been very generous in fulfilling all of our last-minute wishes.

We are also indebted to Mrs. W. M. Van de Ven who not only typed the better part of the book, but also helped in preparing it for the printer, and to Mrs. H. Dohrman who carefully typed many pages. Finally the authors want to thank their wives for all their patience and endurance.

Erlangen/Leiden, February 1984

W. Barth
C. Peters
A. Van de Ven

Table of Contents

Introduction

Historical Note

This book is mainly concerned with the classification of smooth compact complex surfaces, i.e., of compact 2-dimensional complex manifolds, which in the introduction we shall always assume to be connected *).

Surface theory has its roots on the one hand in projective geometry and on the other hand in Riemann's theory of algebraic functions of a single variable. As to projective geometry, around the middle of the 19th century an extensive study was made of (smooth as well as singular) low-degree surfaces in complex-projective 3-space $\mathbb{P}_3$. The twenty-seven lines on a smooth cubic and names such as Cayley cubic and Steiner quartic remind us of that period. The extension of Riemann's work, in geometric form, will always be associated with mathematicians like Clebsch and M. Noether, whereas the topological and transcendental approach is linked to Poincaré and others, in particular Picard.

Soon attention was focused on a classification of algebraic surfaces with respect to birational equivalence. The classical geometers clearly had in mind something similar to what was known for curves: a coarse classification according to the value of some numerical invariants, and then a finer classification. At the beginning of the 20th century Castelnuovo, Enriques and many others had succeeded in creating an impressive, essentially geometric theory of birational classification of smooth algebraic surfaces. (This was in fact a birational classification of *all* algebraic surfaces, smooth or not, since every algebraic surface is birationally equivalent to a smooth one; but a rigorous proof of this fact was given for the first time by R. Walker in 1935.) Among the main birational invariants, it was discovered, are the irregularity or, equivalently, the first Betti number $b_1(X)$ and the plurigenera $P_n(X)$ of a smooth algebraic surface X. For any $n \geq 1$ the n-th plurigenus $P_n(V)$ of a smooth algebraic variety V is defined as the dimension of the space of sections $\Gamma(V, \mathcal{K}_V^{\otimes n})$, where $\mathcal{K}_V = \bigwedge^n \mathcal{T}_V^{\vee}$ is the canonical bundle of V. Traditionally the first plurigenus $P_1(V)$ is denoted by $p_g(V)$, and called the geometric genus of V.

*) From Chap. II on the meaning of the word "surface" in a given chapter is defined at the very beginning of that chapter. In Chap. I there is no danger of confusion.

Given any smooth algebraic surface X, there a four possibilities:
1) all P_n vanish;
2) not all P_n vanish, but are all either 0 or 1;
3) P_n grows linearly in n;
4) P_n grows quadratically in n.

Nowadays this fact is expressed by saying that the Kodaira dimension kod(X) of X is either $-\infty$, 0, 1, or 2. (For a precise definition of this concept, due to Iitaka, we refer to Chap. I, Sect.7.) For curves the corresponding classification is the division into the rational curve $\mathbb{P}_1$ (Kodaira dimension $-\infty$), elliptic curves (Kodaira dimension 0), and curves of higher genus (Kodaira dimension 1).

It was known at the time which surfaces are in class 1), namely those surfaces which are birationally equivalent to the product of $\mathbb{P}_1$ and another curve. This includes in particular the rational surfaces, i.e., those birationally equivalent to $\mathbb{P}_2$. A key stone of the proof was Castelnuovo's criterion: a smooth surface X is rational if and only if its first Betti number $b_1(X)$ and its bi-genus $P_2(X)$ vanish.

As to class 2), the classical geometers knew that there is a subdivision into four types, distinguished by the values of the first Betti number and the plurigenera, namely into surfaces, birationally equivalent to respectively algebraic tori, bi-elliptic surfaces, algebraic K 3-surfaces and Enriques surfaces (the names are the modern ones). The precise classification of the first two types was known, but not that of the last two types.

It had also been established that all surfaces in class 3) are elliptic (i.e., admitting a map onto a curve such that all but a finite number of fibres are elliptic curves); however, not much was known about their further classification.

Finally, the surfaces in class 4), which are analogous to curves of genus $\geq$ 2, were (and still are) called surfaces of general type. The classical geometers certainly had the right idea how to classify them, but – contrary, say, to the case of Castelnuovo's criterion or the case of bi-elliptic surfaces – they never arrived at precise results or even precise statements. Before we explain a little bit the present state of this classification, we first have to make a few remarks of a more general nature.

Today it is standard to look at the above classification of algebraic surfaces (the Enriques classification for algebraic surfaces) in a slightly different way. The basic idea: first a classification according to Kodaira dimension, and then a finer classification, remains the same, but it is seen as a *biregular* classification of *minimal* smooth algebraic surfaces, i.e., surfaces, which cannot be obtained from another *smooth* algebraic surface by blowing up a point. Every smooth surface X can be obtained from such a surface by successive blow-ups. At first sight it might seem that classifying only minimal surfaces is not very satisfactory, because one and the same surface X might be obtained by blowing up different minimal surfaces Y. However, always $\mathrm{kod}(X) = \mathrm{kod}(Y)$ and if $\mathrm{kod}(X) \geq 0$, then Y is determined by X up to an isomorphism. So even from the biregular point of view it is sufficient to classify minimal surfaces, at least in the case of non-negative Kodaira dimension. (If $\mathrm{kod}(X) = -\infty$, then different Y can give the same X, but this case is rather easy to handle.) Furthermore, a birational transformation between minimal surfaces of non-negative Kodaira dimension is always an isomorphism. In other words, for Kodaira dimension ≥ 0 birational classification of all surfaces amounts to biregular classification of minimal surfaces. And, most importantly, whereas from the birational point of view good moduli spaces never exist, they do exist for many of the finer classes in the case of minimal surfaces.

Now let us return to surfaces of general type. We consider minimal ones X with given Chern numbers $c_1^2(X) = p$, $c_2(X) = q$. It turns out that for $n \geq 5$, any n-canonical map (given by the ratios $\gamma_1 : \ldots : \gamma_{N+1}$, where $\gamma_1, \ldots, \gamma_{N+1}$ is a basis for $\Gamma(X, \mathcal{K}_X^{\otimes n})$) is everywhere defined on X and maps this surface birationally onto a surface X' of degree n^2p in $\mathbb{P}_N$ with N depending only on n, p and q. Choosing a different basis for $\Gamma(X, \mathcal{K}_X^{\otimes n})$ yields a surface which is projectively equivalent to X' in $\mathbb{P}_N$. In this way minimal surfaces of general type with fixed c_1^2 and c_2 correspond one-to-one to the points of the quotient of a Zariski-open subset in a Chow variety (or Hilbert scheme) by a projective-linear group. Of course, one wants this quotient to be a variety of moduli (coarse, at least) for the surfaces under consideration. A theorem of Gieseker (1977), based on geometric invariant theory (due to Hilbert and Mumford) says that for n large enough this is indeed the case.

Perhaps it should be mentioned at this point, that the results obtained by the classical geometers, their importance notwithstanding, were in many ways built on sand, for the foundations of algebraic geometry were lacking.

The years 1910–1950 did not bring too much change as far as the classification of surfaces is concerned. In the first two of these decades we see continuing great progress in the general theory of algebraic varieties, from the geometric point of view (Severi) as well as from the transcendental point of view (Lefschetz). But for both directions, a solid basis was still not available. Such a basis was laid in the thirties and forties, on the one hand for geometry by van der Waerden, Zariski, Weil and on the other hand for much of the topological and transcendental theory by de Rham and Hodge.

Once the foundations were present, some of the classification questions were taken up again and also considered for other ground fields: minimal models, Castelnuovo's criterion (Zariski), Enriques surfaces (M. Artin).

Decisive progress came only after the second revolution, i.e., after sheaf theory had been developed, and applied by Serre, Hirzebruch, Grothendieck and many others to analytic and algebraic geometry.

On this basis Kodaira not only extended the classical results on algebraic surfaces in an essential way, but also treated non-algebraic surfaces. For these surfaces the plurigenera and Kodaira dimension can be defined in the same way as for algebraic surfaces, and thus the Enriques classification is extended to the Enriques-Kodaira classification of all compact, complex surfaces.

As to the algebraic surfaces, it will hardly surprise anybody that Kodaira gave the Enriques classification the necessary precision and solid basis. But he went further in many directions. For example, he did the first step towards the classification of K 3-surfaces. A K 3-surface is a compact complex surface with $b_1(X) = 0$ and $\mathcal{K}_X$ trivial. As we have mentioned, the classification of the algebraic ones among them (which form a minority) was already an important problem in older times. Since the fifties they have been studied intensively, the main goal being to prove a conjecture, independently due to Andreotti and Weil about their classification (compare the comments at the end of Chap. VIII). Kodaira verified part of this conjecture, namely that all K 3-surfaces are complex-analytic deformations of each other. The deformation theory of complex manifolds, which Kodaira created together with Spencer, realized at least part of an old ideal of Riemann and Noether: to have a theory of moduli for curves and surfaces. Another contribution of Kodaira of far reaching significance and influence, was his extensive study of (algebraic and non-algebraic) elliptic surfaces, something that had definitely been lacking in the work of the Italian geometers.

Though the concept of an n-dimensional complex manifold had been known implicitly for a long time (certainly since Weyl's *Die Idee der Riemannschen Fläche*), it appeared explicitly only around 1945, in the work of Ehresmann and H. Hopf. In particular Hopf constructed an entirely new class of compact complex surfaces (the first example of what now are called the Hopf surfaces) which are topologically very different from any algebraic surface. A Hopf surface has first Betti number 1 and second Betti number 0, whereas a smooth algebraic surface always has an even first Betti number and a strictly positive second Betti number. So, contrary to tori and K 3-surfaces, a Hopf surface can never be deformed into an algebraic one.

This example shows that there is much more to non-algebraic surfaces than deforming some algebraic ones. It was again Kodaira who started with the classification of non-algebraic surfaces in general, and he completed this task to a considerable degree. In his papers he uses Atiyah-Singer's Riemann-Roch theorem in an essential way.

As to non-algebraic deformations of algebraic surfaces, their significance for a better understanding of algebraic surfaces arises clearly from the Andreotti-Weil conjecture (mentioned before) and Kodaira's work. This point is already obvious from the case of tori, but it gains more weight if K 3-surfaces and elliptic surfaces are taken into consideration.

It would be wrong to think that with Kodaira the theory of surfaces more or less came to its end. On the contrary, the interest in surfaces has only been increasing since the days that Kodaira produced most of his results in the late 1950s and 1960s.

One of the main centres of interest has already been mentioned: the conjecture of Andreotti and Weil on the classification of K 3-surfaces. After important contributions by many mathematicians, in particular Šafarevich and Piateckii-Shapiro, the most important parts of the conjecture could finally be proved, but only with the help of S.-T. Yau's deep differential-geometric results on the Calabi conjecture.

Another centre of attention was the classification of (minimal) surfaces of general type. We spoke already about Gieseker's theorem, saying that for each ordered pair (p, q) of integers there is a (possibly empty) coarse moduli scheme parametrising minimal surfaces of general type X with $c_1^2(X) = p$, $c_2(X) = q$. The next question is of course: when is this scheme non-empty? In this direction an important result was obtained in 1976 by S.-T. Yau and Miyaoka, who independently proved an older conjecture of Van de Ven, saying that for every surface X of general type the inequality $c_1^2(X) \leq 3c_2(X)$ holds. Yau obtained this inequality as a consequence of his famous work on the Calabi conjecture. Miyaoka was very much inspired by Bogomolov, who only proved the weaker inequality $c_1^2(X) \leq 4c_2(X)$, but linked the question in an exciting way to the theory of stable vector bundles. Establishing these inequalities is the first step towards the "geography" of surfaces of general type (in addition to the existence problem this asks the question how special values of the numerical invariants influence the geometry of the surface, such as the existence of special fibrations). In this area Horikawa and Persson were the pioneers, but despite several extensions of their work, it is still not known whether every pair of integers allowed by the preceding inequalities can actually be realized by a minimal surface of general type.

Much research was also done on surfaces which are special, mostly for low values of c_1^2. For example, already many years ago Severi had raised the question, whether there exist surfaces which are homeomorphic, but not algebraically isomorphic to $\mathbb{P}_2$. It took a long time before the final answer was given by S.-T. Yau who proved that these do not exist. However, as was shown by Mumford using p-adic geometry, there does exist at least one "fake projective plane", i.e., a surface different from $\mathbb{P}_2$ but with the same Betti numbers. In spite of many efforts, a direct geometric construction (within the framework of complex algebraic geometry) of such a fake projective plane is still lacking.

As a last example of a classification problem that saw much progress in the years 1970–1980 we mention the case of (minimal) surface without non-constant meromorphic functions. Kodaira had already classified these surfaces, except for surfaces with $b_1 = 1$, $b_2 = 0$ on which there are no curves, and surfaces with $b_1 = 1$, $b_2 \geq 1$. For years no example of either class was known. In 1971 Inoue found some examples of surfaces in the first class, and three years later he showed that also the second one is not empty. Since then Inoue, Bombieri, Kato and Enoki have produced many of these surfaces and have started the classification. This is an active area of research; over the last two decades Bogomolov, Dloussky and several other people have made substantial progress. We refer to the historical remarks at the end of Sect. V.20.

We could go on in this way, but we only wanted to indicate the progress made possible by the introduction of sheaf-theoretic methods and the use of new results in other fields.

As to important developments since the appearance of the first edition of this book (in 1984), we first of all want to mention the spectacular developments concerning the (differential) topology of compact complex surfaces. This started around 1985 with Donaldson's example of two algebraic surfaces (one elliptic, the other rational) which are homeomorphic but not diffeomorphic (the fact that the two surfaces are homeomorphic follows from deep results on the topology of compact 4-manifolds which Freedman had obtained just a few years before). Subsequently Donaldson introduced differentiable invariants, now called the Donaldson polynomials, which can be calculated by algebro-geometric means. The use of these invariants enabled Freedman and Qin to prove the "Van de Ven conjecture": the Kodaira dimension is a differentiable invariant. Differential topology in dimension 4 underwent a second revolution through the work of Seiberg and Witten who produced a new set of invariants whose calculation requires much less algebraic geometry and with which one could even prove more than the Van de Ven conjecture, namely that the plurigenera themselves are differentiable invariants.

A second important development is "Reider's method" for dealing with pluricanonical maps. It simplifies and extends Bombieri's treatment which was based on connectedness properties of pluricanonical divisors. For surfaces of general type much work has also been done on the geography as we already mentioned. But also many moduli spaces have been studied in great detail, mainly by Catanese and his students. This work shows how complicated the behaviour of such moduli spaces can be; for instance the number of components although finite, can be arbitrarily large.

Thirdly we want to mention that recently a direct proof has been found of the fact that a surface with even first Betti number is kählerian. This is due to Lamari and Buchdahl (independent of each other) who use Demailly's deep results on the regularisation of positive currents.

There were, of course, many other developments in the theory of surfaces in the last two decades which we are not able to discuss in this book, such as results obtained by projective methods, the possible number of double points of surfaces in $\mathbb{P}_3$ and their configurations, and the classification of smooth surfaces in $\mathbb{P}_4$ of low degree. For surfaces with many double points, see for instance [Bar] and the references cited there, and [Chm]. As to smooth surfaces in $\mathbb{P}_4$, there exists now, due to the combined effort of many authors, a fairly complete classification up to and including degree 10 with further partial results up to degree 15 (see [D-E-S], [D-S] and the references given there). In particular we know by a result of Ellingsrud and Peskine [E-P] that the degree of any smooth surface in $\mathbb{P}_4$ which is not of general type is bounded. Schreyer conjectures this bound to be 15. It cannot be smaller, since there exist (non-minimal) smooth abelian surfaces and bi-elliptic surfaces of degree 15 in $\mathbb{P}_4$. Braun and Fløysted [B-Fl] proved that this bound is smaller than or equal to 105. This was brought down to 76 by Cook [Ck]. There are abelian surfaces of degree 10 in $\mathbb{P}_4$. These have attracted special attention since they give rise to the Horrocks-Mumford bundle, so far still essentially the only known indecomposable rank 2 bundle on $\mathbb{P}_4$. For a survey article on the rich geometry associated to this bundle see [Hu].

To finish, we mention two developments which do not belong to our subject, but are closely related to it. First, the extension of the Enriques classification to characteristic p by Bombieri and Mumford (some results having been obtained previously by Zariski). In characteristics $\neq 2, 3$ the classification is identical to the complex-algebraic case, but in characteristics 2 and 3 certain 'non-classical' surfaces appear. About the finer classification much less is known than in the complex case, but Cossec and Dolgachev extended many results concerning Enriques surfaces to all characteristics. The structure of the pluricanonical map for surfaces of general type has been studied by Ekedahl and Shepherd-Barron. The results are roughly the same as for the complex-algebraic case.

Secondly, we mention the development by Iitaka, Kawamata, Kollár, Miyaoka, Mori, Reid, Ueno, Viehweg and others of a classification theory for higher dimensional manifolds. Since in this case no unique minimal models exist, it became essential to allow certain singularities and also birational maps such as "flips" and "flops" which are more complicated than blow-ups. The starting point is again a classification according to Kodaira dimension and – at least in dimension 3 – already much is known about the finer division. Here a central role is played by Mori's theorems on the structure of the cone of the so-called nef-divisors (nef is an abbreviation coined by Reid and stands for "numerically eventually free"). A nef-divisor by definition has the property that it has non-negative intersection product with every curve. This concept has indeed become central in modern algebraic geometry. Another major role is played by Iitaka's conjecture $C_{n,m}$: if X and Y are smooth, compact irreducible algebraic varieties, of dimension m and n respectively, and if $f : X \to Y$ is a surjective morphism, then $\mathrm{kod}(X) \geq \mathrm{kod}(Y) + \mathrm{kod}(F)$, where F is a general fibre of f. The conjecture has been proved for several values of m and n, in particular for $m = 2$, $n = 1$ where it follows from the Enriques-Kodaira classification. See also Chap. III where a direct proof is presented.

Some references

Classical results in general: [C-E].

Classical theory of the Enriques classification: [Enr14], [Enr49], [Ge].

Desingularization: [Za71], [Li].

Zariski's work on minimal models and the Castelnuovo criterion: [Za58a], [Za58b].

M. Artin's work on Enriques surfaces can be found in Artin's unpublished thesis [An60]

Enriques classification in characteristic p and subsequent developments: [Mu69], [B-M76], [B-M76], [Co-D], [Ek], [ShB91a].

Classification theory for higher dimensional varieties: [Ue75], [Ue80], [Es], [Cl-K-M], [Vie95].
The other subjects mentioned are treated further on in this book.

The Contents of the Book

As has been explained in the preceding section, the classification of compact, complex surfaces amounts to the classification of minimal surfaces. This is first of all a classification according to Kodaira dimension, which for a surface can assume the values $-\infty$, 0, 1 and 2. A refinement of this very coarse classification is the Enriques-Kodaira classification, a description of which is a first purpose of the book. Some of the classes occurring in the Enriques-Kodaira classification can easily be described in detail, but the others: minimal surfaces of class VII (i.e., minimal surfaces X with $b_1(X) = 0$, $\mathrm{kod}(X) = -\infty$*)), K 3-surfaces, Enriques surfaces, minimal properly elliptic surfaces and minimal surfaces of general type, require further investigation. Apart from the Enriques-Kodaira classification, this book is mainly devoted to a deeper study of some of these classes, namely K 3-surfaces, Enriques surfaces and surfaces of general type. On the other hand, surfaces of class VII and properly elliptic surfaces will not be treated in detail. For elliptic surfaces a number of general properties as well as a classification can be found in Chap. V. We cover only a small part of what is known in this direction, in particular we do not partition elliptic surfaces in families. For this the reader may consult the recent book [F-M94] and the references given there. Surfaces of class VII occur only by way of examples, and neither the beautiful considerations of Kodaira on Hopf surfaces nor most of the work of Enoki, Dloussky, Inoue and Kato on surfaces without non-meromorphic functions can be found in this book.

It goes without saying that in a book like the present one many auxiliary results can only be quoted. As to general theorems on complex and algebraic manifolds or spaces, it has been a difficult question for us to decide whether (special) proofs for the 2-dimensional case should be included or not. Sometimes, when a more elementary treatment is available for the 2-dimensional case, we have explained this in detail. For example, we do not refer to Hironaka for the resolution of surface singularities. The 2-dimensional case is infinitely much simpler and its direct treatment is very rewarding. But at other places we have used a general theorem in spite of the fact that for surfaces an elementary approach exists. For example, in Chapter IV we derive the fundamental projectivity criterion (Theorem IV.6.2) using Grauert's general ampleness theorem, though it would have been possible to avoid this by using a method of Chow and Kodaira. The method we use is shorter, whereas

*) Our definition of a surface of class VII is slightly different from Kodaira's, compare Chap. VI, Sect. 1.

in the elementary proof there is no idea that doesn't already occur elsewhere in the book.

We shall be brief concerning the contents of the different chapters. For more information the reader is advised to consult the introductions that precede each subchapter.

In Chapter I we collect – practically without proofs – most of the definitions and results from topology, algebra, differential geometry, analytic geometry and algebraic geometry which we shall need. Since we would like to make our book as useful as possible for non-specialists, we have thought it better to deviate a little bit from the logical order by collecting some fundamental, but more technical results on complex spaces only after we have dealt with manifolds. This concerns tools like the semi-continuity theorem, the base change theorem and the comparison theorem of Grauert.

Chapter II is devoted to (possibly non-reduced) curves on (not necessarily compact) surfaces and forms a central starting point for the rest of the book. Singularities and their resolutions are treated, the Riemann-Roch theorem is reduced to the smooth case and Serre duality is derived from the reduced projective case. Analytic intersection numbers are defined for divisors and are shown to be the same as the topological ones. In the last section the foundations are laid for Ramanujam's vanishing theorem which will be proved in Chapter IV.

The first part of Chap. III deals with surface singularities, their resolution and the converse of this process, the blowing down of exceptional curves. The results are applied to study bimeromorphic maps and minimal models. The second part of Chap. III is devoted to (proper) curve fibrations over curves and culminates in the proof of Iitaka's conjecture $C_{2,1}$ about the Kodaira dimension of such fibrations. We base it on properties of the period map for stable curves, the Satake compactification and the Torelli theorem for curves.

Chapter IV is not very homogeneous, but contains a number of general results on surfaces which will be needed in subsequent chapters, such as the relations between topological and analytic invariants of compact complex surfaces, the study of the Néron-Severi group and projectivity criteria. The treatment of the invariants is based on the fact that for a compact complex surface the Fröhlicher spectral sequence always degenerates. Combining the consequences of this fact with the topological index theorem we find, following Kodaira, relations between topological and analytical invariants which are crucial in handling non-algebraic surfaces. This also enters in a decisive way in the proof (following Lamari) that a surface with even first Betti number always has a Kähler metric. We also prove the important signature theorem (known as "the algebraic index theorem" for projective surfaces). Crucial for our treatment of the pluricanonical maps in Chap. VII is Reider's theorem which can be found in this chapter. Prior to it we discuss properties of Bogomolov stability for rank 2 vector bundles. From the other subjects treated in this chapter, we mention projectivity criteria (with an application to almost-complex surfaces without any complex structure) and the vanishing theorems of Ramanujam and Mumford.

As to Chap. V (Examples), we first of all have included this chapter as a preparation for the next one, where many examples occur as classes in the Enriques-Kodaira classification. Secondly we have thought that the inclusion of such a list might again be helpful to non-specialists.

In Chap. VI we present the Enriques-Kodaira classification. Our treatment is based on the "rationality theorem" (proved in this chapter) and the fundamental results of Chap. IV. We have chosen this approach, because Mori's approach to the classification of higher dimensional varieties is similar in spirit. For instance, we should mention Mori's detailed study of the cone of nef-divisors through the rationality theorem (which is also valid for higher dimensional projective manifolds). In the previous edition of the book we based the classification on Iitaka's conjecture $C_{2,1}$, whereas we entirely avoid it in the present approach. We are aware of the possibility to prove $C_{2,1}$ as a consequence of the classification. Nevertheless we chose to keep the direct proof of this result given in the first edition, since - together with Mori-theory - this is the only general approach to classification in higher dimensions. At the end of this chapter we discuss the relationship between deformation and classification of surfaces.

Chapter VII is about surfaces of general type. In this edition we base our treatment of pluricanonical maps on Reider's theorem and no longer on Bombieri's method using connectedness of pluricanonical divisors. Since we do not know of any simple proof for Gieseker's theorem, we have chosen to save space by referring to the original paper. In dealing with inequalities for Chern numbers, we present a simplified version of Miyaoka's proof for the inequality $c_1^2 \leq 3c_2$, as well as the more standard proofs for Noether's inequality and the inequalities $c_1^2 > 0$, $c_2 > 0$ for minimal surfaces of general type. In the last part of this chapter we describe various methods of constructing surfaces and we list the results about Chern numbers and Gieseker schemes which these methods have yielded.

Chapter VIII deals with K 3-surfaces and Enriques surfaces. We fully prove some of the main results which have been obtained during the 1970s and the 1980s: the Torelli theorem for marked K 3-surfaces, the surjectivity of the period map for K 3-surfaces, and the bijectivity of the period map for Enriques surfaces. The proof that every K 3-surface is kählerian (based on Yau's results) which we gave in the first edition of the book has now become obsolete since we have presented a direct proof in Chap. IV (based on Demailly's regularity theorem for positive currents) of the fact that a surface with even first Betti number is kählerian. We have added a subchapter with applications of K 3-surfaces which have turned out to be of wider interest: mirror-symmetry, existence of rational and elliptic curves (and the spectacular count of the former inspired by mathematical physics), and, finally, we have given an application to hyperbolic geometry.

In Chapter IX we collect some fundamental results concerning the (differential) topology of complex surfaces. We review Freedman's topological classification of 4-manifolds and we consider which unimodular forms can be represented by simply-connected compact complex surfaces. After this we look at the Donaldson and Seiberg-Witten invariants respectively. We do not explain these fully. Rather, we have chosen two striking results about the differentiable topology of compact complex surfaces whose proofs follow from the theory developed in the rest of this book assuming only a minimal set of axiomatic properties of the Donaldson and Seiberg-Witten invariants.

Standard Notation

$\mathbb{Z}$:	ring of integers
$\mathbb{Z}/m\mathbb{Z}$:	ring of integers mod m
$\mathbb{F}_q$:	field with q elements
$\mathbb{N}$:	set of strictly positive integers
$\mathbb{Q}$:	field of rational numbers
$\mathbb{R}$:	field of real numbers
$\mathbb{R}^n$:	numerical real vector space of dimension n
$\mathbb{1}_n$:	unit matrix of size n by n
S^n:	unit sphere in $\mathbb{R}^{n+1}$
$\mathbb{C}$:	field of complex numbers
$\mathbb{C}^n$:	numerical complex vector space of dimension n
$\mathbb{C}^*$:	multiplicative group of nonzero complex numbers
$\mathbb{P}(V)$:	projective space of the (real or complex) vector space V, i.e., the space of lines in V
$\mathbb{P}_n = \mathbb{P}(\mathbb{C}^{n+1})$:	n-dimensional numerical complex projective space
$\mathcal{O}(1) = \mathcal{O}_{\mathbb{P}^n}(1)$:	the hyperplane bundle on $\mathbb{P}_n$
i:	the square root of -1 with positive imaginary part
$\mathbf{e}(z)$:	$= e^{2\pi i z}$

Chapter I. Preliminaries

Topology and Algebra

In this sub-chapter we collect some basic facts from topology and algebra. Among the less elementary facts needed later on we mention the topological index theorem recalled in §3. The (elementary) facts concerning quadratic forms will be used mainly in the chapter on K 3- and Enriques-surfaces.

1. Notation and Basic Facts

We shall use the standard notations for some algebraic structures which frequently occur. These notations are listed on p. 12.

Let X be a topological space and $\mathcal{S}$ a sheaf of groups on X. We write $\mathcal{S}_x$ for the stalk of $\mathcal{S}$ at $x \in X$ and we shall denote by $H^i(X, \mathcal{S})$ or $H^i(\mathcal{S})$ the i-th cohomology group of X with coefficients in $\mathcal{S}$. For $H^0(X, \mathcal{S})$, the group of sections, we shall also write $\Gamma(X, \mathcal{S})$ or $\Gamma(\mathcal{S})$. If $\mathcal{S}$ is a sheaf of real or complex vector spaces, then $H^i(X, \mathcal{S})$ is also a real or complex vector space. If $\dim H^i(X, \mathcal{S})$ is finite, then $h^i(X, \mathcal{S}) = h^i(\mathcal{S})$ will denote this dimension. If G is an abelian group and G_X the corresponding constant sheaf on X, then we shall write $H^i(X, G)$ for $H^i(X, G_X)$, and $H^i_c(X, G)$ for the i-th cohomology group of G_X with compact supports.

If G is a ring, then $H^*(X, G) = \sum_{i \geq 0} H^i(X, G)$ is made into a graded ring by the cup product. If G has a unit element, then so has $H^*(X, G)$.

We shall often denote the cup product $a \cup b$ of two classes a and b by ab, or $a \cdot b$ or even (a, b). We do this mostly in the case that a and b are of complementary dimension on a connected, compact, oriented manifold, so $ab \in G$.

If X and Y are topological spaces, $\mathcal{S}$ a sheaf of groups on X, and $f : X \to Y$ a continuous map, then we shall write $f_{*i}(\mathcal{S})$ or simply $f_{*i}\mathcal{S}$ for the i-th direct image of $\mathcal{S}$ by f, but mostly just $f_*(\mathcal{S})$ or the 0-th direct image $f_*\mathcal{S}$ for the 0-th direct image. On the other hand, if $\mathcal{T}$ is a sheaf on Y, then $f^{-1}(\mathcal{T})$ will be the inverse image on X (the notation f^* will be reserved for the analytic inverse image, see Sect. 8).

For every continuous map $f : X \to Y$ and every sheaf $\mathcal{S}$ on X there is the Leray spectral sequence ([Go], Chapitre II, Théorème 17.1) with $E_2^{p,q} =$

$H^p(f_{*q}(\mathcal{S})) \Rightarrow H^{p+q}(\mathcal{S})$. The beginning of this spectral sequence leads to the exact sequence

$$0 \to H^1(f_*(\mathcal{S})) \to H^1(\mathcal{S}) \to H^0(f_{*1}(\mathcal{S})) \to H^2(f_*(\mathcal{S})).$$

When we speak of a (mostly complex) manifold, we shall always assume it to be paracompact.

If X is an oriented, connected n-dimensional manifold, then Poincaré duality yields a canonical isomorphism

$$\mathcal{P}_X : H^i_c(X, G) \xrightarrow{\sim} H_{n-i}(X, G)$$

(in particular, $H^n_c(X, G) \simeq H_0(X, G) \simeq G$). This duality induces a product on $H_*(X, G)$, which for $G = \mathbb{Z}$ is nothing but the intersection product ([Dd], p. 336). We shall frequently switch between cohomology with compact supports and homology in this case, without further notice. An element of $H^n_c(X, \mathbb{Z})$ or $H_0(X, \mathbb{Z})$ will be seen as an integer. If $j : Z \hookrightarrow X$ is a compact submanifold of dimension m, the generator $j_*(\mathcal{P}_Z(1)) \in H_m(X, \mathbb{Z})$ is called the fundamental (homology) class, while its Poincaré-dual

$$[Z] = \mathcal{P}_X^{-1} j_* \mathcal{P}_Z(1)) \in H^{n-m}_c(X, \mathbb{Z})$$

is called the fundamental class in cohomology.

If Y is a second connected, oriented manifold of dimension m and $f : X \to Y$ a continuous map, then we shall denote by

$$f^! : H_i(Y, \mathbb{Z}) \to H_{n-m+i}(X, \mathbb{Z})$$

the homomorphism $\mathcal{P}_X f^* \mathcal{P}_Y^{-1}$, whereas the homomorphism $\mathcal{P}_Y^{-1} f_* \mathcal{P}_X$ will be denoted by $f_!$.

(1.1) **Lemma** (Projection formula). *Let X and Y be connected, oriented manifolds, and $f : X \to Y$ a proper continuous map. Then*

$$f_!(x \cdot f^*(y)) = f_!(x) \cdot y \tag{1}$$

for all $x \in H^*_c(X, \mathbb{Z}), y \in H^*_c(Y, \mathbb{Z})$.

For a proof we refer to [Dd], p. 314.

(1.2) **Corollary.** *Let X and Y be compact, connected oriented manifolds of the same dimension. If $f : X \to Y$ is a continuous map of degree different from 0, then $f^* : H^*(Y, \mathbb{R}) \to H^*(X, \mathbb{R})$ is an injective ring homomorphism.*

(1.3) *Remark. If in particular X and Y are complex manifolds and f a surjective holomorphic map, then the degree* $\deg(f)$ *of f is strictly positive and the conclusion of the Corollary applies.*

For certain cases, which play an important role in this book, we shall consider in Chap. II both intersection theory and Poincaré duality more in detail.

When X is a compact oriented n-dimensional manifold, we write $T^i(X)$ for the torsion subgroup of $H^i(X,\mathbb{Z})$ and put $H^i(X,\mathbb{Z})_f = H^i(X,\mathbb{Z})/T^i(X)$. Then we have

$$\operatorname{rank}(H^i(X,\mathbb{Z})_f) = \dim_{\mathbb{R}} H^i(X,\mathbb{R}) = \dim_{\mathbb{C}} H^i(X,\mathbb{C}) = b_i(X),$$

the i-th Betti number of X. Duality implies that $b_i(X) = b_{n-i}(X)$ and $T^i(X) \simeq T^{n-i+1}(X)$ (see [Dd], p. 167).

As for the i-th homotopy group of an arc-wise connected space X, we use the standard notation $\pi_i(X)$ (or $\pi_i(X,p)$ if the base point $p \in X$ plays a role), and in particular we write $\pi_1(X)$ for the fundamental group of X.

2. Some Properties of Bilinear Forms

By a lattice $(L, \langle\ ,\ \rangle)$ we shall mean a finitely generated free $\mathbb{Z}$-module L, endowed with an integral bilinear form $\langle\ ,\ \rangle$ which is either symmetric or skew-symmetric. In the first case we speak of a euclidean lattice, in the second case of a symplectic lattice. Frequently, when confusion is unlikely, we speak of the lattice L instead of $(L, \langle\ ,\ \rangle)$.

If $\{e_1, \ldots, e_n\}$ is a basis for L, then it is easily seen that the determinant of the matrix $(\langle e_i, e_j\rangle)$ is determined uniquely, independent of the choice of basis. This number $d(L)$ is called the discriminant of the lattice. The lattice $(L, \langle\ ,\ \rangle)$ is non degenerate if $\langle\ ,\ \rangle$ is non degenerate, i.e., $d(L) \neq 0$, and $(L, \langle\ ,\ \rangle)$ is unimodular if $d(L) = \pm 1$.

Let $(L, \langle\ ,\ \rangle)$ be a lattice and $L^\vee = \operatorname{Hom}_{\mathbb{Z}}(L,\mathbb{Z})$ be the dual of L. The correlation morphism $\phi : L \to L^\vee$ is defined by

$$\phi(x) = \langle\ , x\rangle.$$

(2.1) **Lemma.** *If $(L, \langle\ ,\ \rangle)$ is a non-degenerate lattice, then*
(i) *the index of $\phi(L)$ in $L^\vee$ is $|d(L)|$,*
(ii) *if M is a submodule of L with* $\operatorname{rank}(M) = \operatorname{rank}(L)$, *then*

$$(L : M)^2 = d(M)d(L)^{-1}.$$

The second assertion follows immediately if you write a basis for M in terms of a basis for L and observe that the determinant of the resulting matrix up to sign is $(L : M)$. Applying the same remark to L and $L^\vee$ yields (i).

The next lemma is a simple exercise in linear algebra over $\mathbb{Q}$.

(2.2) **Lemma.** *If L is a non-degenerate lattice, and M a submodule of L, then*

$$\operatorname{rank} M + \operatorname{rank} M^{\perp} = \operatorname{rank} L. \qquad \square$$

A symplectic form can be non-degenerate only if the rank n of L is even.

(2.3) **Lemma.** *If $(L, \langle\ ,\ \rangle)$ is a non-degenerate, symplectic lattice of rank $2g$, equipped with a non-degenerate symplectic form, then there exists a basis $\{a_1, \dots, a_g, b_1, \dots, b_g\}$ for L and natural numbers $\{d_1, \dots, d_g\}$ with $d_i | d_{i+1} (i = 1, \dots, g-1)$ such that the matrix of $\langle\ ,\ \rangle$ has the form*

$$\begin{pmatrix} 0 & -\Delta \\ \Delta & 0 \end{pmatrix} \quad \textit{with} \quad \Delta = \begin{pmatrix} d_1 & & & \\ & d_2 & & \\ & & \ddots & \\ & & & d_g \end{pmatrix}.$$

For a proof we refer to [Bou59], Chap. 9, 5.1. A basis $\{a_1, \dots, b_g\}$ as in the previous lemma is called canonical. For a unimodular symplectic form all d_i's are necessarily 1 and the matrix in Lemma 2.3 is:

$$\begin{pmatrix} 0 & -1\!\!1_g \\ 1\!\!1_g & 0 \end{pmatrix} \quad \text{(the standard symplectic form).}$$

The submodules of L spanned by $a_1, \dots, a_g$ and by $b_1, \dots, b_g$ are maximal isotropic submodules of L.

A sublattice M of a lattice L is called primitive, if M is a primitive submodule of L, i.e., if L/M is torsion free. If M is primitive, every basis of M can be complemented to a basis of L. In particular this holds if M is a maximal isotropic sublattice of L, since $M = M^{\perp}$ implies that M is primitive. As an application we have:

(2.4) **Lemma.** *If $(L, \langle\ ,\ \rangle)$ is a unimodular, symplectic lattice and $\{a_1, \dots, a_g\}$ a basis for a maximal isotropic subspace of L, then one can always complement $\{a_1, \dots, a_g\}$ to a canonical basis.*

Proof. Since $\langle\ ,\ \rangle$ is unimodular, the correlation morphism is an isomorphism. So, if we complement $\{a_1, \dots, a_g\}$ to a basis of L, every element of the corresponding dual basis for $L^{\vee}$ can be viewed as an element of L. In particular, there are elements $b'_1, \dots, b'_g \in L$ such that $\langle a_i, b'_j \rangle = \delta_{i,j}$. If $\langle b'_i, b'_j \rangle = t_{i,j}$, we set $b_j = b'_j - \sum_{i=1}^{j-1} t_{i,j} a_i$ $(j \geq 2)$ and it is easily checked that $\{a_1, \dots, b_g\}$ yields a canonical basis. $\square$

As we have just remarked, if $M \subset L$ is primitive, every basis of M can be extended to a basis of L and in particular $L^{\vee} \to M^{\vee}$ is surjective. So we can extend every $m^* \in M^{\vee}$ to all of L and restrict the result to $M^{\perp}$, thus obtaining an element of $(M^{\perp})^{\vee}$. If we take different extensions of m^* to L, the difference of two extensions is an element of $L^\vee$, which vanishes identically on M. If in addition L is unimodular, every element of $L^\vee$ is of the form $\langle -, x\rangle$ for some $x \in L$ and we see that the difference of two extensions is of the form $\langle -, y\rangle$ for $y \in M^\perp$. If $\langle\ ,\ \rangle|M$ is non-degenerate then so is $\langle\ ,\ \rangle|M^\perp$ and the correlation morphisms are embeddings. Moreover, we can consider M and $M^\perp$ as submodules of their respective dual modules. In this way we obtain a homomorphism $M^\vee \to (M^\perp)^\vee/M^\perp$, and since it maps M to 0, we get a homomorphism

$$\psi : M^\vee/M \to (M^\perp)^\vee/M^\perp \tag{2}$$

We have:

(2.5) **Lemma.** *If L is a unimodular lattice and M a primitive sublattice such that $\langle\ ,\ \rangle|M$ is non-degenerate, then the homomorphism (2) is an isomorphism.*

Proof. Like any orthogonal complement $M^\perp$ is primitive, and since M is primitive, we have $(M^\perp)^\perp = M$. So we can construct a homomorphism $(M^\perp)^\vee/M^\perp \to M^\vee/M$, which is an inverse for (2). □

(2.6) **Corollary.** *Let L be a unimodular lattice and $M \subset L$ a primitive sublattice, then $|d(M)| = |d(M^\perp)|$. If moreover M is unimodular, then $L = M \oplus M^\perp$.*

Proof. If $d(M) = 0$, clearly $d(M^\perp) = 0$, so we may assume that M is non-degenerate and in that case the previous lemma applies. Since $(M^\vee : M) = |d(M)|$ by Lemma 1.1.4, the first statement follows. The second one is an immediate consequence: if $|d(M)| = |d(M^\perp)| = 1$, then $|d(M \oplus M^\perp)| = 1$ and $M \oplus M^\perp = L$. □

From now on we only consider euclidean lattices $(L, \langle\ ,\ \rangle)$. Associated to the lattice L we have a quadratic form Q defined by $Q(x) = \langle x, x\rangle$. If it takes on even values only, the lattice is called even, otherwise it is called odd. If $Q(x) > 0$ (respectively ≥ 0) for all $x \in L \setminus \{0\}$ it is called positive definite (respectively positive semi-definite) and similar for negative (semi-)definite. A lattice is definite if it is either positive or negative definite. A non-degenerate lattice is called indefinite if it is not definite.

(2.7) **Examples.**

(i) The module $\mathbb{Z}\cdot e$ with $\langle e, e\rangle = \pm 1$ is an odd, unimodular definite lattice. We denote it by $\pm\mathbb{1}$.

(ii) The hyperbolic plane H. As a $\mathbb{Z}$-module it is $\mathbb{Z}^2$ and if e_1, e_2 is the standard basis, the matrix $(\langle e_i, e_j\rangle)$ is just $\begin{pmatrix} 0 & 1 \\ 1 & 0 \end{pmatrix}$. It is even, unimodular and indefinite.

(iii) The root-lattice E_8. As a $\mathbb{Z}$-module, $E_8 = \mathbb{Z}^8$ and on the canonical basis the matrix $(\langle e_i, e_j\rangle)$ is the Cartan matrix of the root system E_8, that is, $2Q(E_8)$ in the notation of Lemma 2.12, or explicitly:

$$E_8 = \begin{pmatrix} 2 & 0 & -1 & 0 & 0 & 0 & 0 & 0 \\ 0 & 2 & 0 & -1 & 0 & 0 & 0 & 0 \\ -1 & 0 & 2 & -1 & 0 & 0 & 0 & 0 \\ 0 & -1 & -1 & 2 & -1 & 0 & 0 & 0 \\ 0 & 0 & 0 & -1 & 2 & -1 & 0 & 0 \\ 0 & 0 & 0 & 0 & -1 & 2 & -1 & 0 \\ 0 & 0 & 0 & 0 & 0 & -1 & 2 & -1 \\ 0 & 0 & 0 & 0 & 0 & 0 & -1 & 2 \end{pmatrix}.$$

The lattice E_8 is even, unimodular and positive definite. Changing all signs yields $-E_8$, a negative definite lattice.

We recall that any quadratic form can be diagonalized over $\mathbb{R}$, that the numbers τ^+, respectively τ^-, of positive, respectively negative entries in every diagonalization over $\mathbb{R}$ are the same, that $\tau^+ + \tau^-$ is called the rank, and $\tau^+ - \tau^-$ the index. Examples of odd unimodular lattices of rank n and index τ are the orthogonal direct sums $\overset{a}{\bigoplus} \mathbb{1} \overset{b}{\bigoplus} -\mathbb{1}$, where $a = \frac{1}{2}(n+\tau)$, $b = \frac{1}{2}(n-\tau)$. Examples of even unimodular lattices of rank n and index τ ($\tau \equiv 0 \bmod 8$ by [M-H] Chap. II, Th. 5.1) are the orthogonal direct sums $\overset{a}{\bigoplus} H \overset{b}{\bigoplus} \pm E_8$, where $a = \frac{1}{2}(n-\tau)$, $b = \pm\tau/8$. The following theorem whose proof can be found e.g. in [M-H], Chap. II (Theor. 5.3) or [Se73], Chap. V (§2.2–2.3), states that these examples give all possible indefinite unimodular lattices and all definite ones of low rank.

(2.8) **Theorem.** *Any indefinite unimodular lattice is up to isometry determined by its rank, index and parity (*i.e., *whether it is odd or even). The same holds for definite unimodular lattices of rank at most 8.*

If an even unimodular lattice L contains k hyperbolic planes, any even lattice of rank $\leq k$ can be realized as a primitive sublattice of L. In fact, we have:

(2.9) **Theorem.** *Let L be an even unimodular lattice containing a sublattice isometric to $\overset{k}{\bigoplus} H$ and let Γ be any even lattice.*

(i) *If* $\operatorname{rank}\Gamma \leq k$, *then there exists a primitive embedding* $i : \Gamma \to L$, i.e., i *is a lattice monomorphism and* $i(\Gamma)$ *is primitive.*

(ii) *If* $\operatorname{rank}\Gamma \leq k-1$, *and* $i, j : \Gamma \to L$ *are primitive embeddings, there exists an isometry* ϕ *of* L *such that* $j = \phi i$.

The proof of (i) is straightforward. Let $\{e_j, f_j\}$ be the standard basis for the j-th copy of H. If $\{c_1, \ldots, c_t\}$ $(t \leq k)$ is a basis of Γ we put

$$i(c_s) = e_s + \frac{1}{2}\langle c_s, c_s\rangle f_s + \sum_{r<s} \langle c_r, c_s\rangle f_r$$

and it is easily checked that i is an isometry. Since the matrix $(\langle i(c_s), f_r\rangle)$ is the identity, $i(\Gamma)$ must be primitive. For the proof of (ii) we refer to [J], [Pi-S], §6, appendix or [L-P], §2.

Finally we state two lemmas, which will turn out to be useful in dealing with intersection forms of curve configurations on surfaces. For a proof of Lemma 2.10 we refer to [Bou68], Chap. 5, § 3.5 and for a short proof of Lemma 2.12, which uses the previous one, we refer to [Dem].

(2.10) **Lemma.** *Let Q be a symmetric bilinear form on $V = \mathbb{R}^n$ or $\mathbb{Q}^n$, given by the matrix $(q_{i,j})$. Suppose*

(i) $Q \geq 0$,

(ii) $q_{i,j} \leq 0$ *for* $i \neq j$,

(iii) *there is no partition $I \cup J$ of $\{1, \dots, n\}$ with $I \neq \varnothing, J \neq \varnothing$ such that for $i \in I, j \in J$ one has $q_{i,j} = 0$.*

Then either $Q > 0$, or its annihilator is 1-dimensional and spanned by a vector $(z_1, \dots, z_n) \in V$ with $z_i > 0$ for all $i = 1, \dots, n$.

If instead of (i), (ii), (iii) *we assume that Q satisfies* (i)', (ii), (iii) *with*

(i)' *the annihilator N of Q contains $z = (z_1, \dots, z_n)$ with $z_i > 0$ for all $i = 1, \dots, n$,*

then $Q \geq 0$, dim $N = 1$ *and N is spanned by z.*

(2.11) **Corollary.** *Let the situation be as in Lemma 2.10 and suppose that in addition to (i), (ii), (iii) we know that $Q > 0$. Then there exists $x = (x_1, \dots, x_n) \in \mathbb{Q}^n$ with $x_i > 0$ such that $Q(x, e_i) > 0$ $(i = 1, \dots, n)$.*

Proof. Let $\lambda \in \mathbb{R}$ be the smallest positive eigenvalue of Q. The form $Q_0 = Q - \lambda 1\!\!1_n$ satisfies (i), (ii) and (iii), so its annihilator is spanned by $y = (y, \dots, y_n)$ with $y_i > 0$. Hence for all $i \in \{1, \dots, n\}$ we have

$$Q(y, e_i) = \lambda y_i > 0.$$

Now by continuity we may find $x \in \mathbb{Q}^n$ close to y such that this inequality holds for x as well. □

(2.12) **Lemma.** *Let Γ be a connected graph with vertices $v_1, \dots, v_n$ such that v_i and v_j are connected by at most one edge $(i \neq j)$. Let $Q(\Gamma)$ be the quadratic form with matrix $(q_{i,j})$ where $q_{i,i} = 1$ $(i = 1, \dots, n)$ and (for $i \neq j$) $q_{i,j} = -\frac{1}{2}$ (respectively 0) if v_i and v_j are connected (respectively are not connected). Then $Q(\Gamma)$ satisfies the conditions (ii) and (iii) of Lemma 2.10. If $Q(\Gamma) \geq 0$* (i.e., *if the condition (i) is also satisfied) only the following series of graphs occur:*

(i) *If $Q(\Gamma) > 0$*

A_n  ($n \geq 1$ vertices)

D_n  ($n \geq 4$ vertices)

E_n 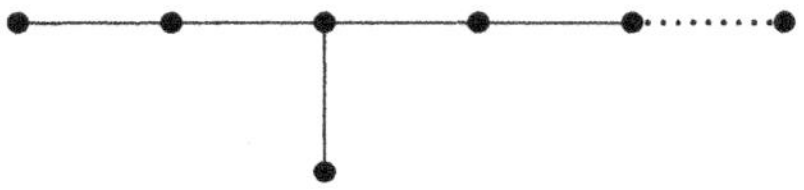 $(n = 6, 7 \text{ or } 8)$

(ii) *If $Q(\Gamma)$ has non-trivial annihilator, the graph Γ is one out of the following list (the weights on the vertices denote the coefficients of a vector spanning the 1-dimensional annihilator):*

$\tilde{A}_n$ 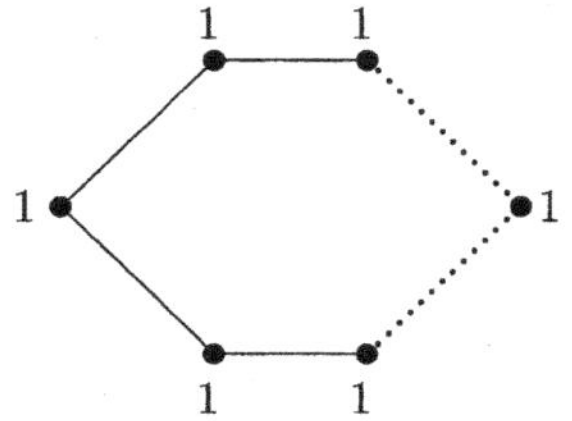

$(n + 1 \text{ vertices}, n \geq 2)$

$\tilde{D}_n$ 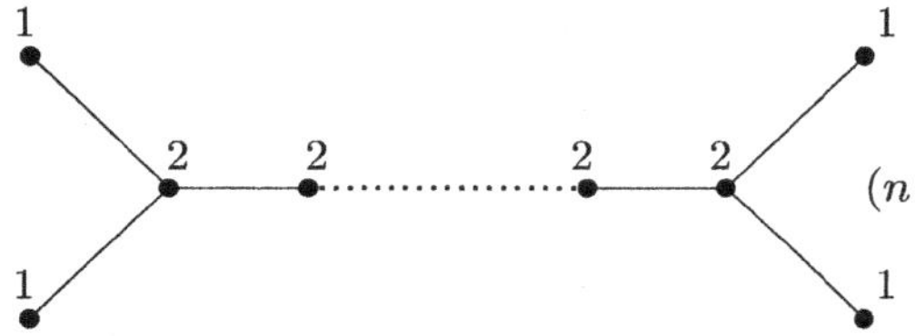

$(n+1 \text{ vertices}, n \geq 4)$

$\tilde{E}_6$ 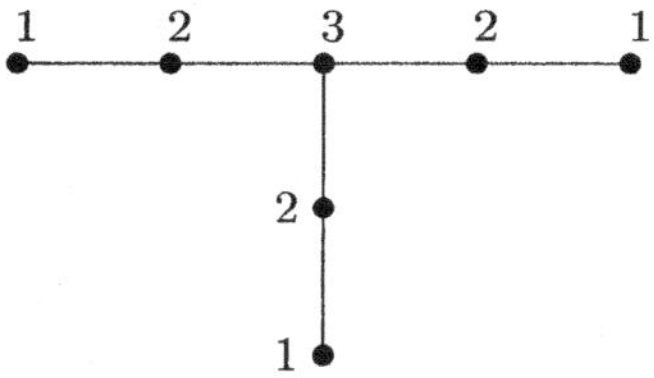

$\tilde{E}_7$ 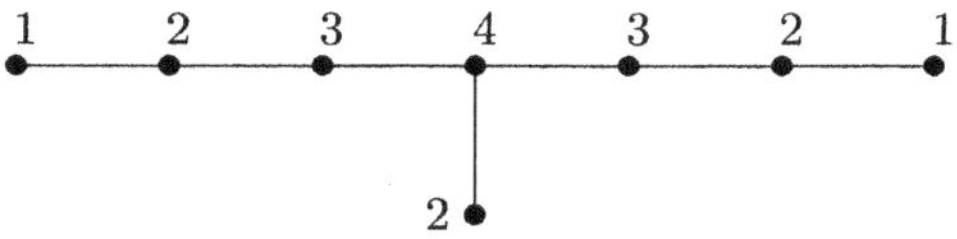

$\tilde{E}_8$ 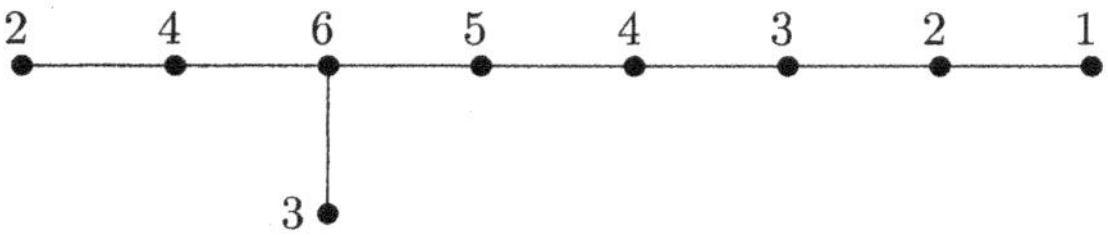

□

3. Vector Bundles, Characteristic Classes and the Index Theorem

Vector bundles, real or complex, will be denoted by $\mathcal{V}, \mathcal{W}, \ldots$, and the fibre of $\mathcal{V}$ over the point x in the base by $\mathcal{V}(x)$. We use of course the standard notations $\mathcal{V} \oplus \mathcal{W}$ for the direct sum of $\mathcal{V}$ and $\mathcal{W}$, $\mathcal{V}^\vee$ for the dual bundle of $\mathcal{V}$, $\mathcal{S}^n\mathcal{V}$ for the n-th symmetric product of $\mathcal{V}$, etc.

If $\mathcal{V}$ is a complex vector bundle of rank d, then $\mathbb{P}(\mathcal{V})$ will denote the associated projective bundle, i.e., if $\rho : \mathrm{GL}(d, \mathbb{C}) \to \mathrm{PGL}(d, \mathbb{C})$ is the canonical epimorphism, and if $\mathcal{V}$ is given by coordinate transformations $\{g_{ij}\}$, then $\mathbb{P}(\mathcal{V})$ is given by the coordinate transformations $\{\rho(g_{ij})\}$. So $\mathbb{P}(\mathcal{V})$ is *not* the associated bundle of $\mathcal{V}^\vee$ as in [Ha66]. If $d = 2$, then $\mathbb{P}(\mathcal{V})$ and $\mathbb{P}(\mathcal{V}^\vee)$ are isomorphic, since in that case $\mathcal{V}^\vee \simeq \mathcal{V} \otimes \bigwedge^2 \mathcal{V}^\vee$.

Let $\mathcal{V}$ be a real r-vector bundle on the paracompact topological space X. Then $w_i(\mathcal{V})$, $i = 0, \ldots, r$ will be its i-th Stiefel-Whitney class, and $w(\mathcal{V}) = \sum_{i=0}^{r} w_i(\mathcal{V}) \in H^*(X, \mathbb{Z}/2\mathbb{Z})$ its total Stiefel-Whitney class. Similarly we have the Pontrjagin classes $p_i(\mathcal{V}) \in H^{4i}(X, \mathbb{Z})$ and the total Pontrjagin class $p(\mathcal{V}) = p_0(\mathcal{V}) + \ldots \in H^*(X, \mathbb{Z})$ if $\mathcal{V}$ is a real r-bundle, as well as the Chern classes $c_i(\mathcal{V}) \in H^{2i}(X, \mathbb{Z})$ and the total Chern class $c(\mathcal{V}) = \sum_{i=0}^{r} c_i(\mathcal{V}) \in H^*(X, \mathbb{Z})$ if $\mathcal{V}$ is a complex r-bundle.

Several times we shall use the fact ([Hir66], p. 73) that if we consider a complex vector bundle $\mathcal{V}$ as a real bundle $\mathcal{V}_\mathbb{R}$, then $w_i(\mathcal{V}_\mathbb{R}) = 0$ for i odd, whereas $w_{2i}(\mathcal{V}_\mathbb{R}) \equiv c_i(\mathcal{V}) \bmod 2$.

Let again $\mathcal{V}$ be any real r-bundle. If we formally put $w(\mathcal{V}) = \prod_{i=1}^{r} (1 + \delta_i)$, then any symmetric power series in $\delta_1, \ldots, \delta_r$ with coefficients in $\mathbb{Z}/2\mathbb{Z}$, yields a polynomial in $w_1, \ldots, w_r$ and hence an element in $H^*(X, \mathbb{Z}/2\mathbb{Z})$. The same can be done, starting from a power series with rational coefficients, with the Pontrjagin classes and with the Chern classes in the case of a complex vector bundle, yielding in both cases an element $e \in H^*(X, \mathbb{Q})$. The component of e which is of dimension i, will be denoted by $t_i(e)$. Thus, once an isomorphism $H^n(X, \mathbb{Q}) \simeq \mathbb{Q}$ is given, $t_n(e)$ becomes a rational number.

Now let $\mathcal{V}$ be an oriented real r-bundle on X. Then we obtain the L-class $L(\mathcal{V}) \in H^*(X, \mathbb{Q})$ by starting from

$$p(\mathcal{V}) = \prod_{i=1}^{r} (1 + \delta_i),$$

and taking the symmetric power series

$$\prod_{i=1}^{r} \frac{\sqrt{\delta_i}}{\tanh \sqrt{\delta_i}}.$$

So

$$L(\mathcal{V}) = 1 + \frac{1}{3}p_1(\mathcal{V}) + \frac{1}{45}(7p_2(\mathcal{V}) - p_1^2(\mathcal{V})) + \dots .$$

Similarly, if $\mathcal{V}$ is a complex r-bundle, and $c(\mathcal{V}) = \prod_{i=1}^{r}(1+\delta_i)$, then the Todd class Todd $(\mathcal{V})$ is obtained from $\prod_{i=1}^{r} \frac{\delta_i}{1-e^{-\delta_i}}$, whereas the Chern character ch$(\mathcal{V})$ is derived from $\sum_{i=1}^{r} e^{\delta_i}$. The beginning of Todd $(\mathcal{V})$ is well known:

$$\mathrm{Todd}(\mathcal{V}) = 1 + \frac{1}{2}c_1(\mathcal{V}) + \frac{1}{12}(c_1^2(\mathcal{V}) + c_2(\mathcal{V})) + \frac{1}{24}c_1(\mathcal{V})c_2(\mathcal{V}) + \dots$$

(so, in our notation, we have for example $t_3(\mathrm{Todd}(\mathcal{V})) = 0$ and $t_6(\mathrm{Todd}(\mathcal{V})) = \frac{1}{24}c_1(\mathcal{V})c_2(\mathcal{V})$).

If X is a differentiable manifold, then we denote by $\mathcal{T}_X$ its tangent bundle. Suppose now that X is compact, connected and oriented. Then $L(\mathcal{T}_X)$ is well-defined, and we set $L(X) = L(\mathcal{T}_X)$.

For any compact, connected, oriented (not necessarily differentiable) manifold X we define the index $\tau(X)$ in the following way. If $\dim X \not\equiv 0 \bmod 4$, we set $\tau(X) = 0$. If $\dim X = 4m$, the cup product form defines on $H^{2m}(X,\mathbb{R})$ a non-degenerate quadratic form $Q(X)$, and we set $\tau(X) = \tau(Q(X))$, i.e., $\tau(X) = b^+(X) - b^-(X)$, where $b^+(X)$ (respectively $b^-(X)$) is the number of positive (respectively negative) eigenvalues of Q (observe that $b_{2m}(X) = b^+(X) + b^-(X)$).

(3.1) **Theorem** (Topological index theorem = Index theorem of Thom-Hirzebruch). *Let X be a compact, connected, oriented differentiable manifold of dimension $4m$. Then*

$$\tau(X) = t_{4m}(L(X)).$$

In particular, if $m = 1$, then $\tau(X) = \frac{1}{3}p_1(X)$, and if in addition X carries an almost-complex structure, then $\tau(X) = \frac{1}{3}p_1(X) = \frac{1}{3}(c_1^2(X) - 2c_2(X))$ by [Hir66], p. 65.

For details we refer to [Hir66], p. 86.

Complex Manifolds

The Riemann-Roch theorem together with the duality theorem of Serre is one of the cornerstones of modern algebraic geometry. We recall these in Sect. 5 in the form needed later on. Useful properties of divisors and line bundles are recalled in Sect. 6. Finally, two other fundamental concepts, the algebraic dimension and the Kodaira dimension are treated in Sect. 7.

4. Basic Concepts and Facts

A 1-dimensional complex manifold will be called a smooth curve or a Riemann surface and a 2-dimensional one a smooth surface.

If X is an n-dimensional*) complex manifold, then we shall denote by

$\mathcal{T}_X$: the (holomorphic) tangent bundle of X;

$c_i(X)$: the i-th Chern class of X, that is $c_i(\mathcal{T}_X)$ (so $c_n(X)$ is equal to the Euler number $e(X)$ if X is compact);

$\mathcal{O}_X$: the structure sheaf of X;

$\mathcal{O}_X^*$: the sheaf of non-vanishing holomorphic function germs on X (with multiplication as group law);

Ω_X^i: the sheaf of germs of holomorphic i-forms on X, i.e., the sheaf of sections in the bundle $\bigwedge^i \mathcal{T}_X^\vee (i \geq 1)$;

$\mathcal{K}_X$: the canonical line bundle on X, i.e., the holomorphic 1-vector bundle $\bigwedge^n \mathcal{T}_X^\vee$;

$\mathcal{N}_{Y/X}$: the normal bundle of the complex submanifold Y in X defined by the normal bundle sequence

$$0 \to \mathcal{T}_Y \to \mathcal{T}_X|Y \to \mathcal{N}_{Y/X} \to 0.$$

We recall that a sheaf of $\mathcal{O}_X$-modules $\mathcal{S}$ is coherent, if locally there always is some exact sequence of sheaves of $\mathcal{O}_X$-modules

$$\mathcal{O}_X^p \to \mathcal{O}_X^q \to \mathcal{S} \to 0.$$

If X is compact and $\mathcal{S}$ coherent, then the complex vector spaces $H^i(X, \mathcal{S})$ are finite-dimensional of dimension $h^i(X, \mathcal{S}) = h^i(\mathcal{S})$, and they vanish for $i > n$. Hence the Euler characteristic $\chi(X, \mathcal{S}) = \sum_{i=0}^{n} (-1)^i h^i(\mathcal{S})$ is well-defined. We put $\chi(X, \mathcal{O}_X) = \chi(X)$. the integer $p_a(X) = (-1)^n(\chi(X) - 1)$ is called the arithmetic(al) genus of X.

Setting for the moment $\Omega_X^0 = \mathcal{O}_X$, we define

*) Complex manifolds are always pure-dimensional (see Sect. 8).

$$h^{p,q}(X) = h^q(\Omega^p_X)$$
$$q(X) = h^{0,1}(X)$$
$$p_g(X) = h^{0,n}(X) \qquad \text{(geometric genus).}$$

5. Holomorphic Vector Bundles, Serre Duality and Riemann-Roch

The sheaf of sections of a holomorphic r-vector bundle on the complex manifold X is locally free, i.e., locally isomorphic to $\mathcal{O}^r_X$; conversely every locally free sheaf $\mathcal{S}$ of $\mathcal{O}_X$-modules is the sheaf of sections of a holomorphic r-vector bundle, which is determined up to isomorphism. Because of this equivalence we shall sometimes use for a holomorphic bundle and its sheaf of sections the same symbol. But tradition does not always permit to do so; thus it prescribes that the sheaf of sections of $\bigwedge^i \mathcal{T}^\vee_X$ is denoted by Ω^i_X (Sect. 4).

Let $\mathcal{V}$ be a holomorphic d-vector bundle on the complex manifold X, let $Y = \mathbb{P}(\mathcal{V})$ and $p : Y \to X$ the projection. Then $p^*(\mathcal{V})$ has a canonical holomorphic subbundle $\mathcal{L}^\vee$ of rank 1, the tautological line bundle of $\mathcal{V}$. The fibre $\mathcal{L}^\vee(y)$ is nothing but the 1-dimensional subspace of $p^*(\mathcal{V})(y) \cong \mathcal{V}(p(y))$ which is represented by y. The quotient bundle is isomorphic to $\mathcal{L}^\vee \otimes \mathcal{W}$, where $\mathcal{W}$ is the bundle along the fibres of Y (see [Hir66], §13). So on Y there is an exact sequence of holomorphic vector bundles

$$0 \to \mathcal{L}^\vee \to p^*(\mathcal{V}) \to \mathcal{L}^\vee \otimes \mathcal{W} \to 0. \tag{3}$$

By eliminating the Chern classes of $\mathcal{W}$ from $p^*(c(\mathcal{V})) = c(\mathcal{L}^\vee) \cdot c(\mathcal{L}^\vee \otimes \mathcal{W})$ we find

$$c_1^d(\mathcal{L}) + p^*(c_1(\mathcal{V})) \cdot c_1^{d-1}(\mathcal{L}) + \ldots + p^*(c_d(\mathcal{V})) = 0.$$

In Chap. VII we shall need the following result from [Ha66], p. 68.

(5.1) **Theorem.** *Let X be a complex manifold, and $\mathcal{V}$ a holomorphic vector bundle on X. Let $Y = \mathbb{P}(\mathcal{V}^\vee), p : Y \to X$ the projection and $\mathcal{L}^\vee$ the tautological line bundle on Y. Then for every coherent sheaf $\mathcal{S}$ on X and for every $n \geq 1$ there are natural isomorphisms of $\mathcal{O}_X$-modules*

$$p_*(p^*(\mathcal{S})) \xrightarrow{\sim} \mathcal{S}$$
$$p_*(\mathcal{L}^{\otimes n} \otimes p^*(\mathcal{S})) \xrightarrow{\sim} S^n\mathcal{V} \otimes \mathcal{S}.$$

Furthermore, $p_{*i}(\mathcal{L}^{\otimes n} \otimes p^*(\mathcal{S})) = 0$ for all $i \geq 1$. Hence by the Leray spectral sequence there are isomorphisms

$$H^i(Y, p^*(\mathcal{S})) \xrightarrow{\sim} H^i(X, \mathcal{S})$$
$$H^i(Y, \mathcal{L}^{\otimes n} \otimes p^*(\mathcal{S})) \xrightarrow{\sim} H^i(X, S^n\mathcal{V} \otimes \mathcal{S})$$

for all $i \geq 0$.

We occasionally use the following geometric interpretation of Chern classes:

(5.2) **Proposition.** *Let $\mathcal{V}$ be a rank r holomorphic vector bundle on a compact complex manifold X. Suppose that $\mathcal{V}$ has holomorphic sections $\sigma_1, \dots, \sigma_{r-k+1}$ such that the subset*

$$D := \{x \in X \mid 0 = \sigma_1(x) \wedge \cdots \wedge \sigma_{r-k+1}(x) \in \Lambda^{r-k+1}\mathcal{V}\}, \qquad (*)$$

defining the locus where these sections become dependent, is a either empty or a subvariety of codimension k, i.e. D is everywhere locally given as the complete intersection of k hypersurfaces meeting transversally. Then the fundamental homology class of D is Poincaré dual to $c_k(\mathcal{V})$.

Remark. In fact, the Proposition also holds if the equations on the right hand side of (*) define an equi-dimensional (singular or even non-reduced) analytic subspace (or subscheme) of X of codimension k provided we use the fundamental cycle in homology which we shall define later (see the remarks just after Theorem 8.8). For a proof of this more general version see [G-H78a], p. 413 for the analytic setting, or [Fult], Example 14.3.2, where an algebro-geometric proof is presented.

A result that we shall use frequently is Serre's duality theorem for manifolds (compare [Se55a], p. 17–20, and Sect. 12 of this chapter.)

(5.3) **Theorem** (Serre's duality theorem for manifolds in its simplest form). *Let X be a compact, connected complex manifold of dimension n and $\mathcal{V}$ a holomorphic vector bundle on X. Then*

$$h^i(\mathcal{V}) = h^{n-i}(\mathcal{V}^\vee \otimes \mathcal{K}_X).$$

As a special case we find $p_g(X) = h^{0,n}(X) = h^{n,0}(X)$.

In Sect. 11 we shall give a more precise formulation and in Chap. II, Sect. 6 we shall give a version for singular *curves*.

An equally fundamental corner-stone of this book is the Riemann-Roch theorem. We do not need it in full generality; the following part of it will be sufficient in our case (compare [Hir66], p. 155 and p. 188).

(5.4) **Theorem** (Hirzebruch-Atiyah-Singer Riemann-Roch theorem). *Let $\mathcal{V}$ be a holomorphic vector bundle on the compact, connected n-dimensional complex manifold X. Then*

$$\chi(X, \mathcal{V}) = t_{2n}(\mathrm{Todd}(X) \cdot ch(\mathcal{V})).$$

In particular, when we take for $\mathcal{V}$ the trivial line bundle, we obtain $\chi(X) = \chi(\mathcal{O}_X) = t_{2n}(\mathrm{Todd}(X))$. The right hand side of this equation is a polynomial with rational coefficients in the Chern classes of X; it is called the Todd genus of X and denoted by $T(X)$.

(5.5) **Theorem** (Todd-Hirzebruch formula). *If X is any compact, connected complex manifold, then*
$$\chi(X) = T(X).$$

For $n = 1$ this gives that $q(X) = g(X)$, where $g(X)$ is the topological genus of X.

For $n = 2$ we find Noether's formula

$$1 - q(X) + p_g(X) = \tfrac{1}{12}(c_1^2(X) + c_2(X)). \tag{4}$$

Applying Theorem 5.4 to a line bundle $\mathcal{L}$ on a compact, connected smooth curve X, we find

$$h^0(X, \mathcal{L}) - h^1(X, \mathcal{L}) = c_1(\mathcal{L}) - g(X) + 1.$$

The integer $c_1(\mathcal{L})$ is called the degree $\deg(\mathcal{L})$ of $\mathcal{L}$, so we can also write

$$h^0(X, \mathcal{L}) - h^1(X, \mathcal{L}) = \deg(\mathcal{L}) - g(X) + 1, \tag{5}$$

which is the classical formula of curves.

Furthermore we find for $\dim X = 2$, rank $\mathcal{V} = 1$:

$$h^0(X, \mathcal{V}) - h^1(X, \mathcal{V}) + h^2(X, \mathcal{V}) = \tfrac{1}{2}c_1(\mathcal{V})(c_1(\mathcal{V}) + c_1(X)) + T(X), \tag{6}$$

and, combining this with Serre duality:

$$h^0(X, \mathcal{V}) - h^1(X, \mathcal{V}) + h^0(X, \mathcal{K}_X \otimes \mathcal{V}^\vee) = \tfrac{1}{2}c_1(\mathcal{V})(c_1(\mathcal{V}) + c_1(X)) + T(X); \tag{7}$$

for $\dim X = 1$, rank $\mathcal{V} = n$:

$$h^0(\mathcal{V}) - h^1(\mathcal{V}) = c_1(\mathcal{V}) - n(g-1); \tag{8}$$

for $\dim X = 2$, rank $\mathcal{V} = n$:

$$h^0(\mathcal{V}) - h^1(\mathcal{V}) + h^2(\mathcal{V}) = \frac{1}{2}(c_1^2(\mathcal{V}) - 2c_2(\mathcal{V})) + \frac{1}{2}c_1(\mathcal{V})c_1(X) + nT(X) \tag{9}$$

and for $\dim X = n$, rank $\mathcal{V} = 1$:

$$\begin{aligned} &h^0(\mathcal{V}) - h^1(\mathcal{V}) + \ldots + (-1)^n h^n(\mathcal{V}) \\ &\quad = \frac{c_1^n(\mathcal{V})}{n!} + \text{terms containing strictly lower powers of } c_1. \end{aligned} \tag{10}$$

6. Line Bundles and Divisors

Let X be a complex manifold of dimension n. With respect to the tensor product, the set of holomorphic line bundles on X forms a group, which is naturally isomorphic to $H^1(X, \mathcal{O}_X^*)$. This group is by definition the Picard group $\mathrm{Pic}(X)$.

The exponential sequence of sheaves

$$0 \to \mathbb{Z}_X \xrightarrow{i} \mathcal{O}_X \xrightarrow{j} \mathcal{O}_X^* \to 0$$

(where i denotes the inclusion and j maps any germ s to $\mathbf{e}(s)$) gives rise to the exponential cohomology sequence

(11) $$\to H^1(\mathbb{Z}_X) \to H^1(\mathcal{O}_X) \to H^1(\mathcal{O}_X^*) \xrightarrow{\delta} H^2(\mathbb{Z}_X) \to H^2(\mathcal{O}_X).$$

We put $\ker(\delta) = \operatorname{Pic}^0(X)$, the identity component of the Picard group, so $\operatorname{Pic}(X)/\operatorname{Pic}^0(X)$ is isomorphic to a subgroup of $H^2(X,\mathbb{Z})$, the Néron-Severi group of X.

(6.1) **Proposition.** *For any holomorphic line bundle $\mathcal{L} \in H^1(\mathcal{O}_X^*)$ we have that $\delta(\mathcal{L}) = c_1(\mathcal{L})$.*

For a proof we refer to [Hir66], Theorem 4.3.1.

A hypersurface $H \subset X$ is any (non-empty) closed subset of X with the following property: every point $p \in H$ has a connected open neighbourhood U, such that $H \cap U$ is the zero set of a non-constant holomorphic function on U. A hypersurface is irreducible if it is not the union of two other hypersurfaces.

A divisor on X is a formal sum $D = \sum_{i=1}^{\infty} d_i D_i$, where $d_i \in \mathbb{Z}$ and $\{D_i\}_{i\in\mathbb{N}}$ a locally finite sequence of irreducible hypersurfaces on X (locally finite means that every point has a neighbourhood which meets only finitely many D_i's). The divisor D is called effective or positive (notation: $D > 0$) if all d_i are non-negative and not all zero; and D is called non-negative if it is effective or the zero divisor.

Let X and Y be connected complex manifolds and $f : X \to Y$ a holomorphic map. If D is any divisor on Y and $f(X) \not\subset \operatorname{supp}(D)$, then $f^*(D)$ is defined in the obvious way by lifting the local equation of its irreducible components.

If $\mathcal{L}$ is a line bundle on X, then every meromorphic section m, different from the zero section, determines a divisor (m) on X, namely its zero divisor minus its polar divisor. Conversely, if any divisor D is given, then there is (up to isomorphism) exactly one $\mathcal{L}$ with a meromorphic section m, such that $(m) = D$. This $\mathcal{L}$ will be denoted $\mathcal{O}_X(D)$. Two divisors D and E are linearly equivalent if and only if $\mathcal{O}_X(D) \cong \mathcal{O}_X(E)$.

The set $|D|$ of effective divisors which are linearly equivalent to a given divisor D, is naturally isomorphic to $\mathbb{P}(\Gamma(\mathcal{O}_X(D)))$ ($\mathbb{P}(0) = \varnothing$).

To explain the notion of a nef-divisor, we need to explain how to intersect divisors and curves. For a divisor D on a compact complex manifold X, the cohomology class $c_1(\mathcal{O}_X(D)) \in H^2(X,\mathbb{Z})$ depends only on the class of D up to linear equivalence. If $C \subset X$ is a smooth irreducible curve, we have defined the fundamental class $[C] \in H^{2n-2}(X,\mathbb{Z})$. If C is any irreducible curve in X, we shall see later (Chap. II, 10) that there is a smooth curve $\tilde{C}$ mapping to to C through a biholomorphism away from the singular points: its desingularization. This gives a holomorphic map, say $f : \tilde{C} \to X$ and then we set

$[C] = f_![\tilde{C}] \in H^{2n-2}(X, \mathbb{Z})$. Hence the intersection number $(c_1(\mathcal{O}_X(D)), [C])$ is well-defined and depends only on the class of D up to linear equivalence. If D happens to be an irreducible divisor, the number $(c_1(\mathcal{O}_X(D)), [C])$ coincides with the topological intersection number $[D] \cdot [C]$. This is a consequence of Prop. 5.2. So, if C intersects D transversally in at least one point, this number is strictly positive, a remark which we will use in the proof of the Lemma below.

A divisor D is called nef if $(c_1(\mathcal{O}_X(D)), [C]) \geq 0$ for all curves C on X. An ample divisor is nef, but the converse need not hold. By the preceding considerations, the property of being nef only depends on the linear equivalence class of the divisor and hence on the associated line bundle. In fact, for any line bundle $\mathcal{L}$ the notion of being nef can be defined just as for divisors.

We need an almost trivial property of nef line bundles :

(6.2) **Lemma** *Let $\mathcal{L}$ be a nef line bundle on a projective manifold such that $\mathcal{L}^{-1}$ has a non-zero section, then $\mathcal{L}$ must be trivial.*

Proof. Suppose that $\mathcal{L}$ is not trivial. Then the section of $\mathcal{L}^{-1}$ must vanish along a non-empty divisor and any curve C transversal to this divisor would satisfy $-(c_1(\mathcal{L}), C) > 0$, contrary to the definition of being nef. □

Canonical divisors on X will be denoted by K_X or K, so $\mathcal{K}_X = \mathcal{O}_X(K_X)$.

Now let Y be an $(n-1)$-dimensional complex submanifold of X; it is a divisor on X in an obvious way. If we denote by $|$ the analytic restriction, we have ([Hir66], Theorem 4.8.1)

(6.3) **Proposition.** $\mathcal{O}_X(Y)|Y \cong \mathcal{N}_{Y/X}$.

If we combine this fact with the isomorphism

$$\bigwedge^{n} \mathcal{T}_X^\vee | Y \cong \bigwedge^{n-1} \mathcal{T}_Y^\vee \otimes \mathcal{N}_{Y/X}^\vee,$$

which follows from the exact sequence

$$0 \to \mathcal{T}_Y \to \mathcal{T}_X|Y \to \mathcal{N}_{Y/X} \to 0$$

of holomorphic vector bundles on Y, we obtain

(6.4) **Theorem** (Adjunction formula). *If Y is a complex submanifold of codimension 1 of the complex manifold X, then*

$$\mathcal{K}_Y = \mathcal{K}_X \otimes \mathcal{O}_X(Y)|Y.$$

7. Algebraic Dimension and Kodaira Dimension

In this section all manifolds are connected.

As to the algebraic dimension, the main result (compare [Rem 56], p. 279) is

(7.1) **Theorem.** *The field of meromorphic functions on a compact connected complex manifold is a finitely generated algebraic function field with a transcendency degree over* $\mathbb{C}$*, that does not exceed the dimension of* X*.*

This transcendency degree is called the algebraic dimension of X; it will be denoted by $\mathrm{a}(X)$.

It follows from GAGA (see Sect. 19) that if X is algebraic, then every meromorphic function is rational, and the field of meromorphic functions is nothing but the field of rational functions on X.

As to the Kodaira dimension, let X be any compact complex manifold. Since there is a pairing

$$\Gamma(X, \mathcal{K}_X^{\otimes m_1}) \otimes \Gamma(X, \mathcal{K}_X^{\otimes m_2}) \to \Gamma(X, \mathcal{K}_X^{\otimes (m_1+m_2)})$$

we can make the direct sum $\mathbb{C} \oplus \sum_{m \geq 1} \Gamma(X, \mathcal{K}_X^{\otimes m})$ into a commutative ring $R(X)$ with unit element. This ring is called the canonical ring of X. It can be proved that $R(X)$ has a finite degree of transcendency, say $\mathrm{tr}(R(X))$, over $\mathbb{C}$. Thus we can define the Kodaira dimension $\mathrm{kod}(X)$ of X as follows:

$$\mathrm{kod}(X) = \begin{cases} -\infty & \text{if } R(X) \cong \mathbb{C} \\ \mathrm{tr}(R(X)) - 1 & \text{otherwise.} \end{cases}$$

We have always that $\mathrm{kod}(X) \leq \mathrm{a}(X) \leq \dim X$. Let $P_m(X) = h^0(\mathcal{K}_X^{\otimes m})$, $m \geq 1$. This number is called the m-th plurigenus of X (so $P_1(X) = p_g(X)$). The Kodaira dimension yields precise information about the behaviour of $P_m(X)$ for $m \to \infty$.

(7.2) **Theorem.** *Let* X *be a compact connected complex manifold. Then:*

$$\mathrm{kod}(X) = -\infty \Leftrightarrow P_m(X) = 0 \quad \text{for all } m \geq 1;$$

$$\mathrm{kod}(X) = 0 \Leftrightarrow P_m(X) = 0 \quad \text{or } 1 \text{ for } m \geq 1, \text{ but not always } 0;$$

$$\left\{ \begin{array}{c} \mathrm{kod}(X) = k \\ 1 \leq k \leq \dim X \end{array} \right\} \Leftrightarrow \left\{ \begin{array}{l} \text{there are real constants } \alpha > 0,\ \beta > 0 \\ \text{such that for } m \text{ large enough} \\ \alpha m^k < P_m(X) < \beta m^k. \end{array} \right\}$$

So for $k \geq 1$ *we have that* P_m *grows like* m^k*.*

For this result we refer to [Ue75], p. 86.

Furthermore we shall need the following two properties of the Kodaira dimension ([Ue75], p. 69 and 73):

(7.3) **Theorem.** *If* X_1 *and* X_2 *are connected compact complex manifolds, then* $\mathrm{kod}(X_1 \times X_2) = \mathrm{kod}(X_1) + \mathrm{kod}(X_2)$.

(7.4) **Theorem.** *Let* X *and* Y *be compact, connected complex manifolds of the same dimension. If there exists a generically finite holomorphic map from* X *onto* Y*, then* $P_n(X) \geq P_n(Y)$ *for* $n \geq 1$*, hence* $\mathrm{kod}(X) \geq \mathrm{kod}(Y)$*. If the map is an unramified covering, then* $\mathrm{kod}(X) = \mathrm{kod}(Y)$.

General Analytic Geometry

Although the book is mainly about non-singular complex manifolds of dimension 2, singularities play an important role. On the one hand these appear when we consider the image of manifolds under holomorphic maps; on the other hand, spaces that parametrise complex structures are often singular, even non-reduced. In this sub-chapter we have gathered all that is needed later on when we deal with such situations. The results are basic, and often deep such as Grauert's base change theorem and Remmert's proper mapping theorem recalled in Sect. 8, and the existence of the Kuranishi family, recalled in Sect. 10.

8. Complex Spaces

By definition a ringed space is a pair $X = (|X|, \mathcal{O}_X)$ consisting of a topological space $|X|$ and a sheaf of commutative rings $\mathcal{O}_X$ on it. Differentiable manifolds and complex manifolds are examples of ringed spaces where the sheaf consists of an appropriate subring of the ring of continuous functions which will then by definition become the ring of differentiable or holomorphic functions on a given open subset of the manifold. We need more general ringed spaces for which the local models are the closed analytic subspaces of an open ball: a closed analytic subspace $Y = (|Y|, \mathcal{O}_Y)$ of an open ball $B \subset \mathbb{C}^n$ is a ringed space which can be obtained in the following way. Let $\{f_i\}_{i \in I}$ be a set of holomorphic functions on B, and $\mathcal{J}$ the subsheaf of $\mathcal{O}_B$, generated (as $\mathcal{O}_B$-module) by the functions f_i. Now let $|Y| = \{z \in B, f_i(z) = 0, i \in I\}$ and $\mathcal{O}_Y = \mathcal{O}_B/\mathcal{J}$. The sheaf $\mathcal{J}$ is the ideal sheaf of Y in B.

A complex space is a Hausdorff ringed space $X = (|X|, \mathcal{O}_X)$, for which $|X|$ is everywhere locally isomorphic to a closed analytic subspace of an open ball in some $\mathbb{C}^n$. The structure sheaf $\mathcal{O}_X$ of a complex space $X = (|X|, \mathcal{O}_X)$ is a sheaf of $\mathbb{C}$-algebras. If $Y = (|Y|, \mathcal{O}_Y)$ is any other complex space, then an analytic or holomorphic map from X into Y is a morphism $f = (|f|, \tilde{f})$ of ringed spaces, such that $\tilde{f} : \mathcal{O}_Y \to |f|_* \mathcal{O}_X$ is a morphism of $\mathbb{C}$-algebras.

Thus the complex spaces form a category, in which direct products and fibre products exist. On any complex space coherent sheaves are defined in exactly the same way as on complex manifolds, i.e., as sheaves $\mathcal{S}$ of $\mathcal{O}_X$-modules, for which there exists locally an exact sequence

$$\mathcal{O}_X^r \longrightarrow \mathcal{O}_X^s \longrightarrow \mathcal{S} \to 0.$$

If $f : X \to Y$ is a holomorphic map and $\mathcal{S}$ a coherent sheaf on Y, then the analytic inverse image or analytic pull-back $f^*(\mathcal{S}) = |f|^{-1}(\mathcal{S}) \otimes_{\mathcal{O}_Y} \mathcal{O}_X$ is a coherent sheaf on X.

Let $X = (|X|, \mathcal{O}_X)$ be an analytic space and $\mathcal{J} \subset \mathcal{O}_X$ a coherent $\mathcal{O}_X$-sheaf of ideals. If $|Y| = \mathrm{supp}(\mathcal{O}_X/\mathcal{J})$, then $(|Y|, \mathcal{O}_Y)$ with $\mathcal{O}_Y = \mathcal{O}_X/\mathcal{J}$ is called a complex subspace of X. The exact sequence

$$0 \to \mathcal{J} \to \mathcal{O}_X \to \mathcal{O}_Y \to 0$$

is called the structure sequence of Y (in X). If there is any danger of confusion we shall write $\mathcal{J}_Y$ or $\mathcal{J}_{Y,X}$ instead of $\mathcal{J}$. If X is smooth and Y locally given by one equation, then $\mathcal{J}_{Y,X} \cong \mathcal{O}_X(-Y)$, and if in addition Y is smooth, then $\mathcal{J}/\mathcal{J}^2 \cong \mathcal{N}^\vee_{Y/X}$, the conormal bundle of Y in X.

A closed embedding of a complex space Y into a complex space X is an isomorphism from Y onto a complex subspace of X.

If $X = (|X|, \mathcal{O}_X)$ is a complex space and $A \subset X$ a closed subset, then A is called a closed analytic subset if there exists a complex subspace Y of X with $|Y| = A$.

Let again $X = (|X|, \mathcal{O}_X)$ be any complex space, and let $\mathcal{R}$ be the coherent sheaf on X, associated to the presheaf which attaches to an open subset $E \subset |X|$ the $\mathcal{O}_X$-module $\{f \in \mathcal{O}_E, f^k = 0 \text{ for some } k \in \mathbb{N}\}$. Then $X_{\text{red}} = \{|X|, \mathcal{O}_X/\mathcal{R}\}$ is again a complex space and the canonical morphism $X_{\text{red}} \to X$ is analytic. This space X_{red} is called the reduction of X, and a space X is called reduced if $X = X_{\text{red}}$. A reduced space X is irreducible if $|X| \neq |X_1| \cup |X_2|$, where X_1, X_2 are closed subspaces of X, with $|X_1|$, $|X_2| \neq |X|$. Every reduced space uniquely decomposes into a locally finite union of irreducible subspaces, the irreducible components. Now let X be reduced. A point $x \in |X|$ is regular or smooth if locally around x the space X is isomorphic to $(B, \mathcal{O}_B)$ for an open ball B in some $\mathbb{C}^N$. A point $x \in |X|$ is singular if it is not smooth; the singular points of X form a proper closed analytic subset of X.

A complex space is a complex manifold if and only if it is reduced and all its points are smooth.

Again, let X be reduced. The dimension $\dim_x X$ at $x \in |X|$ is the Krull dimension of the local ring $\mathcal{O}_{X,x}$. If x is regular, then the dimension is the N above.

If $\dim_x X$ is the same for all $x \in |X|$, then we call this number the dimension of X. This is in particular the case if X is irreducible. The dimension of a reduced complex space is defined as the maximum of the dimensions of its irreducible components, if this maximum exists. If all irreducible components of X have the same dimension d, then we call X of pure dimension d.

We shall say that the complex space X has dimension d if $\dim X_{\text{red}}$ is defined and equal to d. A curve (Riemann surface) is a complex space X with X_{red} of pure dimension 1. Let X be reduced. A point $x \in |X|$ is called normal if its local ring $\mathcal{O}_{X,x}$ is integrally closed in its ring of quotients. A complex space is normal if all its points are normal. The non-normal points always form a thin closed analytic subset, i.e., a closed analytic subset that does not contain any irreducible components of X. The singular locus of a normal complex space has codimension ≥ 2.

A regular point of X is normal, but the converse does not hold if $\dim_x X \geq 2$.

Every reduced complex space X has a normalization, defined uniquely up to analytic isomorphisms, i.e., there exists a normal complex space X_{norm} and a finite, surjective analytic map $\nu_X : X_{\text{norm}} \to X$, such that ν_X maps

$X_{\text{norm}} \setminus \nu_X^{-1}(N)$ isomorphically onto $X \setminus N$, where N is the set of non-normal points on X.

Given any holomorphic map $f : X \to Y$ between reduced complex spaces such that no irreducible component of X is mapped into the locus of non-normal points on Y, then there exists a unique lifting $g : X_{\text{norm}} \to Y_{\text{norm}}$ such that the diagram

$$\begin{array}{ccc} X_{\text{norm}} & \xrightarrow{g} & Y_{\text{norm}} \\ \downarrow{\scriptstyle \nu_X} & & \downarrow{\scriptstyle \nu_Y} \\ X & \xrightarrow{f} & Y \end{array}$$

is commutative.

Now we state a number of fundamental theorems which will be used in this book.

(8.1) **Theorem** (Stein factorization). *Let X, Y be complex spaces and $f : X \to Y$ a proper analytic map. Then f admits a unique factorisation*

$$f : X \xrightarrow{g} Z \xrightarrow{h} Y$$

such that

i) *$g : X \to Z$ is a proper surjective holomorphic map with $g_*\mathcal{O}_X = \mathcal{O}_Z$ (in particular all fibres of g are connected);*

ii) *h is a finite holomorphic map.*

Moreover, if X is normal, then Z is normal, too.

A proof can be found in [G-R77], p. 213.

(8.2) **Theorem** (Grauert's direct image theorem). *Let X, Y be complex spaces, $f : X \to Y$ a proper analytic map. Then for every coherent sheaf $\mathcal{S}$ on X the direct image sheaves $f_{*i}\mathcal{S}, i = 0, 1, 2, \ldots$ are also coherent.*

For this result we refer to [Gr60], [F-K] and [Ki-V]. As an immediate consequence we have the following two facts.

(8.3) **Theorem** (Finiteness theorem of Cartan-Serre). *Let X be a compact complex space and $\mathcal{S}$ a coherent sheaf on X. Then $H^i(X, \mathcal{S})$ has finite dimension for all $i \geq 0$ and $H^i(X, \mathcal{S}) = 0$ for $i >$ dim X.*

(8.4) **Theorem** (Remmert's mapping theorem). *Let X, Y be reduced complex spaces and $f : X \to Y$ a proper analytic map. If $A \subset X$ is a closed analytic subset, then so is $|f|(A) \subset Y$.*

Let X and Y be complex spaces, $f : X \to Y$ a surjective holomorphic map and $y \in |Y|$. The analytic fibre over y, denoted by X_y, is defined as the fibre product $y \times_Y X$. It can be described more explicitly in the following way. Let $\mathcal{J}$ be the ideal sheaf of y in $\mathcal{O}_Y$. Then X_y is the complex subspace

$(|f|^{-1}(y), \mathcal{O}_X/f^*(\mathcal{J}))$ of X. Furthermore, for any coherent sheaf $\mathcal{S}$ on X we denote by $\mathcal{S}_y$ the analytic restriction of $\mathcal{S}$ to X_y (i.e., the analytic inverse image with respect to the embedding of X_y in X).

The next two theorems are e.g. proved in [B-S], Chap. III, § 3.

(8.5) **Theorem.** *Let X, Y be reduced complex spaces and $f : X \to Y$ a proper holomorphic map. If $\mathcal{S}$ is any coherent sheaf on X, which is flat over y (i.e., $\mathcal{S}_x$ is a flat $\mathcal{O}_{|f|(x)}$-module for all $x \in X$), we have*

(i) *the Euler characteristic $\chi(X_y, \mathcal{S}_y)$ is locally constant;*
(ii) (Grauert's semi-continuity theorem) *$h^q(X_y, \mathcal{S}_y)$ is an upper semi-continuous function of y for all $q \geq 0$;*
(iii) *if $h^q(X_y, \mathcal{S}_y)$ is constant, then $f_{*q}\mathcal{S}$ is locally free;*
(iv) (Grauert's base change theorem) *if $h^q(X_y, \mathcal{S}_y)$ is constant, then the "base-change map" $(f_{*q}\mathcal{S})_y/\mathcal{J}_y \cdot (f_{*q}\mathcal{S})_y \to H^q(X_y, \mathcal{S}_y)$ is bijective.*

Property (iv) is also called the base change property.

Let X, Y be reduced complex spaces, $f : X \to Y$ a proper analytic map and $\mathcal{S}$ a coherent sheaf on X. If $y \in Y$, then $(f_{*q}\mathcal{S})_y$ is a finite $\mathcal{O}_{Y,y}$-module by Theorem 8.2; we put

$$\varprojlim_k (f_{*q}\mathcal{S})_y/\mathcal{J}_y^k(f_{*q}\mathcal{S})_y = (\widehat{f_{*q}(\mathcal{S})})_y$$

where $\mathcal{J}_y$ is the maximal ideal in $\mathcal{O}_{Y,y}$. There is a natural homomorphism

$$h^q : (\widehat{f_{*q}(\mathcal{S})})_y \to \varprojlim_k H^q(X_y, \mathcal{S}/(f^*(\mathcal{J}^k))\mathcal{S}).$$

(8.6) **Theorem** (Grauert's comparison theorem). *The homomorphism h^q is an isomorphism.*

A more elementary auxiliary result, that we shall use several times is

(8.7) **Theorem** (Levi's extension theorem). *Let X be an irreducible complex space and $A \subset X$ an analytic subset of codimension at least 2. Then every meromorphic function on $X \setminus A$ extends uniquely to a meromorphic function on X. If X is a complex manifold and the function holomorphic, then the extension is also holomorphic.*

The last part is frequently called Riemann's extension theorem.

For a proof of Levi's theorem we refer to [F], p. 185.

For the proof of the following theorem we refer to [Lj].

(8.8) **Theorem.** *Let X be a complex space and A an analytic subset of X. Then there exists a triangulation of $|X|$ in which A appears as the support of a sub-complex. In particular, there exists an open neighbourhood E of A such that E and $\bar{E}$ can be retracted onto A.*

Let X be a complex space and A an irreducible, compact analytic subset of X. Then it is possible to attach to A an element $a \in H_i(|X|, \mathbb{Z})$, with

$i = 2 \dim A$. This can be done for example by the Borel-Haefliger method ([B-Ha]). In the case that A has a desingularization (e.g., by Hironaka ([Hik64]), when A is a projective-algebraic variety), you can just take for a the image of the fundamental class of the desingularization in $H_i(|X|, \mathbb{Z})$. For curves on surfaces resolution of singularities is much more elementary, as we shall see later (Chap. II, 10).

Finally we shall need

(8.9) **Proposition** (Properness criterion). *Let $f : X \to Y$ be a holomorphic map of complex spaces, $y \in Y$, and $A \subset X$ a connected component of the fibre $f^{-1}(y)$. If A is compact, then there is an open neighbourhood $U \subset X$ of A such that $f|U$ is proper.*

Proof. Let $V \subset X$ be some open neighbourhood of A such that $\bar{V}$ is compact in X. Since $f^{-1}(y)$ is closed, we may take V so small that $V \cap f^{-1}(y) = A$. Then $f(\partial V) \subset Y$ is compact with $y \notin f(\partial V)$. So $U = V \setminus f^{-1}(f(\partial V))$ is an open neighbourhood of A. Whenever $K \subset f(U)$ is compact, then $(f|U)^{-1}(K) = U \cap f^{-1}(K)$ is closed in V, hence compact.

9. The σ-Process

Let $(z_1, \ldots, z_n)$ be the standard coordinates in $\mathbb{C}^n, n \geq 2$, and let $(\xi_1 : \ldots : \xi_n)$ be the standard homogeneous coordinates in $\mathbb{P}_{n-1}$. We take a neighbourhood U of $a = (a_1, \ldots, a_n)$ in $\mathbb{C}^n$, and consider on the product $U \times \mathbb{P}_{n-1}$ the subset $\bar{U}$ given by the equations $(z_i - a_i)\xi_j - (z_j - a_j)\xi_i = 0, i, j = 1, \ldots, n$. Application of the jacobian criterion shows that $\bar{U}$ is an n-dimensional complex submanifold of $U \times \mathbb{P}_{n-1}$; the projection $p : \bar{U} \to U$ maps $\bar{U} \setminus p^{-1}(a)$ biregularly onto $U \setminus a$, whereas $p^{-1}(a)$ is an $(n-1)$-dimensional submanifold of $\bar{U}$, isomorphic to $\mathbb{P}_{n-1}$.

Sometimes we shall refer to this procedure as the σ-process, applied to U in a, or we shall say that we have applied a monoidal transformation to U with centre a. But most frequently we shall say that $\bar{U}$ is obtained from U by blowing up (in) a.

Using local coordinates one can blow up any point x_0 of an n-dimensional complex manifold ($n \geq 2$); up to isomorphism the result is independent of the coordinates used. The resulting complex manifold will be denoted by $\bar{X}(x_0)$ or simply $\bar{X}$, and often we shall loosely speak of the blowing-up $p : \bar{X} \to X$ (at some given point). The divisor $E = p^{-1}(x_0)$, which is isomorphic to $\mathbb{P}_{n-1}$, will be called the exceptional divisor.

In the following theorem we collect some properties of the σ-process, which will be used at one place or another in this book.

(9.1) **Theorem.** *Let X be a complex manifold of dimension ≥ 2 and $p : \bar{X} \to X$ the blowing-up of X at some point. Then $\mathcal{N}_{E/\bar{X}} \cong \mathcal{O}_{\mathbb{P}_{n-1}}(-1)$, and*

(i) *p induces an isomorphism between the fields of meromorphic functions on X and $\bar{X}$. In particular, if X (and hence $\bar{X}$) is compact, then $\mathrm{a}(\bar{X}) = \mathrm{a}(X)$.*

(ii) *$p_*(\mathcal{O}_{\bar{X}}) = \mathcal{O}_X$ and $p_{*i}(\mathcal{O}_{\bar{X}}) = 0$ for $i \geq 1$.*
(iii) *$p^* : H^i(X, \mathcal{O}_X) \to H^i(\bar{X}, \mathcal{O}_{\bar{X}})$ is an isomorphism for all $i \geq 0$.*
(iv) *$p^* : H^i(X, \mathbb{Z}) \to H^i(\bar{X}, \mathbb{Z})$ is bijective for $i = 1$ and injective for $i = 2$. Furthermore,*

$$H^2(\bar{X}, \mathbb{Z}) \cong p^*(H^2(X, \mathbb{Z})) \oplus \mathbb{Z}\{e\}, \qquad \textit{where } e = c_1(\mathcal{O}_X(E)).$$

(v) *For every $a \in H^2(X, \mathbb{Z})$ we have $p_! p^*(a) = a$.*
(vi) *$p^* : H^1(X, \mathcal{O}_X^*) \to H^1(\bar{X}, \mathcal{O}_{\bar{X}}^*)$ is injective, and thus Pic($\bar{X}$) is isomorphic to the product of Pic(X) and the infinite cyclic subgroup, generated by $\mathcal{O}_{\bar{X}}(E)$,*
(vii) *$\mathcal{K}_{\bar{X}} = p^*(\mathcal{K}_X) \otimes \mathcal{O}_{\bar{X}}((\dim X - 1)E)$.*
(viii) *p induces an isomorphism $p^* : \Gamma(X, \mathcal{K}_X^{\otimes m}) \to \Gamma(\bar{X}, \mathcal{K}_{\bar{X}}^{\otimes m})$ for all $m \geq 1$, so if X is compact, $P_m(\bar{X}) = P_m(X)$ for $m \geq 1$ and $\mathrm{kod}(\bar{X}) = \mathrm{kod}(X)$.*

In Chap. II we shall consider blowing-up of points more in detail, but then only for the case that X is a surface.

As to the proof of all these facts, the isomorphism $\mathcal{N}_{E/\bar{X}} \cong \mathcal{O}_{\mathbb{P}_{n-1}}(-1)$ is elementary, whereas (i) and (viii) are direct consequences of Levi's extension theory for meromorphic functions (Theorem 8.7). The Leray spectral sequence combined with (ii) yields (iii). For (iv) we refer to [Ae], p. 269, (v) is a direct consequence of (iv) and Lemma 1.1. Then (vi) is a consequence of (iv), (v) and the exponential sequence. Property (vii) is again elementary.

So it remains to prove (ii). The first statement, namely that p induces an isomorphism $\mathcal{O}_X \xrightarrow{\sim} p_*\mathcal{O}_{\bar{X}}$ follows from (i). As to the vanishing of $p_{*i}\mathcal{O}_{\bar{X}}$, $i \geq 1$, it is clear that $p_{*i}\mathcal{O}_{\bar{X}} = 0$ outside of x_0 (in fact you use that $H^i(\mathcal{O}_B) = 0$ for all $i \geq 1$ and any open ball B). So the only thing left is to prove that the finite-dimensional complex vector space $(p_{*i}\mathcal{O}_{\bar{X}})_{x_0}$ reduces to 0 for $i \geq 1$. By Krull's theorem ([G-R71], p. 211) it is sufficient to show that $(p_{*i}\mathcal{O}_{\bar{X}})_{x_0}/\mathcal{J}_{x_0}^k(p_{*i}\mathcal{O}_{\bar{X}})_{x_0} = 0$ for all $i \geq 1$. Grauert's comparison theorem shows that the claim is proven as soon as we know that $H^i(E, \mathcal{O}_{\bar{X}}/\mathcal{J}_{E,\bar{X}}^k) = 0$ for all k and all $i \geq 1$. Using the exact sequence

$$0 \to \mathcal{J}_E^k/\mathcal{J}_E^{k+1} \to \mathcal{O}_{\bar{X}}/\mathcal{J}_E^{k+1} \to \mathcal{O}_{\bar{X}}/\mathcal{J}_E^k \to 0$$

and the isomorphism $\mathcal{J}_E^k/\mathcal{J}_E^{k+1} \cong \mathcal{N}_{E/\bar{X}}^{\otimes(-k)} \cong \mathcal{O}_{\mathbb{P}_{n-1}}(k)$, this follows by induction.

10. Deformations of Complex Manifolds

A family of compact, complex manifolds $\mathcal{X} = (X, p, S)$ consists of
(i) a pair of connected complex spaces: X and S;
(ii) a surjective, proper holomorphic map $p : X \to S$ which is flat (i.e., $\mathcal{O}_{X,x}$ is a flat $\mathcal{O}_{S,p(x)}$-module for all $x \in X$) and whose fibres X_s are all connected manifolds.

The space S is the base (space) of the family, or the family is a family over S or parametrised by S. If X and S are smooth, then flatness implies that p

is everywhere of maximal rank and the family $\mathfrak{X}$ is called a smooth family. A theorem of Ehresmann (compare [M-K], p. 19) says that a smooth family is differentiably locally trivial; in particular all of its fibres are diffeomorphic.

The simplest examples are given by trivial families, i.e., products $X = V \times S$, where V is a connected compact manifold and S some connected complex space, or by locally trivial families, i.e., holomorphic fibre bundles. (It is by no means true that a smooth family is automatically a fibre bundle. Consider for example the family of elliptic curves in $\mathbb{P}_2$ given by

$$X = \{((x_0 : x_1 : x_2), \lambda) \in \mathbb{P}_2 \times \mathbb{C} \mid x_0^3 + x_1^3 + x_2^3 = 3\lambda x_0 x_1 x_2\}.$$

The family is smooth over $\{\lambda \in \mathbb{C}; \lambda^3 \neq 1\}$ and, up to a multiplicative constant, the j-invariant of X_λ is $\lambda^3(\lambda^3+8)^3(\lambda^3-1)^{-3}$, so not all fibres are isomorphic.)

Let $\mathfrak{X} = (X, p, S)$ be any family, S' a complex space and $f : S' \to S$ a holomorphic map. Then the fibred product $X \times_S S'$ is in a natural way a family over S', called the pull-back of $\mathfrak{X}$ by f. If $\mathfrak{X}_i = (X_i, p_i, S_i), i = 1, 2$, is a pair of families, then a morphism from $\mathfrak{X}_1$ into $\mathfrak{X}_2$ consists of a pair of analytic maps, $g : X_1 \to X_2$ and $f : S_1 \to S_2$, such that $p_2 g = f p_1$. There is always a morphism from the pull-back to the original family.

(10.1) **Theorem** (Local-triviality theorem of Grauert-Fischer). *A smooth family of compact complex manifolds is locally trivial if and only if all fibres are analytically isomorphic.*

This theorem is proved in [F-G].

Let V be a connected compact complex manifold. A deformation of V, parametrised by (or over) the complex space S consists of a connected complex space S, a base point $s_0 \in S$, a family $\mathfrak{X} = (X, p, S)$ and an isomorphism from V onto the fibre X_{s_0}. (Confusingly, the fibres of p are also called deformations of V.) Morphisms between deformations are defined as "base point preserving" morphisms between the families in question which are compatible with the isomorphisms from V onto the fibres over the base points.

In the sequel we shall consider only germs of deformations, tacitly identifying two deformations over S, both with base point s_0, if the coincide in a neighbourhood of this point. Thus if S is smooth in s_0, we can assume all deformations over S (with base point s_0) to be smooth (if the base of a family is smooth, then so is the family). Let again V be a fixed manifold. A deformation $\mathfrak{X} = (X, p, S)$ of V with base point s_0 is called (locally) complete if (locally) every deformation $\mathfrak{X}' = (X', p', S')$ of V with base point s_0' is obtained as the pull-back from $\mathfrak{X}$ by a suitable analytic map $f : S' \to S$ with $f(s_0') = s_0$. If in addition f is always uniquely determined by $\mathfrak{X}'$, the deformation $\mathfrak{X}$ is called (locally) universal. As soon as it exists, the universal deformation of V is unique up to isomorphisms. Although it exists in many cases such as for curves and tori, it does not exist always, as the following example shows. Take

$$X = \{((y_0 : y_1), (x_0 : x_1 : x_2), t) \in \mathbb{P}_1 \times \mathbb{P}_2 \times \mathbb{C} \mid y_0^2 x_1 - y_1^2 x_0 - t y_0 y_1 x_2 = 0\}$$

and consider it as a family over $\mathbb{C}$ via projection onto the t-factor. For $t \neq 0$ the fibre is isomorphic to $\mathbb{P}_1 \times \mathbb{P}_1$, but $X_0 = \Sigma_2$, the Hirzebruch surface (see Chap V, Sect. 4) which is not isomorphic to $\mathbb{P}_1 \times \mathbb{P}_1$. If we take $X = X'$ and if $f(t) = t(1 + \varepsilon(t))$ ($\varepsilon(0) = 0$) is a holomorphic function on a neighbourhood U of $0 \in \mathbb{C}$, we can lift f to a holomorphic map g defined for $t \in U$ by setting

$$g((y_0 : y_1), (x_0 : x_1 : x_2), t) = (((1+\varepsilon(t))y_0 : y_1), ((1+\varepsilon(t))^2 x_0 : x_1 : x_2), f(t)).$$

Since for $\varepsilon(t)$ one can take any holomorphic function with $\varepsilon(0) = 0$ we see that $\{X_t\}_{t \in U}$ can never be universal at 0.

Exactly because of such examples the notion of versal deformation has been introduced. A deformation $\mathcal{X} = (X, p, S)$ of V with base point s_0 is called versal if it has the following property: for every $\mathcal{X}'$ as above there is an f as above which need not be unique, but whose derivative at s_0' is uniquely determined. The preceding example gives a versal deformation of Σ_2. Versal deformations always exist:

(10.2) **Theorem** (Kuranishi's theorem). *Every compact complex manifold has a versal deformation.*

A proof of this basic result can be found in [Dy]. The family constructed by Kuranishi is called the Kuranishi family.

Furthermore there are the following facts, for which we refer to [Wav] and [Dy].

(10.3) *The versal deformation in Theorem 10.2 is complete for any of its fibres, and versal for any of its fibres as soon as $h^1(X_s, \mathcal{T}_{X_s})$ is constant.*

(10.4) *If $H^2(V, \mathcal{T}_V) = 0$, then V admits a smooth versal deformation.*

(10.5) *If $H^0(V, \mathcal{T}_V) = 0$, then V has a universal deformation.*

(10.6) *If a universal deformation exists, then every versal deformation is isomorphic to it.*

Remark. In the case 10.5, i.e., if $H^0(V, \mathcal{T}_V) = 0$, then the universal deformation $\mathcal{X} = (X, p, S)$ has an additional property: any deformation $\mathcal{X}' = (X', p', S')$ of X not only determines the inducing map from S' into S, but also the map from X' into X, i.e., the whole morphism from $\mathcal{X}'$ into $\mathcal{X}$ (see [L-P], p. 170). This does not hold for any universal deformation.

Now let $\mathcal{X} = (X, p, S)$ be a smooth deformation of V with base point s_0. From the exact cohomology sequence of

$$0 \to \mathcal{T}_{p^{-1}(s_0)} \to \mathcal{T}_X | p^{-1}(s_0) \to \mathcal{N}_{p^{-1}(s_0)/X} \to 0$$

we obtain a morphism

$$\Gamma(p^{-1}(s_0), \mathcal{N}_{p^{-1}(s_0)/X}) \to H^1(p^{-1}(s_0), \mathcal{T}_{p^{-1}(s_0)}).$$

Since $\mathcal{N}_{p^{-1}(s_0)/X}$ is trivial, the first of these spaces is naturally isomorphic to $\mathcal{T}_S(s_0)$, and since the second can be identified with $H^1(V, \mathcal{T}_V)$ because of the *given* isomorphism from V onto $p^{-1}(s_0)$, we obtain a homomorphism

$$\rho_{\mathcal{X}} : \mathcal{T}_S(s_0) \to H^1(V, \mathcal{T}_V)$$

which is called the Kodaira-Spencer map. Using the Zariski tangent space (to S at s_0) it can be defined for any family, smooth or not, but we do not need this.

A classical result of Kodaira and Spencer (see [K-S58]) is

(10.7) **Theorem.** *A smooth family is complete if and only if the Kodaira-Spencer map is surjective.*

If $\mathcal{X}'$ is the pull-back of S by an analytic map $f : S' \to S$, then there is a commutative diagram

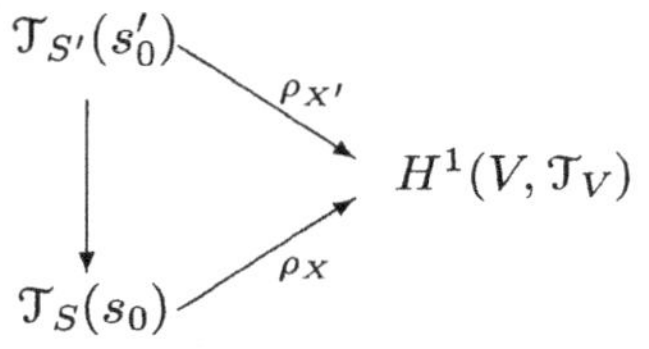

Consequently, if $\mathcal{X}$ is complete and $\rho_{\mathcal{X}}$ bijective, then $\mathcal{X}$ is versal.

Next, suppose we are given a complete smooth deformation $\mathcal{X}$ of V. The Kodaira-Spencer map is surjective, and we may select a smooth subspace S' of the base space of $\mathcal{X}$, such that $\rho_{\mathcal{X}}|\mathcal{T}_S(s_0)$ is an isomorphism. The above diagram shows that this isomorphism coincides with the Kodaira-Spencer map of the restricted family $\mathcal{X}|S'$. By the preceding remark, this family is versal. Consequently: *if V admits a smooth complete deformation, then it admits a versal smooth deformation.*

Differential Geometry of Complex Manifolds

Projective manifolds are examples of Kähler manifolds and several analytic results concerning these will be used in subsequent chapters. This includes the Hodge decomposition theorem and the existence of the Albanese and Picard torus. But also, less classical, the basic results of Yau concerning the existence of particular types of Kähler metrics, the so-called Kähler-Einstein metrics. This all will be reviewed in this sub-chapter.

11. De Rham Cohomology

Let X be a compact, connected, oriented n-dimensional differentiable manifold. We write $\mathcal{D}^p_X$ for the sheaf of real-valued C^∞-forms of rank p on X, and $H^p_{DR}(X)$ for the p-th de Rham group, i.e.,

$$H^p_{DR}(X) = \{\xi \in \Gamma(\mathcal{D}^p_X) \mid d\xi = 0\}/d\Gamma(\mathcal{D}^{p-1}_X).$$

The p-th homology group $H_p(X,\mathbb{Z})$ for the complex of singular chains is the same as the one for the sub-complex of differentiable chains. So Stokes' theorem implies that there is a well-determined homomorphism

$$H^p_{DR}(X) \to \mathrm{Hom}_{\mathbb{Z}}(H_p(X,\mathbb{Z}),\mathbb{R}),$$

defined by sending the class of a closed p-form ξ to the functional

$$c \longmapsto \int_\gamma \xi,$$

where the differentiable chain γ represents the class $c \in H_p(X,\mathbb{Z})$. De Rham's theorem asserts (see [S-T67], p. 147) that in fact this is an isomorphism. The right hand side is of course nothing but the p-th singular cohomology group $H^p(X,\mathbb{R})$. So we obtain the de Rham isomorphism

$$H^p_{DR}(X) \xrightarrow{\sim} H^p(X,\mathbb{R}). \tag{12}$$

The Kronecker pairing

$$H^p(X,\mathbb{R}) \times H_p(X,\mathbb{R}) \to \mathbb{R}$$

translates into integration

$$(x,c) \longmapsto \int_\gamma \xi,$$

where ξ represents x and γ represents c.

If we compare the perfect pairing

$$H^p(X,\mathbb{R}) \times H^{n-p}(X,\mathbb{R}) \to H^n(X,\mathbb{R}) \cong \mathbb{R}$$

given by the cup product with the Kronecker pairing, we obtain an isomorphism between $H_p(X,\mathbb{R})$ and $H^{n-p}(X,\mathbb{R})$ which is nothing but Poincaré duality $\mathcal{P}=\mathcal{P}_X$. In other words, we have

$$x\cdot y=\int_{\mathcal{P}(x)} y, \tag{13}$$

where $x\in H^p(X,\mathbb{R})$ and $y\in H^{n-p}(X,\mathbb{R})$.

If the singular cohomology over $\mathbb{R}$ is naturally identified with the sheaf cohomology over $\mathbb{R}$ (for example by way of a triangulation) then the isomorphism between de Rham cohomology and sheaf cohomology we thus obtain is nothing else but the isomorphism, obtained from the resolution

$$0\to\mathbb{R}_X\to\mathcal{D}_X^0\to\ldots\to\mathcal{D}_X^n\to 0.$$

We occasionally need to work with currents instead of forms. We briefly explain this, referring to [Rh] for details of what follows.

Let X be an n-dimensional differentiable manifold. We endow the vector space of smooth p-forms on X with the following topology. We say that we have a converging sequence $\alpha_i\to\alpha$ of p-forms, if on any coordinate patch any given derivative of a given coordinate function of α_i converges uniformly on compact sets to the corresponding derivative of the coordinate function of α. We denote this topological vector space by $D^p(X)$. Similarly, compactly supported smooth p-forms form a topological vector space $D_c^p(X)$. Here, by definition $\alpha_i\to\alpha$, if for $i\gg 0$ the α_i have support in a common compact subset $K\subset X$ and on this set the convergence is as in $D^p(K)$. The space $D^p(X)$ is an example of a so-called **Fréchet space**. By definition a Fréchet space is a vector space which is complete with respect to a countable family of semi-norms having the property that only for the 0-vector all semi-norms vanish simultaneously. Note that in general $D_c^p(X)$ is not Fréchet.

A **current of degree** $(n-p)$ is a continuous linear function $T: D_c^p(X)\to\mathbb{R}$. This means that T is sequentially continuous, i.e., $\alpha_i\to\alpha$ implies $T(\alpha_i)\to T(\alpha)$. Currents of degree $(n-p)$ form a vector space. We give it the **weak topology** by demanding that $T_n\to T$ if and only if $T_n(\alpha)\to T(\alpha)$ for all $\alpha\in D_c^p(X)$. As such we denote it by $D'^{n-p}(X)$. The corresponding sheaf is denoted $\mathcal{D}'^{n-p}_X$.

Any smooth p-form α defines a p-current by sending any compactly supported $(n-p)$-form β to $\int_X\alpha\wedge\beta$. Any current so obtained is also called a **smooth current**. So the smooth p-currents form a subspace of the space of degree p currents. One can extend the d-operator on forms to currents, by taking the transpose of $d: D_c^p(U)\to D_c^{p+1}(U))$ up to sign $(-1)^p$. Since $d\circ d=0$ on currents, one can consider the corresponding cohomology group

$$H'^p_{\mathrm{DR}}(X):=\{T\in D'^p(X))\mid dT=0\}/dD'^{p-1}(X).$$

Viewing p-forms as degree p currents, the de Rham space $H^p_{\mathrm{DR}}(X)$ naturally maps to this space. By loc. cit. §18 Théorème 14 this is an isomorphism:

(14) $$H^p_{\mathrm{DR}}(X) \xrightarrow{\sim} H'^p_{\mathrm{DR}}(X).$$

As to duality, we first remark that any differentiable $(n-p)$-chain γ on X defines a p-current by the assignment $\alpha \mapsto \int_\gamma \alpha$, $\alpha \in \Gamma_c(X, \mathcal{D}^p_U)$ the current of integration of γ. This current is closed and compactly supported (its support is γ) and hence and defines an element in the de Rham group $H'^p_c(X,\mathbb{R})$ of closed p-currents with compact support modulo exact currents (of compact support). We can identify $H^p_c(X,\mathbb{R})$ and $H'^p_c(X,\mathbb{R})$ through the inclusion $\mathcal{D}^p_c(X) \hookrightarrow \mathcal{D}'^p_c(X)$ as in (14). If we do this, the homology group $H_{n-p}(X,\mathbb{R})$, calculated by means of differentiable $(n-p)$-chains maps to $H^p_c(X,\mathbb{R})$. By loc. cit. §23, in particular Théorème 18 and the discussion following Théorème 19, this is an isomorphism:

$$H_{n-p}(X,\mathbb{R}) \xrightarrow{\sim} H^p_c(X,\mathbb{R}).$$

It extends the (inverse of the) Poincaré duality isomorphism to the case of non-compact manifolds.

12. Dolbeault Cohomology

Let X be a connected compact complex manifold of dimension n. We denote by $\mathcal{D}^{p,q}_X$ the sheaf of $\mathbb{C}$-valued C^∞-forms of type (p,q) on X, and by $H^{p,q}(X)$ the Dolbeault (cohomology) group

$$H^{p,q}(X) = \{\alpha \in \Gamma(\mathcal{D}^{p,q}_X) \mid \bar\partial\alpha = 0\}/\bar\partial\Gamma(\mathcal{D}^{p,q-1}_X).$$

Then we have Dolbeault's isomorphism ([Hir66], p. 119)

$$H^{p,q}(X) \cong H^q(X,\Omega^p_X), \qquad \text{and} \qquad \dim H^{p,q}(X) = h^{p,q}(X).$$

The complexified de Rham-complex $\{\Gamma(\mathcal{D}^\bullet_X \otimes \mathbb{C}), d\}$ is the total complex, associated to the double complex $\{\Gamma(\mathcal{D}^{\bullet,\bullet}_X, \partial, \bar\partial)\}$. Its cohomology groups are $H^k_{DR}(X) \otimes \mathbb{C} = H^k(X,\mathbb{R}) \otimes \mathbb{C} = H^k(X,\mathbb{C})$. Hence the spectral sequence for the double complex reads:

$$E_1^{p,q} = H^{p,q}(X) \Rightarrow H^{p+q}(X,\mathbb{C}).$$

This spectral sequence is called Fröhlicher spectral sequence, and the resulting filtration on $H^{p+q}(X,\mathbb{C})$ is called the Hodge filtration. Explicitly, we have

$$H^k(X,\mathbb{C}) = F^0(H^k) \supset F^1(H^k) \supset \ldots \supset F^k(H^k) \supset 0,$$

where

(15) $$F^p(H^k) = \{[\alpha], \alpha \in \bigoplus_{\substack{p'+q'=k \\ p' \geq p}} \Gamma(\mathcal{D}^{p',q'}_X) \mid d\alpha = 0\}.$$

For later use we need a straightforward criterion for degeneration of spectral sequences obtained by applying [Del71], Prop. 1.3.2.

(12.1) **Lemma.** *The Fröhlicher spectral sequence degenerates at E_1 if and only if d is* strictly compatible with the Hodge filtration, i.e., *every exact form which is a sum of terms of types $(p,q),(p+1,q-1),\ldots,(p+q,0)$ can be written as $d\beta$ where β is decomposable into types $(p,q-1),\ldots,(p+q-1,0)$.*

Now the isomorphism of Dolbeault can be generalized to the case of holomorphic vector bundles $\mathcal{V}$, where it says

$$H^p(X,\mathcal{V}) \cong \{\alpha \in \Gamma(\mathcal{V}\otimes\mathcal{D}^{0,q}) \mid (1\otimes\bar\partial)\alpha = 0\}/(1\otimes\bar\partial)\Gamma(\mathcal{V}\otimes\mathcal{D}^{0,q-1}).$$

In Chap. II we shall need the following explicit description of a special case of Serre duality. As recalled in Theorem 5.3 we have $h^p(\mathcal{V}) = h^{n-p}(\mathcal{V}^\vee\otimes\mathcal{K}_X)$ for any holomorphic vector bundle $\mathcal{V}$.

A duality between $H^i(\mathcal{V})$ and $H^{n-i}(\mathcal{V}^\vee\otimes\mathcal{K}_X)$ can then be obtained in the following way. Considering $H^p(\mathcal{V})$ and $H^{n-p}(\mathcal{V}^\vee\otimes\mathcal{K}_X)$ as Dolbeault groups there is a pairing

$$H^p(\mathcal{V})\otimes H^{n-p}(\mathcal{V}^\vee\otimes\mathcal{K}_X) \to H^n(\mathcal{K}_X) \cong \mathbb{C} \tag{16}$$

and we have

(12.2) **Proposition.** *The pairing (16) of Dolbeault groups is a perfect pairing.*

For this result we refer to Serre's original paper [Se55a], p. 17–20.

13. Kähler Manifolds

Recall that there always exist hermitian metrics on a compact complex manifold X and that any such metric can be given by its associated real $(1,1)$-form. A real $(1,1)$-form α is called positive, denoted $\alpha > 0$, if in local coordinates $(z_1,\ldots,z_n)$ we have $\alpha = \mathrm{i}\sum_{i,j=1}^n \alpha_{ij}dz_i\wedge d\bar z_j$, such that at very point $p\in X$ the hermitian matrix $(\alpha_{ij}(p))$ is positive definite. This condition is independent of the choice of the local coordinates. The $(1,1)$-form associated to a hermitian metric is positive and conversely. A metric whose $(1,1)$-form is closed is called a Kähler metric.

A complex manifold which can be provided with at least one Kähler metric is called a Kähler manifold. Since every submanifold of a Kähler manifold is again a Kähler manifold, and since the Fubini-Study metric on $\mathbb{P}_n$ is kählerian, we have that every projective algebraic manifold is a Kähler manifold. The existence of non-algebraic tori (on which the standard flat metric is a Kähler metric) shows that, even for compact manifolds, the converse does not hold (compare Theorem 19.4.)

A cohomology class $h\in H^2(X,\mathbb{R})$ is called a Kähler class if it can be represented by a Kähler form.

Now let X be a compact connected Kähler manifold and Y a compact, connected complex manifold of dimension k. Suppose there is a holomorphic map $f: Y\to X$ such that $f(Y)$ is a k-dimensional complex subvariety of X. Then if ω is any Kähler form on X, we have

$$\int_Y f^*(\omega^k) > 0.$$

So if $h \in H^2(X, \mathbb{R})$ is the cohomology class of ω, then $f^*(h^k) > 0$. Applying Lemma 1.1 we see that $h^k f_!(1) > 0$, with $f_!(1)$ dual to the homology class of $f(Y)$.

By resolution of singularities ([Hik71]) we thus obtain

(13.1) **Lemma.** *If X is a compact Kähler manifold, h a Kähler class on X and y dual to the homology class of any k-dimensional closed analytic subvariety on X, then $h^k y > 0$.*

Remark. Hironaka's theorem can be avoided at this point by using a suitable integration theory for forms on arbitrary analytic subsets of a complex manifold.

Next, we need a few general facts concerning the existence of a Hodge decomposition. To start, note that the E_1-term of the Fröhlicher spectral sequence is the Dolbeault-space $H^{p,q}(X)$ which by definition maps to $F^p H^{p+q}(X)$ and so to the graded quotient $F^p H^{p+q}(X)/F^{p+1}H^{p+q}(X)$ which is the E_∞-term. So, if the Fröhlicher spectral sequence degenerates, $H^{p,q}(X)$ can be identified with the preceding quotient. If there is a formal Hodge decomposition

$$H^k(X, \mathbb{C}) = \bigoplus_{p+q=k} {}'H^{p,q}, \text{ with } \quad {}'H^{p,q} = F^p(H^k) \cap \overline{F^q(H^k)}$$

then this quotient is isomorphic to ${}'H^{p,q}$ and so $H^{p,q}$ and ${}'H^{p,q}$ get identified in a natural way.

For a compact Kähler manifold this is all true, as follows from the following important fact for which we refer to [Wei71] and [Del68].

(13.3) **Theorem.** *Let X be a compact Kähler manifold. Then*
(i) *the Fröhlicher spectral sequence degenerates at E_1-level;*
(ii) *there is a formal Hodge decomposition as above.*

As a consequence, for compact Kähler manifolds X the spaces $H^{p,q}$ and ${}'H^{p,q}$ can be identified in this way and we obtain the famous Hodge decomposition.

(13.4) **Corollary.** *For any compact Kähler manifold there is a direct sum decomposition*

$$H^k(X, \mathbb{C}) = \bigoplus_{p+q=k} H^{p,q}(X).$$

Furthermore we clearly have

(13.5) **Corollary.** *If X is a compact Kähler manifold, then*
(i) $h^{p,q}(X) = h^{q,p}(X)$
(ii) $b_k(X) = \sum_{p+q=k} h^{p,q}(X)$.

As explained above, the existence of such a Hodge decomposition already follows when the Fröhlicher spectral sequence degenerates and if in addition, one only has a formal Hodge decomposition. This will be an important remark for surfaces, as well as the validity of the $\partial\bar{\partial}$-lemma under these hypotheses, which we now shall explain.

(13.6) **Lemma** ($\partial\bar{\partial}$-Lemma). *Suppose that X is a compact complex manifold such that the following conditions hold:*
(1) *the Fröhlicher spectral sequence degenerates at E_1,*
(2) *there is a formal Hodge decomposition.*
Then the following assertions hold:
(i) *the subspace $'H^{p,q}(X) \subset H^{p+q}(X,\mathbb{C})$ coincides with the subspace representable by closed forms of type (p,q);*
(ii) *any cohomology class can be represented by a form which is ∂- as well as $\bar{\partial}$-closed;*
(iii) *for any d-closed (p,q)-form α on X the following statements are then equivalent:*
 (a) $\alpha = d\beta$ *for some $(p+q-1)$-form β;*
 (b) $\alpha = \bar{\partial}\beta''$ *for some $(p,q-1)$-form β'';*
 (c) $\alpha = \partial\bar{\partial}\gamma$ *for some $(p-1,q-1)$-form γ.*

Proof. (Compare the proof of [Del71], Prop. 4.3.1)
(i) By definition, any class $a \in {}'H^{p,q}(X)$ can be represented by a form α_1, α_2 respectively, whose components of type $(r, p+q-r)$ vanish for $r < p$, $r > p$ respectively. Then $\alpha_1 - \alpha_2 = d\beta$. Splitting the degree $(p+q-1)$-form β into types

$$\beta = \underbrace{\beta^{p+q-1,0} + \cdots + \beta^{p,q-1}}_{\beta_1} + \underbrace{\beta^{p-1,q} + \cdots + \beta^{0,p+q-1}}_{\beta_2},$$

and comparing types, we get a closed form $\alpha_1 - d\beta_1 = \alpha_2 + d\beta_2$ of pure type (p,q) with class a.
(ii) By assumption we have a formal Hodge decomposition and so every complex de Rham class can be written as a sum of classes of pure type and a d-closed form of pure type is automatically ∂- as well as $\bar{\partial}$-closed.
(iii) That (a) and (b) are equivalent follows when we note that having a formal Hodge decomposition implies $F^{p+1} \cap \overline{F^q} = \{0\}$. So $[\alpha] \in {}'H^{p,q} = F^p \cap \overline{F^q}$ is the zero-class if it belongs to F^{p+1}, i.e., if its image in the Dolbeault group $H^{p,q}$ is zero. This is the case if and only if $\alpha = \bar{\partial}\beta''$ for some $(p,q-1)$-form β''.
We now prove that (a) and (c) are equivalent. Clearly (c) implies (a). To prove the converse, assume that α is exact. We know that the Fröhlicher

spectral sequence degenerates so that we can invoke Lemma 12.1. So we can find a $(k-1)$-form $\beta_1 = \beta_1^{k-1,0} + \cdots + \beta_1^{p,k-1-p}$ with $\alpha = d\beta_1$. Since d is also strictly compatible with the conjugate Hodge filtration, there exists a $(k-1)$-form β_2 with $\beta_2 = \beta_2^{p,k-1-p} + \cdots + \beta_2^{0,k-1}$ with $\alpha = d\beta_2$. It follows that $d(\beta_1 - \beta_2) = 0$ and by (ii) we may find

$$\gamma = \underbrace{\gamma^{k-1,0} + \cdots + \gamma^{p,k-1-p}}_{\gamma_1} + \underbrace{\gamma^{p-1,k-p} + \cdots \gamma^{0,k-1}}_{\gamma_2}$$

with $\partial\gamma = \overline{\partial}\gamma = 0$ in the same class as $\beta_1 - \beta_2$. Since the Hodge components of γ are all closed, we have

$$\alpha = d(\beta_1 - \gamma_1).$$

But $(\beta_1 - \gamma_1) - (\beta_2 + \gamma_2) = \beta_1 - \beta_2 - \gamma$ is exact, say $= d\delta$. Let us now consider what happens with the type decomposition of δ under d. We write

$$\delta = \underbrace{\delta^{k-2,0} + \cdots + \delta^{p,k-p-2}}_{\delta_1} + \delta^{p-1,k-p-1} + \underbrace{\delta^{p-2,k-p} + \cdots \delta^{0,k-2}}_{\delta_2}.$$

Clearly, by comparing types, we see that $\beta_1 - \gamma_1 = d\delta_1 + \partial\delta^{p-1,k-p-1}$ and so

$$\alpha = d(\beta_1 - \gamma_1) = d\partial\delta^{p-1,k-p-1} = \overline{\partial}\partial\delta^{p-1,k-1-p}$$

as desired. □

As a corollary of the preceding discussion we indeed have the existence of a Hodge decomposition:

(13.7) **Corollary.** *Under the preceding assumptions on X, the natural map sending a $\overline{\partial}$-closed form of type (p,q) to its De Rham class induces an isomorphism*

$$H^{p,q}(X) \cong {}'H^{p,q} \subset H^{p+q}(X,\mathbb{C})$$

from the Dolbeault group $H^{p,q}(X)$ to the subspace of the de Rham classes representable by closed forms of type (p,q). Identifying these two spaces, we then have a Hodge decomposition

$$H^k(X,\mathbb{C}) = \bigoplus_{p+q=k} H^{p,q}(X).$$

For later reference we need to be able to describe $H^{p,q}(X)$ in terms of Bott-Chern cohomology:

$$H^{p,q}_{\mathrm{BC}}(X) := d\text{-closed forms of type } (p,q)/\partial\overline{\partial}\Gamma(X,\mathcal{D}_X^{p-1,q-1}).$$

(13.8) **Corollary.** *Under the preceding assumptions concerning the $(p+q)$-cohomology for X the natural homomorphism*

$$H^{p,q}_{\mathrm{BC}}(X) \to H^{p+q}_{\mathrm{DR}}(X, \mathbb{C})$$

which sends the class of a d-closed (p,q)-form to its de Rham class is injective with image ${}'H^{p,q}(X) \cong H^{p,q}(X)$.

Proof. This a direct consequence of the $\partial\bar{\partial}$-Lemma. □

Now let X be a connected compact Kähler manifold. To X there are associated two complex tori, both of dimension $g = h^{1,0}(X)$, namely the **Albanese torus** $\mathrm{Alb}(X)$ and the **Picard torus** $\mathrm{Pic}^0(X)$. We define them in a rather primitive way which however is the best one for our purposes.

Firstly, the Albanese torus. Let $\omega_1, \ldots, \omega_g$ be a basis for $\Gamma(X, \Omega^1_X)$. It follows from Corollary 13.4 that $\omega_1, \ldots, \omega_g, \bar{\omega}_1, \ldots, \bar{\omega}_g$ form a basis of $H^1(X, \mathbb{C})$. Furthermore, let $h_1, \ldots, h_{2g}$ be a basis for $H_1(X, \mathbb{Z})$ mod torsion. We consider the vectors

$$v_j = \left(\int_{h_j} \omega_1, \ldots, \int_{h_j} \omega_g \right) \in \mathbb{C}^g, \qquad j = 1, \ldots, 2g,$$

and claim that they are independent over $\mathbb{R}$. Indeed, if they were not, then the $2g$ vectors

$$w_j = \int_{h_j} \omega_1 + \bar{\omega}_1, \ldots, \int_{h_j} \omega_g + \bar{\omega}_g, \qquad \mathrm{i} \int_{h_j} \omega_1 - \bar{\omega}_1, \ldots, \mathrm{i} \int_{h_j} \omega_g - \bar{\omega}_g$$

would be dependent in $\mathbb{R}^{2g}$. Because of (12) there would be real numbers $\lambda_1, \ldots, \lambda_g, \mu_1, \ldots, \mu_g$, not all 0, such that

$$\lambda_1(\omega_1 + \bar{\omega}_1) + \ldots + \lambda_g(\omega_g + \bar{\omega}_g) + \mathrm{i}\mu_1(\omega_1 - \bar{\omega}_1) + \ldots + \mathrm{i}\mu_g(\omega_g - \bar{\omega}_g)$$

would be cohomologous to 0. But this would mean that $\omega_1, \ldots, \omega_g, \bar{\omega}_1, \ldots, \bar{\omega}_g$ would be $\mathbb{C}$-dependent, which is not the case.

So, from the real point of view, the vectors $v_1, \ldots, v_{2g}$ span a lattice in $\mathbb{C}^g$ and thus determine a complex torus. If we replace the h_j's or the ω_k's by another basis, then, up to isomorphism, we obtain the same torus. This torus is $\mathrm{Alb}(X)$.

Fixing a point $x_0 \in X$ define the holomorphic map $f : X \to \mathrm{Alb}(X)$ by $f(x) = \left(\int_{x_0}^{x} \omega_1, \ldots, \int_{x_0}^{x} \omega_g \right)$. Changing x_0 amounts to changing f by a a translation of $\mathrm{Alb}(X)$. Given any holomorphic map g from X into any complex torus T, then if we choose the proper origin on T, g is the composition of f and a unique homomorphism from $\mathrm{Alb}(X)$ into T. In fact, $\mathrm{Alb}(X)$ can be characterized by this property (see [Bl]) which in particular implies that $f(X)$ is never contained in a translate of a proper sub-torus and also that $X = \mathrm{Alb}(X)$ if X is a torus itself.

For later reference we formulate some useful properties of the Albanese torus.

(13.9) **Proposition**

i) *If* $q(X) = 0$ *any map from* X *to a torus must be constant; this holds in particular for* $X = \mathbb{P}_m$.

ii) *Any holomorphic map* $f : X \to Y$ *fits into a commutative diagram*

$$\begin{array}{ccc} X & \xrightarrow{\quad f \quad} & Y \\ \downarrow{\scriptstyle \alpha_X} & & \downarrow{\scriptstyle \alpha_Y} \\ \mathrm{Alb}\, X & \xrightarrow{\quad a(f) \quad} & \mathrm{Alb}\, Y \end{array}$$

iii) *If the image of the Albanese map* α *for* X *is a curve, then* α *is connected,* $\alpha(X)$ *is smooth and has genus* $q(X)$.

Statements i) and ii) follows from the universality property for the Albanese, while iii) follows by looking at the Stein factorisation for the Albanese map

$$\begin{array}{ccc} X & \xrightarrow{\quad \alpha \quad} & C = \alpha(X) \subset \mathrm{Alb}\, X. \\ {\scriptstyle f}\searrow & & \nearrow{\scriptstyle g} \\ & Y & \end{array}$$

Since X and $\mathrm{Alb}(X)$ are smooth, Y is a normal and hence smooth curve. We want to show that the finite map g in fact gives an immersion into the Albanese so that α is indeed connected. The map f fits into a commutative diagram

$$\begin{array}{ccccc} X & \xrightarrow{\quad f \quad} & Y & & \\ \downarrow{\scriptstyle \alpha_X} & & \downarrow{\scriptstyle \alpha_Y} & \searrow{\scriptstyle g} & \\ \mathrm{Alb}\, X & \xrightarrow{\quad a(f) \quad} & \mathrm{Jac}\, Y & \xrightarrow{\quad a(g) \quad} & \mathrm{Alb}\, X \end{array}$$

and since f is surjective, $a(f)$ must be surjective, as remarked before. Moreover since $a(g) \circ a(f) \circ \alpha_X = g \circ f = \alpha_X$, the universal property of α_X implies that $a(g) \circ a(f) = \mathrm{id}$ and so, since we already know that $a(f)$ is surjective, it must be an isomorphism with inverse $a(g)$. Since the Abel-Jacobi map is an embedding, this is also the case for $g = a(g) \circ \alpha_Y$. □

Remark. The Albanese torus has been defined by Blanchard for any compact, connected complex manifold. Observe in any case that as soon as we have a Hodge decomposition in dimension 1, then the preceding construction works also in the non-kählerian case (but there exist compact complex manifolds without such a Hodge decomposition).

The identity component of $\mathrm{Pic}(X)$ is the Picard torus $\mathrm{Pic}^0(X)$. We have

$$\mathrm{Pic}^0(X) = H^1(X, \mathcal{O}_X)/i(H^1(X, \mathbb{Z})),$$

where i comes from the exponential cohomology sequence

$$H^1(X, \mathbb{Z}) \xrightarrow{i} H^1(X, \mathcal{O}_X) \longrightarrow H^1(X, \mathcal{O}_X^*).$$

The point is that if X is a compact Kähler manifold, then $i(H^1(X, \mathbb{Z}))$ is always a lattice in $H^1(X, \mathcal{O}_X)$, and so $\mathrm{Pic}^0(X)$ a complex torus (of dimension $h^{1,0}(X)$). In general this is not the case, even if X is compact.

To prove this property of $\mathrm{Pic}^0(X)$ for a compact Kähler manifold, we observe that there is a factorization of i:

$$\begin{array}{ccc} H^1(X, \mathbb{Z}) & \xrightarrow[i]{\qquad\qquad} & H^1(X, \mathcal{O}_X) \\ & {\scriptstyle j}\searrow \quad \nearrow{\scriptstyle p} & \\ & H^1(X, \mathbb{C}). & \end{array}$$

Now the Hodge decomposition (Corollary 13.4):

$$H^1(X, \mathbb{C}) \cong H^{1,0}(X) \oplus H^{0,1}(X)$$

is such, that the projection onto the second factor becomes the p of the diagram. We shall not prove this here, but we shall prove a similar statement for dimension 2 in Chap. IV (Theorem 2.13). Once this is known, our claim follows immediately: the image of $H^1(X, \mathbb{Z})$ in $H^1(X, \mathbb{C})$ is a lattice in $H^1(X, \mathbb{R})$, and its projection into $H^{0,1}(X)$ remains a lattice, since p maps $H^1(X, \mathbb{R})$ isomorphically onto $H^{0,1}(X)$, if we consider also this last space as a real vector space.

The functional properties of $\mathrm{Pic}^0(X)$ which we shall use, are obvious from the definition.

14. Weight-1 Hodge Structures

Let $H_{\mathbb{C}}$ be a complex vector space of dimension $2g$. By definition, a Hodge structure of weight 1 on $H_{\mathbb{C}}$ consists of

(i) a $\mathbb{Z}$-submodule $H_{\mathbb{Z}} \subset H_{\mathbb{C}}$ of rank $2g$ spanning $H_{\mathbb{C}}$ over $\mathbb{C}$ ($H_{\mathbb{Z}}$ is called the integral lattice);

(ii) a direct sum decomposition $H_{\mathbb{C}} = H^{1,0} \oplus H^{0,1}$ with $H^{1,0} = \overline{H^{0,1}}$.

Here and in the sequel the bar denotes complex conjugation in $H_{\mathbb{C}} = H_{\mathbb{Z}} \otimes \mathbb{C}$ induced by conjugation on $\mathbb{C}$. Its fixed points form the real subvectorspace $H_{\mathbb{R}} = H_{\mathbb{Z}} \otimes \mathbb{R}$ of $H_{\mathbb{C}}$.

(14.1) **Example.** Let X be a compact Kähler manifold. Then $H = H^1(X,\mathbb{C})$ carries a weight-1 Hodge structure given by the Hodge-decomposition (Corollary 13.4) and integral lattice $H^1(X,\mathbb{Z})$.

Denote by $q : H_{\mathbb{C}}^{\vee} \to (H^{1,0})^{\vee}$ the projection dual to the inclusion $H^{1,0} \subset H_{\mathbb{C}}$. The kernel of q is $(H^{0,1})^{\vee}$. By property ii) we have $H_{\mathbb{R}}^{\vee} \cap (H^{0,1})^{\vee} = 0$, so q induces an $\mathbb{R}$-isomorphism $H_{\mathbb{R}}^{\vee} \to (H^{1,0})^{\vee}$ and the image $q(H_{\mathbb{Z}}^{\vee})$ in $(H^{1,0})^{\vee}$ is a lattice of maximal rank. The torus $\operatorname{Alb} H_{\mathbb{C}} = (H^{1,0})^{\vee}/q(H_{\mathbb{Z}}^{\vee})$ is called the Albanese torus of the given Hodge structure.

In the preceding example the map q is nothing but the map $H_1(X,\mathbb{C}) \to (H^{1,0})^{\vee}$ sending the class of the 1-cycle γ to the functional $\omega \to \int_\gamma \omega$, $[\gamma] \in H_1(X,\mathbb{C})$. It follows that the Albanese torus of the Hodge structure on $H^1(X,\mathbb{C})$ coincides with the Albanese torus $\operatorname{Alb}(X)$ defined in Sect. 13.

For tori this has the following useful consequence:

(14.2) **Theorem** (Torelli theorem for tori). *Let T, T' be two tori of the same dimension and let $\varphi : H^1(T,\mathbb{Z}) \to H^1(T',\mathbb{Z})$ be an isomorphism such that its $\mathbb{C}$-linear extension maps $H^{1,0}(T)$ isomorphically onto $H^{1,0}(T')$. Then φ is induced by an isomorphism from T' onto T.*

Proof. The universal property of the Albanese map implies that any torus is isomorphic to its Albanese torus. So there are isomorphisms $\alpha : T \xrightarrow{\sim} \operatorname{Alb}(T)$ and $\alpha' : T' \xrightarrow{\sim} \operatorname{Alb}(T')$. But $\operatorname{Alb}(T) = \operatorname{Alb} H^1(T,\mathbb{C})$ and similarly for T'. So φ induces $\varphi^* : \operatorname{Alb}(T') \xrightarrow{\sim} \operatorname{Alb}(T)$ and $\alpha^{-1} \circ \varphi^* \circ \alpha' : T' \xrightarrow{\sim} T$ induces φ by construction. □

Next, we consider polarized Hodge structures of weight 1. Suppose we are given a Hodge structure of weight 1 on the vector space $H_{\mathbb{C}}$ of dimension $2g$. Given an $\mathbb{R}$-bilinear form Q on $H_{\mathbb{R}}$ we denote its $\mathbb{C}$-bilinear extension to $H_{\mathbb{C}}$ also by Q. We say that a skew form Q on $H_{\mathbb{R}}$ polarises the Hodge structure if
(i) $Q(c,c') = 0$ for all $c, c' \in H^{1,0}$;
(ii) $\mathrm{i}\, Q(c,\bar{c}) > 0$ for all $c \in H^{1,0}$, $c \neq 0$.

We call Q integral if it assumes integral values on $H_{\mathbb{Z}}$. In this case (by Lemma 2.3) $H_{\mathbb{Z}}$ admits a canonical basis in which Q is given by a matrix

$$\begin{pmatrix} 0 & -\Delta \\ \Delta & 0 \end{pmatrix}; \quad \Delta = \begin{pmatrix} \delta_1 & 0 & \dots & 0 \\ 0 & \ddots & & \vdots \\ \vdots & & \ddots & 0 \\ 0 & \dots & 0 & \delta_g \end{pmatrix}.$$

We call Q a polarisation of type Δ, and if all δ_i are 1 we call it a principal polarisation.

(14.3) **Example.** Let X be a compact, connected Riemann surface. The weight-1 Hodge structure on $H^1(X,\mathbb{C})$ considered in Example 14.1 is obviously polarized by the cup product pairing

$$(x,y) = \int_X \xi \wedge \eta$$

(here x, y are the classes represented by the forms ξ, η). The cup product pairing is integral, even unimodular by the Poincaré duality theorem, so defines a principal polarization. A canonical basis consists of classes $\alpha_1, \dots, \alpha_g, \beta_1, \dots, \beta_g \in H^1(X, \mathbb{Z})$ with $(\alpha_i, \alpha_j) = (\beta_i, \beta_j) = 0$ and $(\alpha_i, \beta_j) = \delta_{i,j}$. Let $a_1, \dots, a_g$, $b_1, \dots, b_g \in H_1(X, \mathbb{Z})$ be the Poincaré-duals (see (13)), i.e., $a_i = \mathcal{P}_X \alpha_i$, $b_i = \mathcal{P}_X \beta_i$. Then

$$a_i \cdot a_j = b_i \cdot b_j = 0, \qquad a_i \cdot b_j = \delta_{i,j},$$

so this is a canonical basis for $H_1(X, \mathbb{Z})$ in the classical sense.

Weight-1 Hodge structures on a fixed space $H_{\mathbb{C}}$ of dimension $2g$, with fixed integral lattice $H_{\mathbb{Z}}$ and integral polarisation Q of type Δ, can be classified in the following way. Let $\alpha_1, \dots, \alpha_g$, $\beta_1, \dots, \beta_g$ be a canonical basis for $H_{\mathbb{Z}}$ and $\varphi_1, \dots, \varphi_g$ a basis for $H^{1,0}$. We may express the φ_i in terms of the α_i and β_i

$$(\varphi_1, \dots, \varphi_g) = (\alpha_1, \dots, \beta_g)\Omega,$$

where Ω is a complex $(2g \times g)$-matrix. It is called the **period matrix** of the polarized Hodge structure with respect to the given bases. If we set

$$\Omega = \begin{pmatrix} \Omega_1 \\ \Omega_2 \end{pmatrix}$$

with $g \times g$-matrices Ω_i, properties i) and ii) are equivalent to the matrix conditions:

$$\begin{aligned} &\Omega_1^t \Delta \Omega_2 \qquad \text{symmetric}, \\ &\mathrm{i}(\Omega_1^t \Delta \bar{\Omega}_2 - \Omega_2^t \Delta \bar{\Omega}_1) > 0. \end{aligned}$$

The second condition implies in particular that Ω_1 is invertible. We therefore may substitute

$$(\varphi_1', \dots, \varphi_g') = (\varphi_1, \dots, \varphi_g)\Omega_1^{-1}\Delta^{-1}$$

to obtain a new period matrix

$$\Omega' = \begin{pmatrix} \Delta^{-1} \\ Z \end{pmatrix}.$$

Now the matrix conditions above become

(17) $$Z = Z^t, \qquad \operatorname{Im} Z > 0 \quad \text{(Riemann period relations)}.$$

A basis $\varphi_1', \dots, \varphi_g'$ as above and the corresponding period matrix Ω' are called **normalized** with respect to the canonical basis $\alpha_1, \dots, \beta_g \in H_{\mathbb{Z}}$. The set of complex $g \times g$-matrices Z satisfying (17) is called the **Siegel upper half space** $\mathfrak{H}_g$.

For later use we need another description of the part Z of a normalized period matrix for a principal polarisation. Consider the images $p\alpha_1, \dots, p\beta_g \in H^{0,1}$ under the projection $p : H_{\mathbb{C}} \to H^{0,1}$. Since $p\varphi = 0$ for all $\varphi \in H^{1,0}$, the equation $(\varphi_1, \dots, \varphi_g) = (\alpha_1, \dots, \alpha_g) + (\beta_1, \dots, \beta_g)Z$ defining Z in $H^{0,1}$ becomes

$$(p\alpha_1, \ldots, p\alpha_g) = -(p\beta_1, \ldots, p\beta_g)Z.$$

So $p\beta_1, \ldots, p\beta_g$ form a $\mathbb{C}$-basis of $H^{0,1}$ and $-Z$ is the matrix of coefficients of the vectors $p\alpha_1, \ldots, p\alpha_g$ with respect to this basis.

If we have a *principal polarisation,* then two canonical bases of $H_{\mathbb{Z}}$ are related by

$$(18) \qquad (\alpha_1, \ldots, \beta_g) = (\alpha'_1 \ldots, \beta'_g) \cdot \sigma, \qquad \sigma = \begin{pmatrix} A & B \\ C & D \end{pmatrix} \in Sp(g, \mathbb{Z}).$$

The new period matrix is $\begin{pmatrix} 1\!\!1 \\ Z' \end{pmatrix}$ with

$$(19) \qquad Z' = (DZ + C)(BZ + A)^{-1}.$$

By way of $\sigma : Z \mapsto Z'$ the group $Sp(g, \mathbb{Z})$ operates on $\mathfrak{H}_g$. Since $\pm 1\!\!1$ operates trivially, there is an induced action of the modular group $\Gamma_g = Sp(g, \mathbb{Z})/\{\pm 1\}$. The period domain for principally polarized Hodge structures on a $2g$-dimensional vector space is the quotient $D_g = \mathfrak{H}_g/\Gamma_g$. The group Γ_g acts properly and discontinuously, so this period domain carries the structure of a normal analytic variety. (It is even quasi-projective by [B-B].)

Example. As above let $H = H^1(X, \mathbb{C})$, where X is a connected compact Riemann surface. If $a_1, \ldots, b_g \in H_1(X, \mathbb{Z})$ is a canonical homology basis and $\varphi_1, \ldots, \varphi_g$ is a basis of the space of holomorphic differentials on X, the classical period matrix is

$$\begin{pmatrix} \int_{a_i} \varphi_k \\ \int_{b_i} \varphi_k \end{pmatrix}.$$

This matrix is equivalent with our Ω. In fact, let $\mathcal{P} = \mathcal{P}_X$ and $a_1 = \mathcal{P}\alpha_1, \ldots, b_g = \mathcal{P}\beta_g$. These classes form a canonical cohomology basis with

$$\int_{a_i} \alpha_j = \int_{b_i} \beta_j = 0, \quad \int_{b_i} \alpha_j = -\int_{a_i} \beta_j = \delta_{ij}\ .$$

So the basis $-\beta_1, \ldots, -\beta_g, \alpha_1, \ldots, \alpha_g$, which is also a canonical cohomology basis, is Kronecker dual to $a_1, \ldots, b_g$. Our period matrix Ω for this basis therefore coincides with the classical period matrix, up to the action of the symplectic group. (Instead of the original normalized period matrix $\begin{pmatrix} 1\!\!1 \\ Z \end{pmatrix}$ we have $\begin{pmatrix} 1\!\!1 \\ -Z^{-1} \end{pmatrix}$.)

15. Yau's Results on Kähler-Einstein Metrics

Let X be a Kähler manifold, and ω_g the closed real $(1,1)$-form belonging to a Kähler metric g on X. We can associate to g a second closed real $(1,1)$-form, the Ricci form $\mathrm{Ric}(g)$ of g (see [SemP], p. 79). The metric g is called a Kähler-Einstein metric if there exists a $\lambda \in \mathbb{R}$, such that $\mathrm{Ric}(g) = \lambda\omega_g$.

Around 1955, Calabi formulated several conjectures concerning the existence of such metrics, a (highly non-trivial) part of which was proved by S.-T. Yau in 1976. Yau's results have important implications for the topology of algebraic manifolds. We shall formulate here only those special cases of Yau's results which will be needed in this book. For more details we refer to [Y77], [Y78] and [SemP].

(15.1) **Theorem.** *Let X be a compact complex manifold, of dimension at least 2, with $c_1(X) = 0$. If an element of $H^2(X, \mathbb{R})$ can be represented by a Kähler form, then it can be represented by exactly one Kähler form ω_g such that* $\mathrm{Ric}(g) = 0$.

(15.2) **Theorem.** *Let X be a compact complex manifold of dimension at least 2, such that $-c_1(X)$ can be represented by a Kähler form. Then X admits a Kähler-Einstein metric.*

On the other hand there are the following facts, which were known before. For the first theorem we refer to [K-N], vol. II, p. 171 and for the second to [S-T69], p. 359.

(15.3) **Theorem.** *Let X be an n-dimensional compact, connected Kähler manifold with constant holomorphic sectional curvature. Then the universal covering space of X is either $\mathbb{P}_n$ (with the Fubini metric), or $\mathbb{C}^n$ (with the standard flat metric) or the unit ball in $\mathbb{C}^n$ (with the Bergmann metric).*

(15.4) **Theorem.** *If a 2-dimensional compact, connected complex manifold X with $c_1^2(X) = 3c_2(X)$ admits a Kähler-Einstein metric, then the holomorphic sectional curvature is constant.*

(15.5) **Corollary.** *Assume that X is a 2-dimensional compact connected complex manifold with $c_1^2(X) = 3c_2(X) > 0$ admitting a Kähler-Einstein metric. If X is different from $\mathbb{P}_2$, then the universal covering of X is the unit ball in $\mathbb{C}^2$.*

Proof. If the universal covering of X is $\mathbb{P}_2$, then X is isomorphic to $\mathbb{P}_2$, since every automorphism of $\mathbb{P}_2$ has a fixed point. On the other hand, if the universal covering of X is $\mathbb{C}^2$, and hence the metric on X flat, by Theorems 15.4 and 15.3 we must have $e(X) = c_2(X) = 0$. □

Coverings

Cyclic coverings provide an important method to construct new manifolds which will be used to produce all kinds of surfaces. In order to calculate the invariants some basic computations are given. Coverings will also be used as a technical tool in the guise of "covering tricks" assembled in section 18.

16. Ramification

Apart from the meaning: "system of subsets, covering a whole set", we shall use the word "covering" in two (related) ways.

Firstly in the sense of analytic covering space. This is a triple (X, Y, π) (also to be denoted by $\pi : X \to Y$) where X and Y are connected complex spaces and $\pi : X \to Y$ is a surjective holomorphic map such that all points $y \in Y$ have a connected neighbourhood V_y, with the property that $\pi^{-1}(V_y)$ consists of the union of disjoint open subsets of X, each of which is mapped isomorphically onto V_y by π. Mostly X and Y will be complex manifolds.

If we use the corresponding topological concept we shall speak of a "topological covering". Given any topological covering (X, Y, π) such that Y is a complex manifold, then, up to an isomorphism, there is exactly one complex manifold structure on Y such that (X, Y, π) becomes an analytic covering.

Secondly, we shall use the word "covering" for triples (X, Y, π) (also to be denoted by $\pi : X \to Y$) where X and Y are connected *normal* complex spaces and π a *finite*, surjective proper holomorphic map.

In this last case there exists a proper analytic subset of X, outside of which π is a topological covering. Indeed, on $X' = X \backslash \pi^{-1}(\pi(\operatorname{Sing} X) \cup \operatorname{Sing} Y)$ the map π is a covering between manifolds, so it is a topological covering outside of $\pi^{-1}(\pi(S))$ where $S = \{x \in X' \mid \operatorname{rank}(d\pi)_x \leq \dim X - 1\}$. By definition the degree of π is the degree of $\pi | X' \setminus \pi^{-1}(\pi(S))$. Properness of π implies that for every $x \in X$ there exists at least one connected open neighbourhood V of $\pi(x) \in X$ such that $\pi^{-1}(V)$ is a union of disjoint connected open neighbourhoods U_i of x_i $(i = 1, \ldots, n)$, where $\pi^{-1}\pi(x) = \{x = x_1, x_2, \ldots, x_n\}$. If $V' \subset V$ is another connected neighbourhood of $\pi(x)$, we claim that $\pi^{-1}(V') \cap U_1 = U_1'$ is connected. Indeed, if U_1'' is any connected component of U_1', its image must be V' by Theorem 8.4, so $x \in U_1''$ and $U_1'' = U_1'$. It follows that the degree of $\pi | \pi^{-1}(V') \cap U_1$ is the same for all connected neighbourhoods $V' \subset V$ of x. This degree is called the local degree e_x of π at x or the branching order of π at x. If $e_x \geq 2$ we say that π is ramified at x, and x is called a ramification point. The images of ramification points are called branch points.

Now let us assume that both X and Y are manifolds. The the set of ramification points is the zero divisor R of the canonical section in $\operatorname{Hom}(p^*(\mathcal{K}_Y), \mathcal{K}_X)$, i.e.,

$$\mathcal{K}_X = \pi^*(\mathcal{K}_Y) \otimes \mathcal{O}_X(R) \ . \tag{20}$$

The divisor R is called the **ramification divisor** of π. Formula (20), together with the specification of R, given by Lemma 16.1 below, is called the **Hurwitz-formula**.

We observe that the properness of π implies that $\pi : X \setminus \pi^{-1}(\pi(R)) \to Y \setminus \pi(R)$ is a covering in the first sense; in particular, a covering in the second sense with $R = 0$ is one in the first sense too.

To emphasize the difference we shall frequently call a covering in the first sense an **unbranched, unramified, or étale covering** and one in the second sense a **branched** or **ramified covering**, as soon as $R \neq 0$.

(16.1) **Lemma.** *If $R = \sum r_j R_j$, where R is the ramification divisor of some branched covering and the R_j's its irreducible components, then $r_j = e_j - 1$, where e_j is the branching order at any point $x \in R_j$ which is smooth on R_{red}, and for which $y = \pi(x)$ is smooth on $B_j = \pi(R_j)$.*

Proof. Let $(t_1, \ldots, t_n)$ be local coordinates on Y, centred at y, such that B_j is given by $t_1 = 0$. If $s = 0$ is a local equation for R_j at x, then we have $\pi^*(t_1) = \varepsilon \cdot s^{e_j}$, where ε does not vanish around x, and in fact can be taken to be 1 if s is suitably chosen. If we set $\omega = dt_1 \wedge \cdots \wedge dt_n$, then $\pi^*(\omega) = s^{e_j - 1} ds \wedge d\pi^*(t_2) \wedge \cdots \wedge d\pi^*(t_n)$. This not only shows that $(s, \pi^*(t_2), \ldots, \pi^*(t_n))$ is a local coordinate system at x (so $e_j = e_x$), but also that the zero divisor of $\pi^*(\omega)$ is $(e_j - 1)R_j$. Hence $r_j = e_j - 1$. □

(16.2) **Lemma.** *Let X and Y be compact connected complex manifolds and $f : X \to Y$ a covering of degree d. If $\mathcal{L}$ is a line bundle on Y with $f^*\mathcal{L} = \mathcal{O}_X$, then $\mathcal{L}^{\otimes d} = \mathcal{O}_Y$.*

Proof. Since $f_*\mathcal{O}_X$ is locally free or rank d (compare [Se56a]) this is an immediate consequence of $f_*\mathcal{O}_X = f_*f^*\mathcal{L} = \mathcal{L} \otimes f_*\mathcal{O}_X$. □

17. Cyclic Coverings

Let Y be a connected complex manifold and B a divisor on Y which is either effective or zero. Suppose we have a line bundle $\mathcal{L}$ on Y such that

$$\mathcal{O}_Y(B) = \mathcal{L}^{\otimes n} ,$$

and a section $s \in \Gamma(Y, \mathcal{O}_Y(B))$ vanishing exactly along B (if $B = 0$, we take for s the constant function 1). We denote by L the total space of $\mathcal{L}$ and we let $p : L \to Y$ be the bundle projection. If $t \in \Gamma(L, p^*\mathcal{L})$ is the tautological section, then the zero divisor of $p^*s - t^n$ defines an analytic subspace X in L.

If $B \neq 0$ and reduced, X is an irreducible normal analytic subspace of L, and $\pi = p|X$ exhibits X as an n-fold ramified covering of Y with branch-locus B. We call (X, Y, π) (or X, or π) the n-cyclic covering of Y branched along B, determined by $\mathcal{L}$.

On the other hand, if $B = 0$, we must take n minimal (i.e., $\mathcal{L}$ is exactly of order n in $\text{Pic}(Y)$) in order to obtain a connected manifold X. In this case (X, Y, π) (or X, or π) is called the n-cyclic unramified covering of Y determined by the torsion bundle $\mathcal{L}$.

If $\mathrm{Pic}(Y)$ has no torsion, then B uniquely determines $\mathcal{L}$ and we may speak of *the* n-cyclic covering of Y, branched along B.

It is clear from the above description that X has at most singularities over singular points of B. In particular if B is reduced and smooth, then also X is smooth.

(17.1) **Lemma.** *Let $\pi : X \to Y$ be the n-cyclic covering of Y branched along a smooth divisor B and determined by $\mathcal{L}$, where $\mathcal{L}^{\otimes n} = \mathcal{O}_Y(B)$. Let B_1 be the reduced divisor $\pi^{-1}(B)$ on X. Then*
(i) $\mathcal{O}_X(B_1) = \pi^*\mathcal{L}$;
(ii) $\pi^*B = nB_1$ *(in particular n is the branching order along B_1);*
(iii) $\mathcal{K}_X = \pi^*(\mathcal{K}_Y \otimes \mathcal{L}^{n-1})$.

Proof. If we embed Y as the zero-section in L, then the section $t \in \Gamma(L, p^*\mathcal{L})$ has divisor Y, so $\mathcal{O}_L(Y) = p^*\mathcal{L}$. By construction Y and $X \subset L$ intersect transversally in B_1, so $\mathcal{O}_X(B_1) = \mathcal{O}_L(Y)|X = \pi^*\mathcal{L}$. The identity $\pi^*B = nB_1$ follows from the equation $p^*s - t^n = 0$ for X in L. The formula for $\mathcal{K}_X$ is an application of Lemma 16.1. □

(17.2) **Lemma.** *Let $\pi : X \to Y$ be as in Lemma 17.1. Then $\pi_*\mathcal{O}_X \simeq \bigoplus_{j=0}^{n-1} \mathcal{L}^{-j}$.*

Proof. For an open set $V \subset Y$, any holomorphic function f on $p^{-1}(V)$ has a unique power series expansion $f = \sum_{k=0}^{\infty} a_k t^k$, $a_k \in \Gamma(V, \mathcal{L}^{-k})$. Every function on $\pi^{-1}(V) \subset p^{-1}(V)$ is the restriction of such an f. Using the equation $t_n = \pi^*s$, we obtain a unique expansion $\sum_{k=0}^{n-1} b_k t^k$, $b_k \in \Gamma(V, \mathcal{L}^{-k})$ for holomorphic functions on $\pi^{-1}(V)$. □

18. Covering Tricks

The "*unbranched covering trick*" is nothing but the following remark.

(18.1) **Proposition** (Unbranched covering trick). *Let X be a connected complex manifold.*
(i) *If $b_1(X) \neq 0$, then X admits unbranched coverings of any order;*
(ii) *If $H_1(X, \mathbb{Z})$ contains k-torsion, then X has an unbranched covering of order k.*

Proof. As to i), if $b_1(X) \neq 0$, then $H_1(X, \mathbb{Z})$ is infinite and therefore admits quotient groups of any order. Consequently, also $\pi_1(X)$ has such quotients, and so there exist unbranched coverings of X of any order.

The proof of (ii) is similar. □

The "branched covering trick", though trivial from an algebraic point of view, requires a little bit more care.

(18.2) **Theorem** (Branched covering trick). *We assume a holomorphic* $\mathbb{P}_1$*-bundle over the irreducible, normal complex space* X *is given, with total space* B *and projection* $p : B \to X$*. If* S *is any irreducible divisor on* B*, meeting a general fibre in* n *points, then there exists a normal complex space* Y*, a generically finite surjective map* $f : Y \to X$*, and* n *effective divisors* $S_1, \ldots, S_n$ *on* $B' = B \times_Y X$*, all meeting a general fibre in one point, such that* $g^*(S) = S_1 + \cdots + S_n$*, where* $g : B' \to B$ *is the projection.*

Proof. For $n = 1$ there is nothing to prove. If $n \geq 2$ we consider the normalization $\overline{S}$ of S and observe that, if we put $B_1 = \overline{S} \times_X B$ and denote by $g_1 : B_1 \to B$ the natural projection, we have in a canonical way that $g^*(S) = S_1 + S'$, where S_1, S' respectively, meets a general fibre of $B_1 \to \overline{S}$ in one $((n-1))$ point(s) respectively. Applying the same procedure to $\overline{S}$, and so on, we finally obtain the desired result. □

As a consequence we have (compare [Mi77b], Lemma 11):

(18.3) **Theorem.** *Let* X *be a compact, connected complex manifold and* $\mathcal{L}$ *a holomorphic line bundle on* X *with* $h^0(\mathcal{L}^{\otimes n}) \geq 2$ *for some* $n \geq 1$*. Then there exists a compact complex manifold* Y *and a generically finite-to-one map* $f : Y \to X$*, such that* $h^0(f^*(\mathcal{L})) \geq 2$*.*

Projective-Algebraic Varieties

Projective varieties are important examples of complex spaces. By definition these are the loci where a set of homogeneous polynomials vanish. Any complex subvariety of projective space is a projective variety. This is Chow's theorem, recalled in Sect. 19. In this section we also collect a few a priori tests for projectivity. Some general properties of projective manifolds that are used at various places in the book are stated in Sect. 20.

19. GAGA Theorems and Projectivity Criteria

In the same way as we defined complex-analytic spaces, but using Zariski-topology and affine charts instead of "open ball-charts", we can define algebraic varieties. We shall only need projective-algebraic varieties, i.e., closed algebraic subvarieties of projective spaces $\mathbb{P}^n(\mathbb{C})$, or at most quasi-projective varieties, that is, locally closed algebraic subvarieties of a $\mathbb{P}^n$.

Let us first of all say a word about the results of Serre's classical GAGA-paper [Se56b], of which we shall need only a modest part.

To every (projective-)algebraic variety X we can attach an analytic space X^{an} in an obvious way such that there is a morphism $m : X^{\text{an}} \to X$ of ringed spaces. If $\mathcal{S}$ is any coherent sheaf on X in the algebraic sense, we can make $m^{-1}(\mathcal{S})$ into a coherent sheaf $\mathcal{S}^{\text{an}}$ in the analytic sense by tensoring with $\mathcal{O}_{X^{\text{an}}}$. One has that $m^* : H^i(X, \mathcal{S}) \xrightarrow{\sim} H^i(X^{\text{an}}, \mathcal{S}^{\text{an}})$ for all $i \geq 0$.

(19.1) **Theorem.** *Let X be a projective-algebraic variety and $\widetilde{\mathcal{S}}$ a coherent sheaf on X^{an}. Then there is a coherent sheaf $\mathcal{S}$ on X, unique up to isomorphism, with $\mathcal{S}^{\text{an}} = \widetilde{\mathcal{S}}$. If $\widetilde{\mathcal{S}}$ is locally free, then so is $\mathcal{S}$.*

Roughly we can say: every analytic coherent sheaf on X is algebraic. In particular, every analytic vector bundle on X is algebraic.

An easy consequence of Theorem 19.1 is

(19.2) **Theorem** (Chow's theorem). *Every closed analytic subspace of $\mathbb{P}_n$ is an algebraic subvariety.*

Let X be an analytic space and $\mathcal{L}$ a holomorphic line bundle on X. If for every point $x \in X$ there is at least one section $s \in \Gamma(X, \mathcal{L})$ with $s(x) \neq 0$, then (upon choosing a basis) $\Gamma(X, \mathcal{L})$ yields a holomorphic map into $\mathbb{P}_N$, where $N = \dim \Gamma(X, \mathcal{L}) - 1$. If this map is an isomorphism from X onto its image, then we say that $\mathcal{L}$ is very ample. A line bundle $\mathcal{M}$ is called ample if there exists an $n \geq 1$ such that $\mathcal{M}^{\otimes n}$ is very ample.

The following classical criterion for a line bundle to be ample will play an important role in this book (comparer [Gr62]).

(19.3) **Theorem** (Grauert's criterion). *Let X be a compact complex space and $\mathcal{L}$ a holomorphic line bundle on X. Then $\mathcal{L}$ is ample if and only if the* *following holds: given any irreducible analytic subset Y of strictly positive dimension on X, there exists an $n = n(Y)$, such that $\mathcal{L}^{\otimes n}|Y$ has a section which has at least one zero, but does not vanish identically.*

A projectivity criterion of quite different nature is Kodaira's criterion.

Let X be a complex manifold. A Kähler metric is called a **Hodge metric** if the associated element of $H^2(X, \mathbb{R})$ is integral, i.e., in the image of $H^2(X, \mathbb{Z})$. A Kähler manifold is called a **Hodge manifold** if it admits at least one Hodge metric, and a famous result of Kodaira ([Ko54], Theorem 4) says that this ensures that the manifold is projective:

(19.4) **Theorem** (Kodaira's criterion for Hodge manifolds). *A compact complex manifold is (isomorphic to) a projective-algebraic manifold if and only if it is a Hodge manifold.*

Both Theorems 19.3 and 19.4 are trivial in one direction (as to 19.4 a projective space $\mathbb{P}_n$ obviously has a Hodge metric, and a submanifold of a Hodge manifold is again a Hodge manifold).

Finally, one more fact we shall need:

(19.5) **Proposition.** *The normalization of a projective-algebraic space is again a projective-algebraic space.*

This is, for example, an easy consequence of Grauert's criterion.

20. Theorems of Bertini and Lefschetz

Again we shall need only a limited part of the classical theorems in question.

But first of all we recall a well-known elementary result which is sometimes also called Bertini's theorem, at least in the case that $\dim Y = 1$.

(20.1) **Theorem.** *Let $f : X \to Y$ be a proper, surjective holomorphic map between the complex manifolds X and Y. Then the set of points on X where f is not of maximal rank, is a proper analytic subset of X, the image of which is a proper analytic subset on Y.*

Now the special case of Bertini's theorem which we need (see [G-H78a], p. 137 and [FL], p. 33).

(20.2) **Theorem** (Bertini's theorem). *Let X be a connected n-dimensional compact complex manifold and $f : X \to \mathbb{P}_N$ a holomorphic map such that* $\dim f(X) \geq 2$. *Then for a general hyperplane $H \subset \mathbb{P}_N$ (i.e. for all hyperplanes outside of a proper algebraic subset of $\mathbb{P}_N^\vee$) the (analytic) inverse image of H is a connected $(n-1)$-dimensional complex submanifold on X.*

So, apart from the case that f is constant, we have either the case of the theorem or the case $\dim f(X) = 1$. In this last case there is a Stein factorization $f = h \circ g$ where g is a connected map from X onto a smooth curve C, and $h : C \to \mathbb{P}_N$ finite onto its image. By Theorem 20.1, the general fibre of g is smooth, hence the general fibre of f consists of the disjoint union of finitely many submanifolds on X. In this last case, g is called a pencil (rational if $C \cong \mathbb{P}_1$, irrational otherwise), and f is "composed with" this pencil (compare Chapter IV, Sect. 1).

(20.3) **Corollary.** *Let X be a connected n-dimensional algebraic submanifold of $\mathbb{P}_N$, with $n \geq 2$. Then a general hyperplane H in $\mathbb{P}_N$ intersects X transversally along a smooth irreducible divisor.*

Now let X be any compact, connected complex manifold and $\mathcal{L}$ a line bundle on X. Let $B \subset X$ be the base point set of $\mathcal{L}$, i.e., the analytic set of those points $x \in X$, for which $\gamma(x) = 0$ for all $\gamma \in \Gamma(X, \mathcal{L})$. If $B = \varnothing$, then we have exactly the situation described above if we take for f the map into a projective space, obtained by taking a basis of $\Gamma(X, \mathcal{L})$. If $\dim B \leq \dim X - 2$, we obtain a similar result by considering the desingularization of the graph of the meromorphic map obtained from $\Gamma(X, \mathcal{L})$, after choosing a basis.

Either the image in $\mathbb{P}_N$ has dimension at least 2 and the inverse image of a general hyperplane section (i.e., the zero set of a general member of $\Gamma(X, \mathcal{L})$) on X is irreducible, with singularities at most in B, or f is composed with a "pencil with base points". But we shall have to face this situation for surfaces only, where the desingularization is simple, so we shall return to this point in Chap. IV, Sect. 1.

(20.4) **Theorem** (Lefschetz theorem on hyperplane sections). *Let X be an n-dimensional submanifold of $\mathbb{P}_N$, $n \geq 2$, and let $H \subset \mathbb{P}_N$ be a hyperplane, such that $H \cap X$ is again a complex manifold. Then the inclusion homomorphisms*

$$H_i(X \cap H, \mathbb{Z}) \longrightarrow H_i(X, \mathbb{Z})$$
$$\pi_i(X \cap H, \mathbb{Z}) \longrightarrow \pi_i(X, \mathbb{Z})$$

are isomorphisms for $0 \leq i \leq n-2$.

For a proof we refer to [Mil63].

One can use the Veronese embedding for $X = \mathbb{P}_N$ defined by the linear system $|\mathcal{O}_{\mathbb{P}_N}(d)|$ to obtain a Lefschetz theorem for degree d hypersurfaces in projective space. In fact, one can iterate this procedure for complete intersections:

(20.5) **Corollary** *Let $Y \subset \mathbb{P}_N$ be a smooth complete intersection of dimension $m \geq 1$. Then the inclusion homomorphisms*

$$H_i(Y, \mathbb{Z}) \longrightarrow H_i(\mathbb{P}_N, \mathbb{Z})$$
$$\pi_i(Y, \mathbb{Z}) \longrightarrow \pi_i(\mathbb{P}_N, \mathbb{Z})$$

are isomorphisms for $0 \leq i \leq m-1$. In particular, such a complete intersection is connected and if its dimension is at least 2, also simply connected.

Chapter II. Curves on Surfaces

Unless stated otherwise, X denotes in this chapter a 2-dimensional complex manifold, not necessarily compact or connected.

Embedded Curves

In this sub-chapter two basic technical themes are discussed. The first centres around the Riemann-Roch and Serre duality theorems for any compact curve on a surface and is treated in Sects. 1–6. The second theme, developed in Sects. 7–8 is embedded resolution of singularities of curves and the application to the so-called simple singularities of curves.

1. Some Standard Exact Sequences

A curve on X is a 1-dimensional subspace of X locally defined by one equation, so there is a natural $1-1$ correspondence between curves and effective divisors on X.

If C is a curve on X, $f : C \to X$ the embedding and D a divisor on X, then we denote by $\mathcal{O}_C(D)$ the analytic restriction $f^*(\mathcal{O}_X(D))$. In particular $\mathcal{O}_C(C) = \mathcal{O}_X(C)|C$. The structure sequence (I, Sect. 8)

$$(1) \qquad 0 \to \mathcal{J}_C \to \mathcal{O}_X \to \mathcal{O}_C \to 0$$

yields, in combination with $\mathcal{J}_C \cong \mathcal{O}_X(-C)$, after tensoring with $\mathcal{O}_X(C)$, the sequence

$$(2) \qquad 0 \to \mathcal{O}_X \to \mathcal{O}_X(C) \to \mathcal{O}_C(C) \to 0\ .$$

If C is smooth, then $\mathcal{O}_C(C)$ is the normal bundle of C in X (see Proposition I.6.2). In case C is singular, we still shall call the line bundle $\mathcal{O}_C(C)$ the normal bundle of C in X. If $\{u_i \in \Gamma(\mathcal{O}_X|U_i)\}_{i\in I}$ is a collection of local equations for C, then $\mathcal{O}_C(C)$ is the line bundle defined by the cocycle $(u_i/u_j)|C$. Of course, there is the normal bundle sequence

$$(3) \qquad 0 \to \mathcal{T}_C \to \mathcal{T}_X|C \to \mathcal{N}_{C/X} \to 0,$$

but only when C is smooth.

Let $C = A + B$ be the sum of two effective divisors A, B on X.

The inclusions $\mathcal{J}_C \subset \mathcal{J}_B \subset \mathcal{O}_X$ induce a diagram

$$\begin{array}{ccccccccc}
 & & & & 0 & & 0 & & \\
 & & & & \downarrow & & \downarrow & & \\
 & & 0 & \longrightarrow & \mathcal{O}_X(-C) & \longrightarrow & \mathcal{O}_X(-B) & \longrightarrow & \mathcal{O}_A(-B) \longrightarrow 0 \\
 & & & & \downarrow & & \downarrow & & \\
 & & & & \mathcal{O}_X & = & \mathcal{O}_X & & \\
 & & & & \downarrow & & \downarrow & & \\
0 & \longrightarrow & \mathcal{J}_B/\mathcal{J}_C & \longrightarrow & \mathcal{O}_C & \longrightarrow & \mathcal{O}_B & \longrightarrow & 0. \\
 & & & & \downarrow & & \downarrow & & \\
 & & & & 0 & & 0 & &
\end{array}$$

The first row is the structure sequence for A tensored with $\mathcal{O}_X(-B)$. This shows that $\mathcal{J}_B|\mathcal{J}_C$ is isomorphic to $\mathcal{O}_A(-B)$, a line bundle on A. The resulting exact sequence

$$0 \longrightarrow \mathcal{O}_A(-B) \longrightarrow \mathcal{O}_C \xrightarrow{\text{restr}} \mathcal{O}_B \longrightarrow 0 \tag{4}$$

will be called the decomposition sequence for $C = A + B$. The special case $2C = C + C$ shows that $\mathcal{J}_C/\mathcal{J}_C^2 \simeq \mathcal{O}_C(-C)$ (compare I, Sect. 8). Also in the non-smooth case this sheaf is called the co-normal bundle of C in X.

Now let $C \subset X$ be *reduced.* If $\nu : \tilde{C} \to C$ is the normalization of C (I, Sect. 8), then $\widetilde{C}$ is a (non-singular) Riemann surface. For $x \in C$ singular, $\nu^{-1}(x)$ consists of finitely many points corresponding to the different branches of C through x. On C there is the normalization sequence

$$0 \longrightarrow \mathcal{O}_C \longrightarrow \nu_*\mathcal{O}_{\tilde{C}} \longrightarrow S \longrightarrow 0 \tag{5}$$

with S concentrated at the singularities of C.

For a smooth curve C, we have the adjunction formula (I.6.3)

$$\mathcal{K}_C = \mathcal{K}_X \otimes \mathcal{O}_C(C) \ .$$

If C is singular (and possibly non-reduced) we can formally define a canonical bundle for C by setting

$$\omega_C = \mathcal{K}_X \otimes \mathcal{O}_C(C) \ .$$

This line bundle is mostly called the dualizing sheaf on C. According to our definition it depends on the embedding of C in X, although general theory shows that this is actually not the case. Tensoring (2) with $\mathcal{K}_X$ yields the residue sequence

$$0 \longrightarrow \mathcal{K}_X \longrightarrow \mathcal{K}_X \otimes \mathcal{O}_X(C) \xrightarrow{r} \omega_C \longrightarrow 0 \ . \tag{6}$$

For *reduced* C, we shall give here an explicit description of r. The relation with the residue theorem will be discussed in Sect. 4.

Let (u, v) be local coordinates on X such that u is not constant on any open subset of C. If $f = 0$ is a local equation for C, the partial derivative f_v cannot vanish on any open subset of C, since

$$f_u \nu^*(du) + f_v \nu^*(dv) = \nu^*(df) = 0 .$$

If for a local section $\varphi = h\ du \wedge dv/f$ in $\mathcal{K}_X \otimes \mathcal{O}_X(C)$ we define

$$r'(\varphi) = \nu^*(h\ du/f_v) ,$$

then $r'(\varphi)$ is a meromorphic differential on an open piece of $\widetilde{C}$. It is obvious that $r'(\varphi)$ does not depend on the choice of f. But it also is independent of the choice of the local coordinates (u, v). Namely, let (w, z) be other coordinates, with $\delta = u_w v_z - u_z v_w$. Then

$$\begin{aligned} \nu^*(f_z\ du) &= \nu^*(f_z u_w\ dw + f_z u_z\ dz) \\ &= \nu^*(f_z u_w - f_w u_z)\nu^*(dw) , \\ f_z u_w - f_w u_z &= f_u u_z u_w + f_v v_z u_w - f_u u_w u_z - f_v v_w u_z = f_v \cdot \delta , \\ \nu^*(\delta dw/f_z) &= \nu^*(du/f_v) . \end{aligned}$$

So r' is a globally defined morphism of $\mathcal{K}_X \otimes \mathcal{O}_X(C)$ into the sheaf of meromorphic 1-forms on $\widetilde{C}$. Its kernel consists of those $\varphi = h\ du \wedge dv/f$, for which $\nu^*(h) = 0$, i.e., $\ker(r') = \mathrm{Im}(\mathcal{K}_X)$. So ω_C equals $\nu^*(\mathrm{Im}(r'))$, i.e., ω_C is the sheaf (on C) of meromorphic differentials on $\widetilde{C}$ of the form $r'(\varphi)$ and r can be identified with r'.

2. The Picard-Group of an Embedded Curve

If C is a smooth curve, then there is the exponential sequence (I, Sect. 6) on C

(7) $$0 \longrightarrow \mathbb{Z}_C \longrightarrow \mathcal{O}_C \xrightarrow{\mathbf{e}} \mathcal{O}_C^* \longrightarrow 0 .$$

This sequence can be generalised to the case of any curve C, reduced or not, on a surface X, as we shall now explain. Let $\mathcal{O}_C^* \subset \mathcal{O}_C$ be the subsheaf of those germs f in $\mathcal{O}_C$ for which $f|C_{\text{red}}$ has no zero. Such an f can locally be represented as $f = \tilde{f}|C$, where $\tilde{f}$ is holomorphic without zeros on an open set in X. For $f \in \mathcal{O}_C$ we put $\mathbf{e}(f) = e^{2\pi i f}$, where the exponential e^f is defined in the following way. If locally f is represented as $\tilde{f}|C$, with $\tilde{f}$ holomorphic in a neighbourhood on X of the point on C in question, and if we put $e^f = e^{\tilde{f}}|C$, then this e^f depends on f only, for if $g \in \Gamma(X, \mathcal{J}_C)$, then

$$e^g - 1 = \sum_{m \geq 1} g^m/m! \in \Gamma(X, \mathcal{J}_C)$$

by the closedness property ([G-R71],IV.4.1) of the ideal $\mathcal{J}_C$. To prove exactness in the middle of (7), let $f = \tilde{f}|C$ with $e^f = 1$, i.e., $e^{\tilde{f}} - 1 \in \mathcal{J}_C$. After subtracting the (locally) constant function $f|C_{\mathrm{red}} \in \mathbb{Z}\cdot 2\pi\mathrm{i}$, we may assume $\tilde{f} \in \mathcal{J}_{C_{\mathrm{red}}}$, say $\tilde{f} \in \mathcal{J}^n_{C_i}$, but $\tilde{f} \notin \mathcal{J}^{n+1}_{C_i}$, for some irreducible component $C_i \subset C$. Then also

$$e^{\tilde{f}} - 1 = \sum_{m\geq 1} \tilde{f}^m/m! \in \mathcal{J}^n_{C_i} \setminus \mathcal{J}^{n+1}_{C_i},$$

so n is just the multiplicity of C_i in C, and $\tilde{f} \in \mathcal{J}_C$.
Finally, $\mathbf{e}$ is surjective, because it is surjective on every open set $U \subset X$.

Exactly like in the smooth case, the exponential cohomology sequence

$$H^1(C,\mathbb{Z}) \longrightarrow H^1(\mathcal{O}_C) \longrightarrow H^1(\mathcal{O}^*_C) \longrightarrow H^2(C,\mathbb{Z})$$

describes the group $\mathrm{Pic}(C) = H^1(\mathcal{O}^*_C)$ of line bundles on C.

(2.1) **Proposition.** *If C is compact, the inclusion $\mathbb{Z} \to \mathcal{O}_C$ identifies $H^1(C,\mathbb{Z})$ with a discrete subgroup of $H^1(\mathcal{O}_C)$. So $\mathrm{Pic}^0(C) = H^1(\mathcal{O}_C)/H^1(C,\mathbb{Z})$ carries the structure of a (Hausdorff) abelian complex Lie group.*

Proof. The diagram

$$\begin{array}{ccc} H^1(C,\mathbb{Z}) & \longrightarrow & H^1(\mathcal{O}_C) \\ \| & & \downarrow \scriptstyle\text{restr} \\ H^1(C_{\mathrm{red}},\mathbb{Z}) & \longrightarrow & H^1(\mathcal{O}_{C_{\mathrm{red}}}) \end{array}$$

shows that the assertion holds for C if it holds for C_{red}. So we may assume C reduced.

Since $\Gamma(\mathcal{O}_C)$ and $\Gamma(\mathcal{O}^*_C)$ are spaces of locally constant functions, the map $\mathbf{e} : \Gamma(\mathcal{O}_C) \to \Gamma(\mathcal{O}^*_C)$ is surjective and the injectivity of $H^1(C,\mathbb{Z}) \to H^1(\mathcal{O}_C)$ follows from the exponential cohomology sequence on C.

To prove that the image of $H^1(C,\mathbb{Z})$ is discrete in $H^1(\mathcal{O}_C)$ we consider the normalization sequence (5) for C. From it we obtain a diagram of sheaves

$$\begin{array}{ccccccccc} 0 & \longrightarrow & \mathbb{Z}_C & \longrightarrow & \nu_*\mathbb{Z}_{\widetilde{C}} & \longrightarrow & \Sigma & \longrightarrow & 0 \\ & & \downarrow & & \downarrow & & \downarrow & & \\ 0 & \longrightarrow & \mathcal{O}_C & \longrightarrow & \nu_*(\mathcal{O}_{\widetilde{C}}) & \longrightarrow & S & \longrightarrow & 0 \end{array}$$

and a diagram in cohomology

$$\begin{array}{ccccc} \Gamma(\Sigma) & \longrightarrow & H^1(C,\mathbb{Z}) & \xrightarrow{\nu^*} & H^1(\widetilde{C},\mathbb{Z}) \\ \downarrow & & \downarrow & & \downarrow \\ \Gamma(S) & \longrightarrow & H^1(\mathcal{O}_C) & \xrightarrow{\nu^*} & H^1(\mathcal{O}_{\widetilde{C}}) \,. \end{array}$$

Since $\widetilde{C}$ is non-singular, the right hand vertical arrow has a discrete image. It suffices to prove the same for the left hand vertical arrow. But this can be done locally. If $x \in X$ is a singular point, $\nu^{-1}(x) = \{x_1, \ldots, x_r\}$, $U \subset C$ a small neighbourhood of x, and $\nu^{-1}(U) = U_1 \cup \cdots \cup U_r \subset \widetilde{C}$, then we consider the diagram

$$\begin{array}{ccccccccc} 0 & \longrightarrow & \Gamma(U,\mathbb{Z}) & \xrightarrow{\nu^*} & \bigoplus \Gamma(U_i,\mathbb{Z}) & \longrightarrow & \Sigma_x & \longrightarrow & 0 \\ & & \downarrow & & \downarrow & & \downarrow & & \\ 0 & \longrightarrow & \Gamma(\mathcal{O}_U) & \longrightarrow & \bigoplus \Gamma(\mathcal{O}_{U_i}) & \longrightarrow & S_x & \longrightarrow & 0\,. \end{array}$$

If $e_i \in \Gamma(U_i, \mathbb{Z})$ denotes the constant 1, then $e_1, \ldots, e_r$ generate $\bigoplus \Gamma(U_i, \mathbb{Z})$. The images of these elements in $\bigoplus \Gamma(\mathcal{O}_{U_i})$ are linearly independent over $\mathbb{C}$ and span over $\mathbb{R}$ a vector space V of dimension r. The intersection $V_0 = V \cap \Gamma(\mathcal{O}_U)$ is the one-dimensional real vector space spanned by $1 \in \Gamma(\mathcal{O}_U)$. The image of Σ_x in S_x is a group of rank $r-1$ spanning over $\mathbb{R}$ the $(r-1)$-dimensional vector space V/V_0. So the image of Σ_x is a lattice in $V/V_0 \subset S_x$, hence discrete in S_x. □

If $C = C_1 \cup \cdots \cup C_r \subset X$ is compact and reduced, then ν^* induces an isomorphism $H^2(C, \mathbb{Z}) \to \bigoplus H^2(\widetilde{C}_i, \mathbb{Z}) = r\mathbb{Z}$ (after distinguishing the generator in $H^2(\widetilde{C}_i, \mathbb{Z})$ corresponding to the canonical orientation of $\widetilde{C}_i$). If $\mathcal{L} \in \operatorname{Pic}(C)$, its image $c_1(\mathcal{L}) \in H^2(C, \mathbb{Z})$ can be viewed as an r-tuple $(\ell_1, \ldots, \ell_r)$ of integers. The **degree** of $\mathcal{L}$ is $\deg(\mathcal{L}) = \ell_1 + \cdots + \ell_r$. If $C = n_1C_1 + \cdots + n_rC_r$ is not reduced, we put $\deg(\mathcal{L}) = n_1\ell_1 + \cdots + n_r\ell_r$ with $\ell_i = \deg(\mathcal{L}|(C_i)_{\mathrm{red}})$. For every locally free $\mathcal{O}_C$-sheaf $\mathcal{F}$ we define $\deg(\mathcal{F})$ as $\deg(\det \mathcal{F})$.

3. Riemann-Roch for an Embedded Curve

(3.1) **Theorem** (Riemann-Roch theorem for an embedded curve). *Let $C \subset X$ be a compact curve and $\mathcal{F}$ an $\mathcal{O}_C$-sheaf, locally free of rank r. Then*

$$\chi(\mathcal{F}) = \deg(\mathcal{F}) + r \cdot \chi(\mathcal{O}_C)\ .$$

If C is smooth, this formula already appeared as formula I (9). If C is reduced, it can be proved as follows. Consider the normalization $\nu : \widetilde{C} \to C$ and put $\widetilde{\mathcal{F}} = \nu^*\mathcal{F}$. Then

$$\deg(\widetilde{\mathcal{F}}) = \deg(\mathcal{F}),\quad \chi(\widetilde{\mathcal{F}}) = \chi(\nu_*\widetilde{\mathcal{F}}),\quad \text{and}\quad \chi(\mathcal{O}_{\widetilde{C}}) = \chi(\nu_*\mathcal{O}_{\widetilde{C}})$$

by the Leray spectral sequence (see I, Sect. 1). So Riemann-Roch on $\widetilde{C}$ implies

$$\deg(\mathcal{F}) = \chi(\widetilde{\mathcal{F}}) - r\chi(\mathcal{O}_{\widetilde{C}}) = \chi(\nu_*\widetilde{\mathcal{F}}) - r\chi(\nu_*\mathcal{O}_{\widetilde{C}})\ .$$

Since $\mathcal{F}$ is locally free, $\nu_*\widetilde{\mathcal{F}}/\mathcal{F} \cong r(\nu_*\mathcal{O}_{\widetilde{C}}/\mathcal{O}_C)$ and

$$\begin{aligned} \chi(\mathcal{F}) &= \chi(\nu_*\widetilde{\mathcal{F}}) - \chi(\nu_*\widetilde{\mathcal{F}}/\mathcal{F}) \\ &= \deg(\mathcal{F}) + r\chi(\nu_*\mathcal{O}_{\widetilde{C}}) - r\chi(\nu_*\mathcal{O}_C/\mathcal{O}_C) \\ &= \deg(\mathcal{F}) + r\chi(\mathcal{O}_C)\ . \end{aligned}$$

If C is not reduced, the formula can be proved using a decomposition $C = A + B$ and the corresponding sequences

$$0 \longrightarrow \mathcal{O}_A(-B) \longrightarrow \mathcal{O}_C \longrightarrow \mathcal{O}_B \longrightarrow 0$$

$$0 \longrightarrow \mathcal{F}|A \otimes \mathcal{O}_A(-B) \longrightarrow \mathcal{F} \longrightarrow \mathcal{F}|B \longrightarrow 0\ .$$

From these sequences it follows that

$$\begin{aligned} \chi(\mathcal{O}_C) &= \chi(\mathcal{O}_B) + \chi(\mathcal{O}_A(-B)) \\ \chi(\mathcal{F}) &= \chi(\mathcal{F}|B) + \chi(\mathcal{F}|A \otimes \mathcal{O}_A(-B))\ . \end{aligned}$$

Riemann-Roch on A and B implies

$$\begin{aligned} \chi(\mathcal{F}|B) &= \deg(\mathcal{F}|B) + r\chi(\mathcal{O}_B) \\ \chi(\mathcal{F}|A \otimes \mathcal{O}_A(-B)) &= \deg(\mathcal{F}|A \otimes \mathcal{O}_A(-B)) + r\chi(\mathcal{O}_A) \\ &= \deg(\mathcal{F}|A) + r\deg(\mathcal{O}_A(-B)) + r\chi(\mathcal{O}_A) \\ &= \deg(\mathcal{F}|A) + r\chi(\mathcal{O}_A(-B))\ . \end{aligned}$$

Adding these equations and using $\deg(\mathcal{F}) = \deg(\mathcal{F}|A) + \deg(\mathcal{F}|B)$ we obtain Riemann-Roch for $\mathcal{F}$ on C. □

4. The Residue Theorem

Let X be a complex manifold and $H \subset X$ a non-singular complex hypersurface. We shall say that a C^∞-differentiable complex-valued q-form φ on $X \setminus H$ has **ordinary poles along** H, if in local coordinates $(z_1, \ldots, z_n)$, with $z_1 = 0$ a local equation for H, it can be written

$$\varphi = \varphi_1 \wedge \frac{dz_1}{z_1} + \varphi_2\ , \tag{8}$$

where φ_1 and φ_2 extend over H as C^∞-forms. This notion is independent of the local coordinates chosen. (If the system $(z_1, \ldots, z_n)$ overlaps with $(w_1, \ldots, w_n)$, then $f = z_1/w_1$ and also $1/f$ extend holomorphically over H, so

$$\frac{df}{f} = \frac{dz_1}{z_1} - \frac{dw_1}{w_1}\ ,$$

$$\varphi = \varphi_1 \wedge \frac{dw_1}{w_1} + \varphi_1 \wedge \frac{df}{f} + \varphi_2\ .)$$

We denote by $\mathcal{D}^q_X(H)$ the sheaf on X of such q-forms (more precisely, the direct image on X of that sheaf).

If in the system $(z_1, \ldots, z_n)$ the form $\varphi \in \mathcal{D}^q_X(H)$ has two representations (8), say

$$\varphi = \varphi_1 \wedge \frac{dz_1}{z_1} + \varphi_2 = \psi_1 \wedge \frac{dz_1}{z_1} + \psi_2\ ,$$

then we may write

$$\varphi_1 - \psi_1 = \omega \wedge dz_1 + \sigma \ , \quad \sigma = \sum_{i_k \geq 2} \sigma_{i_2, \ldots, i_q} dz_{i_2} \wedge \cdots \wedge dz_{i_q} \ .$$

So

$$\varphi_1|H - \psi_1|H = \sigma|H \quad \text{with} \quad \sigma \wedge dz_1 = z_1(\psi_2 - \varphi_2) \ .$$

This shows that $\sigma|H$ vanishes and $\varphi_1|H \in \mathcal{D}_H^{q-1}$ is independent of the representation (8) for φ. Since we saw already that $\varphi_1|H$ is independent of the coordinate system, there is a well-defined residue map

$$\text{res} : \mathcal{D}_X^q(H) \to \mathcal{D}_H^{q-1} \ ,$$

defined locally for a form (8) by

$$\text{res}(\varphi) = \varphi_1|H \ .$$

(4.1) **Theorem** (Residue theorem). *The residue map has the following properties.*

(i) $\text{res}(g, \varphi) = g \,\text{res}(\varphi)$ *for any* C^∞*-function* g;

(ii) $\partial(\text{res}\,\varphi) = \text{res}(\partial\varphi)$, $\bar\partial(\text{res}\,\varphi) = \text{res}(\bar\partial\varphi)$;

(iii) *for every real submanifold* $M \subset X$ *of dimension* $q+1$ *intersecting* H *transversally in an oriented submanifold* S *such that* $\text{supp}(\varphi) \cap S$ *is compact, the residue formula*

$$2\pi \mathrm{i} \int_S \text{res}(\varphi) = \lim_{\varepsilon \to 0} \int_{M \cap \{\delta = \varepsilon\}} \varphi \ ,$$

holds, with δ *being some tubular neighbourhood function for* H *on a neighbourhood of* $S \cap \text{supp}(\varphi)$.

A tubular neighbourhood function δ is a real-valued C^∞-function vanishing on H together with its derivatives $\partial\delta/\partial x_1$ and $\partial\delta/\partial x_2$, where $z_1 = x_1 + \mathrm{i}x_2$ is a local equation for H, and the matrix

$$\begin{pmatrix} \dfrac{\partial^2\delta}{\partial x_1^2} & \dfrac{\partial^2\delta}{\partial x_1 \partial x_2} \\ \dfrac{\partial^2\delta}{\partial x_1 \partial x_2} & \dfrac{\partial^2\delta}{\partial x_2^2} \end{pmatrix}$$

is positive definite on H. For small ε, the set $M \cap \{\delta = \varepsilon\}$ is a real manifold of dimension q and it inherits an orientation from S in a natural way.)

Proof. First we prove (i), (ii), (iii) locally where φ is of the form (8) and $\text{res}(\varphi)$ is given by $\varphi_1|H$. Assertions (i),(ii) are obvious. To prove (iii) we put $z_1 = x_1 + \mathrm{i}x_2$ and consider only those φ for which $\text{supp}(\varphi)$ is compact and so small, that $x_1|M$, $x_2|M$ can be extended to a system $x_1, \ldots, x_{q+1}$ of local real coordinates on M in a neighbourhood of $M \cap \text{supp}(\varphi)$. Then (iii) becomes a statement in $\mathbb{R}^{q+1}$, namely

$$2\pi \mathrm{i} \int_{x_1 = x_2 = 0} \varphi_1 = \lim_{\varepsilon \to 0} \int_{\delta(x) = \varepsilon} \varphi \ .$$

But this is just a version with $q-1$ parameters of the usual residue formula in one complex variable:

$$2\pi i h(0) = \lim_{\varepsilon \to 0} \int_{\delta(x_1,x_2)=\varepsilon} h(x_1,x_2) \frac{dx_1 + idx_2}{x_1 + ix_2} ,$$

if $h(x_1,x_2)$ is a C^∞-function near $0 \in \mathbb{R}^2$ and

$$\delta(0) = \frac{\partial \delta}{\partial x_1}(0) = \frac{\partial \delta}{\partial x_2}(0) = 0, \quad \begin{pmatrix} \dfrac{\partial^2 \delta}{\partial x_1^2} & \dfrac{\partial^2 \delta}{\partial x_1 \partial x_2} \\ \dfrac{\partial^2 \delta}{\partial x_1 \partial x_2} & \dfrac{\partial^2 \delta}{\partial x_2^2} \end{pmatrix} > 0 .$$

By a partition of unity argument, the global residue formula is reduced to the local one. □

To conclude, we shall say a word about the case of a smooth curve C on a surface X. Let (u,v) be local coordinates on X with $v=0$ a local equation for C. If $\varphi \in \Gamma(\mathcal{D}_X^2(C))$ is holomorphic, it can be written locally as $\varphi = h\ du \wedge dv/v$ with h holomorphic. So φ is nothing but a section in $\Gamma(\mathcal{K}_X \otimes \mathcal{O}_X(C))$. Also the map $\varphi \mapsto \operatorname{res}(\varphi) = hdu|C$ is the same as the map $r : \mathcal{K}_X \otimes \mathcal{O}_X(C) \to \omega_C$ from Sect. 1.

5. The Trace Map

Let $D \subset X$ be some curve, not necessarily reduced, and ω_D its dualizing sheaf, defined in Sect. 1. In this section we define the **trace map**

$$\operatorname{tr}_D : H_c^1(\omega_D) \to \mathbb{C} .$$

First we recall how to describe $H_c^n(\mathcal{K}_Y)$, where Y is a complex *manifold* of dimension n, in terms of differential forms. Since the sheaves $\mathcal{D}_Y^{n,q}$ are fine, we have $H_c^i(\mathcal{D}_Y^{n,q}) = 0$ for $i \geq 1$, and Dolbeault's $\bar\partial$-resolution of $\mathcal{K}_Y$ gives us an isomorphism

$$H_c^n(\mathcal{K}_Y) = \Gamma_c(\mathcal{D}_Y^{n,n})/\bar\partial\Gamma_c(\mathcal{D}_Y^{n,n-1}) .$$

So each class $[\varphi] \in H_c^n(\mathcal{K}_Y)$ is represented by some (n,n)-form φ with compact support. If we modify φ by $\bar\partial\chi = d\chi$, $\chi \in \Gamma_c(\mathcal{D}_Y^{n,n-1})$, then $\int_Y \varphi$ does not change. So integration over Y defines a $\mathbb{C}$-linear map

$$\operatorname{Tr}_Y : H_c^n(\mathcal{K}_Y) \to \mathbb{C} .$$

If Y is compact connected, then $H_c^n(\mathcal{K}_Y) = H^{2n}(Y)$ is of dimension one and Tr_Y is an isomorphism (I, Sect. 11).

This definition of Tr_Y does not work for singular Y. Therefore, for an arbitrary $D \subset X$ we proceed as follows. Let $\delta : H_c^1(\omega_D) \to H_c^2(\mathcal{K}_X)$ be the map obtained from the sequence (6) and let $\operatorname{tr}_D = \operatorname{Tr}_X \circ \delta$. (This tr_D *a priori* depends on the embedding, but this will not matter in the sequel.)

Of course, we need that for non-singular D this tr_D is essentially the same as Tr_D (Proposition 5.1 below). So we look for an explicit description of δ in the case of *non-singular* D. Recall from Sect. 4 that $\mathcal{D}_X^q(D)$ is the sheaf of q-forms with poles along D. For $q = 0, 1, 2$ we put (in accordance with I, Sect. 12).

$$\begin{aligned} \mathcal{D}_X^{2,q}(D) = &\quad \text{subsheaf of } \mathcal{D}_X^{2+q}(D) \text{ of forms of type } (2,q) \\ \mathcal{D}_X^{1,q}(D) = &\quad \text{sheaf of } C^\infty\text{-forms on } D \text{ of type } (1,q) \\ \mathcal{K}^q = &\quad \text{kernel } \{\text{res} : \ \mathcal{D}_X^{2,q}(D) \longrightarrow \mathcal{D}_D^{1,q}\} \ . \end{aligned}$$

Since res is C^∞-linear, $\mathcal{K}^q$ is a fine sheaf. Furthermore, $\mathcal{K}^0 \subset \mathcal{D}_Y^{2,0}(D)$ is the sheaf of forms $f \ du \wedge dv/u$, where u is a local holomorphic equation of D and the C^∞-function f vanishes on D. If such a form is $\bar\partial$-closed, then f/u is holomorphic, which implies

$$\mathcal{K}^0 \cap \ \text{kernel} \ \{\bar\partial : \mathcal{D}_X^{2,0}(D) \longrightarrow \mathcal{D}_X^{2,1}(D)\} = \mathcal{K}_X \ .$$

So the double complex

$$0 \longrightarrow \mathcal{K}^q \longrightarrow \mathcal{D}_X^{2,q}(D) \xrightarrow{\text{res}} \mathcal{D}_D^{1,q} \longrightarrow 0, \quad q = 0,1,2 \ ,$$

with vertical arrows induced by $\bar\partial$ is a fine resolution of the sequence (6). If now $[\varphi] \in H_c^1(\omega_D)$ is represented by the form $\varphi \in \Gamma_c(\mathcal{D}_D^{1,1})$, then $\delta[\varphi] = [\bar\partial\psi]$ where $\psi \in \Gamma_c(\mathcal{D}_X^{2,1}(D))$ is some pre-image of φ under res and

$$\bar\partial\psi \in \Gamma_c(\mathcal{K}^2) = \Gamma_c(\mathcal{D}_X^{2,2}(D)) \ .$$

(5.1) **Proposition.** *For every class* $[\varphi] \in H_c^1(\omega_D)$ *one has*

$$2\pi \mathrm{i} \int_D [\varphi] = \int_X \delta[\varphi] \ .$$

Proof. Let ψ be as above and let τ be a tubular neighbourhood function for D, valid in a neighbourhood of the support of ψ. The residue formula together with Stokes' theorem gives

$$2\pi \mathrm{i} \int_D \varphi = \lim_{\varepsilon \to 0} \int_{\tau = \varepsilon} \psi = \lim_{\varepsilon \to 0} \int_{\tau \geq \varepsilon} \bar\partial\psi \ .$$

Now $\mathcal{K}^2$ contains $\mathcal{D}_X^{2,2}$ as subsheaf and if $\Phi \in \Gamma_c(\mathcal{D}_X^{2,2})$ is a smooth form representing in $H_c^{2,2}(X)$ the class $\delta[\varphi] = [\bar\partial\psi]$, then $\Phi = \bar\partial\psi + \bar\partial\chi$ with some $\chi \in \Gamma_c(\mathcal{K}^1)$. And if τ is a tubular neighbourhood function also on a neighbourhood of $\mathrm{Supp}\Phi \cap S$ and $\mathrm{Supp}\chi \cap S$, then

$$\begin{aligned} \int_X \delta[\varphi] &= \int_X \Phi = \lim_{\varepsilon \to 0} \int_{\tau \geq \varepsilon} \Phi \\ &= \lim_{\varepsilon \to 0} \left(\int_{\tau \geq \varepsilon} \bar\partial\psi + \int_{\tau \geq \varepsilon} \bar\partial\chi \right) = 2\pi \mathrm{i} \int_D \varphi \ , \end{aligned}$$

because by the residue formula

$$\lim_{\varepsilon \to 0} \int_{\tau \geq \varepsilon} \bar{\partial}\chi = 2\pi \mathrm{i} \int_D \mathrm{res}(\chi) = 0 . \qquad \square$$

6. Serre Duality on an Embedded Curve

In this section we treat in an elementary way Serre-Grothendieck duality for a (not necessarily reduced) compact curve C on a (not necessarily compact) surface X.

(6.1) **Theorem** (Duality theorem for an embedded curve). *For every compact curve C, embedded in a smooth surface X, there is an epimorphism*

$$\mathrm{tr} : H^1(\omega_C) \longrightarrow \mathbb{C} ,$$

such that the cupproduct pairing

$$H^1(\mathcal{F}) \otimes H^0(\mathcal{F}^\vee \otimes \omega_C) \longrightarrow H^1(\omega_C) \xrightarrow{\mathrm{tr}} \mathbb{C}$$

(defined for every $\mathcal{O}_C$-sheaf $\mathcal{F}$) is perfect as soon as $\mathcal{F}$ is locally free.

If C is smooth, then this is nothing but the Serre-duality of Proposition I.12.1. If C is reduced and projective, then the result is again due to Serre ([Se59], Chap. IV). For general projective schemes it is due to Grothendieck (see for example [A-K]), whereas for complex spaces it can be found in [R-R-V]. An elementary treatment, however, does not seem to be available in the literature, so we venture to include one here.

We start by observing that every reduced compact curve C is projective. To see this we remark that there is a line bundle $\mathcal{L}$ on C with a section, vanishing exactly in $p_1 \cup \cdots \cup p_k$, where $p_1, \ldots, p_k$ are smooth points on C, one on every irreducible component. By Grauert's criterion (Theorem I,19.3) the existence of such an $\mathcal{L}$ ensures the projectivity of C.

Next we recall that Serre proved his theorem with ω_C being the sheaf of *rational* Rosenlicht-differentials (see below); it follows from GAGA (I, Sect. 19) that ω_C can also be taken to be the sheaf of *meromorphic* Rosenlicht-differentials as soon as it is clear that this sheaf is coherent.

From here on we proceed as follows

a) we prove that our sheaf ω_C is the same as the sheaf of meromorphic Rosenlicht-differentials. This implies in particular that this last sheaf is coherent and consequently we may indeed apply Serre's result to any reduced C, using our ω_C;

b) we define a trace map for non-reduced C;

c) we prove duality on a non-reduced C by decomposition.

a) Rosenlicht-differentials.

Let C be a reduced curve with normalization $\nu : \widetilde{C} \to C$. A meromorphic differential on C is nothing but a section in $\nu_* \mathcal{M}_{\widetilde{C}}$ with $\mathcal{M}_{\widetilde{C}}$ the sheaf of meromorphic 1-forms on $\widetilde{C}$. Such a meromorphic σ on C is called a **Rosenlicht-differential** if for all $x \in C$, $g \in \mathcal{O}_{C,x}$

$$\sum_{x_k \in \nu^{-1}(x)} \operatorname{res}(x_k, g\sigma) = 0 ,$$

where res denotes the ordinary residue on the Riemann surface $\widetilde{C}$. A Rosenlicht-differential is holomorphic in all regular points of C.

Let C be embedded in a non-singular surface X. In Sect. 1 we defined ω_C and showed how to view it as subsheaf of the sheaf of meromorphic differentials on C.

(6.2) **Proposition.** *$\omega_C \subset \nu_* \mathcal{M}_{\widetilde{C}}$ is the sheaf of Rosenlicht-differentials.*

Proof. Let $x \in C$ be a singular point, (u, v) local coordinates centred at x, and $f = 0$ a local equation for C at x such that $f_v = \partial f / \partial v$ does not vanish identically on any branch of C at x. We put

$$B_\rho = \{|u|^2 + |v|^2 \leq \rho\}, \quad M_\rho = \partial B_\rho, \quad S_\rho = C \cap M_\rho .$$

If ρ is small, B_ρ contains no singularity of C and no zero of $f_v|C$ but for x itself. Also, M_ρ is a compact 3-sphere intersecting C transversally in S_ρ, a union of smooth circles. For every meromorphic function t on C, holomorphic on $(C \cap B_\rho) \setminus \{x\}$, we have

$$\sum_{x_k \mapsto x} \operatorname{res}(x_k, t\, du/f_v) = (2\pi \mathrm{i})^{-1} \int_{S_\rho} t\, du/f_v \tag{9}$$

by the ordinary residue theorem. It t is holomorphic on a neighbourhood in X of $C \setminus \{x\}$, the Residue Theorem 4.1 states:

$$\int_{S_\rho} t\, du/f_v = (2\pi \mathrm{i})^{-1} \lim_{\varepsilon \to 0} \int_{M_\rho \cap \{|f| = \varepsilon\}} t\, du \wedge dv/f . \tag{10}$$

Now we take some $\sigma \in \omega_{C,x}$, which by definition may be written $\sigma = h\, du/f_v$, $h \in \mathcal{O}_{C,x}$. Since h and any other $g \in \mathcal{O}_{C,x}$ is the restriction of a function holomorphic on some B_ρ, ρ small, formula (9) and (10) for $t = gh$ show that

$$\sum_{x_k \to x} \operatorname{res}(x_k, g\sigma) = (2\pi \mathrm{i})^{-2} \lim_{\varepsilon \to 0} \int_{M_\rho \cap \{|f| = \varepsilon\}} gh\, du \wedge dv/f .$$

Since the 2-form $gh\, du \wedge dv/f$ is holomorphic on the complement of C, it is closed on $M_\rho \setminus C$ and by Stokes' theorem

$$\int_{M_\rho \cap \{|f| = \varepsilon\}} gh\, du \wedge dv/f = \int_{M_\rho \cap \{|f| \geq \varepsilon\}} d(gh\, du \wedge dv/f) = 0 .$$

This shows that σ is a Rosenlicht-differential.

Conversely, let σ be some Rosenlicht-differential on C near x. It can be written as $\sigma = s\, du/f_v$ with s meromorphic on C. We show that s is holomorphic. To begin with, for large n, $u^n s$ will be holomorphic, hence the restriction of some $h_n \in \Gamma(B_\rho, \mathcal{O}_X)$. By assumption $\sum \operatorname{res}(x_k, gu^{n-1}\sigma) = 0$ for all holomorphic g. So by formulas (9) and (10) for $t = gh_n/u = gu^{n-1}s$

$$\lim_{\varepsilon \to 0} \int_{M_\rho \cap \{|f| = \varepsilon\}} g\frac{h_n}{u}\, du \wedge dv/f = 0\,.$$

But the integrand is a closed 2-form on $M_\rho \cap \{f \neq 0\} \cap \{u \neq 0\}$. So

$$\lim_{\varepsilon \to 0} \int_{M_\rho \cap \{|u| = \varepsilon\}} g\frac{h_n}{u}\, du \wedge dv/f = 0\,.$$

Now we consider the curve $D = \{u = 0\}$ and apply the residue theorem to $M_\rho \cap D \subset M_\rho$. Then it follows that

$$\operatorname{res}\left(x, \frac{gh_n}{f}\, dv\Big|D\right) = 0$$

for all holomorphic g. But this means h_n/f is holomorphic when restricted to D, i.e., there are holomorphic functions h_{n-1} and a on some B_ρ with $h_n = uh_{n-1} + af$. In other words, $h_n|C$ is divisible by u and we might as well have started with h_{n-1}. Repeating this procedure $(n-1)$ times we find that s itself was already holomorphic. □

b) The trace map.
We consider the trace map $\operatorname{tr}_C : H^1(\omega_C) \to \mathbb{C}$ as defined in Sect. 5.

(6.3) **Proposition.** *If C is an irreducible (hence reduced) compact curve, then $h^1(\omega_C) = 1$ and $\operatorname{tr}_C \neq 0$. So up to a non-zero constant, our tr_C coincides with the algebraic trace map.*

Proof. By duality on the reduced curve C, we have $h^1(\omega_C) = h^0(\mathcal{O}_C) = 1$. If we denote by $S \subset C$ the set of singular points and by D the smooth curve $C \setminus S$, then there is a commutative diagram

$$\begin{array}{ccccc} H^1_c(\omega_D) & \longrightarrow & H^1(\omega_C) & \longrightarrow & H^1(S, \omega_C|S) \\ \downarrow \delta & & \downarrow \delta & & \\ H^2_c(\mathcal{K}_X|X \setminus S) & \longrightarrow & H^2_c(\mathcal{K}_X) & & \end{array}$$

with $H^1(S, \omega_C|S) = 0$ since $\dim S = 0$. So the generator $\xi \in H^1(\omega_C)$ is the image of some $[\eta] \in H^1_c(\omega_D)$ and

$$\operatorname{tr}_C(\xi) = \int_X \delta[\eta] = 2\pi \mathrm{i} \int_D \eta$$

by Proposition 5.1.

Now let $\nu : \widetilde{C} \to C$ be the normalization. By Proposition 6.2 $\nu_* \omega_{\widetilde{C}} \subset \omega_C$. So there is a factorisation

$$H^1_c(\omega_D) \to H^1(\omega_{\widetilde{C}}) = H^1(\nu_* \omega_{\widetilde{C}}) \longrightarrow H^1(\omega_C)$$

showing that $[\eta] \in H^1(\omega_{\widetilde{C}})$ is non-trivial. But this implies

$$\int_D \eta = \int_{\widetilde{C}} \eta \neq 0 . \qquad \square$$

Next we consider a decomposition $C = A + B$ with two curves A, B on the surface X.

(6.4) **Theorem.** *There is a canonical $\mathcal{O}_C$-map $a : \omega_A \to \omega_C$, and* $\mathrm{tr}_C\, H^1(a) = \mathrm{tr}_A$, *where $H^1(a) : H^1(\omega_A) \to H^1(\omega_C)$ is induced by a.*

Proof. The decomposition sequence combines with (6) to a diagram

$$\begin{array}{ccccccccc}
 & & 0 & & 0 & & & & \\
 & & \downarrow & & \downarrow & & & & \\
 & & \mathcal{K}_X & = & \mathcal{K}_X & & & & \\
 & & \downarrow & & \downarrow & & & & \\
0 & \longrightarrow & \mathcal{K}_X \otimes \mathcal{O}_X(A) & \longrightarrow & \mathcal{K}_X \otimes \mathcal{O}_X(C) & \longrightarrow & \omega_B \otimes \mathcal{O}_B(A) & \longrightarrow & 0 \\
 & & \downarrow & & \downarrow & & \| & & \\
0 & \longrightarrow & \omega_A & \xrightarrow[a]{} & \omega_C & \longrightarrow & \omega_B \otimes \mathcal{O}_B(A) & \longrightarrow & 0 \\
 & & \downarrow & & \downarrow & & & & \\
 & & 0 & & 0 & & & &
\end{array}$$

The claim now follows from the definition of tr_A and tr_C. $\square$

c) Duality on non-reduced C.
The duality theorem can also be phrased as follows:
For every locally free $\mathcal{O}_C$-sheaf $\mathcal{F}$ the map

$$\text{(11)} \qquad \mathrm{Hom}_{\mathcal{O}_C}(\mathcal{F}, \omega_C) \longrightarrow H^1(\mathcal{F})^\vee,$$

sending an $\mathcal{O}_C$-morphism $h : \mathcal{F} \to \omega_C$ to the linear form

$$H^1(\mathcal{F}) \longrightarrow H^1(\omega_C) \xrightarrow{\mathrm{tr}} \mathbb{C} ,$$

is an isomorphism.
We show that this statement is true for a compact curve $C = A + B$ if it is true on A and B.

Step 1: injectivity of (11).

Consider the commutative pairing of decomposition sequences

$$\begin{array}{ccccccccc}
0 & \longrightarrow & \mathcal{O}_A(-B) & \longrightarrow & \mathcal{O}_C & \overset{\text{restr}}{\longrightarrow} & \mathcal{O}_B & \longrightarrow & 0 \\
 & & \otimes & & \otimes & & \otimes & & \\
0 & \longleftarrow & \mathcal{O}_A & \overset{\text{restr}}{\longleftarrow} & \mathcal{O}_C & \longleftarrow & \mathcal{O}_B(-A) & \longleftarrow & 0 \\
 & & \downarrow & & \downarrow & & \downarrow & & \\
 & & \mathcal{O}_A(-B) & \longrightarrow & \mathcal{O}_C & \longleftarrow & \mathcal{O}_B(-A) \,. & &
\end{array}$$

Tensoring the first row with $\mathcal{K}_X \otimes \mathcal{O}_C(C)$ we obtain

$$\begin{array}{ccccccccc}
0 & \longrightarrow & \omega_A & \overset{a}{\longrightarrow} & \omega_C & \longrightarrow & \omega_B \otimes \mathcal{O}_B(A) & \longrightarrow & 0 \\
 & & \otimes & & \otimes & & \otimes & & \\
0 & \longleftarrow & \mathcal{O}_A & \longleftarrow & \mathcal{O}_C & \longleftarrow & \mathcal{O}_B(-A) & \longleftarrow & 0 \\
 & & \downarrow & & \downarrow & & \downarrow & & \\
 & & \omega_A & \overset{a}{\longrightarrow} & \omega_C & \overset{b}{\longleftarrow} & \omega_B \,. & &
\end{array}$$

Applying $\mathrm{Hom}(\mathcal{F}, -)$ to the first row and $H^1(\mathcal{F} \otimes -)$ to the second one, we obtain a pairing of exact sequences

$$\begin{array}{ccccccc}
0 & \longrightarrow & \mathrm{Hom}(\mathcal{F}, \omega_A) & \longrightarrow & \mathrm{Hom}(\mathcal{F}, \omega_C) & \longrightarrow & \mathrm{Hom}(\mathcal{F}, \omega_B \otimes \mathcal{O}_B(A)) \\
 & & \otimes & & \otimes & & \otimes \\
0 & \longleftarrow & H^1(\mathcal{F}|A) & \longleftarrow & H^1(\mathcal{F}) & \longleftarrow & H^1(\mathcal{F} \otimes \mathcal{O}_B(-A)) \\
 & & \downarrow & & \downarrow & & \downarrow \\
 & & H^1(\omega_A) & \overset{H^1(a)}{\longrightarrow} & H^1(\omega_C) & \overset{H^1(b)}{\longleftarrow} & H^1(\omega_B) \,.
\end{array}$$

The left hand pairing followed by tr_A is the perfect duality pairing on A and similarly for the right hand pairing. Proposition 5.1 thus gives the diagram

$$\begin{array}{ccccccc}
0 & \longrightarrow & \mathrm{Hom}(\mathcal{F}, \omega_A) & \longrightarrow & \mathrm{Hom}(\mathcal{F}, \omega_C) & \longrightarrow & \mathrm{Hom}(\mathcal{F}, \omega_B \otimes \mathcal{O}_B(A)) \\
 & & \| & & \| & & \| \\
0 & \longrightarrow & H^1(\mathcal{F}|A)^\vee & \longrightarrow & H^1(\mathcal{F})^\vee & \longrightarrow & H^1(\mathcal{F} \otimes \mathcal{O}_B(-A))^\vee \,.
\end{array}$$

It follows that the morphism $\mathrm{Hom}(\mathcal{F}, \omega_C) \to H^1(\mathcal{F})^\vee$ obtained from the column in the middle is injective.

Step 2: equality $h^0(\mathcal{F}^\vee \otimes \omega_C) = h^1(\mathcal{F})$.

We know already $h^0(\mathcal{F}^\vee \otimes \omega_C) \le h^1(\mathcal{F})$ and $h^0(\mathcal{F}) \le h^1(\mathcal{F}^\vee \otimes \omega_C)$. So it suffices to prove $\chi(\mathcal{F}^\vee \otimes \omega_C) = -\chi(\mathcal{F})$. But

$$\begin{aligned}
\chi(\mathcal{F}) &= \chi\big(\mathcal{F}|B\big) + \chi\big(\mathcal{F} \otimes \mathcal{O}_A(-B)\big) \\
\chi(\mathcal{F}^\vee \otimes \omega_C) &= \chi\big(\mathcal{F}^\vee \otimes \omega_B\big) + \chi\big(\mathcal{F}^\vee \otimes \mathcal{O}_A(B) \otimes \omega_A\big) \,,
\end{aligned}$$

and duality on A and B implies $\chi(\mathcal{F}^\vee \otimes \omega_C) = -\chi(\mathcal{F})$. □

7. The σ-process

Let X be a surface and $x \in X$. In I, Sect. 9 we recalled what it means to blow up X at x, or, in other words, to apply the σ-process (σ-transformation) to X at x. Thus a new surface $\overline{X}$ is constructed, together with a regular map $\sigma : \overline{X} \to X$, such that $E = \sigma^{-1}(x)$ is a smooth rational curve, the exceptional curve, whereas σ maps $\overline{X} \setminus E$ biregularly onto $X \setminus x$. If (u, v) are local coordinates on X, centred at x, and $(z_0 : z_1)$ homogeneous coordinates on $\mathbb{P}_1$, then locally $\overline{X}$ can be described as

$$\{(y, (z_0 : z_1)) \in X \times \mathbb{P}_1, u(y)z_0 = v(y)z_1\} ,$$

with $\sigma(y, (z_0 : z_1)) = y$.

On $\overline{X}$, and near E, we can take as local coordinates outside of $z_0 = 0$ the functions $v, \dfrac{z_1}{z_0}$, and outside of $z_1 = 0$ the functions $u, \dfrac{z_0}{z_1}$. So u, v vanish along E to first order. With respect to the covering $E = E_0 \cup E_1$, $E_i = \{z_i \neq 0\}$, the normal bundle $\mathcal{N}_{E|\overline{X}}$ is defined by the cocycle $\dfrac{v}{u} = \dfrac{z_0}{z_1}|E_0 \cap E_1$. This implies $\mathcal{N}_{E/\overline{X}} = \mathcal{O}_{\mathbb{P}_1}(-1)$, the line bundle on E of degree -1.

Now let C be a curve on X, with a point x of multiplicity $\mu_x(C) = \mu$. Let $f = \sum_{p \geq \mu} f_p(u, v) = 0$ be a local equation for C at x, with f_p homogeneous of degree p. Then

$$\sigma^* f_p = f_p\Big(u, \frac{z_1}{z_0}u\Big) = u^p f_p\Big(1, \frac{z_1}{z_0}\Big)$$

where $z_0 \neq 0$ and similarly where $z_1 \neq 0$. This proves that the curve $f^{-1}(C)$ contains E with multiplicity μ. So $\sigma^{-1}(C) = \mu E + \overline{C}$ with $\overline{C}$ the closure in $\overline{X}$ of $\sigma^{-1}(C \setminus \{x\})$. The curve $\sigma^{-1}(C)$ is called the total transform of C and $\overline{C}$ its proper transform. By linearity this definition extends to divisors. If $\tau = \sigma_k \sigma_{k-1} \cdots \sigma_1$ is a succession of finitely many σ-processes, the proper transform under τ is the composition of the proper transforms under the σ_i. The intersections x_i of E with $\overline{C}$ are just the zeros (a^i, b^i) of $f_\mu(z_0, z_1)$ and thus correspond bijectively with the tangents $T_i = \{ub^i = va^i\}$ of C at x. Let $C_i \subset C$ be the union of branches of C at x tangent to T_i, and $\mu_i = \mu(C_i, x)$. Then $\mu = \sum \mu_i$, because there is a factorisation $f = \prod f^i$, with $f^i = 0$ a local equation for C_i. For the multiplicity $\bar{\mu}_i$ of $\overline{C}_i$ at x_i one has the estimate $\bar{\mu}_i \leq \mu_i$. Indeed, if we assume $(a^i, b^i) = (1, 0)$, then a local equation for $\overline{C}_i$ is $\sigma^* f^i / u^{\mu_i} = f^i_{u_i}\Big(1, \dfrac{z_1}{z_0}\Big) + \cdots$, vanishing to an order $\leq \mu_i$. So blowing up does not increase multiplicities of curves in the sense that

$$\mu \geq \sum \bar{\mu}_i .$$

As important application of the σ-process is the desingularization of curves.

(7.1) **Theorem.** *Let X be a (smooth) surface and $C \subset X$ an embedded reduced curve. Then there is a non-singular surface Y and a proper map $\tau : Y \to X$ consisting of a succession of σ-transformations (locally with respect to X, finitely many), such that the proper transform of C in Y is smooth.*

Proof. We may restrict our attention to one singularity x of the curve C. We have to show that the σ-process with centre x transforms C into a "less singular" curve $\overline{C}$. By the estimate above for the multiplicities, this is clear as soon as C has at least two different tangents at x, or if C has only one tangent and if the multiplicity of $\overline{C}$ at the corresponding point is smaller than $\mu_x(C)$. If none of these two favourable cases occurs, we can choose coordinates (u, v) such that the line $v = 0$ is tangent to the singularity, but not a branch of the curve. Then the singularity has the equation $f(u, v) = 0$ with

$$f = av^\mu + \sum_{\ell+m \geq \mu+1} a_{\ell m} u^\ell v^m ,$$

and the proper transform $\overline{C}$ has equation $\bar{f}(u, z) = 0$, $z = \frac{z_0}{z_1}$, with

$$\bar{f} = az^\mu + \sum_{\ell+m \geq \mu+1} a_{\ell m} u^{\ell+m-\mu} z^m .$$

Since the multiplicity of $\overline{C}$ is μ again, the coefficients $a_{\ell m}$ vanish for $\ell + 2m < 2\mu$. Since $\overline{C}$ has one tangent only, $a_{\ell m} = 0$ for $\ell + 2m = 2\mu$ too. This shows that the curve $z = 0$, the proper transform of the line $v = 0$, is tangent to the singularity of $\overline{C}$. By assumption $a_{\ell 0} \neq 0$ for some ℓ. The minimum ℓ_0 of these ℓ is the order to which $f(u, 0)$, the restriction of f to the tangent $v = 0$, vanishes at the origin. Then $\bar{f}(u, 0)$, the restriction to the tangent $z = 0$ of $\bar{f}$, vanishes to the order $\ell_0 - \mu$. We can repeat the σ-process arbitrarily often, but this order of vanishing can decrease only a finite number of times. So after finitely many σ-transformations we shall be in one of the favourable situations described above, and the multiplicity of the proper transform of C will be $< \mu$. Now the assertion follows by induction on μ. □

Sometimes information on the total transform of C is needed.

(7.2) **Theorem.** *Let X be a (smooth) surface and C a reduced curve on the surface X. Then there is a map $\tau : Y \to X$ as in Theorem 7.1 above such that the reduction of the total transform of C has only ordinary double points as singularities.*

Proof. Let $\tau_1 : Y_1 \to X$ be a succession of σ-processes with non-singular proper transform $C_1 \subset Y_1$ of C. The reduction of the total transform $\tau_1^{-1}(C)$ is $C_1 \cup D$ with D the union of all exceptional curves for τ_1. Since the exceptional curve of a σ-process intersects transversally the proper transform of every non-singular curve, it suffices to take $\tau = \tau_1 \circ \tau_2$, with $\tau_2 : Y_2 \to Y_1$ resolving the singularities of $C_1 \cup D$. In fact, a singularity y of the reduction of $\tau^{-1}(C)$ is an ordinary double point unless at y the proper transform of $C_1 \cup D$ meets two exceptional curves E' and E'' for τ_2. Blowing down E' and E'' would create at $\tau_2(y)$ a one-branch singularity for $C_1 \cup D$, a contradiction. □

A curve as in the preceding theorem is an example of a normal crossing divisor, a divisor which locally is given by the vanishing of a function of the form $x^a y^b$ where (x, y) are local holomorphic coordinates and where a and b are non-negative numbers. So the support of such a divisor is a curve having at most ordinary double points. We sometimes use the following straightforward consequence of Theorem 7.2

(7.3) **Corollary** *Let X be a (smooth) surface and $\mathcal{I} \subset \mathcal{O}_X$ a coherent sheaf of ideals. Then there is a non-singular surface Y and a proper map $\tau : Y \to X$ consisting of a succession of locally finitely many σ-transformations, such that the ideal $\tau^*\mathcal{I}$ is the ideal of a normal crossing divisor.*

Proof. The assertion is local. If $\mathcal{I}_x = (f_1 \ldots, f_r)$, $x \in X$, we set $f = f_1 f_2 \cdots f_r$ and we first of all perform successive blow-ups at x say $\tau : Y' \to X$ so that $f \circ \tau = 0$, is a normal crossing divisor at any point $\tilde{x} \in \tau^{-1}(x)$. Replacing X by Y' we may therefore assume that at every point $x \in X$ coordinates exist in which the generators of $\mathcal{I}_x$ are monomial.

If $\mathcal{I}_x$ is not yet principal, by blowing up once at x, we may assume that $\mathcal{I}_x$ has at most one monomial generator in every degree. Indeed, if we look at subset of the generators of $\mathcal{I}_x$ of some degree d, say $u^{n_1} v^{d-n_1}, \ldots, u^{n_k} v^{d-n_k}$, $n_1 < \cdots < n_k$, substituting $u = st, v = t$ replaces these by $t^d s^{n_1}, \ldots, t^d s^{n_k}$, but the first monomial $t^d s^{n_1}$ divides the remaining ones. This takes care of points in one of the two charts above x. For the other chart, we have to make the substitution $u = t, v = st$ and the argument is completely analogous.

So we may assume that $\mathcal{I}_x = (u^{n_1} v^{d_1 - n_1}, \ldots, u^{n_k} v^{d_k - n_k})$, $d_1 < \cdots < d_k$. Again, blowing up once more, we may reduce the number of generators unless $n_1 > \cdots > n_k$. In this case, blowing up at x has the effect of rendering $\mathcal{I}$ principal in the chart (u, v) with $u = t, v = st$. Indeed $\mathcal{I} = (t^{d_1} s^{d_1 - n_1}, \ldots, t^{d_k} s^{d_k - n_k})$ and since $d_1 - n_1 < \cdots < d_k - n_k$ and $d_1 < \cdots < d_k$, the first monomial divides the others. In the other chart $(u, v) = (st, t)$ we find $\mathcal{I} = t^{d_1} s^{n_k} (s^{n_1 - n_k}, \ldots, t^{d_k - d_1})$ and the reduced generators $s^{n_j - n_k} t^{d_j - d_1}$ have degree $d_j + n_j - d_1 - n_k < d_k - n_k$. This is because $n_k < \cdots < n_1 \leq d_1 < \cdots < d_k$ and hence $d_j + n_j < d_k + d_1$. So the maximal reduced degree decreases strictly. We then repeat the preceding procedure. Then either the ideal becomes principal, or the number of generators or the maximal reduced degree decreases. The process thus stops after finitely many blow-ups when the ideal $\mathcal{I}$ has become principal. After this we can once more invoke the previous theorem. □

8. Simple Singularities of Curves

In this section by a curve we always mean a curve on a smooth surface. We give the *A-D-E* classification of simple curve-singularities. A singularity of a reduced curve is called simple, if it is itself a double or triple point, and if when resolving the singularity to a collection of ordinary nodes according to Theorem 7.2, after each blowing up, the (reduced) total transform of our curve again has double or triple points only. We classify simple singularities by classifying all double points, all triple points with two or three different tangents, and simple triple points with one tangent.

The singularity under consideration will always be at the centre $(0,0)$ of a coordinate system (x,y) and will have equation $f(x,y)=0$. By $\mathfrak{m} \subset \mathcal{O}_{(0,0)}$ we denote the maximal ideal.

Classification of double points. These are the singularities with multiplicity $\mu = 2$. The residue $f_2 = f \bmod \mathfrak{m}^3$ is a non-zero homogeneous quadratic polynomial, which after a linear transformation can be put into the form

$$f_2 = x^2 + y^2 \quad \text{or} \quad f_2 = x^2 \ ;$$

In the first case we can write $f = x^2\varphi_1 + y^2\varphi_2$ with $\varphi_1(0,0) \neq 0$, $\varphi_2(0,0) \neq 0$, and we can introduce $x\sqrt{\varphi_1}$ and $y\sqrt{\varphi_2}$ as new coordinates. Then f is in normal form

$$f = x^2 + y^2 \quad \text{(ordinary double point, node, type } A_1) \ .$$

In the second case, we consider the Milnor number $n = \dim_{\mathbb{C}} \mathcal{O}_{(0,0)}/(f_x, f_y)$. It is independent of the choice of coordinates and finite, because we assumed our curve to be reduced. (Indeed, if $n = \infty$, the analytic set $f_x = f_y = 0$ will contain a curve passing through $(0,0)$ and f will vanish on this curve.) We write

$$f(x,y) = x^2 e(x,y) + x\varphi(y) + \psi(y) \tag{12}$$

with $e(0,0) \neq 0$ and φ, ψ vanishing at $y = 0$ to the orders $k \geq 2$, $\ell \geq 3$ respectively. Since up to a unit $\varphi = y \cdot \varphi_y$, we have the inclusion

$$(f_x, f_y) = (2xe + x^2 e_x + \varphi,\ x^2 e_y + x\varphi_y + \psi_y) \subset (x, \varphi_y, \psi_y) \ ,$$

from which we conclude

$$\min\{k, \ell\} \leq n + 1 \ . \tag{13}$$

Replacing x by $x\sqrt{e} + \varphi/2\sqrt{e}$ we find

$$f = x^2 - \varphi^2/4e + \psi \ .$$

Next we may expand

$$-1/4e = c(y) + x\varphi_1(y) + x^2 g(x,y)$$

and put

$$e' = 1 + \varphi^2 g, \quad \varphi' = \varphi^2 \varphi_1, \quad \psi' = \psi + \varphi^2 c\,.$$

Then f appears again in the original form (12), but now k, the vanishing order of φ, has at least doubled. When repeating this procedure arbitrarily often, the finiteness condition (13) prevents ℓ from going to ∞. So it is no loss of generality to assume $k \geq \ell$ and to write $x\varphi(y) + \psi(y) = y^\ell e''(x, y)$ with $e''(0,0) \neq 0$. After introducing new coordinates $x\sqrt{e}$ and $y\sqrt[\ell]{e''}$, the equation appears in normal form ($\ell = n + 1$)

$$f = x^2 + y^{n+1}, \quad n \geq 2 \text{ (type } A_n).$$

Classification of triple points with two or three different tangents. These are the singularities of multiplicity $\mu = 3$ where the cubic homogeneous polynomial $f_3 = f \bmod \mathfrak{m}^4$ has at least two different roots. One of them must be simple and corresponds to a non-singular branch which we can normalize to $y = 0$. Since we exclude the possibility of a non-reduced branch, we have $f = h(x, y) \cdot y$, where $h = 0$ is one of the double points discussed above. Since the quadratic polynomial $h_2 = h \bmod \mathfrak{m}^3$ does not contain the factor y, by a linear transformation, leaving invariant the axis $y = 0$, we can put it into the form $h_2 = x^2 + y^2$ or $h_2 = x^2$. Now we apply the normalizing procedure above to the double point $h = 0$. Noticing that in this procedure we did not change the axis $y = 0$, we find that the triple point can be given by some equation

$$y(x^2 + y^{n-2}) = 0, \quad n \geq 4 \quad \text{(type } D_n).$$

Classification of simple triple points with one tangent. Now $f_3 = f \bmod \mathfrak{m}^4$ has one root only, say $x = 0$. Then there is an expansion

$$f(x, y) = x^3 e(x, y) + x^2 y^2 \varphi_1(y) + x y^3 \varphi_2(y) + y^4 \varphi_3(y)$$

with $e(0,0) \neq 0$. Putting $x = uy$ and dividing by y^3 we obtain the equation

$$\bar{f}(x, y) = u^3 \bar{e}(u, y) + u^2 y \varphi_1(y) + u y \varphi_2(y) + y \varphi_3(y) \tag{14}$$

for the proper transform of our curve after blowing up the origin. This transform is at worst a double point, so one of the following must hold:

$$E_6 : \varphi_3(0) \neq 0$$
$$E_7 : \varphi_3(0) = 0, \quad \varphi_2(0) \neq 0$$
$$E_8 : f_3(0) = \varphi_2(0) = 0, \quad \varphi_3'(0) \neq 0.$$

The case E_6. Replacing y by $y\sqrt[4]{\varphi_3} + x\varphi_2/4\sqrt[4]{\varphi_3^3}$ we bring f into the form (with new e and φ_1)

$$f(x, y) = x^3 e(x, y) + x^2 y^2 \varphi_1(y) + y^4.$$

Substituting $\xi = x\sqrt[3]{e} + y^2\varphi_1/3\sqrt[3]{e^2}$ we change this into

$$\begin{aligned} f(x,y) &= \xi^3 + y^4(1 - x\varphi_1^2/3e - y^2\varphi_1^3/27e^2) \\ &= \xi^3 + y^4 e'(\xi, y) \end{aligned}$$

with $e'(0,0) \neq 0$. Finally, replacing y by $y\sqrt[4]{e'}$, we arrive at the normal form

$$f = x^3 + y^4 \quad (\text{type } E_6).$$

The case E_7. By Eq. (14), our singularity is an ordinary double point of the proper transform. So the original triple point was reducible. Taking for axis $x = 0$ one of the non-singular branches, the equation is put into the form

$$f(x,y) = x(x^2 e(x,y) + xy^2\varphi_1(y) + y^3\varphi_2(y)) \, .$$

Since $\varphi_2(0) \neq 0$, we may replace y by

$$\eta = y\sqrt[3]{\varphi_2} + x\varphi_1/3\sqrt[3]{\varphi_2^2}$$

to find

$$\begin{aligned} f(x,\eta) &= x\left(x^2 e(x,\eta) + \eta^3 - x^2 y\varphi_1^2/3\varphi_2 - x^3\varphi_1^3/27\varphi_2^2\right) \\ &= x\left(x^2 e'(x,\eta) + \eta^3\right) \, . \end{aligned}$$

Since $e'(0,0) \neq 0$ we can replace x by $x\sqrt{e'}$ and divide by a unit to arrive at the normal form

$$f = x(x^2 + y^3) \quad (\text{type } E_7) \, .$$

The case E_8. We write

$$f(x,y) = x^3 e(x,y) + x^2y^2\varphi_1(y) + xy^4\varphi_2(y) + y^5\varphi_3(y)$$

with $\varphi_3(0) \neq 0$. Substituting as above $\xi = x\sqrt[3]{e} + y^2\varphi_1/3\sqrt[3]{e^2}$, we obtain

$$f(\xi,y) = \xi^3 + y^4\left(x\varphi_2(y) - x\varphi_1^2/3e - y^2\varphi_1^3/27e^2\right) + y^5\varphi_3 \, .$$

Expanding the bracket as power series in ξ and y we find (with new functions e and φ_i)

$$f(\xi,y) = \xi^3 e(\xi,y) + \xi^2 y^4\varphi_1(y) + \xi y^4\varphi_2(y) + y^5\varphi_3(y) \, ,$$

where still $\varphi_3(0) \neq 0$. Replacing y by $y\sqrt[5]{\varphi_3} + \xi\varphi_2/5\sqrt[5]{\varphi_3^4}$ we can change this into (again with new e and φ_1)

$$f(\xi,y) = \xi^3 e(\xi,y) + \xi^2 y^3\varphi_1(y) + y^5 \, .$$

With $\zeta = \xi^3\sqrt[3]{e} + y^3\varphi_1/3\sqrt[3]{e^2}$ this becomes

$$\begin{aligned} f(\zeta,y) &= \zeta^3 + y^5(1 - \xi y^2\varphi_1/3e - y^4\varphi_1^3/27e^2) \\ &= \zeta^3 + y^5 e'(\zeta, y) \end{aligned}$$

with $e'(0,0) \neq 0$. Passing from y to $y\sqrt[5]{e'}$ we obtain the normal form

$$f = x^3 + y^5 \quad (\text{type } E_8) \, .$$

Having completed these classifications we can easily prove the basic

(8.1) **Theorem** *The simple singularities of curves are exactly the double points with equations*

$$A_n : x^2 + y^{n+1} = 0, \quad n \geq 1$$

and the triple points with equations

$$D_n : y(x^2 + y^{n-2}) = 0, \quad n \geq 4$$
$$E_6 : x^3 + y^4 = 0$$
$$E_7 : x(x^2 + y^3) = 0$$
$$E_8 : x^3 + y^5 = 0 .$$

Proof. By the classification above, it is clear that all simple singularities must occur in this list. To prove that the A-D-E singularities are all simple, it suffices to compute their total transforms under blowing up the origin and to observe that the new singularities are again of A-D-E type. In fact we have:

singularity	A_2	$A_n,\ n \geq 3$	D_4	D_5	$D_n,\ n \geq 6$	E_6	E_7	E_8
transform	A_3	D_{n+1}	$3A_1$	A_1, A_3	A_1, D_{n-2}	A_5	D_6	E_7

□

Intersection Theory

In this subsection we shall discuss properties concerning the intersection of two divisors, culminating in a study of 1-connected divisors. This will be used later in Chap. III.

9. Intersection Multiplicities

Let x be an isolated intersection of two reduced curves C and D on a (non-singular) surface X. Let $f,\, g \in \mathcal{O}_{X,x}$ be local equations for C, D respectively. We define the following numbers:

$$i_1 = \dim_{\mathbb{C}} \mathcal{O}_{X,x}/(f_x, g_x)$$
$$i_2 = \sum_{k=1}^{r} \operatorname{ord}_{x_k}(g_k) \ ,$$

where $\nu : \widetilde{C} \to C$ is the normalization $g_1,\ \ldots,\ g_r$ are the germs of $g \circ \nu$ in the points $x_1,\ \ldots,\ x_r \in \nu^{-1}(x)$, and $\operatorname{ord}_{x_k}(g_k)$ denotes their order of vanishing.

(9.1) **Proposition** *The numbers i_1 and i_2 are equal.*

The number $i_1 = i_2 = i_x(C, D)$ is called the intersection multiplicity of C and D in x.

Proof of Proposition 9.1. Let $\mathcal{J} \subset \mathcal{O}_C$ be the ideal generated by the function $g|C$. Then $i_1 = \dim_{\mathbb{C}}(\mathcal{O}_C/\mathcal{J})_x$. Consider the exact diagram on C:

$$\begin{array}{ccccccccc}
 & & 0 & & 0 & & & & \\
 & & \downarrow & & \downarrow & & & & \\
0 & \longrightarrow & \mathcal{J} & \longrightarrow & \nu_*\nu^*\mathcal{J} & \longrightarrow & \mathcal{C}_1 & \longrightarrow & 0 \\
 & & \downarrow & & \downarrow & & \downarrow{\scriptstyle p} & & \\
0 & \longrightarrow & \mathcal{O}_C & \longrightarrow & \nu_*\mathcal{O}_{\widetilde{C}} & \longrightarrow & \mathcal{C}_2 & \longrightarrow & 0 \\
 & & \downarrow & & \downarrow & & & & \\
 & & \mathcal{O}_C/\mathcal{J} & \xrightarrow{q} & \nu_*(\mathcal{O}_{\widetilde{C}}/\nu^*\mathcal{J}) & & & & \\
 & & \downarrow & & \downarrow & & & & \\
 & & 0 & & 0 & & & &
\end{array}$$

Since $\mathcal{J}$ is locally isomorphic with $\mathcal{O}_C$, the two quotients $\mathcal{C}_1$ and $\mathcal{C}_2$ will be isomorphic too. Both have support in x, so $\ker p$ and $\operatorname{coker} p$ are sheaves of the same length supported in x. Now the familiar snake lemma shows $\ker p = \ker q$ and $\operatorname{coker} p = \operatorname{coker} q$. Hence

$$\begin{aligned}
i_1 &= \dim_{\mathbb{C}}(\mathcal{O}_C/\mathcal{J})_x = \dim_{\mathbb{C}}(\nu_*(\mathcal{O}_{\widetilde{C}}/\nu^*\mathcal{J}))_x \\
&= \sum_1^r \dim_{\mathbb{C}}(\mathcal{O}_{\widetilde{C},x_k})/(g_k) = i_2 \ . \qquad \square
\end{aligned}$$

The preceding descriptions of $i_x(C, D)$ allow to draw the following conclusions:

1) $i_x(C, D) = 1$ if and only if f and g span $\mathfrak{m}_x \subset \mathcal{O}_{X,x}$, i.e., if C and D are non singular and do not touch at x.
2) $i_x(C, D)$ is symmetric in C and D. Since $\operatorname{ord}_{x_k}(g_k h_k) = \operatorname{ord}_{x_k}(g_k) + \operatorname{ord}_{x_k}(h_k)$, the intersection number depends bilinearly on C and D. So by linearity the definition can be extended to divisors.
3) Let $\sigma : \overline{X} \to X$ be the σ-process with centre x. Then i_2 does not change if one replaces C by $\overline{C}$, the proper transform, and D by $\nu^{-1}(D)$, the total transform. So one has

$$i_x(C, D) = \sum_{y \in C \cap \sigma^{-1}(x)} i_y(\overline{C}, \sigma^{-1}(D)) \ .$$

10. Intersection Numbers

Let X be a connected surface. Then $H_c^4(X,\mathbb{Z}) \cong H_0(X,\mathbb{Z}) \cong \mathbb{Z}$, in a canonical way, and we have the cupproduct pairing

$$\begin{array}{rcl} H_c^2(X,\mathbb{Z}) \times H^2(X,\mathbb{Z}) & \longrightarrow & H_c^4(X,\mathbb{Z}) \cong \mathbb{Z} \\ \xi \times \eta & \longrightarrow & (\xi, \eta) \,. \end{array}$$

We shall denote (ξ, η) sometimes by $\xi\eta$.

Definition If D is a divisor on X with compact support, E some other divisor and $\mathcal{M}$ a line bundle on X, we put

$$(D, E) = DE = ([D], c_1(\mathcal{O}_X(E)))$$

$$(D, \mathcal{M}) = ([D], c_1(\mathcal{M})) \,.$$

(Recall that $[D] \in H_c^2(X,\mathbb{Z})$ is the fundamental cohomology class of D, cf. page 14). If X is compact we set

$$(\mathcal{L}, \mathcal{M}) = \big(c_1(\mathcal{L}), c_1(\mathcal{M})\big)$$

for any two line bundles $\mathcal{L}$, $\mathcal{M}$ on X. All the integers thus defined are called intersection numbers.

The intersection product has the following properties.

(i) *It is bilinear with respect to the tensor product operation on line bundles and with respect to addition of divisors.*

(ii) *It is symmetric.*

(iii) *If $\pi : Y \to X$ is a proper map from another surface Y onto X, then the product formula I(1) implies*

$$(\pi^*(\mathcal{L}), \pi^*(\mathcal{M})) = (\deg(\pi))(\mathcal{L}, \mathcal{M}).$$

(iv) *If C is any compact curve, then*

$$(C, \mathcal{M}) = \deg(\mathcal{M}|C) \,. \tag{15}$$

Proof. If C is a smooth compact curve, then the homology class of C is dual to $c_1(\mathcal{O}_X(C))$ (Proposition I.6.4). Therefore, if $\mathcal{M}$ is any line bundle on X, we have (15) again by the product formula.
If C is singular, but still reduced and irreducible, and if $\nu : \widetilde{C} \to C$ is the desingularization (Theorem 7.1) then $\nu^* : H^2(C,\mathbb{Z}) \to H^2(\widetilde{C},\mathbb{Z})$ is an isomorphism, and consequently (15) holds in this case too. The general case follows by linearity. □

(v) *If D_1 and D_2 are two compact divisors, then D_1D_2 is the intersection product of their homology classes.*
To see this it is sufficient to show that for any compact divisor D the class $c_1(\mathcal{O}_X(D))$ is dual to the homology class of D. By linearity this can be reduced to the case of an irreducible D. We take a succession of σ-processes

$\tau : \overline{X} \to X$ such that the proper transform $\overline{D} \subset \overline{X}$ of D is smooth. Using Proposition I.6.4 and linearity of c_1 we see that $c_1(\tau^*(\mathcal{O}_X(D))$ is dual to $\overline{D}+$ (sum of exceptional curves). Applying Lemma I.1.1, we find the result we want.

(vi) *If the compact divisors D, E have no common component (i.e., if they intersect in a finite set of points), then*

$$DE = \sum_{x \in D \cap E} i_x(D, E) \, .$$

To prove this, by linearity we may assume D and E to be effective, even irreducible. Let $\nu : \widetilde{D} \to D$ be the normalization, then by *(iv)* above

$$DE = \deg \mathcal{O}_D(E) = \deg \nu^*(E) \, .$$

But the canonical section in $\mathcal{O}_{\widetilde{D}}(\nu^*(E))$ vanishes exactly in the points $y \in \widetilde{D}$ with $x = \nu(y) \in D \cap E$ and there to the order $\mathrm{ord}_y(\nu^*(f_x))$, with $f_x \in \mathcal{O}_{X,x}$ a local equation for E. So

$$\deg \nu^*(E) = \sum_{x \in D \cap E} \sum_{y \in \nu^{-1}(x)} \mathrm{ord}_y(\nu^*(f_x))$$

and this proves the assertion, because

$$\sum_{y \in \nu^{-1}(x)} \mathrm{ord}_y(\nu^*(f_x)) = i_x(D, E) \, .$$

11. The Arithmetical Genus of an Embedded Curve

Let C be a compact irreducible curve. The genus $g(\widetilde{C})$ of its normalization $\widetilde{C}$ is called the geometric(al) genus of C. If C is embedded in a non-singular surface X, then the arithmetic(al) genus of C (compare I, Sect. 4) is

$$g(C) = 1 - \chi(\mathcal{O}_C) = 1 + \chi(\omega_C)$$

(the second equality is due to duality). This $g(C)$ has the following properties:

a) *If C is irreducible non-singular, then arithmetic and geometric genus are the same. If $C = \bigcup_1^k C_i$ is non-singular with C_i connected, then $g(C) = \sum_1^k g(C_i) - (k-1)$. If $C \subset X$ is reduced with $\nu : \widetilde{C} \to C$ its normalization, then the normalization sequence (5) shows that $\chi(\mathcal{O}_C) = \chi(\nu_* \mathcal{O}_{\widetilde{C}}) - h^0(\nu_* \mathcal{O}_{\widetilde{C}} / \mathcal{O}_C)$. Hence*

$$g(C) = g(\widetilde{C}) + \delta(C)$$

with

$$\delta = \sum_{x \in C} \dim_{\mathbb{C}} (\nu_* \mathcal{O}_{\widetilde{C}} / \mathcal{O}_C)_x \ .$$

b) *For every $C \subset X$ we have the adjunction formula or genus formula*

$$g(C) = 1 + \frac{1}{2} \deg(\mathcal{K}_X \otimes \mathcal{O}_X(C) \mid C) \ . \tag{16}$$

To prove this, we start from

$$\deg(\omega_C) = \deg(\mathcal{K}_X \otimes \mathcal{O}_X(C) \mid C) \ .$$

Riemann-Roch plus duality show

$$\deg(\omega_C) = \chi(\omega_C) - \chi(\mathcal{O}_C) = 2\chi(\omega_C),$$

yielding (16).

(If C is smooth then formula (16) is indeed equivalent to the "old" adjunction formula of Theorem I.6.3.) By this formula we can also define the arithmetical genus for divisors C. If $C = A + B$, then

$$g(A + B) = g(A) + g(B) + AB - 1 \ . \tag{17}$$

c) *If C is reduced and connected, then $g(C) \geq 0$, and $g(C) = 0$ implies that C is a tree of non-singular rational curves.*
Here $C = \sum R_i$ is a tree if
 (i) $R_i R_j \leq 1$ for $i \neq j$,
 (ii) there is no cycle $R_{i_1}, \ \ldots, \ R_{i_n} \subset C$, $n \geq 3$, with $R_{i_j} R_{i_{j+1}} \neq 0$ for $j = 1, \ \ldots, \ n-1$ and $R_{i_1} R_{i_n} \neq 0$,
 (iii) three different curves never have a point in common.
Proof of c). Since $h^0(\mathcal{O}_C) = 1$ for every reduced connected C, it follows that $g(C) = 1 - h^0(\mathcal{O}_C) + h^1(\mathcal{O}_C) \geq 0$. Conversely, if $g(C) = 0$, then the assertion can be proved by induction on the number of irreducible components: if $C = A + B$, then $AB \geq 1$, so $g(C) \geq g(A) + g(A)$, and $g(C)$ vanishes only if $g(A) = g(B) = 0$ and $AB = 1$. □

12. 1-Connected Divisors

We shall call an effective divisor C connected, if supp(C) is connected. For compact reduced C, connectedness implies $h^0(\mathcal{O}_C) = 1$, but this is no longer true if C is not reduced. The following lemma deals with this situation. For $\mathcal{L} = \mathcal{O}_C$, it is due to Ramanujam ([Ram], Lemma 3).

(12.1) **Lemma.** *Let C be a compact curve on the non-singular surface X and $\mathcal{L}$ a line bundle on C, the restriction of which to any irreducible component of C has degree 0. If $h \in H^0(\mathcal{L})$, and $C = C_1 + C_2$, with $C_1 \leq C$ a maximal divisor satisfying $h|C_1 \equiv 0$, then*

$$C_1 C_2 \leq 0 \, .$$

Proof. By assumption $h \in H^0(\mathcal{J}_{C_1} \cdot \mathcal{L}) = H^0(\mathcal{O}_{C_2}(-C_1) \otimes \mathcal{L})$. The map $h : \mathcal{O}_{C_2} \to \mathcal{O}_{C_2}(-C_1) \otimes \mathcal{L}$ is injective and its cokernel $\mathcal{Q}$ has finite support. The exact sequence

$$0 \longrightarrow \mathcal{O}_{C_2} \longrightarrow \mathcal{O}_{C_2}(-C_1) \otimes \mathcal{L} \longrightarrow \mathcal{Q} \longrightarrow 0 \tag{18}$$

and Riemann-Roch on C_2 then show

$$\begin{aligned} -C_2 C_1 &= \deg \left(\mathcal{O}_{C_2}(-C_1) \otimes \mathcal{L}\right) \\ &= \chi(\mathcal{O}_{C_2}(-C_1) \otimes \mathcal{L}) - \chi(\mathcal{O}_{C_2}) = h^0(\mathcal{Q}) \geq 0 \, . \end{aligned}$$ □

Definition A compact effective divisor C on a surface is called m- connected, if $C_1 C_2 \geq m$ for each effective decomposition $C = C_1 + C_2$.

Of course, every 1-connected ("numerically connected") divisor C has connected support, but the converse is false. (Take for example an irreducible curve D with negative self-intersection. Then for $n \geq 2$ the curve $C = nD$ is connected but not 1-connected.)

(12.2) **Lemma.** *Let C be 1-connected and $\mathcal{L}$ a line bundle on C as in Lemma 12.1 above. Then $h^0(\mathcal{L}) \leq 1$ and $h^0(\mathcal{L}) = 1$ if and only if $\mathcal{L} = \mathcal{O}_C$.*

Proof. For $h \in H^0(\mathcal{L})$ define $C = C_1 + C_2$ as above. If $h \neq 0$, then $C_2 \neq 0$, and so $C_1 = 0$ by 1-connectedness. But now the exact sequence (18) and Riemann-Roch show $h^0(\mathcal{Q}) = 0$, i.e., $\mathcal{Q} = 0$, and $h : \mathcal{O}_C \to \mathcal{L}$ is an isomorphism. □

As a special case we note

(12.3) **Corollary.** *If C is a 1-connected effective divisor, then $h^0(\mathcal{O}_C) = 1$.*

We also note the following auxiliary result which is often used in technical applications.

(12.4) **Lemma.** *Let D be a 1-connected effective divisor on X and $E \not\subset$ supp(D) an irreducible curve on X with $DE = 1$. Then $h^0(\mathcal{O}_D(E)) = 1$, unless the component R of D intersecting E is non-singular rational. (Notice that because of $DE = 1$ the curve E intersects only one component E of D, which is necessarily reduced.)*

Proof. If we write $D = C + R$, then $E \cap \text{supp}(C) = \varnothing$ and we have two decomposition sequences.

$$\begin{array}{ccccccccc} 0 & \longrightarrow & \mathcal{O}_C(E-R) & \longrightarrow & \mathcal{O}_D(E) & \longrightarrow & \mathcal{O}_R(E) & \longrightarrow & 0 \\ & & \| & & & & & & \\ 0 & \longrightarrow & \mathcal{O}_C(-R) & \longrightarrow & \mathcal{O}_D & \longrightarrow & \mathcal{O}_R & \longrightarrow & 0\,. \end{array}$$

Since D is 1-connected, by Corollary 12.3 the restriction $H^0(\mathcal{O}_D) \to H^0(\mathcal{O}_R)$ is injective, so $h^0(\mathcal{O}_C(E-R)) = h^0(\mathcal{O}_C(-R)) = 0$. And if $h^0(\mathcal{O}_D(E)) > 1$, the line bundle $\mathcal{O}_R(E)$ of degree 1 will have two independent sections, defining an isomorphism from R onto $\mathbb{P}_1$. □

Historical Remarks

The contents of Sects. 1–7 are absolutely standard. In algebraic formulation, most of it is contained in Serre's book [Se59], in particular chapter I. Serre duality for reduced projective curves is formulated and proved there for the first time. Duality on projective schemes is due to Grothendieck, see the lecture notes [A-K]. There is also an analytic version of this duality, see e.g. [R-R]. A simple treatment of the curve case in analytic language however does not seem available. Of course, every compact analytic curve is projective, but the easiest way to prove this is using duality.

It should be mentioned that the "fundamental class" in analytic and algebraic duality differs by a power of $2\pi\mathrm{i}$. If X is a compact algebraic manifold of dimension n, then under the identification

$$H^n_{\mathrm{alg}}(X, \Omega^n) = H^n_{\mathrm{an}}(X, \Omega^n)$$

we have

$$\text{algebraic fund. class} \;=(2\pi\mathrm{i})^n \text{ analytic fund. class}\,.$$

For our purposes, such differences are not important.

The relation between simple singularities and simple Lie groups is one of the most beautiful discoveries in mathematics. It is impossible to attribute it to a single author. Any list of references will contain the names V.I. Arnold, E. Brieskorn, P. Du Val, F. Klein, and it will lead back to the classification of platonic solids. Section 8 contains the classification of simple singularities of curves in the most elementary way we know of. Dynkin diagrams will appear in III. Sect. 7.

Sections 9, 10, 11 are again standard knowledge, although references in the analytic case are scarce. Section 12 is essentially due to Ramanujam (see [Ram]).

Chapter III. Mappings of Surfaces

In this chapter a surface is a reduced 2-dimensional complex space, unless specified otherwise. We draw in particular attention to the convention valid for Sects. 8–18.

Bimeromorphic Geometry

In the framework of analytic geometry bimeromorphic maps play the same role as birational maps in algebraic geometry. The first aspect concerning these is the process of desingularization. We have seen (Sect. I.8) that the normalization can be seen as a first step in making the singularities less complicated. So we may assume that the surfaces under consideration are normal. For these there is indeed a desingularization as shown in Sect. 6. The proof is by reduction to the case of Hirzebruch-Jung singularities treated in Sect. 5. The existence of a *minimal* resolution is very particular for dimension 2 and depends on the fact that one can contract a (-1)-curve, i.e., a smooth rational curve with self-intersection -1, to a *smooth* point. This and its applications to the existence of minimal models is treated in Sect. 4. Curves contractible to (singular) points are treated more generally in Sect. 2.

The simplest kind of singularities are rational singularities. We shall consider these in Sect. 3, while an important subclass, namely the simple singularities, will be treated in Sect. 7.

1. Bimeromorphic Maps

Let X, Y be irreducible surfaces. A proper holomorphic surjective map $\pi : X \to Y$ is called bimeromorphic if there are proper analytic subsets $T \subset X$ and $S \subset Y$ such that $\pi : X \setminus T \to Y \setminus S$ is biholomorphic. A basic example of such a map is provided by the normalization $\nu : \widetilde{X} \to X$, introduced in I, Sect. 8.

In the sequel we consider only bimeromorphic maps $\pi : X \to Y$ with X and Y *normal*. Then there is a discrete set $S \subset Y$ such that $\pi|\pi^{-1}(Y \setminus S)$ is biholomorphic and such that $\pi^{-1}(y)$ is a curve in X for every $y \in S$ (see [G-R55]). The points $y \in S$ are called fundamental points for π and the curves $\pi^{-1}(y)$ exceptional curves for π. It is a consequence of the Riemann and Levi extension theorems on the normal surface Y (Theorem I.8.7) that π^* and π_* give isomorphisms between the rings of holomorphic functions and the fields of meromorphic functions. This implies in particular that the exceptional curves $\pi^{-1}(y)$ are *connected.*

A special type of bimeromorphic map – the σ-process – appeared already before (I, Sect. 9 and II, Sect. 7). Like in that special case we define for a bimeromorphic map $\pi : X \to Y$ between smooth surfaces X and Y the **total transform** of any divisor D on Y to be the divisor $\pi^*(D)$ on X, whereas the **proper transform** $\overline{D}$ of D is the divisor obtained from $\pi^*(D)$ by removing all components which are exceptional for π.

If X and Y are normal projective surfaces, then by Chow's theorem (Theorem I.19.2) a bimeromorphic map is nothing but a birational map.

Let X, Y be an ordered pair of normal surfaces. A **bimeromorphic correspondence** between X and Y is a triple (Z, π_1, π_2), where Z is a third irreducible and normal surface and $\pi_1 : Z \to X$, $\pi_2 : Z \to Y$ are bimeromorphic maps. Such a correspondence induces a **bimeromorphic transformation** $\tau : X \to Y$ given by $\tau = \pi_2 \circ \pi_1^{-1}$. Given a bimeromorphic transformation (Z_1, π_1, π_2) between X_1 and X_2 and a bimeromorphic transformation (Z_2, π_3, π_4) between X_2 and X_3, there exists a diagram

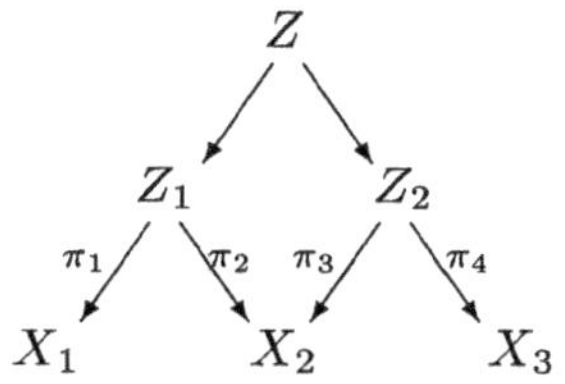

where Z is a properly chosen component of the normalization of $Z_1 \times_{X_2} Z_2$. This diagram provides a bimeromorphic transformation from X_1 onto X_3. So, if we call two normal surfaces **bimeromorphically equivalent** as soon as there is at least one bimeromorphic equivalence between them, we see that this equivalence is indeed an equivalence relation.

Remark. When two surfaces are bimeromorphically equivalent, then their fields of meromorphic functions are isomorphic. The converse does not hold in general, but it is true in the case of algebraic surfaces, for which bimeromorphic equivalence is the same as birational equivalence.

2. Exceptional Curves

A compact, reduced, connected curve C on a *nonsingular* surface X is called **exceptional**, if there is a bimeromorphic map $\pi : X \to Y$ such that C is exceptional for π, i.e., if there is an open neighbourhood U of C in X, a point $y \in Y$, and a neighbourhood V of y in Y, such that π maps $U \setminus C$ biholomorphically onto $V \setminus \{y\}$, whereas $\pi(C) = y$. We shall express this situation also by saying that C is **contracted** to y. Since by agreement we assume $y \in Y$ to be a normal point, this singularity is uniquely determined by the embedding $C \subset X$ up to biholomorphic equivalence. (Compare the proof of Proposition 8.5.) The following characterisation is due to Grauert ([Gr62], p. 367) and its earlier algebraic version to Mumford (see [Mu61])

(2.1) **Theorem** (Grauert's criterion). *A reduced, compact connected curve C with irreducible components C_i on a smooth surface is exceptional if and only if the intersection matrix (C_iC_j) is negative definite.*

The following examples of exceptional curves are the most important ones:

i) *Exceptional curves of the first kind.* These are nonsingular rational curves with self-intersection -1. Frequently we shall call such curves (-1)-curves. A very useful characterisation of (-1)-curves is given by

(2.2) **Proposition.** *An irreducible curve $C \subset X$ is a (-1)-curve if and only if*

$$C^2 < 0 \quad \textit{and} \quad (\mathcal{K}_X, C) < 0.$$

Proof. If C is a (-1)-curve then by definition $C^2 = -1$, hence $(\mathcal{K}_X, C) = -1$ by the adjunction formula. Conversely, if the inequalities of the proposition hold, then $\deg(\omega_C) < 0$, hence C is nonsingular rational (II, Sect.11), and $C^2 = -1$, again by the adjunction formula. □

(2.3) **Proposition.** *Let X be a smooth compact connected surface with* $\mathrm{kod}(X) \geq 0$, *and D an effective divisor on X such that $(\mathcal{K}_X, D) < 0$ Then D contains a (-1)-curve.*

Proof. It is sufficient to show that if D is an irreducible curve with $(\mathcal{K}_X, D) < 0$, then D is a (-1)-curve. But by assumption, for some $n \geq 1$ there is a non-negative n-canonical divisor $K = \sum c_iC_i$, $c_i \geq 0$. Since $KD < 0$, the curve D must be one of the C_i's, say $D = C_0$ Hence $D(K - c_0D) \geq 0$ and $D^2 < 0$ The assertion thus follows from Proposition 2.2. □

We can formulate the above proposition in terms of nef divisors as follows.

(2.4) **Corollary** *Let X be a smooth compact connected surface with $\mathcal{K}_X$ not nef. Then either* $\mathrm{kod}(X) = -\infty$ *or X contains a (-1)-curve. In particular, the canonical bundle is nef for any minimal surface with non-negative Kodaira dimension.*

ii) *Hirzebruch-Jung strings.* These are unions $C = \bigcup_1^r C_i$ of smooth rational curves C_i such that

$$\begin{aligned} C_i^2 &\leq -2 \ \text{ for all } \ i, \\ C_iC_j &= 1 \ \text{ if } \ |i-j| = 1, \\ C_iC_j &= 0 \ \text{ if } \ |i-j| \geq 2. \end{aligned}$$

If $e_i = C_i^2$, then this configuration is visualized by the dual graph

The intersection matrix

$$\begin{pmatrix} e_1 & 1 & 0 & \cdots \\ 1 & e_2 & 1 & \ddots \\ 0 & 1 & e_3 & \ddots \\ \vdots & \ddots & \ddots & \ddots \end{pmatrix}$$

is negative definite. Concrete examples of such curves are easily constructed: let U_i be a small neighbourhood of the zero-section in $\mathcal{O}_{\mathbb{P}_1}(e_i)$, the line bundle over $C_i = \mathbb{P}_1$ of degree e_i. Then patch the U_i together like this:

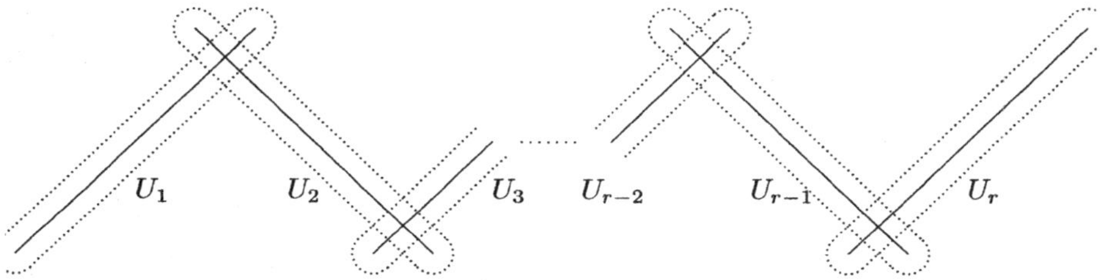

The simplest Hirzebruch-Jung string is a smooth rational curve with self intersection -2. We shall call such a curve a (-2)-curve.

iii) *A-D-E curves.* These are the exceptional curves $C = \bigcup C_i$ or which all irreducible components are (-2)-curves. Because of

$$(C_i + C_j)^2 = 2(C_iC_j - 2) < 0,$$

for all $i \neq j$ one has $C_iC_j \leq 1$, i.e., two such curves can intersect in at most one point and then transversally. The intersection form of C being negative definite, it must be one of the forms described by Dynkin diagram A_n with $n \geq 1$, D_n with $n \geq 4$, or E_6, E_7, or E_8 (I, Sect. 2). So these Dynkin diagrams are the dual graphs of our curves. Notice that the curves A_n are Hirzebruch-Jung strings.

A-D-E-curves can be recognized by the following criterion.

(2.5) **Proposition.** *Let $C \subset X$ be an exceptional curve with $(\mathcal{K}_X, C_i) = 0$ for each irreducible component C_i of C. Then C is an A-D-E curve.*

Proof. From $(\mathcal{K}_X, C_i) = 0$ and $C_i^2 < 0$ it follows that $g(C_i) \leq 0$. Hence each C_i is nonsingular rational and $C_i^2 = -2$ by the adjunction formula. □

Grauert has shown in ([Gr62], p. 357) that an exceptional curve $C \subset X$ possesses arbitrarily small strictly pseudo-convex neighbourhoods $U \subset X$ with the following property: for every locally free $\mathcal{O}_U$-sheaf $\mathcal{S}$ there is some $k_0 > 0$ such that for all $k \geq k_0$

$$\text{restr} : H^i(U, \mathcal{S}) \to H^i(\mathcal{S}|kC) \quad (i \geq 1)$$

is injective. C being of dimension one, this shows in particular $H^2(U, \mathcal{S}) = 0$. Using $H^2(U, \mathcal{S} \otimes \mathcal{O}_U(-kC)) = 0$, one even finds that

$$\text{restr} : H^1(U, \mathcal{S}) \to H^1(\mathcal{S}|kC) \tag{1}$$

is bijective for $k \gg 0$.

3. Rational Singularities

Let $y \in Y$ be the singularity obtained by contracting the exceptional curve $C \subset X$ via $\pi : X \to Y$. The singularity is called rational if $\pi_{*1}\mathcal{O}_X$ vanishes. By Grauert's result (1) and the cohomology sequences of the "decomposition series"

$$0 \to \mathcal{O}_C(-kC) \to \mathcal{O}_{(k+1)C} \to \mathcal{O}_{kC} \to 0$$

$k \geq 1$, we see that this is the case if and only if $h^1(\mathcal{O}_{kC}) = 0$ for all $k \geq 1$. From this criterion we can now deduce that the examples i)–iii) of the preceding section are rational.

(3.1) **Proposition.** *A (-1)-curve or a Hirzebruch-Jung string gives rise to a rational singularity.*

Proof. In view of the decomposition sequences above it is sufficient to show that

$$h^1(\mathcal{O}_C(-kC)) = h^0(\omega_C \otimes \mathcal{O}_C(kC)) = 0 \quad \text{for all} \quad k \geq 0.$$

But if C is a (-1)-curve, then $\mathcal{O}_C(C) = \mathcal{O}_{\mathbb{P}_1}(-1)$ and $h^1(\mathcal{O}_C(-kC)) = h^1(\mathcal{O}_{\mathbb{P}_1}(k))$ obviously vanishes.

If $C = \bigcup C_i$ is a Hirzebruch-Jung string, then

$$\begin{aligned} &\deg(\omega_C \otimes \mathcal{O}_C(kC)|C_i) \\ &\quad = \deg(\omega_{C_i}) + C_i(C - C_i) + kCC_i \\ &\quad = -2 - e_i + (k+1)CC_i \\ &\quad = \begin{cases} k - 1 + ke_i & \text{if } C_i \text{ is an end of the string,} \\ 2k + ke_i & \text{if } C_i \text{ is not an end.} \end{cases} \end{aligned}$$

This degree is always ≤ 0 and strictly negative if C_i is an end curve. So $\omega_C \otimes \mathcal{O}_C(kC)$ admits the trivial section only. □

To deal with the curves of type D_n, E_6, E_7, E_8, we use (compare [An62], Proposition 1).

(3.2) **Theorem** (Artin's criterion). *Let $C = \bigcup C_i$ be an exceptional curve. The following assertions are equivalent*

i) *C contracts to a rational singularity;*

ii) *each divisor $Z = \sum r_i C_i$, $r_i \geq 0$, has arithmetical genus $g(Z) \leq 0$;*

iii) $H^1(\mathcal{O}_Z) = 0$.

Proof. Firstly, let the singularity be rational. For each divisor Z there is some $k > 0$ and an epimorphism $\mathcal{O}_{kC} \to \mathcal{O}_Z$. Since supp$(C)$ is of dimension one, $H^1(\mathcal{O}_Z) = 0$ as soon as $H^1(\mathcal{O}_{kC})$ vanishes. But this last group always vanishes for rational singularities. So i) $\Rightarrow$ iii). Clearly iii) $\Rightarrow$ ii) so that only ii) $\Rightarrow$ i) remains to be shown. Assume now that $g(Z) \leq 0$ for all Z. Putting in particular $Z = C_i$, we find that all C_i are nonsingular rational (II, Sect.11). Now we use induction on $r = \sum r_i$ to show $h^1(\mathcal{O}_Z) = 0$. So let $r \geq 2$ and $h^1(\mathcal{O}_{Z'}) = 0$ for all $Z' = \sum r'_i C_i$ with $\sum r'_i = r - 1$. Let C_0 be a component of Z, and $Z_0 = Z - C_0$. Because of the decomposition sequence

$$0 \to \mathcal{O}_{C_0}(-Z_0) \to \mathcal{O}_Z \to \mathcal{O}_{Z_0} \to 0,$$

it suffices to prove $h^1(\mathcal{O}_{C_0}(-Z_0)) = 0$, which is equivalent to $Z_0 C_0 \leq 1$, that is

$$ZC_0 \leq 1 + C_0^2.$$

As we may choose C_0 arbitrarily, we have finished unless $ZC_i \geq 2 + C_i^2$ for each C_i. But if this were the case, then

$$\begin{aligned} \deg(\omega_Z) &= (\mathcal{O}_Z(Z) \otimes \mathcal{K}_X, Z) \\ &= \sum r_i \{ZC_i + \deg(\mathcal{K}_X | C_i)\} \\ &= \sum r_i \{ZC_i - 2 - C_i^2\} \geq 0, \end{aligned}$$

and we would have the contradiction $g(Z) = 1 + \frac{1}{2}\deg(\omega_Z) > 0$. □

By the adjunction formula $\omega_Z = \mathcal{O}_Z(\mathcal{K}_X + Z)$ and since $h^0(\omega_Z) = h^1(\mathcal{O}_Z)$, using the exact sequence

$$0 \to \mathcal{K}_X \to \mathcal{K}_X(Z) \to \omega_Z \to 0 \tag{2}$$

we deduce the following result which we need later:

(3.3) **Corollary.** *Let X be a compact surface and let $Z \geq 0$ be a divisor supported on a curve which contracts onto a rational singularity. Then the inclusion $\mathcal{O}_X \to \mathcal{O}_X(Z)$ induces an isomorphism $H^0(\mathcal{K}_X) \xrightarrow{\sim} H^0(\mathcal{K}_X(Z))$.*

(3.4) **Proposition.** *The A-D-E curves contract to rational singularities.*

Proof. Since

$$(\mathcal{K}_X, C_i) = -2 - C_i^2 = 0,$$

for any effective divisor $Z = \sum r_i C_i$ we have

$$\deg(\omega_Z) = Z^2 = \left(\sum r_i C_i\right)^2 < 0,$$

because C is exceptional So $g(Z) = 1 + \frac{1}{2}\deg(\omega_Z) \leq 0$. □

Rational singularities are easy to deal with, because they admit arbitrarily small neighbourhoods $U \subset X$ with $H^1(U, \mathcal{O}_U^*) = H^2(U, \mathbb{Z})$, and

$$H^2(U, \mathbb{Z}) = H^2(C, \mathbb{Z}) = \bigoplus H^2(C_i, \mathbb{Z})$$

is the free group with one generator for each component C_i. Since for an exceptional A-D-E curve $C = \bigcup C_i \subset X$ we have $(\mathcal{K}_X, C_i) = 0$ for all i, this implies

(3.5) **Proposition.** *Any A-D-E curve has a neighbourhood U with $\mathcal{K}_U = \mathcal{O}_U$.*

Coming back to the exact sequence (2), tensoring it with any multiple of the canonical bundle we deduce the next result, needed in Chapter VII:

(3.6) **Corollary.** *Let X be a compact surface and let $Z \geq 0$ be a divisor supported on an A-D-E curve. Then for all $k \in \mathbb{Z}$ the inclusion $\mathcal{O}_X \to \mathcal{O}_X(Z)$ induces an isomorphism $H^0(\mathcal{K}_X^{\otimes k}) \xrightarrow{\sim} H^0(\mathcal{K}_X^{\otimes k}(Z))$.*

An effective divisor D on U is the divisor (f) of a holomorphic function if and only if $DC_i = 0$ for all i. Any effective D is of the form $D = Z + \sum U_j$ where $Z = \sum r_i C_i$ has support on C and the U_j intersect C in finitely many points at most. Since $((f), C_i) = 0$ for all i and $U_j C_i \geq 0$, we see that Z satisfies

$$ZC_i \leq 0 \quad \text{for all} \quad i. \tag{3}$$

Conversely, if Z satisfies (3), then (perhaps after shrinking U) one may pick U_j's such that $C_i(Z + \sum U_j) = 0$ for all i. Thus we have

(3.7) **Proposition.** *The divisors $Z = \sum r_i C_i$, $r_i \geq 0$, satisfying (3) are exactly the parts contained in C of divisors of holomorphic functions defined on some neighbourhood of C.*

Given two divisors $Z = \sum r_i C_i$ and $Z' = \sum r_i' C_i$ which are the parts in C of the principal divisors of two holomorphic functions f, f', then for general α, $\alpha' \in \mathbb{C}$ the part of $(\alpha f + \alpha' f')$ in C is $\sum \min\{r_i, r_i'\} C_i$, which therefore also satisfies (3). So there is a minimal effective divisor Z satisfying (3). In [An62] Artin has called this Z the fundamental cycle of the singularity.

(3.8) **Proposition.** *Let $\mathfrak{m} \subset \mathcal{O}_{Y,y}$ denote the maximal ideal. Then for all $k \geq 1$*

$$\pi^*(\mathfrak{m}^k) = \mathcal{O}_X(-kZ) \subset \mathcal{O}_X.$$

Proof. If suffices to show $\pi^*(\mathfrak{m}) = \mathcal{O}_X(-Z)$ Now if $g \in \mathfrak{m}$, then $g \circ \pi$ vanishes on C and $(g \circ \pi) \geq Z$ by construction of the fundamental cycle. This shows $\pi^*(\mathfrak{m}) \subset \mathcal{O}_X(-Z)$

To prove the converse, let $x \in C$. As pointed out above, there is on a neighbourhood of C a holomorphic function f such that $(f) = Z + \sum U_j$. The U_j's may be chosen not to contain x. So in a neighbourhood of x, the sheaf $\mathcal{O}_X(-Z)$ is generated by f. But $\pi_* f$ is holomorphic on a neighbourhood of $y \in Y$. So $f = \pi^*(\pi_* f) \in \pi^*(\mathfrak{m})$. □

The fundamental cycles for the different types are the following (Lemma I, 2.12):

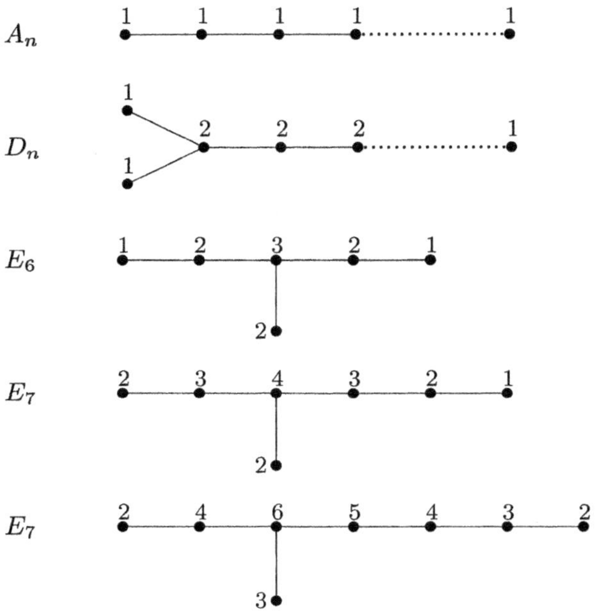

As an immediate consequence we have

(3.9) **Proposition.** *For the fundamental cycle Z of an exceptional A-D-E curve we have*

$$Z^2 = -2.$$

The fundamental cycles can also be used for constructing functions, which make A-D-E singularities locally double coverings of the plane.

(3.10) **Lemma.** *Let $C \subset X$ be an exceptional A-D-E curve and $y \in Y$ the singularity arising from contracting it. Then locally Y can be realized as a double covering of a nonsingular surface.*

Proof. We map a neighbourhood of y into $\mathbb{C}^2$ by constructing two functions f, g on a neighbourhood of C. Since the singularity is rational, it suffices to give effective divisors (f), (g) satisfying $((f), C_i) = ((g), C_i) = 0$ for all irreducible components $C_i \subset C$. We want f and g to satisfy the following conditions:

(i) $\{f = g = 0\}$ is the set C,

(ii) $(f) = \sum r_i C_i + R$ where R is a nonsingular curve intersecting C either in one point, at which $g|R$ vanishes to the second order, or in two different points, at which $g|R$ has simple zeros.

Property (i) implies that the map $Y \to \mathbb{C}^2$ is proper in a neighbourhood of y, and (ii) forces the degree of this covering to be 2. For finding the divisors (f) and (g), we distinguish between two cases:

a) In the case of an A_n-singularity, let R_1 and R_2 be two curves intersecting only the end curves C_1 and C_n, and both of them transversally in one point. Let S_1 and S_2 be two curves with the same properties, but disjoint from R_1 and R_2. Then we put $(f) = Z + R_1 + R_2$ and $(g) = Z + S_1 + S_2$.

b) In the remaining cases there is a distinguished component C_{i_0} of multiplicity 2 in Z with $ZC_{i_0} = -1$. Let R and S be two disjoint curves intersecting no C_i but C_{i_0}, and C_{i_0} transversally in one point. Then we put $(f) = Z + R$ and $(g) = Z + S$. □

4. Exceptional Curves of the First Kind

The contraction of a (-1)-curve is always the converse of a σ-transform. This will follow from Theorems 4.1 and 4.2 below.

(4.1) **Theorem.** *Let X be a nonsingular surface, $E \subset X$ a (-1)-curve and $\pi : X \to Y$ the map contracting E. Then $y = \pi(E)$ is nonsingular on Y.*

Proof. We may assume X to be so small a neighbourhood of E that there are disjoint nonsingular curves U, $V \subset X$ intersecting E transversally in two distinct points:

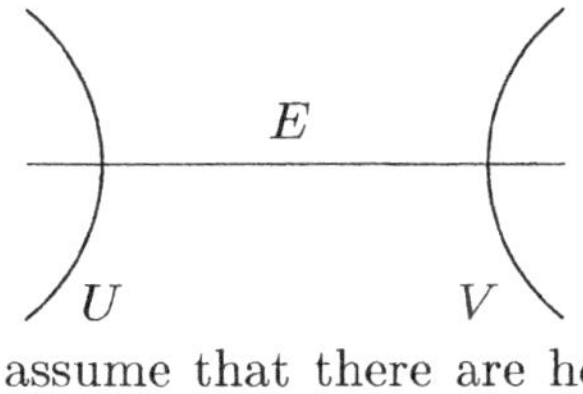

By Sect. 3, we also may assume that there are holomorphic functions u, v on X with divisors $(u) = U + E$, $(v) = V + E$. If $\rho : X \to \mathbb{C}^2$ is the map $x \to (u(x), v(x))$, then $\rho^{-1}(0) = \{u = v = 0\} = E$. Since ρ is proper on a neighbourhood of E (see I.8.9), and since $\rho(E) = 0$, there is a factorization $\rho = \gamma\pi$ with $\gamma : Y \to B$ a (perhaps ramified) covering of a neighbourhood $B \subset \mathbb{C}^2$ of the origin. The curve U is mapped into the v-axis $\{u = 0\} \subset \mathbb{C}^2$ and since v vanishes on U to the first order at $U \cap E$, the map $\rho|U$ is injective near this point. Since u also vanishes to the first order only along U, there is no ramification along this curve. This proves that the degree of the covering γ is one, so $Y = B$ and $\rho = \pi$. □

(4.2) **Theorem** (Uniqueness of the σ-process). *Let X and Y be smooth surfaces and $\pi : X \to Y$ a bimeromorphic map. If $E = \pi^{-1}(y)$ is an irreducible curve, then near E, the map π is equivalent to the σ-process with centre y.*

Proof. Let u, v be local coordinates on Y centred at y. Put $U = \{u = 0\}$, $V = \{v = 0\} \subset Y$ and let $\overline{U}, \overline{V} \subset X$ be the proper transforms of these curves. If $\overline{U} \cap E = \{x\}$, then $i_x(\overline{U}, \pi^*(V)) = \mathrm{ord}_x(\pi^*(v)|\overline{U}) = \mathrm{ord}_y(v|U) = 1$. So $\overline{U} \cap \overline{V} = \varnothing$ and $\overline{U}E = \overline{V}E = 1$. This implies $(\pi^*(u)) = \overline{U} + E$ and $(\pi^*(v)) = \overline{V} + E$. So the meromorphic function $\pi^*(u/v)$ on X has no points of indeterminacy and defines a map $\lambda : X \to \mathbb{P}_1$ such that $\lambda|E : E \to \mathbb{P}_1$ is an isomorphism. Then $\pi \times \lambda : X \to Y \times \mathbb{P}_1$ is an isomorphism of X onto the result $\overline{Y} \subset Y \times \mathbb{P}_1$ of the σ-process with centre y. □

(4.3) **Lemma** (Factorization lemma). *Let $\pi : X \to Y$ be a bimeromorphic map with X, Y nonsingular surfaces. Unless π is an isomorphism, there is a factorization $\pi = \pi' \circ \sigma$, where $\sigma : X \to X'$ is a σ-process.*

Proof. Take some $y \in Y$ with $C = \pi^{-1}(y) \subset X$ an exceptional curve. We may assume Y so small that there are global coordinates u, v on Y. Then $\pi^*(du \wedge dv)$ is an effective canonical divisor on X, say $K = \sum k_i C_i$, with $k_i > 0$ and C_i irreducible components of C. Then $K^2 < 0$ by Grauert's Criterion (Theorem 2.1) , and there will be some C_i with $KC_i < 0$ and $C_i^2 < 0$. This C_i is a (-1)-curve by Proposition 2.2. By Theorem 4.2 the contraction $\sigma : X \to X'$ of this curve is a σ-transform. There is a factorization $\pi = \pi' \circ \sigma$ with $\pi' : X' \to Y$ holomorphic outside of $x' = \sigma(C_i)$. Since however $u \circ \pi'$ and $v \circ \pi'$ extend holomorphically over x', this map π' is holomorphic in x' too. □

(4.4) **Corollary** (Decomposition of bimeromorphic maps). *Let X, Y be nonsingular and $\pi : X \to Y$ a bimeromorphic map. Then π is equivalent to a succession of σ-transforms, which locally (with respect to Y) are finite in number.*

Proof. Induction on the number of irreducible components of each exceptional curve. □

A smooth surface is called minimal, if it does not contain any (-1)-curve.

A minimal surface X can also be characterized by the property that any birational morphism $X \to X'$ with X' smooth, is an isomorphism.

A nonsingular surface $X_{\min}$ is called a minimal model of the nonsingular surface X, if $X_{\min}$ is minimal itself, and if there is a bimeromorphic map (i.e., a succession of σ-transforms) from X onto $X_{\min}$.

(4.5) **Theorem.** *Every compact nonsingular surface X has a minimal model.*

Proof. Suppose that X contains a (-1)-curve and let X_1 be obtained from X by contracting it. If X_1 contains another (-1)-curve, the process can be repeated, and so on. This must lead to a surface without (-1)-curves after a finite number of blowing downs, since each time the second Betti number diminishes by 1 (Theorem I.9.1, (iv)). □

(4.6) **Proposition.** *If X is a nonsingular compact connected surface with* $\mathrm{kod}(X) \geq 0$, *then all minimal models of X are isomorphic.*

To prove this, we invoke Corollary 2.4. So it suffices to show the slightly more general

Claim Let X, Y be two compact connected nonsingular surfaces and $f : X \to Y$ a bimeromorphic map. If $\mathcal{K}_Y$ is nef, f is a morphism. If in addition $\mathcal{K}_X$ is nef, f is an isomorphism.

Proof. This statement follows quite easily from the fact that if $\sigma : \tilde{X} \to X$ is the blowing up in $p \in X$ and $\tilde{C} \subset \tilde{X}$ is any irreducible curve mapping onto a curve $C \subset X$, (having p as a point of multiplicity m) then

$$(\mathcal{K}_{\tilde{X}}, \tilde{C}) = (\sigma^*\mathcal{K}_X + E, \sigma^*C - mE) = (\mathcal{K}_X, C) + m \geq (\mathcal{K}_X, C).$$

So first of all the number $(\mathcal{K}_X, C)$ does not increase under blowing down and secondly, if $\mathcal{K}_X$ is nef, any curve $\tilde{C}$ on $\tilde{X}$ with $(\mathcal{K}_{\tilde{X}}, \tilde{C}) \leq -1$ must be mapped to a point. If the bimeromorphic map $X \to Y$ would not be a morphism, one has to blow up X in order to get a morphism, say $f' : X' \to Y$. This last morphism is a sequence of σ-transforms and any curve in X' arising from blowing up X will be mapped to a curve C in Y and hence to a curve $\tilde{C} \subset \tilde{Y}$, where $\tilde{Y}$ is the first σ-process $\tilde{Y} \to Y$ into which f' decomposes. Such a curve $\tilde{C}$ satisfies the inequality $(\mathcal{K}_{\tilde{Y}}, \tilde{C}) \leq -1$ which is impossible if $\mathcal{K}_Y$ is nef. □

Remark. If $\mathrm{kod}(X) = -\infty$, then the statement is no longer true, for $\mathbb{P}_1 \times \mathbb{P}_1$ blown up in one point can be blown down to $\mathbb{P}_2$. (Compare VI, Sect. 7.)

5. Hirzebruch-Jung Singularities

Let $C = \sum_{i=1}^{r} C_i$ with $C_i^2 = e_i \leq -2$ as in Sect. 2, Example ii). For sufficiently small $X \supset C$ there is a (closed, but not necessarily compact) smooth curve C_0 which intersects C_1 transversally in one point, without meeting any of the other curves C_i. Similarly there is such a curve C_{r+1} intersecting C_r transversally in one point which does not intersect any of the curves $C_0, \ldots, C_{r-1}$. Thus we have a dual diagram

Let $n_i \in \mathbb{Z}$, $n_i \geq 0$, $i = 1, \ldots, r+1$. By Sect. 3 there is a holomorphic function φ on X with divisor $(\varphi) = \sum_{i=0}^{r+1} n_i C_i$ if and only if for $k = 1, \ldots, r$

$$n_{k-1} + e_k n_k + n_{k+1} = \sum_{i=0}^{r+1} n_i C_i C_k = 0.$$

Given n_0 and n_1, the coefficients n_k, $k = 2, \ldots, r+1$ are determined uniquely by the recursion formula

$$n_k = |e_{k-1}| \cdot n_{k-1} - n_{k-2}. \tag{4}$$

If $n_0 < n_1$ (respectively $n_0 \leq n_1$), then it follows by induction that $n_k < n_{k+1}$ (respectively $n_k \leq n_{k+1}$) for $k = 1, \ldots, r$. So if we determine integers μ_k, ν_k from (3) starting with initial data

$$\mu_0 = 0, \quad \mu_1 = 1 \rightsquigarrow \mu_k,$$
$$\nu_0 = 1, \quad \nu_1 = 1 \rightsquigarrow \nu_k,$$

then for $k \geq 1$ these integers will be positive. Therefore, we have holomorphic functions g, h on X with divisors

$$(g) = \sum_{i=0}^{r+1} \mu_i C_i, \quad (h) = \sum_{i=0}^{r+1} \nu_i C_i.$$

Notice that the integers μ_i satisfy

$$\mu_1 = 1, \quad \mu_2 = |e_1|,$$
$$\frac{\mu_3}{\mu_2} = |e_2| - 1/|e_1|$$
$$\frac{\mu_{k+1}}{\mu_k} = |e_k| - 1\rfloor\overline{|e_{k-1}|} - \cdots - 1\rfloor\overline{|e_1|},$$

where the last symbol means expansion as a continued fraction. The recursion formula (3) implies

$$\text{g.c.d.}(\mu_{k+1}, \mu_k) = \text{g.c.d.}(\mu_k, \mu_{k-1}) = \cdots = \text{g.c.d.}(\mu_2, \mu_1) = 1.$$

It follows that μ_k and μ_{k+1} are coprime, and so they might also have been defined by the above expansion.

We put in particular $n' = \mu_{r+1}$, $g' = \mu_r$. Then the expansion

$$\frac{n'}{q'} = |e_r| - 1\rfloor\overline{|e_{r-1}|} - \cdots - 1\rfloor\overline{|e_1|}$$

shows that the self-intersections e_i are determined by the two integers n' and q'.

Finally, we define a divisor

$$(f) = \sum_{i=0}^{r+1} \lambda_i C_i$$

by integers λ_i satisfying (3) and

$$\lambda_{r+1} = 0, \quad \lambda_r = 1.$$

There λ's are the μ's we would have obtained when starting with our index i at the other end of the string. So

$$\lambda_1 = q, \quad \lambda_0 = n,$$

with

$$\frac{n}{q} = |e_1| - 1\rfloor\overline{|e_2|} - \cdots - 1\rfloor\overline{|e_r|}. \tag{5}$$

By induction we obtain

$$\lambda_k + (n-q)\mu_k = n\nu_k \tag{6}$$

$$\lambda_k \mu_{k+1} - \lambda_{k+1}\mu_k = n. \tag{7}$$

For $k = r$, Eq. (6) implies

$$n' = n,$$

whereas Eq. (5) shows that q' is the integer determined uniquely by

$$0 < q' < n, \quad qq' \equiv 1 \bmod n.$$

For the functions f, g and h, Eq. (5) means

$$(fg^{n-q}) = (f) + (n-q)(g) = n(h) = (h^n).$$

So fg^{n-q}/h^n is a function in $\Gamma(\mathcal{O}_X^*)$, which can be absorbed into f for example. Then we have the relation

$$h^n = fg^{n-q}.$$

In other words: by

$$w = h, \quad z_1 = f, \quad z_2 = g,$$

X is mapped into the surface

$$W = \{(w, z_1, z_2) \subset \mathbb{C}^3 \;:\; w^n = z_1 z_2^{n-q}\} \subset \mathbb{C}^3 \quad \text{(the singularity } A_{n,q}\text{)}.$$

(5.1) **Theorem.** *For $0 < q < n$, n and q coprime, let $C \subset X$ be the Hirzebruch-Jung string (see Sect. 2 ii) with self-intersection numbers e_i satisfying (4), and let $y \in Y$ be the singularity resulting from contracting C. Then this singularity is isomorphic to the unique singularity lying over $0 \in \mathbb{C}^3$ in the normalization of the surface W above.*

Remark. This theorem shows in particular that the singularity $y \in Y$ (hence the embedding $C \subset X$) depends on n and q only. We therefore call it *the* singularity $A_{n,q}$.

Proof of the Theorem. We denote by $\pi : X \to Y$ the contraction of C and by $\rho : X \to \mathbb{C}^2$ the map defined by $z_1 = f$, $z_2 = g$. Since the divisors (f) and (g) intersect in C only, we have $\rho^{-1}(0) = C$. By Proposition I.8.9 we may assume ρ proper. There is a factorization $\rho = \gamma\pi$, where $\gamma : Y \to \mathbb{C}^2$ is proper and finite over a neighbourhood $Z \subset \mathbb{C}^2$ of the origin, so a (ramified) covering. Since γ factors through the normalization $\widetilde{W}$ of W, the assertion follows as soon as we know that $\deg(\gamma) = n$. But consider for example the curve $\pi(C_0) = \gamma^{-1}\{z_1 = 0\}$. At $C_0 \cap C_1$ the function g vanishes to first order only, and f vanishes along C_0 to n-th order. So γ is ramified cyclically of order n along C_0, hence $\deg(\gamma) = n$ near $y \in Y$. □

Next we describe some frequently occurring situations which give rise to $A_{n,q}$ singularities.

i) *Coverings branched over the coordinate axes.*
Let Z denote the unit dicylinder $\{(z_1, z_2) \in \mathbb{C}^2 : |z_1| < 1, |z_2| < 1\}$. We consider here coverings $\gamma : B \to Z$, with B a connected normal surface, which are only branched over $z_1 \cdot z_2 = 0$. Put $Z^* = Z \setminus \{z_1 z_2 = 0\}$ and $B^* = \gamma^{-1}(Z^*)$. Then γ and B are determined uniquely by the unramified covering $B^* \to Z^*$ ([St], Satz 1), i.e., by the subgroup $\Gamma = \gamma_* \pi_1(B^*) \subset \pi_1(Z^*)$. If $w_i \in \pi_1(Z^*)$ is the class of a positively oriented little loop around the z_i-axis, then $\pi_1(Z^*) = \mathbb{Z} \times \mathbb{Z}$ with generators $w_1 = (1, 0)$ and $w_2 = (0, 1)$.

a) First we notice that if over one axis, say $z_1 = 0$, no ramification takes place, then Γ is generated by $(0, 1)$ and some element $(b, 0)$, $b > 0$. But this is the case also for the covering $Z \to Z$, $z_1 \mapsto z_1^b$, $z_2 \mapsto z_2$.

b) Next we determine the subgroup Γ in case $B = Z$ is smooth and $\gamma(z_1, z_2) = (z_1^{m_1}, z_2^{m_2})$. Then obviously $\gamma_*(w_i) = m_i w_i$, so Γ is the subgroup generated by $(m_1, 0)$ and $(0, m_2)$.

c) Now we identify Γ if $B = Y$ is a neighbourhood of the Hirzebruch-Jung singularity $A_{n,q}$ mapped onto Z by $z_1 = f$, $z_2 = g$ as above. Let $\delta_k \in \pi_1\left(X \setminus \bigcup_{i=1}^{r+1} C_i\right)$ be the class of a positively oriented little loop around the curve C_k. Since f and g vanish along C_k to order λ_k and μ_k respectively, we find $(\gamma\pi)_*(\delta_k) = \lambda_k w_1 + \mu_k w_2$. In particular, $\Gamma \subset \mathbb{Z} \times \mathbb{Z}$ contains the elements

$$(\gamma\pi)_*(\delta_0) = (n, 0),$$
$$(\gamma\pi)_*(\delta_1) = (q, 1).$$

But these elements generate in $\mathbb{Z} \times \mathbb{Z}$ a subgroup of index n, which must coincide with Γ.

d) Finally we treat the general case. Let $\Gamma = \gamma_*(\pi_1(B^*))$ be an arbitrary subgroup of finite index. We pick generators for Γ as follows: $\Gamma \cap (\mathbb{Z}, 0)$ is non-trivial, so there is some $(n', 0) \in \Gamma$ generating this intersection, $n' > 0$. The quotient $\Gamma/\mathbb{Z}\cdot(n', 0)$ is isomorphic to $\mathbb{Z}$, so there is some $(q', m_2) \in \Gamma$, $0 \le q' < n'$, $m_2 > 0$, such that $(n', 0)$ is isomorphic to $\mathbb{Z}$, so there is some $(q', m_2) \in \Gamma$, $0 \le q' < n'$, $m_2 > 0$, such that $(n', 0)$ and (q', m_2) generate Γ. We may assume $q' > 0$, because $q' = 0$ is case b), above. Let $m_1 = \text{g.c.d.}(n', q')$ and $n' = nm_1$, $q' = qm_1$. Then Γ is contained in the subgroup $\Gamma_1 = \mathbb{Z}(m_1, 0) + \mathbb{Z}(0, m_2)$. By b) above. this corresponds to a covering $\gamma_1 : Z_1 \to Z$ with Z_1 nonsingular. The original covering factors as $\gamma = \gamma_1\gamma_2$. If we identify again $\pi_1(Z_1^*)$ with $\mathbb{Z} \times \mathbb{Z}$, then $\gamma_2 : B \to Z_1$ corresponds to the subgroup with generators $(n, 0)$ and $(q, 1)$. So B has an $A_{n,q}$ singularity over the origin. This proves

(5.2) **Theorem.** *Let $\gamma : Y \to Z$ be a covering with Z a nonsingular surface and Y normal. If γ is branched over a curve in Z with at worst nodes, then the singularities of Y can only be of Hirzebruch-Jung type. If in particular the ramification curve is nonsingular, then Y is nonsingular too.*

ii) *Riemann existence domain of the function* $\sqrt[n]{z_1^a z_2^b}$.
By this existence domain we mean the normalization $\widetilde{W}$ of the surface

$$W = \{(w, z_1, z_2) \in \mathbb{C}^3 \ ; \ w^n = z_1^a z_2^b\}.$$

The projection $(w, z_1, z_2) \mapsto (z_1, z_2)$ exhibits $\widetilde{W}$ as an n-fold covering of $\mathbb{C}^2$, branched only over $z_1 \cdot z_2 = 0$. So $\widetilde{W}$ will be smooth, except at the points lying over $(0,0)$, where singularities of $A_{n,q}$-type appear. Explicitly, the situation is as follows:

a) If we put $d = \text{g.c.d.}(n, a, b)$, $n = \nu d$, $a = \alpha d$ and $b = \beta d$, then

$$w^n - z_1^a z_2^b = \prod_{j=1}^{d} \left(w^\nu - z_1^\alpha z_2^\beta \mathbf{e}(\frac{d}{j})\right)$$

and $\widetilde{W}$ decomposes into d different coverings, all of which are isomorphic to the existence domain of $\sqrt[\nu]{z_1^\alpha z_2^\beta}$.

b) Assume now that $\text{g.c.d.}(n, a, b) = 1$ and let

$$\begin{aligned} d_a &= \text{g.c.d.}(n, a), \quad a = \alpha d_a \ , \\ d_b &= \text{g.c.d.}(n, b), \quad b = \beta d_b \ , \\ n &= \nu d_a d_b. \end{aligned}$$

In $\mathbb{C}^3$ we consider the surface

$$U = \{(u, y_1, y_2) \in \mathbb{C}^3 : \ u^\nu = y_1^\alpha y_2^\beta\}.$$

It is mapped to W by

$$w = u, \quad z_1 = y_1^{d_b}, \quad z_2 = y_2^{d_a} \ .$$

Using g.c.d.$(\alpha, d_b) =$ g.c.d.$(\beta, d_a) =$ g.c.d.$(d_a, d_b) = 1$, we find that the map $U \to W$ is injective over $\mathbb{C}^2 \setminus \{z_1 z_2 = 0\}$. Since both the coverings $\widetilde{W}$ and $\widetilde{U}$ over $\mathbb{C}^2(z_1, z_2)$ *] are of degree n, there is an isomorphism $\widetilde{U} \to \widetilde{W}$. In other words: the covering $\widetilde{W} \to \mathbb{C}^2(z_1, z_2)$ factors through $\mathbb{C}^2(y_1, y_2) \to \mathbb{C}^2(z_1, z_2)$ and over $\mathbb{C}^2(y_1, y_2)$ the surface $\widetilde{W}$ is the existence domain of $\sqrt[\nu]{y_1^\alpha y_2^\beta}$.

c) Finally, let g.c.d.$(n, a) =$ g.c.d.$(n, b) = 1$. We define the integer q, $0 < q < n$, by

$$aq \equiv -b \bmod n,$$

say $aq = rn - b$ with $0 < r \leq a$. Then q and n are coprime because g.c.d.$(n, b) = 1$. Consider in $\mathbb{C}^3(u, z_1, z_2)$ the surface $u^n = z_1 z_2^{n-q}$. Since a and n are coprime, via $v = u^a$ this surface is mapped bijectively onto the surface

$$v^n = z_1^a z_2^{a(n-q)} = z_1^a z_2^b \cdot z_2^{n(a-r)} .$$

Outside of the axis $z_2 = 0$, the function $w = v/z_2^{a-r}$ is holomorphic and maps this last surface bijectively onto W. This shows that $\widetilde{W}$ is the existence domain of $\sqrt[n]{z_1 z_2^{n-q}}$. Now the functions f, g, h from Theorem 5.1 define explicitly an isomorphism between $A_{n,q}$ and a neighbourhood of the singularity in this domain.

iii) *Cyclic quotients.*
In this section we denote the elements of $\mathbb{Z}/n\mathbb{Z}$, by integers $k \in \mathbb{Z}$ (identified $\bmod\, n$). Every *linear* operation of $\mathbb{Z}/n\mathbb{Z}$ on $\mathbb{C}^2$ can be given with respect to suitable coordinates (u_1, u_2) by

$$k\begin{pmatrix} u_1 \\ u_2 \end{pmatrix} = \begin{pmatrix} \mathbf{e}(q_1 k/n) \cdot u_1 \\ \mathbf{e}(q_2 k/n) \cdot u_2 \end{pmatrix}$$

with integers q_1, q_2 satisfying $0 \leq q_i < n$. Except for their ordering these integers q_i are determined uniquely by the operation. We call them the **weights** of the operation. If one of them vanishes, the operation is essentially 1-dimensional and the quotient smooth. From here on we exclude this possibility.

Furthermore, if $c =$ g.c.d.$(n, q_1, q_2) > 1$, then the action of $\mathbb{Z}/n\mathbb{Z}$ can be considered as an action of $\mathbb{Z}/(n/c)\mathbb{Z}$. So without loss of generality we may assume that g.c.d.$(n, q_1, q_2) = 1$.

We use the following notation ($i = 1, 2$):

$$\begin{aligned} & d_i = \text{g.c.d.}(n, q_1), \quad n = n_i d_i, \quad q_i = p_i d_i \ , \\ & m = \text{g.c.d.}(n_1, n_2) \ , \\ & p'_i \quad \text{the integer with } p_i p'_i \equiv 1 \bmod m, \quad 0 < p'_i < m \ , \\ & \ q \quad \text{the integer with } q \equiv p_1 p'_2 \bmod m, \quad 0 < q < m \ . \end{aligned}$$

Notice that $n =$ l.c.m.(n_1, n_2) and $n_1 n_2 = mn$.

---

*] $\mathbb{C}^n(z_1, \ldots, z_n)$ stands for $\mathbb{C}^n$ with coordinates $(z_1, \ldots, z_n)$.

(5.3) **Proposition.** *The image of* $(0,0) \in \mathbb{C}^2$ *in the quotient* $\mathbb{C}^2/(\mathbb{Z}/n\mathbb{Z})$ *is a singularity of type* $A_{m,q}$.

Proof. Let $\mathbb{Z}/n\mathbb{Z}$ be embedded in $\mathbb{Z}/n_1\mathbb{Z} \times \mathbb{Z}/n_2\mathbb{Z}$ as the subgroup generated by the element $(p_1 \bmod n_1,\ p_2 \bmod n_2)$. By way of this embedding the action of $\mathbb{Z}/n\mathbb{Z}$ is induced by the action

$$\begin{pmatrix} k_1 \bmod n_1 \\ k_2 \bmod n_2 \end{pmatrix} \begin{pmatrix} u_1 \\ u_2 \end{pmatrix} = \begin{pmatrix} \mathbf{e}(k_1/n_1)u_1 \\ \mathbf{e}(k_2/n_2)u_2 \end{pmatrix}$$

of the group $\mathbb{Z}/n_1\mathbb{Z} \times \mathbb{Z}/n_2\mathbb{Z}$ on $\mathbb{C}^2$. The quotient map γ of this action can be identified with the covering $z_1 = u_1^{n_1}$, $z_2 = u_2^{n_2}$ of $\mathbb{C}^2$ branched over $\{z_1 \cdot z_2 = 0\}$ only. It is determined by the subgroup $\Gamma \subset \mathbb{Z} \times \mathbb{Z} = \pi_1(\mathbb{C}^2 \setminus \{z_1 \cdot z_2 = 0\})$ which is generated by $(n_1, 0)$ and $(0, n_2)$.

If we denote by γ_0 the quotient map $\mathbb{C}^2 \to \mathbb{C}^2/(\mathbb{Z}/n\mathbb{Z})$, then there is a factorization $\gamma = \delta\gamma_0$, where δ displays our quotient $\mathbb{C}^2/(\mathbb{Z}/n\mathbb{Z})$ as a covering of $\mathbb{C}^2$ of degree $n_1 n_2/n = m$, branched over $\{z_1 \cdot z_2 = 0\}$ only. We have to determine the subgroup $\Delta \subset \mathbb{Z} \times \mathbb{Z} = \pi_1(\mathbb{C}^2 \setminus \{z_1 \cdot z_2 = 0\})$ which corresponds to δ.

Under the canonical epimorphism $\mathbb{Z} \times \mathbb{Z} \to \mathbb{Z} \times \mathbb{Z}/\Gamma = \mathbb{Z}/n_1\mathbb{Z} \times \mathbb{Z}/n_2\mathbb{Z}$ the group Δ is mapped onto $\mathbb{Z}/n\mathbb{Z}$, embedded in $\mathbb{Z}/n_1\mathbb{Z} \times \mathbb{Z}/n_2\mathbb{Z}$ as above. So $\Delta \subset \mathbb{Z} \times \mathbb{Z}$ is generated by

$$(n_1, 0), \quad (0, n_2) \quad \text{and} \quad (p_1, p_2).$$

Since g.c.d.$(n_1, n_2 p_1) = m$, there are integers a and b with

$$a(n_1, 0) + b(n_2 p_1, n_2 p_2) = (m, b n_2 p_2) = (m, 0) + b p_2 (0, n_2).$$

This proves $(m, 0) \in \Delta$. Similarly one finds $(0, m) \in \Delta$. Moreover,

$$p_2'(p_1, p_2) \equiv (q, 1) \operatorname{mod}(m, m) \ ,$$

hence $(q, 1) \in \Delta$. On the other hand, the elements $(m, 0)$ and $(q, 1)$ generate a subgroup of index m in $\mathbb{Z} \times \mathbb{Z}$, which must coincide with Δ. □

By a result of H. Cartan ([Car], Lemma 2, p. 98) every action of a finite group on a manifold can locally be linearized. Applying this result together with Proposition 5.3 we find

(5.4) **Theorem.** *If the finite cyclic group* G *acts on the smooth surface* X, *then the quotient* X/G *has only singularities of Hirzebruch-Jung type.*

6. Resolution of Surface Singularities

In this section we shall show that the preceding considerations quickly lead to a proof of a central and classical result: every surface is bimeromorphically equivalent to a nonsingular surface.

(6.1) **Theorem.** *For every normal surface X there is a bimeromorphic map $\pi : Y \to X$ with Y nonsingular.*

Proof. The singularities of a normal surface are isolated, so we may concentrate on one point $x \in X$ and proceed locally on X. There is a (local) embedding $X \subset \mathbb{C}^n$ and a linear projection $\gamma : \mathbb{C}^n \to \mathbb{C}^2$ with $x = \gamma^{-1}(\gamma(x)) \cap X$, the Remmert-Stein projection (see [G-R58], Satz IV, p. 256). This γ makes X (locally) into a covering of some domain $B \subset \mathbb{C}^2$ ramified over some curve $C \subset B$. Singularities of X will lie over singularities of B only (Theorem 5.2). By Theorem II.7.2 we can choose a bimeromorphic map $\beta : \overline{B} \to B$ such that $\beta^{-1}(C) \subset \overline{B}$ has normal crossings only. Forming $X \times_B \overline{B}$ and (if necessary) normalizing this surface, we obtain a surface X_1 which is provided with two proper projections, a bimeromorphic one onto X and a second one onto $\overline{B}$, which is a covering with ramification locus contained in $\beta^{-1}(C)$. This shows that the only singularities of X_1 are those which appear on coverings of $\mathbb{C}^2$ ramified over the coordinate axes. By Theorem 5.2 these singularities can be resolved by Hirzebruch-Jung strings. □

A resolution of singularities $\pi : Y \to X$ is called minimal, if π does not contract any (-1)-curve in Y.

(6.2) **Theorem.** *Every normal surface X admits a minimal resolution of singularities which is determined uniquely by X.*

Proof. The existence of a minimal resolution is shown as in the proof of Theorem 4.5: you start with some resolution, the existence of which is assured by Theorem 6.1, then you blow down all (-1)-curves contained in the exceptional curves and repeat this process if necessary.

Now assume that there are two minimal resolutions $\mu_i : M_i \to X$, $i = 1, 2$. Let $M \subset M_1 \times_X M_2$ be the 2-dimensional component mapped bimeromorphically onto X. After normalizing and resolving the singularities of M we obtain a nonsingular surface $\overline{M}$. The maps $\overline{M} \to M_i$, $i = 1, 2$, are bimeromorphic, and by Corollary 4.4 both consist of successions of σ-processes. So M_1 and M_2 are isomorphic, unless at some stage of blowing down $\overline{M}$ to M_1, say, we would meet an exceptional curve E containing two intersecting (-1)-curves C_1 and C_2. But this would imply $(C_1 + C_2)^2 \geq 0$, contradicting Theorem 2.1. □

(6.3) **Theorem** (Structure theorem for bimeromorphic transformations). *Let $\tau : X \to Y$ be a bimeromorphic transformation of nonsingular surfaces. Then τ is the composition of σ-processes (in both directions),* i.e., *there is a nonsingular surface Z, two maps $\pi_1 : Z \to X$, $\pi_2 : Z \to Y$ which are compositions of σ-processes such that $\pi = \pi_2 \circ \pi_1^{-1}$.*

Proof. By definition, there is a normal surface Z' such that $\tau = \pi_2' \circ (\pi_1')^{-1}$ with $\pi_1' : Z' \to X$, $\pi_2' : Z' \to Y$ bimeromorphic. If Z is the desingularization of Z' and π_1, π_2 the maps induced by π_1', π_2', then by Corollary 4.4 both maps π_1 and π_2 are compositions of σ-processes. □

Theorem I.9.1, (iii) and (viii) yield

(6.4) **Corollary.** *The plurigenera P_n and the irregularity q of smooth, compact, connected surfaces are invariant under bimeromorphic transformations.*

7. Singularities of Double Coverings, Simple Singularities of Surfaces

Let X be a normal surface and Y a smooth surface, both connected, and let $X \to Y$ be a double covering, ramified over $B \subset Y$. Singularities of X (lying over the singularities of B) can be resolved by the method used in the proof of Theorem 6.1. But sometimes a variation, the canonical resolution, is more economical, though in general it does not lead to a minimal resolution. It goes as follows. Let μ_y be the multiplicity of $y \in B$, and let $\sigma_1 : Y_1 \to Y$ be the σ-process simultaneously applied in all singularities $y \in B$. If $E_y = \sigma_1^{-1}(y)$ is the exceptional curve over y, then $\sigma^*(B) = \overline{B} + \sum \mu_y E_y$ is the total transform of B. Let X_1 be the normalization of $X \times_Y Y_1$. It is a double covering of Y_1, ramified over

$$B_1 = \overline{B} + \sum_{\mu_y \text{ odd}} E_y.$$

Unless B_1 is nonsingular we repeat this construction, and so on. Since the new ramification curves $B_1, B_2, \ldots,$ are contained in the total transforms of B, Theorem II.7.2 implies that after finitely many steps (locally with respect to Y) we arrive at a ramification curve B_k with at worst nodes. So all singularities of B_k have multiplicities $\mu_y = 2$, hence $B_{k+1} = \overline{B}_k$ is a nonsingular curve, and X_{k+1} is a resolution of the singularities of X.

We demonstrate this method by resolving simple surface singularities. These are the singularities of double coverings branches over a curve B having an A-D-E singularity. In II, Sect. 8, we gave explicit equations for simple curve singularities. Thus we obtain the following explicit equations for simple surface singularities:

$$\begin{array}{lll} A_n \ (n \geq 1): & w^2 + x^2 + y^{n+1} & = 0 \\ D_n \ (n \geq 4): & w^2 + y(x^2 + y^{n-2}) & = 0 \\ E_6: & w^2 + x^3 + y^4 & = 0 \\ E_7: & w^2 + x(x^2 + y^3) & = 0 \\ E_8: & w^2 + x^3 + y^5 & = 0. \end{array}$$

(Here the covering is $(w, x, y) \mapsto (x, y)$ and the singularity lies in the origin.)

The result is shown in Table 1. This table shows the total transform of B after the canonical resolution has been taken, and the exceptional curve is shown in the covering lying over it. Straight lines are symbols for copies of $\mathbb{P}_1$, mapped biregularly onto their images, the little fat curves form the proper transform of B, and the symbol

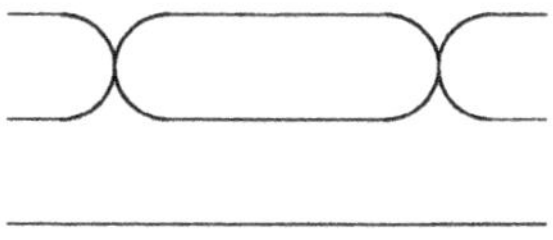

means a projective line lying over its image ramified in two points. Numbers without brackets are multiplicities and numbers within brackets denote self intersections.

Table 1 shows that all the irreducible components of the exceptional curve in the covering surface are (-2)-curves. So the canonical resolution in these cases is minimal. By inspection one even finds:

(7.1) **Theorem.** *A simple surface singularity is resolved by an exceptional A-D-E curve of the corresponding type.*

As we shall presently show (Theorem 7.3), the converse of this theorem is also true. In other words, *A-D-E* surface singularities and simple surface singularities are one and the same thing. But first we need some preparations.

(7.2) **Theorem.** *Let $\gamma : X \to Y$ be a double covering with X normal and Y nonsingular, ramified over the (reduced) curve $B \subset Y$. Let $\mathcal{L}$ be the line bundle on Y, satisfying $\mathcal{L}^{\otimes 2} = \mathcal{O}_Y(B)$, which determines the covering as in I, Sect. 17. Consider the canonical resolution diagram*

$$\begin{array}{ccc} \overline{X} & \xrightarrow{\pi} & X \\ {\scriptstyle\bar{\gamma}}\downarrow & & \downarrow{\scriptstyle\gamma} \\ \overline{Y} & \xrightarrow{\tau} & Y \end{array}$$

where τ is a sequence of σ-processes and $\overline{X}$ is nonsingular. Then there is a divisor $Z \geq 0$ on $\overline{X}$, with supp(Z) *contained in the union of the exceptional curves for π such that*

$$\mathcal{K}_{\bar{X}} = (\gamma\pi)^*(\mathcal{K}_Y \otimes \mathcal{L}) \otimes \mathcal{O}_{\bar{X}}(-Z). \tag{8}$$

Furthermore, $Z = 0$ if and only if the singularities of B (hence of X) are simple.

Proof. Let $\tau = \sigma\tau'$, where $\sigma : Y_1 \to Y$ is the σ-transform in one singularity $y \in B$. Since the covering $X \to Y$ is obtained as inverse image of the canonical section in $\mathcal{O}_Y(B)$ under the squaring map $\mathcal{L} \to \mathcal{O}_Y(B)$, the normalized covering $X_1 \to Y_1$ is obtained from the canonical section in $\mathcal{O}_{Y_1}(B_1)$ under the squaring map $\mathcal{L}_1 \to \mathcal{O}_{Y_1}(B_1)$, where B_1 is either $\overline{B}$ or $\overline{B} + E_y$ and

$$\mathcal{L}_1 = \sigma^*(\mathcal{L}) \otimes \mathcal{O}_{Y_1}\left(-\left[\frac{\mu_y}{2}\right]E_y\right).$$

So we have

$$\mathcal{K}_{Y_1} \otimes \mathcal{L}_1 = \sigma^*(\mathcal{K}_Y \otimes \mathcal{L}) \otimes \mathcal{O}_{Y_1}\left(\left(1 - \left[\frac{\mu_y}{2}\right]\right)E_y\right).$$

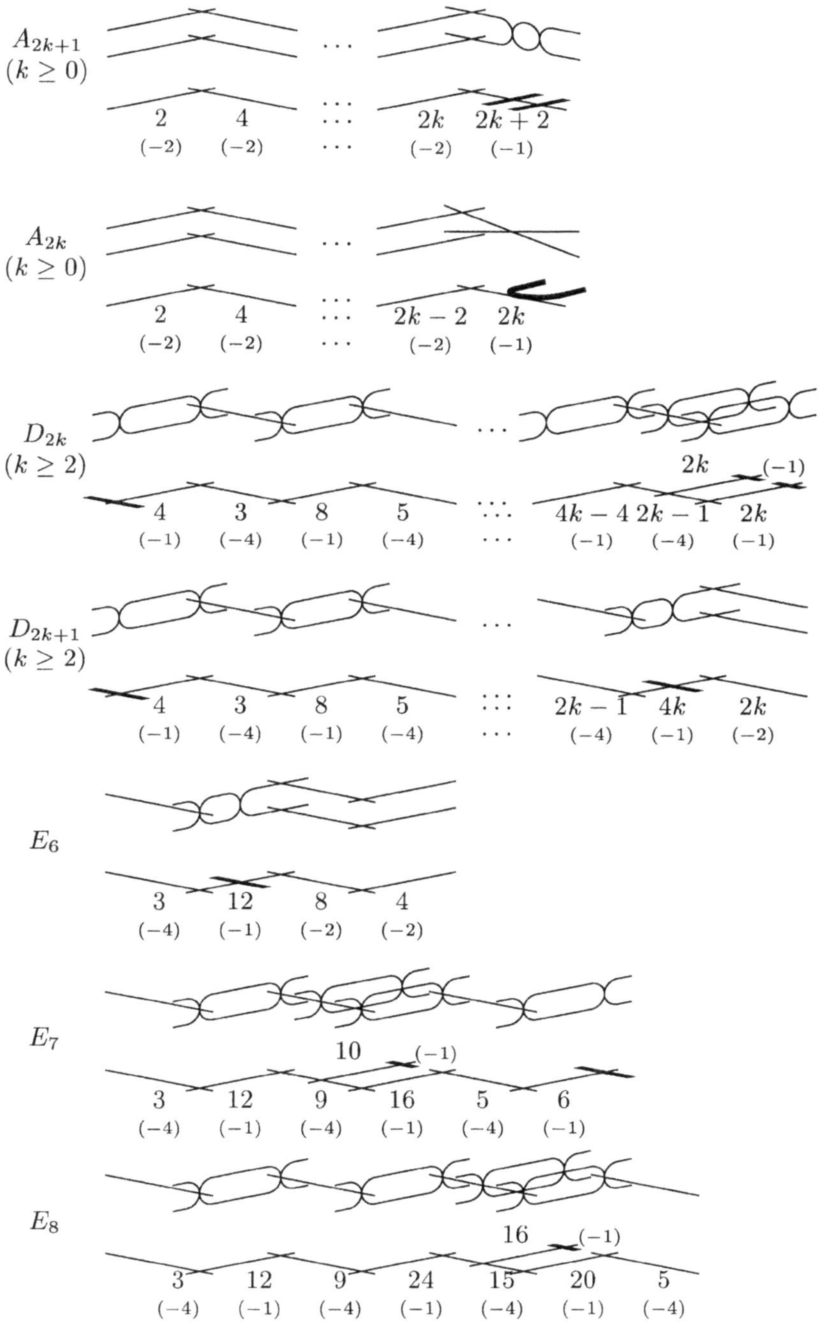

A_{2k+1} $(k \geq 0)$
2 4 … 2k 2k + 2
(−2) (−2) … (−2) (−1)
A_{2k} $(k \geq 0)$
2 4 … 2k − 2 2k
(−2) (−2) … (−2) (−1)
D_{2k} $(k \geq 2)$
2k (−1)
4 3 8 5 … 4k − 4 2k − 1 2k
(−1) (−4) (−1) (−4) … (−1) (−4) (−1)
D_{2k+1} $(k \geq 2)$
4 3 8 5 … 2k − 1 4k 2k
(−1) (−4) (−1) (−4) … (−4) (−1) (−2)
E_6
3 12 8 4
(−4) (−1) (−2) (−2)
E_7
10 (−1)
3 12 9 16 5 6
(−4) (−1) (−4) (−1) (−4) (−1)
E_8
16 (−1)
3 12 9 24 15 20 5
(−4) (−1) (−4) (−1) (−4) (−1) (−4)

Table 1.

Now the adjunction formula for coverings (I. Sect. 16) shows $\mathcal{K}_{\bar{X}} = \bar{\gamma}^*(\mathcal{K}_{\bar{Y}} \otimes \overline{\mathcal{L}})$, where $\overline{\mathcal{L}}$ is the line bundle on $\overline{Y}$ satisfying $\overline{\mathcal{L}}^{\otimes 2} = \mathcal{O}_{\bar{Y}}(\overline{B})$ and determining the covering $\bar{\gamma}$. By repeating the blowing up procedure above we see that $\mathcal{K}_{\bar{Y}} \otimes \overline{\mathcal{L}} = \tau^*(\mathcal{K}_Y \otimes \mathcal{L}) \otimes \mathcal{O}_Y(-\overline{Z})$ with $\overline{Z} \geq 0$ a divisor on $\overline{Y}$ contained in the union of exceptional curves. Putting $Z = (\bar{\gamma})^*(\overline{Z})$ we obtain formula (7).

Furthermore, $Z = 0$ if and only if $1 - [\mu_y/2]$ vanishes for all singularities of all the ramification curves $B, B_1, B_2, \ldots$ i.e., if all these singularities are only double or triple points. By II, Sect. 8 this holds if all the singularities of B are simple. Conversely, if a singularity $y \in B$ is a double point, it is of type A_n, by Theorem II.8.1 hence simple. If it is a triple point with two different tangents, it is simple of type D_n. If it is a triple point with one tangent, and if $B_1 = \overline{B} + E_y$ has at most triple points, then the singularity is simple of type E_6, E_7 or E_8. □

(7.3) **Theorem.** *An exceptional A-D-E curve contracts to a simple surface singularity with the corresponding equation. In particular, the singularity (and the embedding of the exceptional curve) is determined uniquely by the corresponding dual graph.*

Proof. Let $x \in X$ be the singularity in question and $\gamma : X \to Y$ the double covering constructed in Lemma 3.10. By Proposition 3.5 and Theorem 7.2, the ramification curve $B \subset Y$ has a simple singularity at $y = \gamma(x)$. □

Fibrations of Surfaces

We prove Zariski's lemma about the intersection of fibre components of a fibration. Subsequently we prove the stable reduction theorem, a formula for the topological Euler characteristics of a fibration, and, finally, the relative duality theorem.

In Sects. 8–18, X will always be a connected smooth surface (not necessarily compact), S a smooth connected curve and $f : X \to S$ a proper surjective holomorphic map. Unless stated otherwise, the map f is also assumed to be connected. Sometimes we write $X \to S$ instead of $f : X \to S$.

8. Generalities on Fibrations

A point $x \in X$ is a critical point of f, if $df = 0$ at x. The critical points form an analytic set. By Remmert's Theorem I.8.4, the critical values of f (i.e., the images $f(x) \in S$ of critical points $x \in X$) form in S an analytic subset of dimension 0, i.e., a discrete subset. If $s \in S$, with $\mathfrak{m}_s \subset \mathcal{O}_S$ its maximal ideal, then the fibre X_s is the curve $f^{-1}(s)$ on X with sheaf of ideals $f^*(\mathfrak{m}_s)$. This fibre is singular, if and only if s is a critical value. So almost all fibres are smooth. All smooth fibres are diffeomorphic, hence they all have the same genus.

If S is not compact, $H^1(\mathcal{O}_S) = H^2(S, \mathbb{Z}) = 0$, so the bundle $\mathcal{O}_S(s)$, $s \in S$, is trivial. Hence $\mathcal{O}_X(X_s)$ is trivial too. We have

(8.1) **Lemma.** *Unless X is compact, all line bundles $\mathcal{O}_X(X_s)$, $s \in S$, are trivial. Whether X is compact or not, the normal bundle $\mathcal{O}_{X_s}(X_s)$ of each fibre is trivial.*

(8.2) **Lemma** (Zariski's lemma). *Let $X_s = \sum n_i C_i$, $n_i > 0$, $C_i \subset X$ irreducible, be a fibre of the fibration $X \to S$. Then we have*

(9) $\quad C_i X_s = 0 \quad$ *for all i.*

(10) $\quad$ *If $D = \sum m_i C_i$, $m \in \mathbb{Z}$, then $D^2 \leq 0$.*

(11) $\quad$ *$D^2 = 0$ holds in (9) if and only if $D = rX_s$, $r \in \mathbb{Q}$ (i.e., $pD = qX_s$ with $p, q \in \mathbb{Z}$, $p \neq 0$).*

Proof. Property (8) follows from Lemma 8.1. Consider the sub vector-space $V \subset H_2(X, \mathbb{Q})$, spanned by the classes of the curves C_i. The class of X_s belongs to the annihilator of the bilinear form on V, induced by the intersection product on X. Furthermore, let W be the abstract vector space $\bigoplus \mathbb{Q}C_i$. The canonical homomorphism from W onto V induces a bilinear form Q on W. We can apply Lemma I.2.10 to $-Q$: conditions (i') and (ii) are obviously satisfied, whereas (iii) is true since X_s is connected. It follows that Q is semi-negative definite and that $\sum n_i C_i$ spans its annihilator. Thus we obtain (9) and (10). $\square$

A singular fibre $X_s = \sum n_i C_i$ is called multiple fibre (of multiplicity n) if $n = \text{g.c.d.}\{n_i\} > 1$. Then $X_s = n \cdot F$ with F another effective divisor on X. Lemma 8.2 shows that F is 1-connected in the sense of Ramanujam (II, Sect. 12).

(8.3) **Lemma.** *Let $S = \Delta \subset \mathbb{C}$ be the unit disc and $X_0 = n \cdot F$ a multiple fibre with multiplicity n. Then $\mathcal{O}_X(F)$ on X and $\mathcal{O}_F(F)$ are both torsion bundles of order n.*

Proof. Since $\mathcal{O}_X(nF) = \mathcal{O}_X(X_0)$ is the trivial bundle, the order of $\mathcal{O}_X(F)$ is finite and does not exceed n. If $\mathcal{O}_X(F)$ would be of order $< n$ then there would be some holomorphic function on X vanishing along X_0 of lower order than $z \circ f$, z being the coordinate function on Δ. But this is impossible, because any function on X is the pull-back of some function on Δ.

To prove that ord $\mathcal{O}_F(F)$ is also n, we first shrink Δ a little bit (compare Theorem I.8.8) so that the restriction $H^i(X, \mathbb{Z}) \to H^i(F, \mathbb{Z})$ is bijective, $i = 1, 2$. Then we consider the exponential diagram

$$\begin{array}{ccccccc}
H^1(X,\mathbb{Z}) & \longrightarrow & H^1(\mathcal{O}_X) & \longrightarrow & H^1(\mathcal{O}_X^*) & \longrightarrow & H^2(X,\mathbb{Z}) \\
\| & & \downarrow & & \downarrow & & \| \\
H^1(F,\mathbb{Z}) & \longrightarrow & H^1(\mathcal{O}_F) & \longrightarrow & H^1(\mathcal{O}_F^*) & \longrightarrow & H^2(F,\mathbb{Z}) . \\
 & & & & & & \| \\
 & & & & & & \mathbb{Z}^N
\end{array}$$

Since $\mathcal{O}_X(F) \in H^1(\mathcal{O}_X^*)$ has finite order and since $H^2(X,\mathbb{Z})$ is torsion free, there is a pre-image $\xi \in H^1(\mathcal{O}_X)$ of $\mathcal{O}_X(F)$. If m is the order of $\mathcal{O}_F(F)$, then $m|n$, and there is some class $c \in H^1(X,\mathbb{Z})$ mapped onto $m \cdot (\xi|F)$. So $(n/m)c$ and $n\xi$ both have the same image in $H^1(\mathcal{O}_F)$. Since the map $H^1(F,\mathbb{Z}) \to H^1(\mathcal{O}_F)$ is injective by Proposition II.2.1, the class $m\xi$ equals the image of c in $H^1(\mathcal{O}_X)$, i.e., $\mathcal{O}_X(F)^{\otimes m} = 0$. So $m = n$. □

Let $f : X \to S$, $g : Y \to S$ be two (connected) fibrations. We call them **bimeromorphically equivalent**, if there exists a bimeromorphic correspondence between X and Y respecting the fibrations, i.e., if there is a surface Z and bimeromorphic maps $\xi : Z \to X$, $\eta : Z \to Y$ such that $f \circ \xi = g \circ \eta$.

(8.4) **Proposition.** *If the genus of its general fibre is strictly positive, then $X \to S$ factors through a unique nonsingular surface Y containing in none of its fibres a (-1)-curve.*

Proof. We blow down all (-1)-curves contained in fibres X_s, and repeat this until a surface Y is obtained containing no more such curves (at least locally, this stage is reached after a finite number of blowing downs, and this is enough). This Y is uniquely determined unless some fibre X_s contains two intersecting (-1)-curves C_1, C_2. Then $(C_1 + C_2)^2 \geq 0$ and $(C_1 + C_2)^2 = 0$ only if $(C_1, C_2) = 1$. By Lemma 8.2, (10), the fibre X_s containing both C_1 and C_2 must be of the form $n(C_1 + C_2)$. Then for the (connected) general fibre X_t we have

$$(\mathcal{K}_X, X_t) = n(\mathcal{K}_X, C_1 + C_2) = -2n.$$

So $n = 1$ and the general fibre is rational. □

A fibration which has no (-1)-curves in any of its fibres is called **relatively minimal**.

(8.5) **Proposition.** *Let $f : X \to S$, $g : Y \to S$ (with also Y smooth, connected and g connected) be relatively minimal. Furthermore let $s \in S$, and $C \subset X_s$, $D \subset Y_s$ be curves with $X_s \setminus C \neq \varnothing$ and $Y_s \setminus D \neq \varnothing$. Then the existence of a fibre-preserving biholomorphic map $X \setminus C \to Y \setminus D$ implies that the fibrations f and g are biholomorphically equivalent. If the relative minimality is dropped, then the conclusion is that the fibrations are bimeromorphically equivalent.*

Proof. By Lemma 8.2, (10), the curves C and D are exceptional. Let $(X, C) \to (\overline{X}, x)$ and $(Y, D) \to (\overline{Y}, y)$ be the maps contracting them, and let $\varphi : \overline{X} \setminus$

$\{x\} \to \overline{Y}\backslash\{y\}$ be the induced map. Then $\lim_{\nu\to\infty}\varphi(x_\nu) = y$ for each sequence $(x_\nu)_{\nu=1}^{\infty}$ in $X \setminus \{x\}$ converging to x. So φ extends continuously and for every open neighbourhood V of y in $\overline{Y}$ there exists an open neighbourhood U of x in $\overline{X}$ with $\varphi(U\backslash x) \subset V$. We may take V so small that it is an analytic subvariety of some open ball in a $\mathbb{C}^N$. Since x is normal, the coordinate functions of φ extend holomorphically over x and so φ extends to a biholomorphic map from $\overline{X}$ onto $\overline{Y}$. The assertion now follows from Theorem 6.2. □

9. The n-th Root Fibration

Let $f : X \to \Delta$ be a fibration over the unit disc Δ. We define its n-th root as follows. Let $\delta_n : \Delta \to \Delta$ be the map $z \mapsto z^n$, $X' = \Delta \times_\Delta X$ the fibre product with respect to δ_n, X'' the normalization of X', and $X^{(n)}$ the minimal desingularization of X''. Then there is a diagram

$$\tag{12} \begin{array}{ccccccc} & & & \overset{\tau^{(n)}}{\curvearrowright} & & & \\ X^{(n)} & \xrightarrow{\tau''} & X'' & \xrightarrow{\tau'} & X' & \xrightarrow{\tau} & X \\ \downarrow{\scriptstyle f^{(n)}} & & \downarrow{\scriptstyle f''} & & \downarrow{\scriptstyle f'} & & \downarrow{\scriptstyle f} \\ \Delta & = & \Delta & = & \Delta & \xrightarrow{\delta_n} & \Delta \end{array}$$

and we call $f^{(n)} : X^{(n)} \to \Delta$ the n-th root fibration of f.

Remark. The smooth surface $X^{(n)}$ does not contain any (-1)-curves over the singular points of X'', but in general the singular fibres of $f^{(n)}$ do contain (-1)-curves.

(9.1) **Proposition** *Let X_0 be a multiple fibre of multiplicity n as in Lemma 8.3. Then X'' is nonsingular, so $X^{(n)} = X''$. Furthermore $\tau^{(n)} = \tau\tau'$: $X^{(n)} \to X$ is an unramified covering, and $f^{(n)} : X^{(n)} \to \Delta$ has no multiple fibre over $0 \in \Delta$.*

Proof. If f is locally given by the function g^n (with g in general reducible), then a local equation for $X' \subset \Delta \times X$ is

$$z^n - g^n = (z-g)(z-\varepsilon g)\cdots(z-\varepsilon^{n-1}g) = 0$$

with $\varepsilon = \mathbf{e}(1/n)$. The normalization X'' decomposes into n pieces with local equations $z = g, \ldots, z = \varepsilon^{n-1}g$. So X'' is nonsingular and the multiplicities n_i'' of the components of X_0'' are n_i/n, where n_i is the multiplicity of the corresponding component in X_0.

(9.2) **Proposition** *Let $X_0 = \sum n_i C_i$ with C_i irreducible and such that $(X_0)_{\mathrm{red}} = \sum C_i$ has only normal crossings. If n is chosen such that $n_i | n$ for all i, then $(X^{(n)})_0$ is reduced and has only normal crossings.*

Proof. Every $x \in X_0$ is the centre of a local coordinate system (z_1, z_2) with $f = z_1^a z_2^b$. Let us first assume that x is a regular point of $(X_0)_{\text{red}}$. Then, say, $b = 0$ and f is given by z_1^a, where a is one of the n_i's. Locally $X^{(n)} = X'$ is defined as $\{(w, z_1, z_2) : w^n = z_1^a\}$ with $f^{(n)}$ given by w. If we write $n = \nu \cdot a$, then X' decomposes into a surfaces $w^\nu = z_1 \mathbf{e}(k/a)$, $k = 1, \ldots, a$. These surfaces are all nonsingular and w vanishes on X'_0 to first order.

In the general case, i.e., if $ab \neq 0$, we put $d = \text{g.c.d.}(a, b)$, $a = \alpha d$, $b = \beta d$, $n = r\alpha\beta d$. Then X', defined by $z^n = z_1^a z_2^b$, decomposes into d copies of the existence domain of $w = \sqrt[r\alpha\beta]{z_1^\alpha z_2^\beta}$. In X'' these copies become separated and f'' is given by w. Following the procedure of Sect. 5, ii), one finds that X'' is locally isomorphic to the existence domain of $w = \sqrt[r]{y_1 y_2}$ where $z_1 = y_1^\beta$, $z_2 = y_2^\alpha$. Now Sect. 5; ii), shows that X'' has a singularity of type $A_{r,r-1} = A_{r-1}$ and w becomes the function h from Theorem 5.1. But if $q = n - 1$ in (4), all e_k equal 2 and all $\nu_k = 1$. So w vanishes on the exceptional curve only to first order. □

10. Stable Fibrations

Let $f : X \to S$ be a fibration with $g > 0$, where $g = h^1(\mathcal{O}_{X_s})$ is the genus of the general fibre X_s.

Definition. A fibre X_s, $s \in S$, is stable, if it has the following three properties:
i) X_s is reduced,
ii) the only singularities of X_s are nodes,
iii) X_s contains no (-1)-curves.
The fibration f is stable, if all fibres X_s are stable.

Remarks. a) For any irreducible component C of X_s, the conditions i) and ii) imply that $-C^2$ equals the number of intersections of C with other components of X_s. So iii) can be replaced by

iii') every nonsingular rational component (of a fibre) has at least two points in common with the union of the remaining components.

b) When Deligne and Mumford ([D-M]) first defined stability of curves, instead of iii') they imposed the stronger condition that each nonsingular rational component should have at least three points in common with the union of the other components. In other words: X_s should not contain (-2)-curves. Of course one can achieve this by blowing down exceptional curves in X_s consisting of (-2)-curves. Since this produces singularities on X, we prefer to work with the above definition. Our stable curves usually are called semi-stable, but for the sake of brevity we drop the prefix "semi".

Example. The only stable fibres with $g(X_s) = 1$ are of the type I_b, $b > 0$, where following Kodaira in [Ko60], part. II we define

I_0 : nonsingular elliptic,

I_1 : irreducible rational with one node,

I_b : cycle of b different (-2)-curves $(b \geq 2)$.

This is an easy consequence of II, (17)

$$g(C) + g(X_s - C) + C(X_s - C) - 1 = g(X_s) = 1$$

for each irreducible component $C \subset X_s$.

The following results are crucial in the sense that they help to reduce the study of fibrations in general to the study of stable fibrations, which are much easier to handle.

(10.1) **Theorem** (Local stable reduction theorem). *Let $f : X \to \Delta$ be such that $X_0 = \sum n_i C_i$, the only singular fibre, has a reduction $(X_0)_{\mathrm{red}} = \sum C_i$ with at worst nodes. If $n \in \mathbb{N}$ is a multiple of all n_i, then, after blowing down the (-1)-curves contained in its fibres, the n-th root fibration $f^{(n)} : X^{(n)} \to \Delta$ is a stable fibration.*

This theorem follows at once from Proposition 9.2 because blowing down a (-1)-curve in $(X^{(n)})_0$ does not affect the remaining components of $(X^{(n)})_0$.

Theorem 10.1 makes it possible to define for every fibration $f : X \to \Delta$ and arbitrarily big $n \in \mathbb{N}$ a stable reduction. This is a diagram

(13)

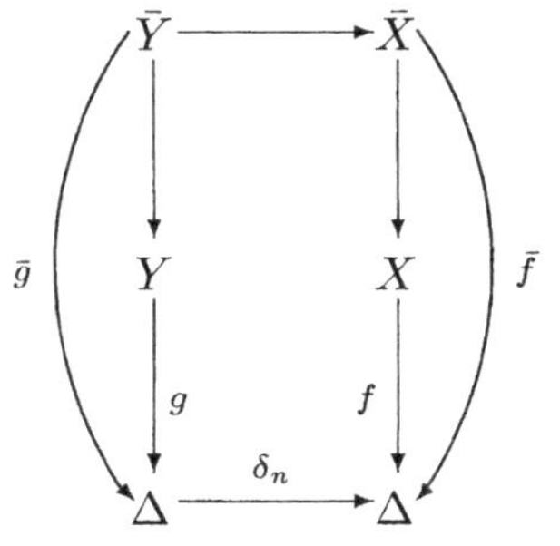

where
- the map $\overline{X} \to X$ is bimeromorphic and resolves $(X_0)_{\mathrm{red}}$ to a curve $(\overline{X}_0)_{\mathrm{red}}$ with only nodes (always possible by Theorem II.7.2),
- n is some multiple of all multiplicities of components in $\overline{X}_0$,
- $\bar{g}$ is the n-th root fibration for $\bar{f}$ (Proposition 9.2),
- $\overline{Y} \to Y$ blows down all (-1)-curves in the fibres of $\bar{g}$,
- $g : Y \to \Delta$ is stable.

Morally, X is the quotient $Y/(\mathbb{Z}/n\mathbb{Z})$, but unfortunately this holds only bimeromorphically:

(10.2) **Proposition.** *In the situation* (12), *the cyclic group* $\mathbb{Z}/n\mathbb{Z}$ *operates on* Y *covering its action* $z \mapsto \mathbf{e}(1/n)z$ *on* Δ. *The quotient* $Y/(\mathbb{Z}/n\mathbb{Z})$ *is bimeromorphically equivalent to* X.

Proof. If $X' = \Delta \times_\Delta \overline{X}$ with respect to $\delta : \Delta \to \Delta$, $z \mapsto z^n$ and X'' is its normalization as in (11), then $\mathbb{Z}/n\mathbb{Z}$ acts on X'' compatibly with the fibrations and such that the quotient is $\overline{X}$. This action lifts to $\overline{Y}$ and as a consequence of Proposition 8.5, $\overline{Y}/\mathbb{Z}/n\mathbb{Z}$ is bimeromorphically equivalent to $\overline{X}$. The $\mathbb{Z}/n\mathbb{Z}$-operation descends to Y, the relatively minimal model of $\overline{Y}$, and $Y/(\mathbb{Z}/n\mathbb{Z})$ is bimeromorphically equivalent to $\overline{Y}/(\mathbb{Z}/n\mathbb{Z})$. □

So diagram (12) can be extended as to contain $Y/(\mathbb{Z}/n\mathbb{Z}$ in the following way ($\overline{Y/(\mathbb{Z}/n\mathbb{Z})}$ denotes the desingularization of $Y/(\mathbb{Z}/n\mathbb{Z})$):

$$
\begin{array}{ccccc}
\overline{Y} & \longrightarrow & & & \overline{X} \\
\Big\downarrow & & & \overline{Y/(\mathbb{Z}/n\mathbb{Z})} & \Big\downarrow \\
 & & \swarrow & & \searrow \\
Y & \longrightarrow & Y/(\mathbb{Z}/n\mathbb{Z}) & & X \\
{\scriptstyle g}\Big\downarrow & & \Big\downarrow & & \Big\downarrow{\scriptstyle f} \\
\Delta & \xrightarrow{\delta_n} & \Delta & = = = = & \Delta\,.
\end{array}
\tag{14}
$$

Finally, it is easy to patch local stable reductions together to obtain (compare [D-M]):

(10.3) **Theorem** (Stable reduction theorem). *Consider a fibration* $f : X \to S$, *with* S *compact. Let* $\overline{X} \to X$ *be a bimeromorphic map resolving all fibres* X_s *to fibres* $\overline{X}_s$ *with* $(\overline{X}_s)_{\text{red}}$ *having at worst nodes. Then there exists a cyclic covering* $\delta : T \to S$, *ramified only over the critical values* $s_1, \ldots, s_k \in S$ *of* f *and one more point* $s_0 \in S$ *such that* $T \times_S \overline{X}$ *is bimeromorphically equivalent to a stable fibration.*

Proof. We choose some multiple $N \in \mathbb{N}$ of all multiplicities n_i occurring as multiplies of fibre components. Then we take $s_0 \in S \backslash \{s_1, \ldots, s_k\}$ arbitrarily. If, furthermore, $\ell \in \mathbb{N}$ is chosen such that $N | k + \ell$, then there is at least one line bundle $\mathcal{L} \in \operatorname{Pic}(S)$ with $\mathcal{L}^{\otimes N} = \mathcal{O}_S(\ell s_0 + s_1 + \cdots + s_k)$. Such an $\mathcal{L}$ admits an N-valued section T, the pre-image under $\mathcal{L} \to \mathcal{L}^{\otimes N}$, $t \mapsto t^N$, of the canonical section in $\mathcal{O}_S(\ell s_0 + s_1 + \cdots + s_k)$. This T can be interpreted as a Riemann surface, lying cyclically of order N over S. Locally at $s_1, \ldots, s_k$, the map $T \to S$ is just δ_N from Sect. 9, so the assertion follows now from the local version of the theorem. □

11. Direct Image Sheaves

Every fibration $f : X \to S$ of the type we consider is flat (see e.g. [F], p. 154), so Theorem I.8.5 on direct image sheaves applies.

(11.1) **Lemma.** *Let $f : X \to S$ be a fibration as before, but not necessarily connected. Then $h^i(\mathcal{O}_{X_s})$, $i = 0, 1$ is independent of s. ($h^0(\mathcal{O}_{X_s})$ equals the number of components of a nonsingular fibre.)*

Proof. Firstly, let f be connected. Then $h^0(\mathcal{O}_{X_s}) = 1$ is s is not a critical value and $h^0(\mathcal{O}_{X_s}) \geq 1$ for all $s \in S$. Assume that there is a critical value $s \in S$ with $h^0(\mathcal{O}_{X_s}) > 1$. Ramanujam's Lemma II.12.2 shows that the divisor X_s cannot be 1-connected. So $X_s = nF$ must be a multiple fibre with F 1-connected by Zariski's Lemma 8.2. Now consider for $1 \leq \nu \leq n-1$ the exact sequences

$$0 \longrightarrow \mathcal{O}_F(-\nu F) \longrightarrow \mathcal{O}_{(\nu+1)F} \longrightarrow \mathcal{O}_{\nu F} \longrightarrow 0.$$

By Lemma 8.3 all the bundles $\mathcal{O}_F(-\nu F)$ are non-trivial of finite order. Ramanujam's lemma shows that $h^0(\mathcal{O}_F(-\nu F))$ vanishes. By induction on ν it follows that

$$h^0(\mathcal{O}_{X_s}) = h^0(\mathcal{O}_{nF}) \leq h^0(\mathcal{O}_F)$$

with $h^0(\mathcal{O}_F) = 1$, because F is 1-connected. So in this case we have $h^0(\mathcal{O}_{X_s}) = 1$ for all $s \in S$.

If f is not connected, let $f = \gamma \circ \rho$ be its Stein factorization, so $\gamma : \widetilde{S} \to S$ is a ramified covering and ρ is connected. Then $h^0(\mathcal{O}_{\rho^{-1}(t)}) = 1$ for all $t \in \widetilde{S}$. If $s \in S$ is not a ramification point, then $h^0(\mathcal{O}_{X_s}) = d$, with $d = \deg(\gamma)$. If however $s \in S$ is a ramification point, say $\gamma^{-1}(s) = \{t_1, \ldots, t_k\}$ with ν_k the ramification order at t_k, then $\nu_1 + \cdots + \nu_k = d$ and $X_s = \nu_1\rho^{-1}(t_1) + \cdots + \nu_k\rho^{-1}(t_k)$. So it suffices to show $h^0(\mathcal{O}_{\nu\cdot\rho^{-1}(t)}) \leq \nu$ for every fibre of the connected fibration ρ. But this follows as usual from the exact sequences

$$0 \longrightarrow \mathcal{O}_{\rho^{-1}(t)}\big((1-\nu)\rho^{-1}(t)\big) \longrightarrow \mathcal{O}_{\nu\cdot\rho^{-1}(t)} \longrightarrow \mathcal{O}_{(\nu-1)\cdot\rho^{-1}(t)} \longrightarrow 0,$$

because $\mathcal{O}_{\rho^{-1}(t)}\big(\rho^{-1}(t)\big) = \mathcal{O}_{\rho^{-1}(t)}$ by Lemma 8.1.

So we have completed the proof for $i = 0$. The case $i = 1$ follows from Theorem I.8.5, (i). □

From Theorem I.8.5, (iii) we now obtain

(11.2) **Corollary.** *If $f : X \to S$ is a not necessarily connected fibration, then the sheaves $f_*\mathcal{O}_X$ and $f_{*1}\mathcal{O}_X$ are locally free and have the base-change property.*

(11.3) **Proposition.** *Let $g : Y \to T$ be a stable reduction of $f : X \to S$ with $S = T/(\mathbb{Z}/N\mathbb{Z})$ as in Theorem 10.3. Then, if $\delta : T \to S$ is the quotient map, there is a natural injection $\delta^*(f_{*1}\mathcal{O}_X) \to g_{*1}\mathcal{O}_Y$. If δ is ramified at $t \in T$ cyclically of order n, $n|N$, then $\big(\delta^*(f_{*1}\mathcal{O}_X)\big)_t$ is the submodule of $\mathbb{Z}/n\mathbb{Z}$-invariants of $(g_{*1}\mathcal{O}_Y)_t$.*

Proof. Since δ is ramified at t cyclically of order n, we can identify a suitable neighbourhood of t with Δ and also a neighbourhood of s with Δ such that locally the situation (13) occurs (with $\delta|\Delta = \delta_n$). Let $f' : \overline{Y/(\mathbb{Z}/n\mathbb{Z})} \to X \to$

Δ and $f'' : Y/(\mathbb{Z}/n\mathbb{Z}) \to \Delta$ be the induced maps. By Proposition 10.2, the map $\overline{Y/(\mathbb{Z}/n\mathbb{Z})} \to X$ is bimeromorphic, hence a composition of σ-processes (Theorem 4.4) and so by Proposition 3.1, the natural morphism $f_{*1}\mathcal{O}_X \to f'_{*1}\mathcal{O}_{\overline{Y/(\mathbb{Z}/n\mathbb{Z})}}$ is bijective. The natural map $f''_{*1}\mathcal{O}_{Y/(\mathbb{Z}/n\mathbb{Z})} \to f'_{*1}\mathcal{O}_{\overline{Y/(\mathbb{Z}/n\mathbb{Z})}}$ is bijective too, because $Y/(\mathbb{Z}/n\mathbb{Z})$ has only cyclic quotient singularities, which are of type $A_{n,q}$ by Theorem 5.4, hence rational by Proposition 3.1. The assertion now follows from [Gk57b], p. 202, Cor. of Prop. 5.2.3: $H^1(\mathcal{O}_{Y/(\mathbb{Z}/n\mathbb{Z})})$ is the subspace of $\mathbb{Z}/n\mathbb{Z}$-invariants in $H^1(\mathcal{O}_Y)$. □

(11.4) **Proposition.** *Let $f : X \to S$ be a fibration and X_{gen} a nonsingular fibre. Then*

i) *$e(X_s) \geq e(X_{\text{gen}})$ for all fibres X_s;*

ii) *if X is compact, then*

$$e(X) = e(X_{\text{gen}}) \cdot e(S) + \sum_{s \in S} \big(e(X_s) - e(X_{\text{gen}})\big).$$

Proof.

i) Since $h^1(\mathcal{O}_{X_s})$ is independent of s by Lemma 11.1, we have to show that $e(X_s) \geq 2-2h^1(\mathcal{O}_{X_s})$. Since $e(X_s) = e\big((X_s)_{\text{red}}\big)$ and $h^1(\mathcal{O}_{X_s}) \geq h^1\big(\mathcal{O}_{(X_s)_{\text{red}}}\big)$, it suffices to prove that $e(C) \geq 2 - 2h^1(\mathcal{O}_C)$ for a connected compact reduced curve C. But for such a curve we have $H^2(C, \mathbb{Z}) = \mathbb{Z}^N$, where N is the number of irreducible components, whereas $b_1(C) = \operatorname{rank} H^1(C, \mathbb{Z}) \leq 2h^1(\mathcal{O}_C)$ by Proposition II.2.1.

ii) Let $s_1, \ldots, s_k \in S$ be the critical values of f. Then we have

$$\begin{aligned} e(X) &= e\Big(X \setminus \bigcup_{i=1}^{k} X_{s_i}\Big) + e\Big(\bigcup_{i=1}^{k} X_{s_i}\Big) \\ &= \big(e(S) - k\big)e(X_{\text{gen}}) + \sum_{i=1}^{k} e(X_{s_i}) \\ &= e(S)e(X_{\text{gen}}) + \sum_{i=1}^{k} \big(e(X_{s_i}) - e(X_{\text{gen}})\big). \end{aligned}$$ □

(11.5) *Remark.* Assertion i) can easily be sharpened. In fact, for singular X_s one always has $e(X_s) > e(X_{\text{gen}})$ unless X_s is a multiple fibre with $(X_s)_{\text{red}}$ nonsingular elliptic.

(11.6) **Corollary.** *If X is a compact surface and $f : X \to S$ a fibration with fibre genus g_1 and base genus g_2, then $e(X) \geq 4(g_1 - 1)(g_2 - 1)$.*

12. Relative Duality

The differential *df* of $f : X \to S$ can be viewed as an injection of sheaves $f^*(\mathcal{K}_S) \to \Omega^1_X$. Its cokernel $\Omega_{X/S}$ is called the sheaf of relative differentials.

In general, $\Omega_{X/S}$ is very far from being locally free. Indeed, if some fibre has a multiple component, then *df* vanishes along this component and $\Omega_{X/S}$ contains a torsion subsheaf with one-dimensional support. However, away from the singularities of f, there is an exact sequence *of vector bundles*

$$0 \longrightarrow f^*(\mathcal{K}_S) \longrightarrow \Omega^1_X \longrightarrow \Omega_{X/S} \longrightarrow 0$$

inducing outside of the singularities an isomorphism between $\Omega_{X/S}$ and $\mathcal{K}_X \otimes f^*(\mathcal{K}_S^\vee)$.

Definition. The line bundle $\omega_{X/S} = \mathcal{K}_X \otimes f^*(\mathcal{K}_S^\vee)$ on X is called the dualizing sheaf of f.

By definition (see II, Sect.1) and Lemma 8.1 the restriction $\omega_{X/S}|X_s$ is isomorphic to ω_{X_s} for all $s \in S$. So by duality on X_s the dimensions $h^0(\omega_{X/S}|X_s) = h^1(\mathcal{O}_{X_s})$ and $h^1(\omega_{X/S}|X_s) = h^0(\mathcal{O}_{X_s})$ are independent of $s \in S$. Because of Theorem I.8.5, (iii) and Lemma 11.1, this proves

(12.1) **Proposition.** *The sheaves $f_*\omega_{X/S}$ and $f_{*1}\omega_{X/S}$ on S are locally free and have the base-change property.*

The following result is the heart of this section.

(12.2) **Proposition.** *There is an epimorphism* $\mathrm{Tr} : f_{*1}\omega_{X/S} \to \mathcal{O}_S$ *which on every fibre restricts to the trace morphism* tr_{X_s} *from* II, *Sect.* 5. *If f is connected,* Tr *is an isomorphism.*

Proof. Let X be embedded in $X \times S$ by way of $j = (\mathrm{id}_X, f)$, and let $X' = j(X)$. Then we have a diagram

$$\begin{array}{ccccc} X & \xrightarrow{j} & X \times S & \xrightarrow{q} & X \\ & {}_{f}\searrow & \downarrow{\scriptstyle p} & & \\ & & S & & \end{array}$$

where p, q are the projections.

Tensoring the structure sequence

$$0 \longrightarrow \mathcal{O}_{X\times S} \longrightarrow \mathcal{O}_{X\times S}(X') \longrightarrow \mathcal{O}_{X'}(X') \longrightarrow 0$$

with $q^*(\mathcal{K}_X)$, and using the adjunction formula I.6.4 as well as $\mathcal{K}_{X\times S} \cong p^*(\mathcal{K}_S) \otimes q^*(\mathcal{K}_X)$, we obtain

$$(15) \qquad 0 \longrightarrow q^*(\mathcal{K}_X) \longrightarrow q^*(\mathcal{K}_X) \otimes \mathcal{O}_{X\times S}(X') \longrightarrow \omega_{X/S} \longrightarrow 0.$$

Now, for any sheaf $\mathcal{F}$ on $X \times S$, let $p_*^c(\mathcal{F})$ be the direct image sheaf with compact supports in X-direction (see [Bre], p. 135). Then, by Dolbeault a germ in $p^c_{*2}(q^*(\mathcal{K}_X))_{s_0}$, $s_0 \in S$, is represented by a $\bar\partial$-closed $(2,2)$-form Φ on a neighbourhood of $X \times \{s_0\} \subset X \times S$. This can be written as

$$\Phi = \varphi(s) + \psi(s) \wedge d\bar{s} + \psi'(s) \wedge ds\ .$$

Here $\varphi(s) = \Phi|X \times \{s\}$ is a $(2,2)$-form and $\psi(s)$, $\psi'(s)$ are forms on $X \times \{s\}$ of type $(2,1)$, $(1,2)$ respectively. If we denote by $\bar{\partial}_X$ the derivative in X-direction, then

$$\frac{\partial\varphi}{\partial\bar{s}} \wedge d\bar{s} + \bar{\partial}_X\psi(s) \wedge d\bar{s} + \frac{\partial\psi'}{\partial\bar{s}} \wedge d\bar{s} \wedge ds = \bar{\partial}\Phi = 0\ .$$

So

$$\begin{aligned}\frac{\partial}{\partial\bar{s}}\int_{X\times\{s\}}\Phi &= \frac{\partial}{\partial\bar{s}}\int_X \varphi(s) = \int_X \frac{\partial\varphi(s)}{\partial\bar{s}}\\ &= -\int_X \bar{\partial}_X\psi(s) = -\int_X d\psi(s) = 0,\end{aligned}$$

because $\psi(s)$ has compact support. This shows that the function $s \mapsto \int_{X\times\{s\}}\Phi$ is holomorphic near s_0, i.e., integration over the p-fibres defines a morphism $p^c_{*2}(q^*(\mathcal{K}_X)) \to \mathcal{O}_S$.

Now we define Tr as the composition of this morphism with the morphism

$$\delta : f_{*1}\omega_{X/S} = p^c_{*1}\omega_{X/S} \longrightarrow p^c_{*2}(q^*\mathcal{K}_X)$$

induced by the sequence (14). Evaluated at every $s \in S$, this Tr is the trace morphism tr_{X_s} (base-change for $\omega_{X/S}$). So the surjectivity of Tr follows from the surjectivity of all morphisms tr_{X_s}, $s \in S$. □

Next let $\mathcal{F}$ be some locally free $\mathcal{O}_X$-sheaf. The cup-product (of classes defined in a neighbourhood of f-fibres) defines a pairing of $\mathcal{O}_S$-sheaves

$$f_*(\mathcal{F}^\vee \otimes \omega_{X/S}) \otimes_{\mathcal{O}_S} f_{*1}\mathcal{F} \longrightarrow f_{*1}\omega_{X/S} \xrightarrow{\mathrm{Tr}} \mathcal{O}_S\ .$$

For every $s \in S$, there is a commutative diagram

$$\begin{array}{ccccc} f_*(\mathcal{F}^\vee \otimes \omega_{X/S})/\mathfrak{m}_s f_*(\mathcal{F}^\vee \otimes \omega_{X/S}) & \otimes_{\mathcal{O}_S} & f_{*1}\mathcal{F}/\mathfrak{m}_s f_{*1}\mathcal{F} & \longrightarrow & \mathbb{C}\\ \downarrow{\scriptstyle\varphi_0} & & \downarrow{\scriptstyle\varphi_1} & & \|\\ H^0(\mathcal{F}^\vee \otimes \omega_{X_s}) & \otimes_{\mathbb{C}} & H^1(\mathcal{F}|X_s) & \longrightarrow & \mathbb{C}\end{array}$$

relating this pairing of sheaves with the duality pairing on the curve X_s. The induced morphism

$$(16) \qquad f_*(\mathcal{F}^\vee \otimes \omega_{X/S}) \longrightarrow (f_{*1}\mathcal{F})^\vee$$

is called the relative duality morphism.

(12.3) Theorem (Relative duality theorem). *If $f : X \to S$ is a fibration and $\mathcal{F}$ some locally free $\mathcal{O}_X$-sheaf, then the relative duality morphism* (15) *is bijective.*

Proof. We consider the restriction morphisms φ_0, φ_1 in the commutative diagram of pairings above.

Since $\dim S = 1$, we have $\mathfrak{m}_s f_*(\mathcal{F}^\vee \otimes \omega_{X/S}) = f_*(\mathcal{I}_{X_s} \cdot \mathcal{F}^\vee \otimes \omega_{X/S})$, so φ_0 is injective. We denote its image by E. It is the subspace in $H^0(\mathcal{F}^\vee \otimes \omega_{X_s})$

of sections extending to a neighbourhood of X_s and its dimension is $h^0(\mathcal{F}^\vee \otimes \omega_{X_{\text{gen}}})$, where X_{gen} is a general fibre of f.

Since the fibre dimension if 1, φ_1 is surjective. We denote by $(f_{*1}\mathcal{F})_{\text{tors}}$ the torsion submodule of $(f_{*1}\mathcal{F})_s$ and by $T \subset H^1(\mathcal{F}|X_s)$ its image $\varphi_1(f_{*1}\mathcal{F})_{\text{tors}}$. The free quotient $(f_{*1}\mathcal{F})^{\vee\vee}$ then maps onto $H^1(\mathcal{F}|X_s)/T$, so the codimension of T is at most $h^1(\mathcal{F}|X_{\text{gen}})$, which equals $h^0(\mathcal{F}^\vee \otimes \omega_{X_{\text{gen}}})$ by duality on X_{gen}.

Since $f_*(\mathcal{F}^\vee \otimes \omega_{X/S}) \otimes (f_{*1}\mathcal{F})_{\text{Tors}}$ goes to zero in $\mathcal{O}_S$, the spaces E and T annihilate each other under the duality pairing on the curve X_s. Since this pairing is perfect by Theorem II.6.1, we find $\operatorname{codim} T = \dim E$, and the composed map

$$f_*(\mathcal{F}^\vee \otimes \omega_{X/S})/\mathfrak{m}_s f_*(\mathcal{F}^\vee \otimes \omega_{X/S}) \longrightarrow E \longrightarrow$$
$$\longrightarrow \left(H^1(\mathcal{F}|X_s)/T\right)^\vee \longrightarrow \left((f_{*1}\mathcal{F})^{\vee\vee}/\mathfrak{m}_s(f_{*1}\mathcal{F})^{\vee\vee}\right)^\vee = (f_{*1}\mathcal{F})^\vee/\mathfrak{m}_s f_{*1}\mathcal{F})^\vee$$

is an isomorphism.

This proves that the reduction of (15) modulo $\mathfrak{m}_s$ is bijective, and so (15) is an isomorphism too. □

The Period Map of Stable Fibrations

The period matrix of a smooth fibre of a stable fibration defines a point in the period domain $\mathfrak{h}_g/\Gamma_g$. We show that the period map extends over the Satake compactification. In a subsequent section we then show that for non-constant period maps the direct image of the relative canonical bundle has positive degree. This is a crucial step in the proof of Iitaka's conjecture $C_{2,1}$. In higher dimensions the analogue of this conjecture is central in classification theory, but, except from some special cases, still remains a conjecture.

13. Period Matrices of Stable Curves

For a nonsingular connected compact Riemann surface F we have defined in I, Sect. 14 what we mean by a canonical basis of $H^1(F,\mathbb{Z})$ and by the corresponding period matrix. Here we generalise this to the case where F is a reduced, connected, compact curve with no other singularities but ordinary double points.

Let $\nu : \widetilde{F} \to F$ be the normalization. We put $g = h^1(\mathcal{O}_F)$ and $\tilde{g} = h^1(\mathcal{O}_{\widetilde{F}})$, the sum of the genera of the irreducible components $\widetilde{F}_k \subset \widetilde{F}$. For all α, $\beta \in H^1(F,\mathbb{Z})$ we can define a product $(\alpha, \beta) \in \mathbb{Z}$ by mapping

$$\alpha \cup \beta \in H^2(F,\mathbb{Z}) \simeq \bigoplus_k H^2(\widetilde{F}_k,\mathbb{Z}) \simeq \bigoplus_k \mathbb{Z} \longrightarrow \mathbb{Z}, \quad \left((\varphi_k)_k \mapsto \sum \varphi_k\right)$$

(both isomorphisms are canonical).

(13.1) **Proposition.** *Let $p : H^1(F,\mathbb{Z}) \to H^1(\mathcal{O}_F)$ be induced by the inclusion $\mathbb{Z} \hookrightarrow \mathcal{O}_F$. There is a basis $\alpha_1, \ldots, \alpha_{\tilde{g}}, \beta_1, \ldots, \beta_g \in H^1(F,\mathbb{Z})$ such that*
i) *the images $p\beta_1, \ldots, p\beta_g \in H^1(\mathcal{O}_F)$ span $H^1(\mathcal{O}_F)$ over $\mathbb{C}$,*
ii) $(\alpha_i, \alpha_j) = (\beta_i, \beta_j) = 0$, $(\alpha_i, \beta_j) = \delta_{ij}$.

Proof. We choose canonical bases for all components $\widetilde{F}_k$. They combine to a set $\tilde{\alpha}_1, \ldots, \tilde{\alpha}_{\tilde{g}}, \tilde{\beta}_1, \ldots, \tilde{\beta}_{\tilde{g}} \in H^1(\widetilde{F},\mathbb{Z})$ with $p\tilde{\beta}_1, \ldots, p\tilde{\beta}_{\tilde{g}}$ generating $H^1(\mathcal{O}_{\widetilde{F}})$ and

$$(\tilde{\alpha}_i, \tilde{\alpha}_j) = (\tilde{\beta}_i, \tilde{\beta}_j) = 0, \quad (\tilde{\alpha}_i, \tilde{\beta}_j) = \delta_{ij} \; . \tag{17}$$

Now the cohomology diagram from II, Sect 2

$$\begin{array}{ccccccccccc} 0 \to & H^0(F,\mathbb{Z}) & \to & H^0(\widetilde{F},\mathbb{Z}) & \to & H^0(\Sigma) & \to & H^1(F,\mathbb{Z}) & \overset{\nu^*}{\to} & H^1(\widetilde{F},\mathbb{Z}) & \to 0 \\ & \downarrow & & \downarrow & & \downarrow & & \downarrow p & & \downarrow p & \\ 0 \to & H^0(\mathcal{O}_F) & \to & H^0(\mathcal{O}_{\widetilde{F}}) & \to & H^0(S) & \to & H^1(\mathcal{O}_F) & \to & H^1(\mathcal{O}_{\widetilde{F}}) & \to 0 \end{array}$$

shows that there are pre-images $\alpha_1, \ldots, \alpha_{\tilde{g}}, \beta_1, \ldots, \beta_{\tilde{g}} \in H^1(F,\mathbb{Z})$ for the classes $\tilde{\alpha}_1, \ldots, \tilde{\beta}_{\tilde{g}}$. Since ν^* commutes with cup products, they have the same products (16) as their images on $\widetilde{F}$.

Since each singularity $x_i \in F$ is an ordinary double point, we have $\Sigma_{x_i} = \mathbb{Z}$ and $S_{x_i} = \mathbb{C}$. This implies rank $H^1(F,\mathbb{Z}) = g + \tilde{g}$. So the image of $H^0(\Sigma)$ in $H^1(F,\mathbb{Z})$ will be a direct summand of rank $g - \tilde{g}$. We pick a basis $\beta_{\tilde{g}+1}, \ldots, \beta_g$ for this $\mathbb{Z}$-module. Since $\nu^*\beta_{\tilde{g}+1} = \cdots = \nu^*\beta_g = 0$, cup products with $\beta_{\tilde{g}+1}, \ldots, \beta_g$ annihilate all of $H^1(F,\mathbb{Z})$. This proves ii) for $\alpha_1, \ldots, \beta_g$.

Obviously the image of $H^0(\Sigma)$ in $H^0(S)$ spans this $\mathbb{C}$-vector space. This implies that $p\beta_1, \ldots, p\beta_g$ span $H^1(\mathcal{O}_F)$. □

A basis as in the lemma will be called a canonical basis for $H^1(F,\mathbb{Z})$. We may express $p\alpha_1, \ldots, p\alpha_{\tilde{g}} \in H^1(\mathcal{O}_F)$ in the basis $p\beta_1, \ldots, p\beta_g$. The result is a $g \times \tilde{g}$ matrix Z with

$$(\alpha_1, \ldots, \alpha_{\tilde{g}}) = -(\beta_1, \ldots, \beta_g)Z \; .$$

After applying ν^*, the first $\tilde{g} \times \tilde{g}$ square block of Z becomes the direct sum $\widetilde{Z}$ of the period matrices for the components $\widetilde{F}_k$. So the Riemann period relations I, (16) hold for this block: $\widetilde{Z} = \widetilde{Z}^t$, $\operatorname{Im} \widetilde{Z} > 0$.

14. Topological Monodromy of Stable Fibrations

Let $f : X \to \Delta$ be a fibration with only one singular fibre X_0, which is reduced and has no singularities but ordinary double points. For more details of the following discussion we refer to [Le] or [La].

The pair (X, X_0) is triangulable, so X_0 is a deformation retract of arbitrarily small neighbourhoods in X (Theorem I.8.8). The function $|f|$ can be used as a Morse-function on $X \setminus X_0$. This shows that there is a deformation retraction $\pi : X \to X_0$. It follows that the canonical maps

$$\text{restr} : H^i(X, \mathbb{Z}) \longrightarrow H^i(X_0, \mathbb{Z}) \quad \text{and} \quad H_i(X_0, \mathbb{Z}) \longrightarrow H_i(X, \mathbb{Z})$$

are isomorphisms. Near a double point $x_i \in X_0$ we can find local coordinates (u, v) on X such that $f = u^2 + v^2$. For every $\varepsilon > 0$ we can choose π such that outside of the ball $B_i = \{|u|^2 + |v|^2 < 2\varepsilon\}$ the map $\pi|X$ is a diffeomorphism for all s with $|s| < \varepsilon$. For $s \neq 0$ the circle $S_i = \{(u,v) \in B_i,\ u^2 + v^2 = s,\ \operatorname{Im} U = \operatorname{Im} v = 0\}$ is a deformation retract of $B_i \cap X_s$ (in fact $B_i \cap X_s$ is homeomorphic to $S_i \times (0,1)$). For $s = 0$ this circle becomes the points x_i. The retraction π can be chosen such that it retracts S_i to x_i and induces a map $X_s \setminus \bigcup_i S_i \to X_0 \setminus \bigcup x_i$, homotopic to a diffeomorphism. This explains why the circle S_i, or rather its class $e_i \in H_1(X_s, \mathbb{Z})$ (determined up to sign) is called the vanishing cycle.

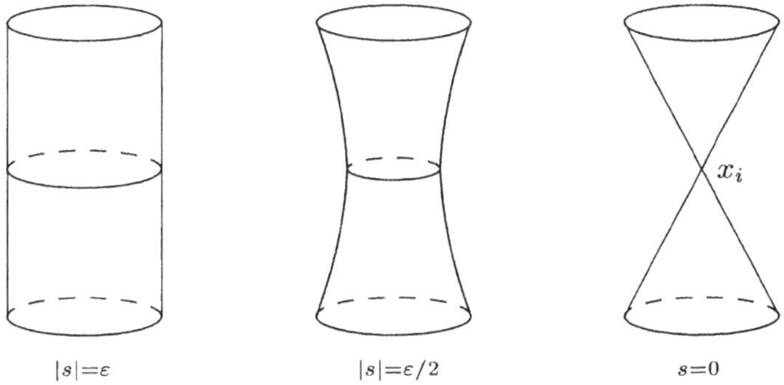

Part of X_s in B_i for various values of s

How to obtain $B_i \cap X_s$ $(0 < |s| < \varepsilon)$ from the neighbourhood $(B_i \cap X_0)^\sim$ of y_i, y_i' in $\tilde{X}_0$

Fig. 2

Also it should be clear that X_s is homeomorphic to a connect sum, obtained from the components of $\widetilde{X}_0$, the normalization of X_0, in the following way. Each time X_0 has a double point x_i with inverse images y_i, $y_i' \in \widetilde{X}_0$ one must remove two discs about y_i and y_i' and close up with a cylinder (compare IV, Sect. 9). We have visualized this above in Fig. 2. From this it follows that in homology π induces a diagram

$$\begin{array}{ccccccc} \bigoplus_i H_1(S_i,\mathbb{Z}) & \longrightarrow & H_1(X_s,\mathbb{Z}) & \longrightarrow & H_1(X_s,\bigcup S_i;\mathbb{Z}) & \longrightarrow & \bigoplus_i H_0(S_i,\mathbb{Z}) \\ \downarrow & & \downarrow{\scriptstyle \pi_*} & & \| & & \| \\ 0 & \longrightarrow & H_1(X_0,\mathbb{Z}) & \longrightarrow & H_1(X_0,\bigcup x_i;\mathbb{Z}) & \longrightarrow & \bigoplus_i H_0(x_i,\mathbb{Z}) \ . \end{array}$$

If we denote by $V \subset H_1(X_s,\mathbb{Z})$ the subspace spanned by the vanishing cycles, then the sequence

$$(18) \qquad 0 \longrightarrow V \longrightarrow H_1(X_s,\mathbb{Z}) \xrightarrow{\pi_*} H_1(X_0,\mathbb{Z}) \longrightarrow 0$$

is exact.

Differentiably, f is locally trivial over Δ^*. Circling the origin $0 \in \Delta$ once in counter-clockwise direction we obtain an automorphism

$$T : H_1(X_s,\mathbb{Z}) \longrightarrow H_1(X_s,\mathbb{Z}),$$

the Picard-Lefschetz monodromy. It is determined by the vanishing cycles e_i by way of the classical formula (compare [P-Si], [Le], [La])

$$T(a) = a - \sum_i (a \cdot e_i) e_i \ , \quad a \in H_1(X_s,\mathbb{Z}) \ .$$

On $H^1(X_s,\mathbb{Z}) = \mathrm{Hom}_{\mathbb{Z}}(H_1(X_s,\mathbb{Z}),\mathbb{Z})$ the monodromy T operates canonically by

$$(T\alpha)(b) = \alpha(T^{-1}b), \quad \alpha \in H^1, \quad b \in H_1 \ .$$

T respects the intersection pairing on H_1 and this implies that T commutes with $P : H^1 \to H_1$, the inverse of Poincaré duality.

(14.1) **Theorem** (Local invariant cycle theorem). *The image of*

$$\mathrm{restr} : H^1(X,\mathbb{Z}) \longrightarrow H^1(X_s,\mathbb{Z})$$

is exactly the subgroup of classes invariant under T.

Proof. The sequence (17) dualises to

$$(19) \qquad 0 \longrightarrow H^1(X,\mathbb{Z}) \xrightarrow{\mathrm{restr}} H^1(X_s,\mathbb{Z}) \longrightarrow \mathrm{Hom}(V,\mathbb{Z}) \ ,$$

and so the image of restr consists exactly of the classes $\alpha \in H^1(X_s,\mathbb{Z})$ vanishing on V.

Now let $\alpha = Pa$. From $TP = PT$ it follows that

$$T\alpha = TPa = P\big(a - \sum (a\cdot e_i)e_i\big)$$
$$= \alpha - P\big(\sum \alpha(e_i)e_i\big).$$

So $T\alpha = \alpha$ if and only if $\sum \alpha(e_i)e_i$ vanishes.

Let v_k be a basis for V and $e_i = \sum c_{ik}v_k$. Then the two integral matrices c_{ik} and $\big(\sum_i c_{ik}c_{i\ell}\big)$ are of maximal rank. So

$$\sum_i \alpha(e_i)e_i = \sum c_{ik}c_{i\ell}\alpha(v_k)v_\ell$$

vanishes if and only if $\sum_k c_{ik}c_{i\ell}\alpha(v_k) = 0$ for all ℓ, i.e., $\alpha(v_k) = 0$ for all k, in other words, if α vanishes on V. □

15. Monodromy of the Period Matrix

We fix a canonical basis $\alpha_1, \ldots, \alpha_{\tilde g}, \beta_1, \ldots, \beta_g \in H^1(X_0, \mathbb{Z})$ in the sense of Lemma 13.1. Because of the isomorphism $H^1(X,\mathbb{Z}) \to H^1(X_0,\mathbb{Z})$ these classes can be thought of as existing on all of X. So they restrict to classes $\alpha_i(s), \beta_j(s) \in H^1(X_s,\mathbb{Z})$, $s \neq 0$. The map $\pi^* : H^2(X_0,\mathbb{Z}) \to H^2(X_s,\mathbb{Z}) = \mathbb{Z}$ is just the map from Sect. 13. So the intersection-product formulae in Proposition 13.1, (ii) hold also for $\alpha_1(s), \ldots, \alpha_{\tilde g}(s), \beta_1(s), \ldots, \beta_g(s)$, if $s \neq 0$. The exact sequence (18) shows that $H^1(X,\mathbb{Z})$ is a direct summand in $H^1(X_s,\mathbb{Z})$. Therefore we can find, for fixed $s \neq 0$, classes $\alpha_{\tilde g+1}(s), \ldots, \alpha_g(s)$ such that $\alpha_1(s), \ldots, \alpha_g(s), \beta_1(s), \ldots, \beta_g(s)$ form a basis in $H^1(X_s,\mathbb{Z})$. The intersection product being unimodular on the space spanned by $\alpha_1(s), \ldots, \alpha_{\tilde g}(s)$, $\beta_1(s), \ldots, \beta_{\tilde g}(s)$, by Lemma I.2.3 we may choose $\alpha_{\tilde g+1}(s), \cdots, \alpha_g(s)$ such that $\big(\alpha_1(s), \ldots, \alpha_g(s), \beta_1(s), \ldots, \beta_g(s)\big)$ is a canonical basis for $H^1(X_s,\mathbb{Z})$.

On Δ we consider the following direct image sheaves:

$\mathcal{E} = f_{*1}\mathcal{O}_X$ which is locally free of rank g by Lemma 11.2, and $L = f_{*1}\mathbb{Z}_X$.

The exponential sequence on X induces an exact sequence

$$f_*\mathcal{O}_X \xrightarrow{\mathbf{e}} f_*\mathcal{O}_X^* \longrightarrow L \xrightarrow{p} \mathcal{E}$$

making L into a subsheaf of $\mathcal{E}$. If E is the vector bundle corresponding to $\mathcal{E}$, then L defines a lattice in each $E(s)$, $s \neq 0$. The natural morphism $H^1(X,\mathbb{Z}) \to H^0(\Delta, f_{*1}\mathbb{Z}_X) = H^0(\Delta, L)$ is bijective (consider the beginning of Leray's spectral sequence), so the classes $p\alpha_1, \ldots, p\alpha_{\tilde g}, p\beta_1, \ldots, p\beta_g$ are holomorphic sections in $L \subset \mathcal{E}$. The additional classes $\alpha_{\tilde g+1}(s), \ldots, \alpha_g(s)$ extend locally on Δ^* to holomorphic sections in $L \subset \mathcal{E}$, but globally on Δ^* we obtain only multivalued sections, because of non-trivial monodromy.

The period matrix $Z(s)$, $s \in \Delta^*$, is the holomorphic multivalued matrix defined by

$$(20) \qquad (p\alpha_1(s), \ldots, p\alpha_g(s)) = -\big(p\beta_1(s), \ldots, p\beta_g(s)\big)Z(s) .$$

It satisfies the Riemann period relations I, (16) in each $s \in \Delta^*$. If we write it in blocks

$$Z(s) = \begin{pmatrix} Z_{11}(s) & Z_{12}(s) \\ Z_{12}^t(s) & Z_{22}(s) \end{pmatrix} \begin{matrix} \} \tilde{g} \\ \} g - \tilde{g} \end{matrix}$$

then Z_{11} is a symmetric $\tilde{g} \times \tilde{g}$ matrix, holomorphic and univalued on all of Δ. Also Z_{12} is holomorphic and univalued on Δ, but Z_{22} changes under monodromy, say $T : Z_{22}(s) \to Z_{22}(s) + \Lambda$.

(15.1) **Lemma.**
i) *The matrix* Λ *is constant with integral entries.*
ii) $\Lambda = \Lambda^t$.
iii) $\det \Lambda \neq 0$.

Proof.
i) Consider for $j = \tilde{g} + 1, \ldots, g$ the "variation" $T\alpha_j - \alpha_j$. Since $T\beta_i = \beta_i$, we have $(T\alpha_j, \beta_j) = (\alpha_j, \beta_j)$ for all i, so $T\alpha_j - \alpha_j$ must be an integral linear combination of $\beta_1, \ldots, \beta_g$ in $H^1(X_s, \mathbb{Z})$.
ii) Z and TZ are both symmetric.
iii) If Λ were not of maximal rank $g - \tilde{g}$, there would be an integral linear combination $\alpha \neq 0$ of $\alpha_{\tilde{g}+1}(s), \ldots, \alpha_g(s)$ invariant under monodromy. By Theorem 14.1, this α would come from $H^1(X, \mathbb{Z})$ and it would thus be an integral linear combination of $\alpha_1, \ldots, \alpha_{\tilde{g}}, \beta_1, \ldots, \beta_g$, a contradiction!
□

Next we consider on Δ^* the single-valued holomorphic matrix

$$Z'_{22}(s) = Z_{22}(s) - (2\pi \mathrm{i})^{-1} \log s \cdot \Lambda .$$

(15.2) **Theorem.** Z'_{22} *extends holomorphically over* 0.

Proof. $\operatorname{Im} Z_{22}(s)$ is positive definite. So for each diagonal entry z_{jj} of this matrix, we have $|\mathbf{e}(z_{jj})| < 1$, and by Riemann's extension theorem $\mathbf{e}(z_{jj})$ extends to a holomorphic function h on all of Δ. If we write $h(s) = s^m g(s)$ with $m \in \mathbb{N}$ and $g(0) \neq 0$, then $z_{jj} - m(2\pi\mathrm{i})^{-1} \log s$ is holomorphic and single-valued on Δ. The integer m must be the diagonal entry λ_{jj} of Λ. So the diagonal entries of Z'_{22} extend over 0.

If $i \neq j$, then applying the form $\operatorname{Im} Z_{22}$ to the difference $e_i - e_j$ of basis vectors yields

$$\operatorname{Im}(z_{ii} + z_{jj}) > 2 \operatorname{Im} z_{ij} .$$

Hence the preceding argument also applies to the function $z_{ii} + z_{jj} - 2z_{ij}$ showing that $z_{ij} - \lambda_{ij}(2\pi\mathrm{i})^{-1} \log s$ extends too. □

(15.3) **Corollary.** $\Lambda > 0$.

Proof. $\operatorname{Im} Z_{22}(s) = \operatorname{Im} Z'_{22}(s) - \frac{1}{2\pi}\log|s| \cdot \Lambda$ is positive definite and $Z'_{22}(s)$ is bounded near $s = 0$. So necessarily $\Lambda \geq 0$, but $\det \Lambda \neq 0$, as was already observed (Lemma 15.1). □

Finally we consider a fibration $f : X \to X$ *without singular fibres*. We choose a base point $0 \in S$ and a canonical basis $\alpha_1, \ldots, \alpha_g, \beta_1, \ldots, \beta_g \in H^1(X_0, \mathbb{Z})$. If S is simply-connected, the direct image sheaf $L = f_{*1} \cdot \mathbb{Z}_X$ is trivial, and the canonical basis extends uniquely to a basis of sections in L all over S. But then the period matrix Z is globally a well-defined holomorphic map $S \to \mathfrak{H}_g$.

(15.4) **Theorem.** *Let $f : X \to S$ be a fibration without singular fibres for which S is compact and either rational or elliptic. Then f is locally trivial.*

Proof. If S is simply connected and compact, the period map $S \to \mathfrak{H}_g$ is constant. It S is elliptic, we still have a well-defined period map $\mathbb{C} \to \mathfrak{H}_g$ on this universal covering. Since $\mathfrak{H}_g$ is a bounded domain, this map is constant. It follows in both cases, that all curves X_s have the same period point in $\mathfrak{H}_g$, so they are biholomorphically equivalent by Torelli's theorem (compare [Mar]). The assertion then follows from Theorem I.10.1. □

Remark. If S is rational and the fibration of genus ≥ 2, then it is even *globally* trivial.

16. Extending the Period Map

By assigning to $s \in \Delta^*$ the class of $Z(s)$ in the period domain $\mathfrak{H}_g/\Gamma_g$, we obtain a holomorphic map $\Delta^* \to \mathfrak{H}_g/\Gamma_g$, the period map.

Satake's compactification $\overline{\mathfrak{H}_g/\Gamma_g}$ of the period domain ([B-B]) is a normal projective variety, which is stratified by subvarieties

$$\overline{\mathfrak{H}_g/\Gamma_g} = \mathfrak{H}_g/\Gamma_g \cup \mathfrak{H}_{g-1}/\Gamma_{g-1} \cdots \mathfrak{H}_1/\Gamma_1 \cup pt\ .$$

To describe the topology of Satake's compactification we need the notion of Siegel sets. Every real symmetric matrix $Y > 0$ has a unique decomposition $Y = W^t D W$ with D diagonal and

$$W = \begin{pmatrix} 1 & w_{12} & \cdots & w_{1g} \\ 0 & \ddots & \ddots & \vdots \\ \vdots & \ddots & 1 & w_{g-1,g} \\ 0 & \cdots & 0 & 1 \end{pmatrix}.$$

It is easy to see that the entries of D and W depend continuously on Y. For real u, let us denote by $\mathfrak{H}_g(u) \subset \mathfrak{H}_g$ the set of all matrices

$$Z = X + \mathrm{i}Y, \quad Y = W^t D W$$

satisfying

$$\begin{array}{lll} |x_{ij}| < u, & |w_{ij}| < u & 1 \le i,j \le g, \\ 1 < ud_{11}, & d_{ii} < ud_{i+1,i+1} & 1 \le i \le g-1. \end{array}$$

For fixed $\tilde{g} < g$ we write

$$Z = \tilde{g}\{\overbrace{\begin{pmatrix} Z_{11} & Z_{12} \\ Z_{12}^t & Z_{22} \end{pmatrix}}^{\tilde{g}}, \quad D = \begin{pmatrix} D_1 & 0 \\ 0 & D_2 \end{pmatrix}, \quad W = \begin{pmatrix} W_1 & W_{12} \\ 0 & W_2 \end{pmatrix}.$$

A sequence converges in Satake's compactification to the point represented by $Z_0 \in \mathfrak{H}_{\tilde{g}}$, if there is a representing sequence of matrices $Z^{(\nu)} \in \mathfrak{H}_g$ such that

a) there is some u with $Z^{(\nu)} \in \mathfrak{H}_g(u)$ for all ν,
b) $Z_{1,1}^{(\nu)} \to Z_0$ in the usual sense,
c) $D_2^{(\nu)} \to \infty$, i.e., all entries go to infinity.

(16.1) **Theorem.** *The period map $\Delta^* \to \mathfrak{H}_g/\Gamma_g$ extends holomorphically to a map $\Delta \to \overline{\mathfrak{H}_g/\Gamma_g}$. If*

$$Z(s) = \begin{pmatrix} Z_{11}(s) & Z_{12}(s) \\ Z_{12}^t(s) & Z_{22}'(s) \end{pmatrix} + (2\pi \mathrm{i})^{-1} \log s \begin{pmatrix} 0 & 0 \\ 0 & \Lambda \end{pmatrix},$$

then the class of $Z_{11}(0)$ in $\mathfrak{H}_{\tilde{g}}/\Gamma_{\tilde{g}} \subset \overline{\mathfrak{H}_g/\Gamma_g}$ is the image of $0 \in \Delta$ under the extended map.

Proof.

Let $(2\pi\mathrm{i})^{-1} \log s = \xi + \mathrm{i}\eta$. We may restrict our attention to one branch of $\log s$, say the one given by $|\xi| \le 1$, because the other branches of $Z(s)$ represent the same class in $\mathfrak{H}_g/\Gamma_g$. In the strip $|\xi| \le 1$ the matrix $X = \operatorname{Re} Z(s)$ is bounded and for $Y = \operatorname{Im} Z$ we have

$$Y = \begin{pmatrix} Y_{11} & Y_{12} \\ Y_{12}^t & Y_{22} \end{pmatrix} + \eta \begin{pmatrix} 0 & 0 \\ 0 & \Lambda \end{pmatrix}$$

with Y_{11}, Y_{12}, Y_{22} extending continuously for $\eta \to \infty$. In fact $Y_{11}(0) > 0$, because $Z_{11}(0)$ is the direct sum of the period matrices of all components of the normalization $\widetilde{X}_0$ (see Sect. 13).

Comparing Y with

$$W^t D W = \begin{pmatrix} W_1^t D_1 W_1 & W_1^t D_1 W_{12} \\ W_{12}^t D_1 W_1 & W_{12}^t D_1 W_{12} + W_2^t D_2 W_2 \end{pmatrix}$$

we see that W_1, D_1 extend continuously for $s \to 0$ as is the case for $W_{12} = D_1^{-1}(W_1^t)^{-1} Y_{12}$. Hence $W_2^t \cdot \frac{1}{\eta} D_2 \cdot W_2$ converges to Λ if $\eta \to \infty$. So W_2 is bounded and $D_2 \to \infty$. Conditions b) and c) are fulfilled and for u sufficiently large condition a) also holds. This shows that the period map extends continuously, hence holomorphically for $s \to 0$. □

17. The Degree of $f_*\omega_{X/S}$

Now let $f : X \to S$ be a stable fibration with S compact and connected. We cover S by open sets $U_i \cong \Delta$ on which there is a canonical basis $\alpha_1^i(s), \ldots, \beta_g^i(s)$, $s \in U_i$, as in Sect. 15. Either U_i contains no critical value, and this basis is univalued, or U_i contains one critical value and $\alpha_{\bar{g}_i}^i(s), \ldots, \alpha_g^i(s)$ are multi-valued. In any case, the sections $p\beta_1^i, \ldots, p\beta_g^i \in \Gamma(U_i, f_{*1}\mathcal{O}_X)$ form a basis, and the (multi-valued, holomorphic) period matrix $Z^i(s)$ on U_i is defined by (19). If we denote by Λ^i the $g \times g$ monodromy-matrix (vanishing if U_i contains no critical value), then monodromy acts as (*cf.* Sect. 15):

$$Z^i \longmapsto Z^i + \Lambda^i$$

$$(\alpha_1^i, \ldots, \beta_g^i) \longmapsto (\alpha_1^i, \ldots, \beta_g^i) \begin{pmatrix} 1 & 0 \\ -\Lambda^i & 1 \end{pmatrix} .$$

Next let $\sigma^{ji} = \begin{pmatrix} A^{ji} & B^{ji} \\ C^{ji} & D^{ji} \end{pmatrix} \in Sp(g, \mathbb{Z})$ be a matrix transforming the canonical bases over $U_i \cap U_j$ by

$$(\alpha_1^i, \ldots, \beta_g^i) = (\alpha_1^i, \ldots, \beta_g^j)\sigma^{ji}.$$

By I, (18) we have on $U_i \cap U_j$:

$$Z^j = (D^{ji}Z^i + C^{ji})(B^{ji}Z^i + A^{ji})^{-1} .$$

Notice that monodromy changes σ^{ij} into

$$\begin{pmatrix} 1 & 0 \\ \Lambda^j & 1 \end{pmatrix} \sigma^{ji} \begin{pmatrix} 1 & 0 \\ -\Lambda^i & 1 \end{pmatrix} = \begin{pmatrix} A^{ji} - B^{ji}\Lambda^i & B^{ji} \\ * & * \end{pmatrix} ,$$

but $B^{ji}Z^i + A^{ji}$ is unchanged. So

$$c^{ji} = \det(B^{ji}Z^i + A^{ji}) \in \Gamma(U_i \cap U_j, \mathcal{O}_S)$$

is a well-defined function.

(17.1) **Proposition.** *The system $\{c^{ji}\}$ is a cocycle defining the line bundle* $\det(f_*\omega_{X/S})$, i.e., *a section h in this bundle is given by functions $h^i \in \Gamma(U_i, \mathcal{O}_S)$ satisfying $h^i = c^{ij}h^j$.*

Proof. Consider the basis $p\beta_1^i, \ldots, p\beta_g^i$ in $\Gamma(U_i, f_{*1}\mathcal{O}_X)$. Its dual basis (under relative duality) in $\Gamma(U_i, f_*\omega_{X/S})$ consists of sections $f_*\omega_1^i, \ldots, f_*\omega_g^i$ with $\omega_1^i, \ldots, \omega_g^i \in \Gamma(f^{-1}(U_i), \omega_{X/S})$. Then $c^i = f_*\omega_1^i \wedge \cdots \wedge f_*\omega_g^i$ generates $\det(f_*\omega_{X/S})$ over U_i. We have

$$\sigma^{ij} = (\sigma^{ji})^{-1} = \begin{pmatrix} D^{ji} & -C^{ji} \\ -B^{ji} & A^{ji} \end{pmatrix}^t ,$$

$$(p\beta_1^j, \ldots, p\beta_g^j) = (p\alpha_1^i, \ldots, p\beta_g^i)\begin{pmatrix} B^{ij} \\ D^{ij} \end{pmatrix} = (p\beta_1^i, \ldots, p\beta_g^i)(B^{ji}Z^i + A^{ji})^t ,$$

$$(f_*\omega_1^j, \ldots, f_*\omega_g^j) = (f_*\omega_1^i, \ldots, f_*\omega_g^i) \cdot (B^{ji}Z^i + A^{ji})^{-1} ,$$

$$c^{ji} = c^i/c^j . \qquad \square$$

(17.2) **Proposition.** *Unless the period map maps S to a point in $\mathfrak{H}_g/\Gamma_g$, the line bundle* $\det(f_*\omega_{X/S})$ *on S has degree > 0.*

Proof. We use modular forms of some weight $m > 0$ to construct a non-trivial section $h \in \Gamma(\det(f_*\omega_{X/S})^{\otimes m})$ with zeros. By definition a modular form of weight m is a holomorphic function φ on $\mathfrak{H}_g$ satisfying for all $Z \in \mathfrak{H}_g$, $\begin{pmatrix} A & B \\ C & D \end{pmatrix} \in Sp(g, \mathbb{Z})$ the relation

$$\varphi\left((AZ + B)(CZ + D)^{-1}\right) = \det(CZ + D)^m \cdot \varphi(Z).$$

(If $g = 1$, it is additionally required that φ stays bounded for $\operatorname{Im} z \to \infty$.) The modular forms form a finitely generated graded ring. Its associated projective variety is just the Satake compactification $\overline{\mathfrak{H}_g/\Gamma_g}$, see [B-B]. So there is a very ample line bundle $\mathcal{O}(1)$ on $\overline{\mathfrak{H}_g/\Gamma_g}$, whose sections come from modular forms of some weight m. If $\varphi_0, \ldots, \varphi_N$ form a basis for the vector space of these forms, then the projective embedding of $\overline{\mathfrak{H}_g/\Gamma_g}$ is on $\mathfrak{H}_g/\Gamma_g$ defined by

$$Z \operatorname{mod} \Gamma_g \mapsto \left(\varphi_0(Z) : \cdots : \varphi_N(Z)\right) \in \mathbb{P}_N(\mathbb{C}) .$$

Assume now that the image of S in $\overline{\mathfrak{H}_g/\Gamma_g}$ has dimension one. Then φ_0 may be chosen in such a way that the hyperplane $\mathbb{P}_{N-1} \subset \mathbb{P}_N$ corresponding to $\varphi_0 = 0$ intersects the image of S in a discrete set, not only consisting of critical values for f. A non-trivial section h in $(\det f_*\omega_{X/S})^{\otimes m}$ is now constructed as follows.

If we put $h^i(s) = \varphi_0(Z^i(s))$ on U_i, then $h^i(s)$ is holomorphic for s not a critical value. Since $\begin{pmatrix} D & C \\ B & A \end{pmatrix}$ belongs to $Sp(g, \mathbb{Z})$ if $\begin{pmatrix} A & B \\ C & D \end{pmatrix}$ does, the relation between Z^j and Z^i and the transformation property of φ_0 show that

$$\begin{aligned} h^j &= \varphi_0\big((D^{ji}Z^i + C^{ji})(B^{ji}Z^i + A^{ji})^{-1}\big) \\ &= \det(B^{ji}Z^i + A^{ji})^m \varphi_0(Z^i) = (c^{ji})^m h^i. \end{aligned}$$

So the h^i patch together and yield a section in $(\det f_*\omega_{X/S})^{\otimes m}$ over $S \setminus \{$critical values$\}$. If $s \in U_i$ converges to some critical value, we have seen in the proof of Theorem 16.1, that we may choose $Z^i(s)$ with all its values in a Siegel set $\mathfrak{H}_g(u)$. It is a rather elementary fact, that on such a set any modular form is bounded (see [Maa], Theorem 1 on p. 185). So h^i extends holomorphically to all of U_i. $\square$

We recall that the degree of a vector bundle on a curve is defined as the degree of its determinant bundle (II, Sect. 2).

(17.3) **Theorem.** *Let $f : X \to S$ be a fibration whose singular fibres are reduced with at worst nodes and do not contain (-1)-curves. Then* $\deg(f_*\omega_{X/S}) > 0$, *unless f is locally trivial (and hence all fibres of f are nonsingular and isomorphic).*

Proof. If the period map is not constant, $\deg(f_*\omega_{X/S})$ is positive by Proposition 17.2 above. So let us assume that the image of the period map $S \to \overline{\mathfrak{H}_g/\Gamma_g}$ is a point. This point, contained in $\mathfrak{H}_g/\Gamma_g$ will be the period point of all the nonsingular fibres. By Torelli's theorem (see [Mar]) all nonsingular fibres will be isomorphic. The assertion then follows from Theorem I.10.1, provided that we prove the absence of singular fibres.

So let $X_s = \sum C_i$, $s \in S$ be a singular fibre. By Theorem 16.1 it has the same period matrix as all the nonsingular fibres. If g_i is the genus of the normalization $\widetilde{C}_i$, then $\tilde{g} = \sum g_i$ equals g, the genus of a nonsingular fibre. So Theorem 11.1 shows $\tilde{g} = h^1(\mathcal{O}_{X_s})$.

The cohomology sequence of II, (5) implies that the components C_i are nonsingular and also that they form a tree. This tree cannot have a rational curve at one of its ends, because such a curve would be a (-1)-curve. So there must be at least two different irrational components C_i, and the period matrix Z of X_s decomposes as direct sum of at least two blocks. The nonsingular fibres have the same period matrix, so their jacobian will be a product of at least two factors and the theta-divisor on this jacobian will be reducible. But for a nonsingular curve, the theta-divisor, an image of the $(g-1)$-fold symmetric product of the curve, is always irreducible. This contradiction proves the theorem. □

18. Iitaka's Conjecture $C_{2,1}$

The aim of this section is to prove Iitaka's conjecture $C_{2,1}$, which will appear as Theorem 18.4 below. But first we need some preliminaries.

(18.1) **Theorem.** *Let F be a nonsingular, compact, connected Riemann surface of genus $g > 0$. If there is a non-trivial finite subgroup $G \subset \mathrm{Aut}(F)$ operating trivially on $H^1(\mathcal{O}_F)$, then F must be elliptic and G must consist of translations.*

Proof. The quotient $Q = F/G$ is nonsingular with $H^1(\mathcal{O}_Q) \subset H^1(\mathcal{O}_F)$ the subgroup of G-invariants (compare [Gk57b]). If G acts trivially on $H^1(\mathcal{O}_F)$, then $g(Q) = g(F)$ and from Hurwitz' formula

$$2g(F) - 2 \geq |G|(2g(Q) - 2)$$

we deduce that not only both F and Q are elliptic, but also that the map $F \to Q$ is unramified. So G consists of translations only, because all other automorphisms of F have fixed points. □

This lemma and the Stable Reduction Theorem 10.3 now enable us to generalise Theorem 17.3 to the case of arbitrary fibrations.

(18.2) **Theorem.** *Let X and S be compact, and $f : X \to S$ a relatively minimal fibration with strictly positive fibre genus. Then*

$$\deg(f_*\omega_{X/S}) \geq 0$$

and this degree vanishes if and only if
- *either f is locally trivial (hence smooth)*
- *or $g = 1$ and the only singular fibres are multiples of nonsingular elliptic curves.*

Proof. Instead of $f_*\omega_{X/S}$ we consider its dual $f_{*1}\mathcal{O}_X$. We take a stable reduction

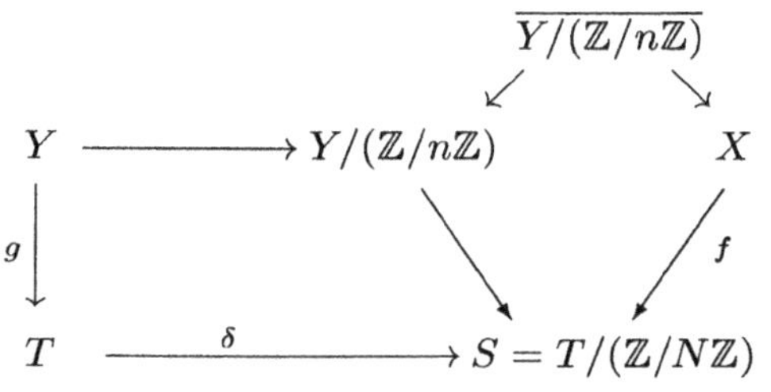

as constructed in Sect. 10. In Proposition 11.3 we observed that there is an injection of $\mathcal{O}_T$-sheaves $\iota : \delta^*(f_{*1}\mathcal{O}_X) \to g_{*1}\mathcal{O}_Y$ which is an isomorphism outside of the ramification points. Both sheaves are locally free of the same rank, so Theorem 17.1 shows that

$$\deg(f_*\omega_{X/S}) = -\deg(f_{*1}\mathcal{O}_X) = -\frac{1}{n}\deg\delta^*(f_{*1}\mathcal{O}_X) > 0$$

unless $\deg(g_{*1}\mathcal{O}_Y) = 0$ and ι is an isomorphism. By Theorem 17.3, g is locally trivial.

Now we consider some ramification point $t \in T$ lying over a critical value $s \in S$ with local ramification group $\mathbb{Z}/\nu\mathbb{Z}$, $\nu | n$. Then $\mathbb{Z}/\nu\mathbb{Z}$ acts non-trivially on Y_t, because otherwise $Y/(\mathbb{Z}/\nu\mathbb{Z})$ were nonsingular over s, hence $X = \overline{Y/(\mathbb{Z}/\nu\mathbb{Z})} = Y/(\mathbb{Z}/\nu\mathbb{Z})$ is locally trivial near X_s. Since ι is an isomorphism in t, all germs in $(g_{*1}\mathcal{O}_Y)_t$ are invariant under $\mathbb{Z}/\nu\mathbb{Z}$. By base-change it follows that $\mathbb{Z}/\nu\mathbb{Z}$ acts trivially on $H^1(\mathcal{O}_{Y_t})$. So by Lemma 18.1 above, Y_t is elliptic and $\mathbb{Z}/\nu\mathbb{Z}$ acts by translations. This implies that X_s is a multiple of a nonsingular elliptic curve. □

(18.3) **Proposition.** *If $\deg(f_*\omega_{X/S}) = 0$ in the situation of Theorem 18.2, then there is a finite unramified covering $\sigma : T \to S$ such that the pull-back $\sigma^*(f_*\omega_{X/S})$ is trivial (hence $\det(f_*\omega_{X/S})$ is a torsion bundle).*

Proof. If $g \geq 2$, then the automorphism group $\mathrm{Aut}(X_s)$ of the fibre is finite. So f is locally trivial with $\mathrm{Aut}(X_s)$ as structure group and the structure group of the bundle $f_*\omega_{X/S}$ can be reduced to this same group.

If $g = 1$, we take a cyclic covering $\delta_1 : T_1 \to S$ such that the pull-back $f' : X_1 = X \times_S T_1 \to T_1$ is locally trivial (stable reduction). The automorphism group of the fibre is an extension

$$0 \to \mathbb{C}/\Gamma \to \mathrm{Aut}(X_s) \to \mathbb{Z}/n\mathbb{Z} \to 0$$

with $\mathbb{C}/\Gamma$ operating trivially on $H^0(\omega_{X_s})$. So the structure group of $f'_*\omega_{X_1/T_1} = \delta_1^*(\omega_{X/S})$ can be reduced to $\mathbb{Z}/n\mathbb{Z}$, and there is an unramified n-fold covering $\delta_2 : T_2 \to T_1$ such that the pull-back of $f_*\omega_{X/S}$ to T_2 is trivial. The line bundle $f_*\omega_{X/S}$ represents an element in the kernel of $\mathrm{Pic}^0(S) \to \mathrm{Pic}^0(T_2)$. This kernel must be a proper subgroup of $\mathrm{Pic}^0(S)$ and by Lemma I.16.2 it consists of torsion elements only. Applying the unbranched covering trick (I.18.1), the result follows. □

(18.4) **Theorem** (Iitaka's conjecture $C_{2,1}$) *Let $f : X \to S$ be a fibration with X minimal and compact. Then the inequality*

$$\mathrm{kod}(X_s) + \mathrm{kod}(S) \leq \mathrm{kod}(X)$$

holds for a general fibre X_s.

Proof. Without loss of generality we can assume $g = g(X_s) > 0$ and $g(S) > 0$. Then Theorem 18.2 and Proposition 18.3 leave us with the following cases:

a) $\deg(f_*\omega_{X/S}) > 0$. After passing to an unramified double covering we may assume $\deg(f_*\omega_{X/S}) \geq 2$. So by Riemann-Roch on S

$$h^0(f_*\mathcal{K}_X) - h^1(f_*\mathcal{K}_X) = \deg\left(f_*\omega_{X/S}\right) + g(g(S) - 1),$$

and we conclude $h^0(\mathcal{K}_X) \geq 2$, and in particular $\mathrm{kod}(X) \geq 1$. There is a pencil $C_\lambda + F \in |K_X|$ of effective canonical divisors, with F the fixed part and C_λ, $\lambda \in \mathbb{P}_1$ variable. By Proposition 2.3 and minimality we have

$$K_X^2 = C_\lambda^2 + FC_\lambda + K_XF \geq 0.$$

Since $K_X^2 > 0$ together with $p_g(X) \geq 2$ would imply $\mathrm{kod}(X) = 2$, we may assume $K_X^2 \leq 0$, i.e., $C_\lambda^2 = FC_\lambda = K_XF = 0$. So $K_XC_\lambda = 0$, and using Proposition 2.3 again we conclude $K_XD = 0$ for each irreducible component $D \subset C_\lambda$. This, the vanishing of C_λ^2, and Stein factorization imply that there exists a connected holomorphic map from X onto a curve T such that the general fibre is elliptic. Then either the general fibre X_s of f is elliptic, or (by Lüroth's theorem) S is elliptic. In both cases the asserted inequality holds.

b) $g > 1$ and f is locally trivial. Since $\mathrm{Aut}(X_s)$ is finite, there is some finite unramified covering $T \to S$ such that the pull-back of X becomes the product $X_s \times T$. Then

$$\begin{aligned} \mathrm{kod}(X) &= \mathrm{kod}(X_s \times T) && \text{(by Theorem I.7.4)} \\ &= \mathrm{kod}(X_s) + \mathrm{kod}(T) && \text{(by Theorem I.7.3)} \\ &= \mathrm{kod}(X_s) + \mathrm{kod}(S) && \text{(by Theorem I.7.4).} \end{aligned}$$

c) $g = 1$ and $f_*\omega_{X/S}$ is a torsion bundle. Consider an unramified covering $\gamma : T \to S$, where the pull-back of $f_*\omega_{X/S}$ becomes trivial. If $g : Y = X \times_S T \to T$, then $g_*\omega_{Y/T} = \gamma^*(f_*\omega_{X/S})$ is trivial, and there will be a non-zero map $g^*\omega_T \to \mathcal{K}_Y$. This implies $\mathrm{kod}(Y) \geq \mathrm{kod}(T)$. But $\mathrm{kod}(Y) = \mathrm{kod}(X)$ and $\mathrm{kod}(T) = \mathrm{kod}(S)$ by Theorem I.7.4. □

Historical Remarks

The methods dealing with rational singularities, described in Sect. 3, are essentially due to M. Artin, see [An62], [An66]. In the treatment of Hirzebruch-Jung singularities (Sect. 5) we follow Hirzebruch in [Hir53]. For the history of the famous question how to desingularize a surface we refer to Lipman's survey [Li].

For stable fibrations we already referred to [D-M]. Relative duality (Sect. 12) is due to Grothendieck. A simple treatment of the analytic case in relative dimension one does not seem to be available (see however [R-R-V]).

The results in Sects. 15 and 6 are due to A. Mayer, Mumford, Griffiths and others. We follow the treatment given in the unpublished seminar [Cl]. Theorems 15.2 and 16.1 are due to A. Mayer, the proof of Theorem 15.2 follows Jambois in [Cl].

Theorem 17.3 is due to Paršin ([Pa]) and Arakelov ([Ar]). For the algebraic case compare also Astérisque 86, Exp. 3. Theorem 18.2 is contained in [Ue77]. Proofs of Iitaka's conjecture $C_{2,1}$ which are independent of the classification of surfaces were first given by Ueno ([Ue77]) and Viehweg ([Vie77]). Our proof is a mixture of these two proofs.

Chapter IV. Some General Properties of Surfaces

In this chapter a surface is always a connected 2-dimensional complex manifold.

In the fundamental Sect. 1 we treat meromorphic maps induced by line bundles. Then, in Sect. 2, we deal with special features for differential forms on compact surfaces. The main point is that for surfaces the Fröhlicher spectral sequence always degenerates. Combining the consequences of this fact with the topological index theorem we find, following Kodaira, relations between topological and analytical invariants which are crucial in handling non-algebraic surfaces. For instance, in Sect. 3 we give a direct proof (due to Lamari) that a compact surface is Kähler if and only if its first Betti number is even.

We also prove (Sect. 2) the important signature theorem (known as algebraic index theorem in the case of algebraic surfaces). Lefschetz theorem on $(1,1)$-classes is also proved in this section and leads to the introduction of the Néron-Severi-group. For algebraic surfaces the ample and nef divisors define cones in the associated real vector space and properties of these, derived in Sect. 7 play an important role in the proof of the classification theorem which we give in Chap. VI.

Bogomolov's theorem is proved in Sect. 10 and is used to derive Reider's theorem which plays a dominant role in our treatment of the pluricanonical maps of surfaces of general type (Chap. VII). It is also used in the final Sect. 12.

From the other subjects treated in this chapter, we mention projectivity criteria (with an application to almost-complex surfaces without any complex structure, see Sect. 6 and 8) and the vanishing theorems of Ramanujam and Mumford (Sect. 12).

In Sect. 4, we introduce the period map and prove a result on deformations which will be used in Sect. VI.8; likewise, Sect. 8 will be applied in Chap. VI.

1. Meromorphic Maps, Associated to Line Bundles

The considerations in this section apply mutatis mutandis to complex manifolds of any dimension, but since we shall use resolution of singularities we shall restrict ourselves to the 2-dimensional case which is the only one we shall need.

Let X be a compact surface, and $\mathcal{L} = \mathcal{O}_X(D)$ a holomorphic line bundle on X with $h^0(\mathcal{L}) \geq 2$. There exists a maximal divisor $V \geq 0$ with $D - V \geq 0$ for all $D \in |D|$. This divisor V is called the fixed part of $|D|$. There is an obvious isomorphism between $|D-V|$ and $|D|$. The line bundle $\mathcal{M} = \mathcal{O}_X(D-V)$ has the property that there exists a finite number of base points $b_1, \ldots, b_k$

on X such that $B = \bigcup_{=1}^{k} b_i$ is exactly the subset of X where all sections of $\Gamma(\mathcal{M})$ vanish.

Let $\gamma_1, \ldots, \gamma_{N+1}$ be a base for $\Gamma(\mathcal{M})$. Each point $x_0 \in X$ has a neighbourhood U with local coordinates $u = (u_1, u_2)$ on U such that, with respect to a trivialization of $\mathcal{M}$ over U, the section γ_i can be given by a holomorphic function $f_i(u)$. There exists an analytic subset $A \subset X \times \mathbb{P}_N$ which on each $U \times \mathbb{P}_N$ is given by the equations

$$f_i(u)z_j - f_j(u)z_i = 0 \ ,$$

where $(z_1 : \cdots : z_{N+1})$ are homogeneous coordinates on $\mathbb{P}_N$ and $1 \leq i, j \leq N+1$.

Over $X \backslash B$ the set A is nothing but a section s in $X \times \mathbb{P}_N$, i.e., a holomorphic map $s : X \backslash B \to \mathbb{P}_N$ which has the property that $s^*\big(\mathcal{O}_{\mathbb{P}_N}(1)\big) = \mathcal{M}|X \backslash B$, as is easily checked. Let Y be the irreducible component of A which contains S. Then Y is a 2-dimensional reduced complex space which can be seen as a "meromorphic map" from X into $\mathbb{P}_N$. This is by definition the **meromorphic map associated to** $\mathcal{M}$ and the base $\gamma_1, \ldots, \gamma_{N+1}$. Replacing this base by another one, can be interpreted as a change of homogeneous coordinates in $\mathbb{P}_N$. Since for many purposes such a change is irrelevant, we shall often speak of *the* meromorphic map $f_{\mathcal{M}}$ associated to $\mathcal{M}$. The meromorphic map $f_{\mathcal{L}}$ associated to the original line bundle $\mathcal{L}$ is by definition $f_{\mathcal{M}}$ again. The image $f_{\mathcal{L}}(X)$ is defined as the projection of Y in $\mathbb{P}_N$ which is the same as $\overline{s(X \setminus B)}$.

Remarks. 1) If X is a projective algebraic surface and $\mathcal{L}$ an algebraic line bundle on X (by GAGA every vector bundle on an algebraic variety is analytically equivalent to an algebraic one), then Y is algebraic and $f_{\mathcal{L}}$ is the **rational map, associated to** $\mathcal{L}$.

2) The preceding considerations apply equally well to the case that one does not take all of $\Gamma(X, \mathcal{L})$, but only a subspace of dimension at least 2.

Let $\overline{Y}$ be the minimal desingularization of Y (compare Theorem III.6.2 and let $\pi : \overline{Y} \to X$, $\rho : \overline{Y} \to \mathbb{P}_N$ be the natural projections. There are two possibilities:

(i) $\dim \rho(\overline{Y}) = 2$. In this case Bertini's Theorem I.20.2 yields that for a general section $\beta \in \Gamma\big(\mathbb{P}_N, \mathcal{O}_{\mathbb{P}_N}(1)\big)$ the induced section $\rho^*(\beta)$ vanishes on a smooth irreducible curve. We claim that the images by π of the zero divisors of such sections $\rho^*(\beta)$ are exactly the divisors in $|D|$. Indeed, if E is such an image, then certainly $\mathcal{O}_{X \backslash B}(E) = \mathcal{O}_{X \backslash B}(D)$, but if a line bundle is trivial outside B, it is trivial on X (the section 1 can be extended by Riemann's extension theorem to a section on X which nowhere vanishes). Conversely, every element of $|D|$ appears in this way for dimension reasons. We conclude that in this case the general member of $|D|$ is irreducible and smooth outside of the base points.

(ii) $\dim \rho(\overline{Y}) = 1$, i.e., $\rho(\overline{Y})$ is an irreducible curve Z. We can apply Stein factorization to ρ, i.e., $\rho = \sigma\tau$ where $\tau : \overline{Y} \to W$ is a connected map onto the smooth irreducible curve W and where $\sigma : W \to Z$ is finite. There is an integer $\ell \in \mathbb{N}$ such that $\rho^*\big(\mathcal{O}_{\mathbb{P}_N}(1)\big) = \mathcal{O}_{\overline{Y}}(F_1 + \cdots + F_\ell)$, where $F_1, \ldots, F_\ell$ are general fibres of τ. On the other hand, by construction $\rho^*\big(\mathcal{O}_{\mathbb{P}_N}(1)\big) = \mathcal{O}_{\overline{Y}}\big(\pi^*(D)\big)$ (mod exceptional divisors over B). The images on X of the fibres F form a 1-dimensional irreducible system of effective divisors P, all connected and in general irreducible. If W is rational, then this system is easily seen to be a pencil in the usual sense (1-dimensional linear system), otherwise it is called an irrational pencil (of genus equal to genus (W)). Multiplicity taken into account, the fibres D are sums of ℓ curves P. The situation is described by saying that D is composed with a (rational or irrational) pencil. In the case of an irrational pencil all exceptional curves over B must be contained in fibres of ρ by Lüroth's theorem, i.e., $B = \varnothing$ in this case.

In particular, if we take $\mathcal{L} = \mathcal{K}_X^{\otimes m}$, then the meromorphic map $f_\mathcal{L}$ is called the m-th pluricanonical map of X and denoted by f_m (it need not exist of course). If $f_m(X)$ is a surface, it is called the m-th canonical model of X. If $P_m(X) \geq 2$ for at least one m, it is known that

$$\mathrm{kod}(X) = \max_{m\geq 1} \dim f_{\mathcal{K}^{\otimes m}}(X)$$

(see [Ue75], p. 86). This remains true if X is replaced by a connected compact complex manifold of any dimension.

2. Hodge Theory on Surfaces

Let X be a compact Kähler manifold. If we consider $H^k(X, \mathbb{C})$ as a de Rham space, then the subspace of those elements which can be represented by a d-closed form of type (p, q) is naturally isomorphic to the Dolbeault group $H^{p,q}(X)$ and thus there is a decomposition $H^k(X, \mathbb{C}) \cong \bigoplus_{p+q=k} H^{p,q}(X)$ (see Corollary I.13.4). If X is not kählerian, in general such a decomposition does not exist. However, if X is a non-kählerian *surface*, then there always exists such a decomposition for $k = 2$, whereas for $k = 1$ the existence of such a decomposition is assured as soon as $b_1(X)$ is even. All this will be explained in the present section. The main point is the fact that, like for Kähler manifolds, the Fröhlicher spectral sequence degenerates at E_1-level for any compact surface.

(2.1) **Lemma.** *Every holomorphic differential form on a compact surface X is closed.*

Proof. The assertion for 0- and 2-forms being trivial, we consider a holomorphic 1-form ω. By Stokes' theorem we have

$$\int_X d\omega \wedge d\bar{\omega} = \int_X d(\omega \wedge d\bar{\omega}) = 0 .$$

Upon writing locally $d\omega = f dz_1 \wedge dz_2$ and

$$d\omega \wedge d\bar{\omega} = -|f|^2 \, dz_1 \wedge d\bar{z}_1 \wedge dz_2 \wedge d\bar{z}_2 = |f|^2 \, dx_1 \wedge dy_1 \wedge dx_2 \wedge dy_2$$

(with $z_j = x_j + \mathrm{i}y_j$, $j = 1, 2$) we see that the vanishing of the integral implies $d\omega = 0$. □

Remark. The same proof works for holomorphic $(n-1)$-forms on an n-dimensional compact complex manifold. But if the manifold is non-kählerian, then holomorphic forms of lower dimension need not be closed (in fact there exist 3-manifolds with non-closed holomorphic 1-forms, compare [Ue75], §17).

We need a well known auxiliary result:

(2.2) **Lemma** *Let X be a compact connected complex manifold and f a smooth complex-valued function with $\bar{\partial}\partial f = 0$. Then f is constant.*

Proof. The set where f assumes its maximum is closed. It is also open, since locally in any coordinate patch $\bar{\partial}\partial f = 0$ means that f is harmonic in each variable separately. So f is harmonic with respect to the standard euclidean metric and hence obeys a maximum principle. □

Remark A function with the property of the previous lemma is called pluriharmonic since its restriction to each embedded 1-disk is harmonic.

(2.3) **Lemma.**
(i) *If for $\omega \in H^{1,0}$ there is some $f \in C^\infty(X)$ with $\omega = \partial f$, then $\omega = 0$;*
(ii) *If for $\sigma \in H^{2,0}$ there is some $\varphi \in \Gamma(\mathcal{D}^{1,0})$ with $\sigma = \partial\varphi$, then $\sigma = 0$.*

Proof.
(i) Since $\bar{\partial}\partial f = \bar{\partial}\omega = 0$, the function f is constant by the previous Lemma.
(ii) Since $\bar{\partial}\varphi \wedge \bar{\sigma} = 0$, we have $\sigma \wedge \bar{\sigma} = d\varphi \wedge \bar{\sigma} = d(\varphi \wedge \bar{\sigma})$, so by Stokes $\int_X \sigma \wedge \bar{\sigma} = 0$, and arguing as in the proof of Lemma 2.1 we find $\sigma = 0$. □

Using Lemma 2.1 we have first of all homomorphisms

$$H^{1,0}(X) \longrightarrow H^{0,1}(X) : \omega \to \bar{\omega}$$
$$H^{2,0}(X) \longrightarrow H^{0,2}(X) : \sigma \to \bar{\sigma}$$

which are injective by Lemma 2.3 (so the second map is an isomorphism by Serre duality). Next we define $H^{1,0}(X) \to H^1(X, \mathbb{C})$ and $H^{2,0}(X) \to H^2(X, \mathbb{C})$ by sending a holomorphic form to its de Rham class. Again by Lemma 2.3 these homomorphisms are injective, and we shall identify $H^{1,0}(X)$ and $H^{2,0}(X)$ with their images.

On $H^k(X, \mathbb{C}) = H^k(X, \mathbb{R}) \otimes \mathbb{C}$ ($k = 1, 2$) one has complex conjugation. As usual, we denote by $\overline{V}$ the conjugate of any subspace $V \subset H^k(X, \mathbb{C})$ with respect to the complex conjugation on $H^k(X, \mathbb{C}) = H^k(X, \mathbb{R}) \otimes \mathbb{C}$.

(2.4) **Proposition.** $H^{k,0}(X) \cap \overline{H^{k,0}(X)} = 0$ *($k = 1, 2$).*

Proof. A class in $H^{1,0}(X) \cap \overline{H^{1,0}(X)}$ can be written as $[\omega_1] = [\bar{\omega}_2]$, with ω_1, $\omega_2 \in \Gamma(\Omega^1_X)$. So $\omega_1 - \bar{\omega}_2 = df$ for some $f \in C^\infty(X)$, hence $\omega_1 = \partial f = 0$ by Lemma 2.3. As to the case $k = 2$, any class in $H^{2,0}(X) \cap \overline{H^{2,0}(X)}$ can be written as $[\sigma_1] = [\bar{\sigma}_2]$, with $\sigma_1, \sigma_2 \in \Gamma(\Omega^2_X)$. So $\int_X \sigma_1 \wedge \bar{\sigma}_2 = \int_X \bar{\sigma}_2 \wedge \bar{\sigma}_1 = 0$, and exactly as in the proof of Lemma 2.1 we find $\sigma_1 = 0$. □

(2.5) *Remark.* Let V be a real vector space and W a complex subspace of $V \otimes_{\mathbb{R}} \mathbb{C}$, which as a subspace is invariant under conjugation. Then W is the complexification of $W \cap V$.

Proof. The space W is the direct sum of the subspace of elements, invariant under conjugation and the subspace of elements, which are anti-invariant under conjugation. Since the first of these subspaces is just $W \cap V$ and since the second one can be obtained from the first one by multiplication with i, the statement follows. □

(2.6) **Lemma.** *For every compact surface X the following inequalities hold:*
(i) $2h^{1,0} \leq h^{0,1} + h^{1,0} \leq 2h^{0,1}$
(ii) $2p_g \leq b^+$.

Proof. The inequality $2h^{1,0} \leq b_1$ is an immediate consequence of Proposition 2.4.

If for a moment we let $\mathcal{S}$ be the sheaf of closed holomorphic 1-forms on X, then there is an exact sequence

$$0 \longrightarrow \mathbb{C}_X \longrightarrow \mathcal{O}_X \xrightarrow{d} \mathcal{S} \longrightarrow 0.$$

Using Lemma 2.1 we obtain from this an exact sequence

$$(1) \qquad 0 \longrightarrow H^0(\Omega^1_X) \longrightarrow H^1(X, \mathbb{C}) \longrightarrow H^1(\mathcal{O}_X),$$

which proves $h^{1,0} + h^{0,1} \geq b_1$, hence $b_1 \leq 2h^{0,1}$ because of Lemma 2.1.

To prove (ii) we consider the linear subspace of $H^2(X, \mathbb{R})$, consisting of the classes $[\sigma + \bar{\sigma}]$ with $\sigma \in H^{2,0}$, i.e., by Lemma 2.4 $(H^{2,0} \oplus \overline{H^{2,0}}) \cap H^2(X, \mathbb{R})$. On this space the intersection form is

$$(\sigma_1 + \bar{\sigma}_1, \sigma_2 + \bar{\sigma}_2) = \int_X (\sigma_1 + \bar{\sigma}_1) \wedge (\sigma_2 + \bar{\sigma}_2) = \int_X \sigma_1 \wedge \bar{\sigma}_2 + \bar{\sigma}_1 \wedge \sigma_2,$$

so this restriction is positive definite and (ii) follows by Remark 2.5 ($H^{2,0} \oplus \overline{H^{2,0}}$ is the complexification of its intersection with $H^2(X, \mathbb{R})$, so this intersection is of dimension $2h^{2,0} = 2p_g$ by Lemma 2.4). □

(2.7) **Theorem.** *Let X be a compact surface. Then*
(i) $b_1(X) = h^{1,0} + h^{0,1}$;
(ii) *if $b_1(X)$ is even, then $h^{1,0} = h^{0,1}$ and $b^+(X) = 2p_g(X) + 1$;*
(iii) *if $b_1(X)$ is odd, then $h^{1,0} = h^{0,1} - 1$ and $b^+(X) = 2p_g(X)$;*

(iv) *$q(X)$ and $p_g(X)$ are topological invariants, $q(X)$ of the non-oriented, and p_g of the oriented underlying manifold.*

Proof. Eliminating $c_1^2(X)$ from the index theorem of Thom-Hirzebruch (Theorem I.3.1)

$$\tau(X) = \frac{1}{3}\left(c_1^2(X) - 2c_2(X)\right)$$

and the Riemann-Roch formula

$$1 - q + p_g = \frac{1}{12}\left(c_1^2(X) + c_2(X)\right)$$

we obtain

$$4 - 4q + 4p_g - b^+ + b^- = c_2(X) = e(X) = 2 - 2b_1 + b^+ + b^-,$$

which means

$$(b^+ - 2p_g) + (2q - b_1) = 1.$$

Since by Lemma 2.6 both terms within brackets are non-negative, we are left with only two possibilities, namely (ii) and (iii) of the theorem. Both imply (i). □

Remark. Noether's formula, together with Theorem 2.7, (iv) shows that $c_1^2(X)$ is a topological invariant of the underlying oriented manifold, but this is of course already clear from the index theorem.

(2.8) **Theorem.** *For any compact surface the Fröhlicher spectral sequence degenerates at E_1-level.*

Proof. If we consider $H^p(\Omega_X^q)$ as a Dolbeault group, then at E_1-level the derivatives are given by the ∂-operator:

$$\begin{aligned} H^0(\mathcal{O}_X) &\xrightarrow{\partial} H^0(\Omega_X^1) \xrightarrow{\partial} H^0(\Omega_X^2),\\ H^1(\mathcal{O}_X) &\xrightarrow{\partial} H^1(\Omega_X^1) \xrightarrow{\partial} H^1(\Omega_X^2),\\ H^2(\mathcal{O}_X) &\xrightarrow{\partial} H^2(\Omega_X^1) \xrightarrow{\partial} H^2(\Omega_X^2). \end{aligned}$$

The first map in the first row clearly vanishes, and the second map in the first row vanishes because of Lemma 2.1. Furthermore, Theorem 2.7 and sequence (1) imply that $H^1(\mathbb{C}) \to H^1(\mathcal{O}_X)$ is surjective, so $H^1(\mathcal{O}_X) \to H^1(\Omega_X^1)$ vanishes. The compatibility of Serre duality with exterior products (Proposition I.12.2) then implies the vanishing of all other derivatives at E_1-level. Similarly it can be shown that also at E_2-level all derivatives vanish. □

Once we have this result we can try to mimic the procedure of Sect. I.13. So we start from the Hodge filtrations defined on $H^1(X, \mathbb{C})$ and $H^2(X, \mathbb{C})$ by the Fröhlicher spectral sequence:

$$F^1(H^1) = \{[\omega] \mid \omega \in \Gamma(\mathcal{D}_X^{1,0}), d\omega = 0\}$$
$$F^1(H^2) = \{[\sigma] \mid \sigma \in \Gamma(\mathcal{D}_X^{2,0} \oplus \mathcal{D}_X^{1,1}), d\sigma = 0\}$$
$$F^2(H^2) = \{[\sigma] \mid \sigma \in \Gamma(\mathcal{D}_X^{2,0}), d\sigma = 0\} ,$$

and show

(2.9) **Proposition.** *There are formal Hodge decompositions*
(i) $H^1 = F^1(H^1) \oplus \overline{F^1(H^1)}$ *(if b_1 is even);*
(ii) $H^2 = F^2(H^2) \oplus \big(F^1(H^2) \cap \overline{F^1(H^2)}\big) \oplus \overline{F^2(H^2)}$ *(always).*

Proof. The case (i) being very similar, we restrict ourselves to the case (ii). The degeneracy of the Fröhlicher spectral sequence implies that

$$\dim F^1(H^2) = \dim \overline{F^1(H^2)} = h^{2,0} + h^{1,1}$$

and also that $\dim H^2 = h^{2,0} + h^{1,1} + h^{0,2}$. Hence $\dim(F^1(H^2) \cap \overline{F^1(H^2)}) \geq h^{1,1}$. So (ii) follows as soon as we have proved that the three summands have intersection 0, in other words that $F^2(H^2) \cap \overline{F^1(H^2)} = 0$. But a class in $F^2(H^2) \cap \overline{F^1(H^2)}$ can be represented by some $\alpha^{2,0} \in \Gamma(\Omega_X^2)$ and at the same time by a form $\sigma^{1,1} + \sigma^{0,2} \in \Gamma(\mathcal{D}_X^{1,1}) \oplus \Gamma(\mathcal{D}_X^{0,2})$. Then $\sigma^{1,1} + \sigma^{0,2} = \alpha^{2,0} + d\omega$ for some 1-form ω. Writing $\omega = \omega^{1,0} + \omega^{0,1}$ one finds $\alpha^{2,0} = \partial\omega^{1,0} = 0$ by Lemma 2.3. □

These formal decompositions are in fact the usual Hodge decompositions. This is clear for the first one. As to (ii), by definition $F^2(H^2)$ consists of those de Rham classes which can be represented by a form of type $(2,0)$. Consequently, $\overline{F^2(H^2)}$ consists of classes of $(0,2)$-forms. We claim that $F^1(H^2) \cap \overline{F^1(H^2)}$ consists of the de Rham classes which are representable by a (closed) form of type $(1,1)$. Indeed, by definition $[\alpha] \in F^1(H^2) \cap \overline{F^1(H^2)}$ if and only if there are closed forms α_1 and α_2, with $[\alpha_1] = [\alpha_2] = [\alpha]$, such that $\alpha_1 = \alpha_1^{1,1} + \alpha_1^{2,0}$ and $\alpha_2 = \alpha_2^{1,1} + \alpha_2^{0,2}$ with $\alpha_k^{i,j} \in \Gamma(\mathcal{D}_X^{i,j})$. Now $\alpha_1 - \alpha_2 = d\beta$, for some 1-form β. If $\beta = \beta^{1,0} + \beta^{0,1}$ is the type decomposition of β, we have that $\alpha_1 - d\beta^{1,0} = \alpha_2 + d\beta^{0,1}$ is a d-closed $(1,1)$-form of class $[\alpha]$. Since on the other hand every element of $H^2(X,\mathbb{C})$ which can be represented by a closed $(1,1)$-form lies in $F^1(H^2) \cap \overline{F^1(H^2)}$, we have proved our claim.

From the degeneration of the Fröhlicher spectral we obtain natural isomorphisms

$$H^{2,0} = E_\infty^{2,0} \cong F^2(H^2)$$
$$H^{1,1} = E_\infty^{1,1} \cong F^1(H^2)/F^2(H^2) \cong F^1(H^2) \cap \overline{F^1(H^2)}$$
$$H^{0,2} = E_\infty^{0,2} \cong H^2(X,\mathbb{C})/F^1(H^2) \cong \overline{F^2(H^2)}.$$

So we finally obtain

(2.10) **Theorem.** *Let X be a compact surface. Then, always for $p+q=2$ and if $b_1(X)$ is even also for $p+q=1$, the Dolbeault group $H^{p,q}(X)$ is naturally isomorphic to the subspace of $H^{p+q}(X,\mathbb{C})$ whose elements can be* *represented by a d-closed form of type* (p,q). *In this way one obtains natural decompositions* $H^k(X,\mathbb{C}) = \bigoplus_{p+q=k} H^{p,q}(X)$, $k = 1,2$.

In particular, every class in $H^2(X,\mathbb{C})$ has a unique decomposition into types.

hree applications of the preceding considerations; the first one
Albanese torus, the second one is about Lefschetz's theorem on
ınd the third one is about the index theorem.
seen that $H^1(X,\mathbb{C})$ is generated by holomorphic 1-forms and
ıes as soon as $b_1(X)$ is even. This gives, exactly like in the
(I, Sect. 14), a Hodge structure of weight 1 and

ary. *If X is a compact surface with $b_1(X)$ even, then* Alb(X)
limension $q(X) = h^{1,0}(X) = \frac{1}{2}b_1(X)$.

hall prove Lefschetz's theorem on $(1,1)$-classes for any compact
entify the sheaf cohomology group $H^2(X,\mathbb{C})$ with the de Rham
d 2-forms in the canonical way, and we also identify $H^2(X,\mathcal{O}_X)$
ace of $(0,2)$-classes in $H^2(X,\mathbb{C})$ in the same way as described
ve have

sition. *In 2-cohomology the injection* $\mathbb{C}_X \to \mathcal{O}_X$ *induces the*
o the $(0,2)$*-component.*

ʒ for $k = 0,1,2$ a complex k-form to its $(0,k)$-component defines
f complexes

$$j^\bullet : \Gamma(\mathcal{D}^\bullet_X) \to \Gamma(\mathcal{D}^{0,\bullet}_X)$$

the injection of sheaves $j : \mathbb{C}_X \to \mathcal{O}_X$. In view of Theorem 2.10
ıap $j^* : H^2(X,\mathbb{C}) \to H^2(X,\mathcal{O}_X)$ is of course nothing but the
□

Lefschetz theorem on $(1,1)$-classes answers the following ques-
$(X,\mathcal{O}^*_X) \to H^2(X,\mathbb{Z})$ is the boundary homomorphism in the
ohomology sequence, and if $i^* : H^2(X,\mathbb{Z}) \to H^2(X,\mathbb{C})$ is in-
embedding $i : \mathbb{Z}_X \to \mathbb{C}_X$, then what is $i^* \circ \delta(H^1(X,\mathcal{O}^*_X))$,
ıe image of Pic(X) in $H^2(X,\mathbb{C})$?

em (Lefschetz theorem on $(1,1)$-classes). *Let X be a compact*
the image of Pic(X) *in* $H^2(X,\mathbb{C})$ *is* $H^{1,1}(X) \cap i^*(H^2(X,\mathbb{Z}))$.
: an element of $H^2(X,\mathbb{C})$ *is in the image of* Pic(X) *if and only*
al" and can be represented by a real closed $(1,1)$*-form.*

$_X \to \mathcal{O}_X$ is the natural injection, then we have a commutative

We giv
concerns th
(1,1)-classe
We ha
their conju
kählerian c

(2.11) **Cor**
is a torus

Next w
surface. We
group of cl
with the su
above. The

(2.12) **Pro**
projection

Proof. Senc
a morphisr

which exte
the induce
projection.

Now t
tion: if δ :
exponentia
duced by
i.e., what i

(2.13) **The**
surface. T
In other w
if it is "int

Proof. If k
diagram

$$\begin{array}{ccccc} \mathrm{Pic}(X) = H^1(X, \mathcal{O}_X^*) & \xrightarrow{\delta} & H^2(X, \mathbb{Z}) & \xrightarrow{k^*} & H^2(X, \mathcal{O}_X) \\ & & {\scriptstyle i^*}\searrow & & \nearrow{\scriptstyle j^*} \\ & & & H^2(X, \mathbb{C}) & \end{array}$$

It follows that $\mathrm{Im}(i^* \circ \delta) = (\ker j^*) \cap (\mathrm{Im}\, i^*)$, hence the image of $\mathrm{Pic}(X)$ in $H^2(X, \mathbb{C})$ coincides with

$$(\ker j^*) \cap (\mathrm{Im}\, i^*) = (\ker j^*) \cap (\overline{\ker j^*}) \cap \mathrm{Im}(i^*) = H^{1,1}(X) \cap \mathrm{Im}\, i^*.$$

As to the second formulation, we have to prove that every element of $H^{1,1}(X) \cap H^2(X, \mathbb{R})$ can be represented by a *real* d-closed $(1,1)$-form. But such an element can be represented by a d-closed $(1,1)$-form $\rho = \sigma + d\tau$, with $[\sigma] \in H^2(X, \mathbb{R})$. So it is also represented by the real d-closed $(1,1)$-form $\frac{1}{2}(\rho + \bar{\rho}) = \sigma + d(\tau + \bar{\tau})$. □

Remark. The group $H^{1,1}(X) \cap i^*(H^2(X, \mathbb{Z}))$ is the Néron-Severi group of X (I, Sect. 6); its rank is the Picard number $\rho(X)$.

Actually, $H^{1,1}(X)$ is the complexification of $H^{1,1}_{\mathbb{R}}(X) = H^{1,1}(X) \cap H^1(X, \mathbb{R})$. This follows from Remark 2.5.

We come to our last application the important

(2.14) **Theorem** (Signature theorem). *Let X be a compact surface. Then the cup product-form on $H^2(X, \mathbb{R})$, restricted to $H^{1,1}_{\mathbb{R}}(X)$, is non-degenerate of type $(1, h^{1,1} - 1)$ if $b_1(X)$ is even and of type $(0, h^{1,1})$ if $b_1(X)$ is odd.*

Proof. The complex subspace $H^{2,0}(X) \oplus H^{0,2}(X) \subset H^2(X, \mathbb{C})$ is the complexification of $(H^{2,0}(X) \oplus H^{0,2}(X)) \cap H^2(X, \mathbb{R})$ by Remark 2.5. On this $2p_g$-dimensional subspace of $H^2(X, \mathbb{R})$ the cup-product-form is positive definite. Using Theorem 2.10 we see that its orthogonal complement in $H^2(X, \mathbb{R})$ is nothing but $H^{1,1}_{\mathbb{R}}(X)$. So the result follows from Theorem 2.7, (ii) and (iii). □

(2.15) **Corollary.** *Let X be a compact surface with $b_1(X)$ even, and let h be an element of $H^{1,1}_{\mathbb{R}}(X)$ with $h^2 > 0$. Then the cup product-form is negative definite on the orthogonal complement in $H^{1,1}_{\mathbb{R}}(X)$ of the line, determined by h.*

The signature theorem is mostly used in the following form:

(2.16) **Corollary** (Algebraic index theorem or Hodge index theorem). *Let D, E be divisors with rational coefficients on the algebraic surface X. If $D^2 > 0$ and $DE = 0$, then $E^2 \leq 0$ and $E^2 = 0$ if and only if E is homologous to 0 (in rational homology).*

This follows from Corollary 2.15, Theorem 2.13 and the fact that an algebraic surface, being kählerian, has always an even first Betti number.

Warning: The reader should not confuse the algebraic index theorem above with the topological index theorem of Thom-Hirzebruch (Theorem I.3.1). Unfortunately, both names are well-established.

3. Existence of Kähler Metrics

The aim of this section is to present a direct proof of the following statement.

(3.1) **Theorem.** *A compact complex surface is Kähler if and only if its first Betti number is even.*

This theorem will be proved in several steps, but first we need some preliminaries. We start by recalling that a (complex or real) current of degree k on a complex surface X is a (complex or real) k-form with distributional coefficients. We can likewise speak of currents of type (p, q). So in a coordinate chart with coordinates (z_1, z_2), a real current of type $(2, 2)$ can be written $T dz_1 \wedge dz_2 \wedge d\bar{z}_1 \wedge d\bar{z}_2$ with T a real distribution. If in addition $T(f) \geq 0$ for all non-negative test-functions f, we speak of a non-negative measure. If $T(f) > 0$ for non-zero test functions $f \geq 0$, we say that T is a positive measure. Any volume form gives an example of a positive measure.

We shall mainly work with *real* $(1, 1)$-forms and distributions:

$$\begin{aligned} D^{1,1}(X, \mathbb{R}) &= \{\text{real (global) forms of type } (1,1) \text{ on } X\} \\ (D')^{1,1}(X, \mathbb{R}) &= \{\text{real (global) distributions of type } (1,1) \text{ on } X\} \end{aligned}$$

Recall (Chapter I, Sect. 13) that a real $(1, 1)$-form α is called positive, denoted $\alpha > 0$ if in local coordinates (z_1, z_2) valid on $U \subset X$ we have

$$(2)\quad \alpha = \mathrm{i} \sum_{i,j=1}^{2} \alpha_{ij} dz_i \wedge d\bar{z}_j, \text{ the matrix } (\alpha_{ij}(p)) \text{ is positive definite, } \forall p \in U.$$

Equivalently, α is the $(1, 1)$-form associated to a hermitian metric. This notion is clearly punctual. Replacing strict inequalities by inequalities in the larger sense, we arrive at the notion of a non-negative form $\alpha \geq 0$. These form the cone

$$D^{1,1}_{\geq 0} = \{\alpha \in D^{1,1} \mid \alpha \geq 0.\}$$

Definition.

1 A real current T of type $(1, 1)$ on a complex surface X is non-negative denoted $T \geq 0$, if $T(\alpha) \geq 0$ for all decomposable test forms $\alpha = \mathrm{i}\beta \wedge \bar{\beta}$ of type $(1, 1)$ where β is any $(1, 0)$-form. These currents form the cone

$$(D')^{1,1}_{\geq 0} = \{T \in (D')^{1,1} \mid T \geq 0.\}$$

2 If in addition T is closed, X is compact and for some hermitian metric on X with associated $(1,1)$-form ω_0 there exists a constant $C > 0$ with $T - C \cdot \omega_0 \geq 0$, we say that T is a Kähler current. This notion is independent of the choice of the hermitian metric since X is compact.

Considering a non-negative real $(1,1)$-form ω as a current, $\omega \geq 0$ precisely when ω is non-negative in the previous sense. Indeed, more generally, a current $T = \mathrm{i} \sum_{i,j=1}^{2} T_{ij} dz_i \wedge d\bar{z}_j$ of type $(1,1)$ is non-negative if and only if for every vector $(\lambda_1, \lambda_2) \in \mathbb{C}^2$ the distribution $\tilde{T} := \sum_{i,j=1}^{2} \lambda_i \bar{\lambda}_j T_{ij}$ yields the non-negative measure $\tilde{T} dz_1 \wedge dz_2 \wedge d\bar{z}_1 \wedge d\bar{z}_2$. Also ω is a Kähler current precisely when it is a Kähler form.

In the sequel, we need the *real* operator

$$d^c = \frac{\mathrm{i}}{2}(\bar{\partial} - \partial).$$

So, instead of $\bar{\partial}\partial$ we shall use the real operator dd^c.

(3.2) **Examples.**

1. A $(1,1)$-form α is positive if $T(\alpha) > 0$ for all non-zero $T \geq 0$. Indeed, writing locally $\alpha = \sum_{i,j} \alpha_{ij} dz_i \wedge d\bar{z}_j$, we may assume that the hermitian matrix (α_{ij}) is diagonal at p and testing against $\delta(p) dz_i \wedge d\bar{z}_i$, $p \in X$ with $\delta(p)$ the Dirac delta function at p, we see that $\alpha(p) > 0$.
2. A distribution u for which $dd^c u \geq 0$ is called plurisubharmonic. Plurisubharmonic *functions* are easily seen to obey a maximum principle. A plurisubharmonic distribution can be shown to be equal (in the sense of distributions) to a locally integrable plurisubharmonic function. For a proof of this fact, well known since the appearance of [Lel68] see [Dema97], Theorem 5.8. As a consequence, on a compact complex manifold X, plurisubharmonic distributions are constant.
3. A real $(1,1)$-current T is non-negative if and only if $T(\alpha) \geq 0$ for all real $(1,1)$-forms $\alpha \geq 0$. The reason is that first of all, as observed before, positivity of forms is a punctual notion, and secondly every positive definite hermitian matrix such as $(\alpha_{ij}(p))$, $p \in X$ as in (2) can be diagonalized. So testing on decomposable non-negative forms is equivalent to testing on all non-negative forms.
4. The current of integration of a curve, or more generally of an effective divisor $D = \sum n_i C_i$ on a surface is a non-negative current denoted by $[D]$ and defined by the formula

$$([D], \alpha) = \sum n_i \int_{C_i'} \alpha, \qquad \alpha \in D^{1,1}(X),$$

where the integration is over the smooth part C_i' of the curve C_i. That this is indeed a converging integral was shown by Lelong in [Lel]. He also shows that the current $[D]$ is closed and non-negative. Of course this remains true for any analytic subset D of a manifold X.

We can now proceed to the **Proof of Theorem 3.1.**
Step 1: reduction to the construction of a Kähler current. For this step the hypothesis "b_1 even" is not needed. Still, this is the hardest step since it uses deep results due to Demailly ([Dema]). In order to restrict technicalities to a minimum, we present the reduction for surfaces only, but the proofs are valid for any dimension. The main result is as follows.

(3.3) **Proposition.** *Let X be a compact complex surface admitting a Kähler current. Then some (repeated) blow-up of X admits a Kähler metric.*

We give the proof of this proposition as communicated to us by Demailly. We need the regularity theorem from [Dema] in the following guise:

(3.4) **Theorem.** *Let T be a Kähler $(1,1)$-current on a compact complex manifold X. There is a distribution f on X such that $T = T_0 + dd^c f$ with T_0 a Kähler current with at most logarithmic singularities. This means that every point in X has an open neighbourhood U such that*

$$T_0|U = \frac{1}{\pi} dd^c \varphi_U, \quad \varphi_U = \frac{\lambda}{2}\log(\sum_{j=1}^{\infty} |g_j|^2) + \psi_U$$

for suitable holomorphic functions $g_j \in \mathcal{O}_X(U)$, a positive constant λ and a C^∞ function ψ_U.

Moreover, there exists a sheaf of ideals $\mathcal{J}$ on X and a finite open cover $\{U_\alpha\}$, $\alpha \in J$ of X such that (i) *the above representation of T_0 is valid over each U_α and* (ii) *the ideal generated by the g_j in $\mathcal{O}_X(U_\alpha)$ is $\mathcal{J}(U_\alpha)$.*

In fact, the main steps of the proof of this result are in loc. cit. p. 376–380. Further details can be found in [Dema-P], §3.

Suppose that in the above theorem in each open set U_α we need only one holomorphic function $g = g_1$ which then is the divisor of zeros of a globally defined divisor D. By the Poincaré-Lelong formula (see [G-H78a], p. 388) the current $\frac{1}{\pi} dd^c \log(g)$ equals the current $[D]$ defined by this divisor and so, globally $\omega = T_0 - \lambda[D]$ is a smooth closed $(1,1)$-form with $T_0 = \lambda[D] + \omega$. We have $\omega \geq C\omega_0$ outside the support of D. By continuity this inequality remains true also on D. So ω is a positive form, i.e., a Kähler form.

Note that in the preceding theorem in general there may be infinitely many g_j. But the ideal $\mathcal{J}(U)$ generated by the $g_j \in \mathcal{O}_X(U)$ is globally defined. For surfaces X we can now apply the theorem of the resolution of ideals (Chap. II, 7.3) to replace the ideal $\mathcal{J}$ by the ideal of a normal crossing divisor with support in the common zero set of the pull back of the g_j. Precisely, we can blow successively up X, say

$$\mu : \widetilde{X} = X_N \xrightarrow{\mu_{N-1}} X_{N-1} \to \cdots \to X_{j+1} \xrightarrow{\mu_j} X_j \to \cdots \to X_2 \xrightarrow{\mu_1} X_1 = X,$$

where μ_j is the blow up of X_j in some point p_j, in such a way that $\mu^*\mathcal{J}$ becomes an invertible ideal sheaf $\mathcal{O}_{\widetilde{X}}(-D)$, where $D = \sum n_j D_j$ is a normal crossing divisor on $\widetilde{X}$. We can locally write $g_j \circ \mu = g \cdot h_j$ where the h_j have no common zeroes and where g is a defining equation for D. By construction φ_U then lifts to

$$\varphi_U \circ \mu = \frac{\lambda}{2} \log(|g|)^2 + \widetilde{\psi},$$

where $\widetilde{\psi}$ is a smooth form. By the Poincaré-Lelong formula we discussed before, the globally defined current $\widetilde{T} = \mu^* T_0 - \lambda[D]$ is a smooth $(1,1)$-form on $\widetilde{X}$.

Although $\widetilde{T} \geq 0$, unfortunately it ceases to be positive along the exceptional divisor, but this can be remedied by successively applying the following lemma.

(3.5) **Lemma** *Let X be a compact complex surface and let S be a Kähler current on X. If $\sigma : Y \to X$ is the blow-up of X at p, there is a smooth $(1,1)$-form u on Y in the same cohomology class as the exceptional curve E and $\varepsilon > 0$ such that $\sigma^* S - \varepsilon u$ is a Kähler current.*

Proof. The proof is similar to the proof of [Ko54], Lemma 1 and so we give a sketch only. The line bundle $\mathcal{O}_Y(-E)|E$ is ample and so there is a hermitian metric on $\mathcal{O}_Y(-E)$ whose curvature form $-u$ is positive in the neighbourhood of E, while the pull back of S is positive on the complement of E. The result then follows by compactness. □

Let us apply this Lemma to $S = (\mu_1 \circ \cdots \circ \mu_{j-1})^* T_0$ on $X = X_j$, $p = p_j$. We find $\varepsilon_{j+1} > 0$ and a smooth $(1,1)$-form u_{j+1} in the cohomology class of $[E_{j+1}]$ such that $T_{j+1} = \mu_j^* T_j - \varepsilon_{j+1} u_{j+1}$ is a Kähler current on X_{j+1}. If $\widetilde{u}_j$ is the pull-back of u_j to the final blow-up $\widetilde{X}$, we conclude inductively that indeed $\mu^* T_0 - \sum \varepsilon_j \widetilde{u}_j$ is a Kähler current. We write the Kähler current $\mu^* T_0 - \sum \varepsilon_j \widetilde{u}_j$ as a sum $\lambda[D] + (\mu^* T_0 - \lambda[D] - \sum \varepsilon_j \widetilde{u}_j)$ where the second term $\omega = \mu^* T_0 - \sum \varepsilon_j \widetilde{u}_j - \lambda[D]$ is smooth. The same argument as in the case where $\mathcal{J}$ was already principal shows that $\omega > 0$, i.e., ω is a Kähler form on $\widetilde{X}$.

We now complete the first reduction step, using the remark that a surface bimeromorphic to a Kähler surface is Kähler because of a well known extension theorem (see for instance [Mi74a]):

(3.6) **Lemma.** *Let ω be a Kähler form on the punctured unit ball B in $\mathbb{C}^n$. There exists a Kähler form on the entire ball coinciding with ω outside an arbitrary small ball centred at the puncture.*

(3.7) **Corollary.** *A compact complex surface admitting a Kähler current is Kähler.*

Step 2: a criterion for Kähler currents. This is the most elementary step, using only basic properties from functional analysis. As for step 1, there exist generalizations to arbitrary dimensions.

Let us introduce the following real vector spaces

$$Z_d^{1,1}(X,\mathbb{R}) = \{d\text{-closed real forms of type } (1,1) \text{ on } X\}$$
$$Z_{dd^c}^{1,1}(X,\mathbb{R}) = \{dd^c\text{-closed real forms of type } (1,1) \text{ on } X\}.$$

A similar notation with accented letters will be used to denote the corresponding spaces of currents.

In I. Sect.11 we have introduced the (complex) Bott-Chern spaces $H_{\mathrm{BC}}^{p,q}(X)$. The relevant space $H_{\mathrm{BC}}^{1,1}(X)$ is the complexification of

$$(3)\quad H_{\mathrm{BC}}^{1,1}(X,\mathbb{R}) = Z_d^{1,1}(X,\mathbb{R})/dd^c D^0(X,\mathbb{R}) \xrightarrow{\sim} Z'^{1,1}_d(X,\mathbb{R})/dd^c D'^0(X,\mathbb{R}).$$

The isomorphism is induced by the inclusion j of $(1,1)$-forms in the space of $(1,1)$-currents. That it is indeed an isomorphism follows upon inspecting the commutative diagram of sheaves

$$\begin{array}{ccccccccc} 0 & \to & \mathcal{H} & \to & \mathcal{D}^0_{\mathbb{R}} & \xrightarrow{dd^c} & \mathcal{D}^{1,1}_{\mathbb{R}} & \xrightarrow{d} & \mathcal{D}^3_{\mathbb{R}} \\ & & \downarrow{\scriptstyle \mathrm{id}} & & \downarrow{\scriptstyle j} & & \downarrow{\scriptstyle j} & & \downarrow{\scriptstyle j} \\ 0 & \to & \mathcal{H} & \to & (\mathcal{D}')^0_{\mathbb{R}} & \xrightarrow{dd^c} & (\mathcal{D}')^{1,1}_{\mathbb{R}} & \xrightarrow{d} & (\mathcal{D}')^3_{\mathbb{R}}, \end{array}$$

where $\mathcal{H}$ is the sheaf of real pluriharmonic functions on X (see the remark after Lemma 2.2 for the notion of pluriharmonic). Indeed, from this diagram we see that both spaces compute $H^1(X,\mathcal{H})$. In particular they are finite-dimensional.

To arrive at a useful criterion for the existence of Kähler currents, we shall more generally consider the convex cone $(D')^{1,1}_{\geq 0}(X)$ of non-negative $(1,1)$-currents. This cone is closed in the weak topology, i.e. the topology we introduced on page 40. To get a weakly compact set, we use some hermitian metric with $(1,1)$-form ω_0 on X and form the convex set

$$D^1_+(X) = \{T \in (D')^{1,1} \mid T \geq 0,\ T(\omega_0) = 1\}.$$

This set cannot intersect $dd^c(D')^0(X)$, since a distribution u with $dd^c u \geq 0$ is plurisubharmonic and hence constant by the maximum principle (3.2, 2)). By the Hahn-Banach theorem, there is some continuous functional, given by a smooth "test"-form α on which the distributions in $D^1_+(X)$ are positive and on which all distributions from the subspace $dd^c D^0(X)$ vanish. By Example 3.2 1)) we have $\alpha > 0$ and so α is a form associated to a hermitian metric. On the other hand, $dd^c\alpha = 0$ since by assumption $u(dd^c\alpha) = 0$ for all distributions u. Summarizing, we have shown:

(3.8) **Proposition.** *On any compact complex surface there exist hermitian metrics whose associated form is dd^c-closed.*

Remark Metrics having this property are called Gauduchon. A general existence and uniqueness proof can be found in [Ga].

We need some topological properties of the operator dd^c. The spaces $E = D^{1,1}(X, \mathbb{R})$ and $F = D^4(X, \mathbb{R})$ are Fréchet spaces (see Chapt. I, Sect. 11) and $E' = (D')^{1,1}(X, R)$ and $F' = (D')^0(X, \mathbb{R})$ are their weak duals. The map dd^c defines a continuous linear operator $v : E \to F$ with adjoint $v' : F' \to E'$ which is also given by dd^c. By the open mapping theorem for Fréchet spaces ([Schae], Ch. IV, §8) the fact that the range of v has finite codimension implies that this range is closed. Next, applying [Schae], Ch. IV, 7.7, one sees that also v' has closed range. For all $\phi \in F'$ and $e \in E$ we have $v'(\phi)e = \phi(v(e))$ and so the annihilator of the image of v' in E is $\ker(v)$. Since $\mathrm{Im}(v')$ is closed in E', the discussion in loc. cit. after Ch. IV, 1.5 Coroll. 4 implies that we may take annihilators once more, i.e., the annihilator of $\ker(v)$ in E' is $\mathrm{Im}(v')$. Hence:

(3.9) **Lemma.** *Any real* $(1,1)$*-current* T *which vanishes on the subspace of real* dd^c*-closed forms of type* $(1,1)$ *is of the form* $T = dd^c u$. *If* T *is smooth, also* u *is smooth.*

Proof. We only need to prove the last clause. It follows from the isomorphism (3). Indeed, if T is smooth such that $T = dd^c u$ for some distribution u, the class T defines is zero on the right hand side of (3) and hence also on the left. So there is a smooth function u with $T = dd^c u$. Since $dd^c u = 0 = \bar{\partial}\partial u$ the function u is constant by Lemma 2.2 and so if T is smooth, all solutions of $T = dd^c u$ are smooth. □

As a first application we have:

(3.10) **Lemma.** *Let* (X, ω_0) *be a compact hermitian surface. Suppose that we have a* $(1,1)$*-form* α *with*

$$T(\alpha) \geq 0 \quad \forall T \in (D')^{1,1}_{\geq 0} \cap (Z')^{1,1}_{dd^c}. \tag{4}$$

Then we can find a smooth function f_ϵ *on* X, *depending on the choice of* $\epsilon > 0$ *such that* $\alpha + dd^c f_\epsilon \geq -\epsilon\omega_0$.

If (4) is only true for T *varying over the subset of non-negative* dd^c*-closed* forms, *we can find a* distribution u *such that* $\alpha + dd^c u \geq 0$ *as currents.*

Proof. The form α in both cases is non-negative on the cone C of positive dd^c-closed forms of type $(1,1)$. If it is zero on this cone, it vanishes on the entire subspace $Z_{dd^c}(X, \mathbb{R})$ in which the cone lies and so, by Lemma 3.9 $\alpha = -dd^c f$ for some smooth function f. So this case is settled.

If however $T_0(\alpha) > 0$ for some $T_0 \in C$ the argument is more involved. The previously introduced weakly compact set $D^1_+(X) \subset (D')^{1,1}(X, \mathbb{R})$ meets the dd^c-closed currents in

$$K' = D^1_+(X) \cap (Z')^{1,1}_{dd^c}(X, \mathbb{R})$$

which is non-empty because of the previous Proposition. After rescaling we may suppose that $T_0 \in K'$ and consider

$$D^1_\epsilon = D^1_+(X) + \epsilon T_0 \subset (D')^{1,1}(X,\mathbb{R})$$
$$K'_\epsilon = D^1_\epsilon \cap (Z')^{1,1}_{dd^c}(X,\mathbb{R}) = K' + \epsilon T_0 \subset (Z')^{1,1}_{dd^c}(X,\mathbb{R}).$$

The form α is positive on D^1_ϵ and so the closed linear subspace $E = \{T \in (D')^{1,1}(X) \mid T(\alpha) = 0\} \cap (Z')^{1,1}_{dd^c}(X,\mathbb{R})$ does not meet the weakly compact set K'_ϵ and, by Hahn-Banach, there is a smooth $(1,1)$-form β annihilated by E but with $T(\beta) > 0$ for $T \in K'_\epsilon$. So writing $T_0(\alpha) = cT_0(\beta)$, the number c must be positive.

We have $T_0(\alpha - c\beta) = 0$ and, by construction, all currents in the hyperplane E vanish on $\alpha - c\beta$. It follows that $\alpha - c\beta$ is annihilated by $(Z')^{1,1}_{dd^c}(X,\mathbb{R})$ and so, again by Lemma 3.9, $\alpha - c\beta = -dd^c g_\epsilon$ for some smooth function g_ϵ. If $S \in D^1_\epsilon$ then $S(\beta) \geq 0$ and so $S(\alpha + dd^c g_\epsilon) = cS(\beta) \geq 0$. If now $T \neq 0$ is any non-negative current of type $(1,1)$, we scale it into $D^1_+(X,\mathbb{R})$ by dividing by $T(\omega_0)$ so that $S := T/T(\omega_0) + \epsilon T_0 \in D^1_\epsilon$. The inequality $S(\alpha + dd^c g_\epsilon) \geq 0$, using the fact that T_0 is dd^c-closed, translates into:

$$T(\alpha + dd^c g_\epsilon) \geq -\epsilon T_0(\alpha) T(\omega_0).$$

Replacing ϵ by $\epsilon' = \epsilon / T_0(\alpha)$ and setting $f_\epsilon = g_{\epsilon'}$, Example 3.2 (ii) then implies the required inequality $\alpha + dd^c f_\epsilon \geq -\epsilon\omega_0$.

For the second assertion, suppose first that $\alpha > 0$ on the entire interior of the cone C. This means that for u in the interior of the cone, $\alpha(u) = \int_X \alpha \wedge u > 0$. So it does not meet

$$F := \{\beta \in D^{1,1}(X,\mathbb{R}) \mid \int_X \beta \wedge \alpha = 0\} \cap Z^{1,1}_{dd^c}(X,\mathbb{R}),$$

a closed subspace of $Z^{1,1}_{dd^c}(X,\mathbb{R})$. By Hahn-Banach, there exists a $(1,1)$-current τ such that $\tau(\beta) = 0$ for all $\beta \in F$, while $\tau(c) > 0$ for all u in the interior of C. Since also $\alpha(c) > 0$, we must have $\alpha(c) = \lambda\tau(c)$ with $\lambda > 0$. Then the current $\alpha - \lambda\tau$ vanishes on the hyperplane F of $Z^{1,1}_{dd^c}(X,\mathbb{R})$ as well as on a point $c \in F$ outside this hyperplane. So it is identically zero on the space $Z^{1,1}_{dd^c}(X,\mathbb{R})$ and by Lemma 3.9 we have $\alpha - \lambda\tau = -dd^c u$. The current $\alpha + dd^c u$ is non-negative by construction.

If $\alpha(c) = 0$ for some interior point $c \in C$, we claim that $\alpha = 0$ on $Z^{1,1}_{dd^c}(X,\mathbb{R})$ so that we are in the situation at the start of the proof and so we are also ready. To substantiate this claim, consider for all non-negative $c' \in Z^{1,1}_{dd^c}(X,\mathbb{R})$ the affine linear function $\beta : t \mapsto \alpha((1-t)c + tc') = t\alpha(c')$. We have $\beta(0) = 0$ and since for $|t|$ small $(1-t)c + tc'$ is in the interior of C for such t we have $\beta(t) \geq 0$. But then β must be identically zero and so $\alpha = 0$ on the entire cone C. Thus β must vanish on $Z^{1,1}_{dd^c}(X,\mathbb{R})$ as claimed. □

Definition. A real current $T \in (D')^{1,1}(X,\mathbb{R})$ is called weakly positive if it is the (weak) limit

$$T = \lim_{k\to\infty} \alpha_k$$

of non-negative real forms α_k of type $(1,1)$. If in addition the forms α_k are pluriharmonic, i.e., $dd^c\alpha_k = 0$, we call T weakly positive pluriharmonic.

Let us now dualise the complexes we used previously to compute the Bott-Chern spaces. Recall that the latter is the quotient of the space of d-closed real forms of type $(1,1)$ by the subspace which is the image of dd^c. The adjoint of the d-operator $d : D^{1,1}(X,\mathbb{R}) \to D^3(X,\mathbb{R})$ is the operator $d' : (D')^1(X,\mathbb{R}) \to (D')^{1,1}(X,\mathbb{R})$ which is the d-operator on currents followed by projection onto currents of type $(1,1)$. The adjoint of the map $dd^c : D^0(X,\mathbb{R}) \to D^{1,1}(X,\mathbb{R})$ is the map $dd^c : (D')^{1,1}(X,\mathbb{R}) \to (D')^2(X,\mathbb{R})$ on the level of currents. Hence, the following variant of the Bott-Chern space

$$\tilde{H}^{1,1}_{\mathrm{BC}}(X,\mathbb{R}) := (Z')^{1,1}_{dd^c}(X,\mathbb{R})/d'(D')^1(X,\mathbb{R}).$$

is the dual of the Bott-Chern space $H^{1,1}_{\mathrm{BC}}(X,\mathbb{R})$; in particular it is finite dimensional. As before, we deduce that d' has closed range (see also [H-L] p. 174]) and that $Z^{1,1}_d(X,\mathbb{R}) \subset D^{1,1}(X,\mathbb{R})$ is the annihilator of the image of d' (see Lemma 3.9).

We then have

(3.11) **Lemma.** *Suppose that the cone of weakly positive pluriharmonic* $(1,1)$ *currents meets the subspace of* d'*-exact currents only in zero. Then there exists a* d*-closed* $(1,1)$*-form* β *such that* $\beta-\omega_0$ *is non-negative on the set of positive* dd^c*-closed forms of type* $(1,1)$.

Proof. By assumption, the compact set $K := D^1_+(X,\mathbb{R}) \cap (Z')^{1,1}_{dd^c}(X,\mathbb{R})$ is disjoint from the closed set of d'-boundaries and so by Hahn-Banach there is a real $(1,1)$-form β', zero on the d'-boundaries but positive on K. By the above remark ($Z^{1,1}_d(X,\mathbb{R})$ is the annihilator of $\mathrm{Im}(d')$) we know that β' is d-closed.

Let $C > 0$ be the maximal value of β' on the compact set K. Then $\beta := \beta'/C$ satisfies the positivity requirement on the set of dd^c-closed forms τ of type $(1,1)$. Indeed $\tau/\tau(\omega_0) \in K$ and so $\tau(\beta) \geq \tau(\omega_0)$. □

Now we are able to formulate:

(3.12) **Proposition.** (Criterion for Kähler currents). *Suppose that* X *is a compact complex surface with the property that the set of weakly positive pluriharmonic* $(1,1)$ *currents meets the subspace of* d'*-exact currents only in zero. Then* X *admits a Kähler current.*

Proof. The conclusion of the previous lemma can serve as the input of Lemma 3.10 with $\alpha := \beta - \omega_0$ and $\epsilon = \frac{1}{n}$. So there exist closed currents S_n type $(1,1)$, $n = 1,2,\dots$ in the same class as α such that $S_n - \omega_0 \geq \frac{1}{n}\omega_0$ as currents. A subsequence converges then to a closed current S of type $(1,1)$ with $S - \omega_0 \geq 0$, i.e., S is a Kähler current. □

Step 3: applying the criterion to construct a Kähler current. Here we are finally using the hypothesis that $b_1(X)$ is even.

According to the criterion, we need to show that a weakly positive pluriharmonic $(1,1)$-current T which is also of the form $d'u := \overline{\partial}u + \partial\bar{u}$ must be the zero current. In fact it suffices to show that T is closed:

(3.13) **Lemma.** *Let $b_1(X)$ be even. Suppose that T is a positive d-closed real current of type $(1,1)$ which is also a d'-boundary. Then $T = 0$.*

Proof. The Fröhlicher spectral sequence degenerates and if b_1 is even, there are formal Hodge decompositions for the cohomology (Theorem 2.10). We have seen (Lemma I. 13.6) that this implies that the $\partial\overline{\partial}$-Lemma holds so that the first Bott-Chern space injects onto the subspace $H^{1,1}_{\mathbb{R}}(X)$ of de Rham classes representable by real $(1,1)$-forms. By duality, sending a d-closed real form of type $(1,1)$ to its corresponding class in $\tilde{H}^{1,1}_{\mathrm{BC}}(X,\mathbb{R})$ identifies the latter also with $H^{1,1}_{\mathbb{R}}(X)$. So we have natural isomorphisms

$$H^{1,1}_{\mathrm{BC}}(X,\mathbb{R}) \xrightarrow{\sim} H^{1,1}_{\mathbb{R}}(X) \xrightarrow{\sim} \tilde{H}^{1,1}_{\mathrm{BC}}(X,\mathbb{R}).$$

Since we may compute the first of these with currents as well, T defines a class in the first space. Viewing this class then as a class in the last space, it is zero by definition. So it must be zero as a class in the first space as well. This means that $T = dd^c u$ for some distribution u. Since $T \geq 0$, this distribution is plurisubharmonic. By the maximum principle (see 3.2, 2)) u must be the constant function. So $T = 0$. □

This final step of the proof requires a substantial strengthening of the $\partial\overline{\partial}$-Lemma. First we note that the space $\tilde{H}^{1,1}_{\mathrm{BC}}(X,\mathbb{R})$ can be computed using *forms* and we then consider the homomorphism

$$p : H^2(X,\mathbb{R}) \to \tilde{H}^{1,1}_{\mathrm{BC}}(X,\mathbb{R}) = Z^{1,1}_{\mathrm{BC}}(X,\mathbb{R})/d'D^1(X,\mathbb{R})$$

dual to the inclusion $H^{1,1}_{\mathrm{BC}}(X,\mathbb{R}) \hookrightarrow H^2(X,\mathbb{R})$. It is induced by the linear map

$$p^{1,1} : d\text{-closed real 2-forms} \to Z^{1,1}_{dd^c}(X,\mathbb{R})$$

which sends a form to its $(1,1)$-component (one may easily verify that this component indeed is dd^c-closed).

(3.14) **Proposition.** *For a surface with b_1 even, there is a unique $\mathbb{R}$-linear map*

$$Q : Z^{1,1}_{dd^c}(X,\mathbb{R}) \to \{d\text{-closed real 2-forms on } X\}$$

such that

i) *the $(1,1)$-part of $Q(\omega)$ is ω,*
ii) *the $(2,0)$-part of $Q(\omega)$ is ∂-exact.*

Proof. Let ω be a real dd^c-closed $(1,1)$-form. Consider the $(2,1)$-form $\partial\omega$. It is $\overline{\partial}$-closed by assumption and thus defines a class in the Dolbeault group $H^{2,1}(X)$. When b_1 is even, complex conjugation $H^{0,1}(X) \to H^{1,0}(X)$ is an $\mathbb{R}$-linear isomorphism (this follows from Proposition 2.9, (i)). Serre duality implies that this holds likewise for the complex conjugation $H^{2,1}(X) \to$

$H^{1,2}(X)$. But the conjugate of $\partial\omega$ is equal to $\bar\partial\bar\omega$ and hence represents the zero class. So this is also true for $\partial\omega$, which means that $\partial\omega = -\bar\partial\beta$ for some $(2,0)$-form β. For type reasons, this form is ∂-closed and so the conjugate form $\bar\beta$ is $\bar\partial$-closed. Hence it defines a Dolbeault-class $\bar b$ in $H^{0,2}(X)$. Via complex conjugation we get a Dolbeault-class $b \in H^{2,0}(X)$ represented by a holomorphic 2-form β'. The form $\overline{\beta'}$ represents $\bar b$ since complex conjugation $H^{2,0}(X) \to H^{0,2}(X)$ is an isomorphism. So $\bar\beta = \overline{\beta'} + \bar\partial\bar\gamma$ with γ a form of type $(1,0)$. Now introduce the 2-form θ whose components are given by

$$\begin{aligned}\theta^{2,0} &= \beta - \beta' \\ \theta^{1,1} &= \omega \\ \theta^{0,2} &= \overline{\theta^{2,0}}.\end{aligned}$$

This is a real 2-form and by construction $d\theta^{1,1} = d\omega = -(\bar\partial\beta + \partial\bar\beta) = -d\theta^{2,0} - d\theta^{0,2}$, i.e., θ is closed. Moreover, $\theta^{2,0} = \partial\gamma$. Now θ satisfies all requirements to define the map Q. We only need to verify that these properties together with the fact that θ be closed determine γ and hence θ uniquely. To show this, assume that $\partial\gamma = \partial\gamma' = \theta^{2,0}$. Then since $\bar\partial\theta^{2,0} = \partial\theta^{1,1}$, we have $\bar\partial\partial(\gamma-\gamma') = 0$. Hence $\partial(\gamma-\gamma')$ is a ∂-exact holomorphic $(2,0)$-form and so it must vanish by Lemma 2.3.

So the map $\omega \mapsto \theta$ is well-defined and we set $Q(\omega) = \theta$. That this map Q is linear follows from the construction. □

(3.15) **Corollary** *The map Q descends to an $\mathbb{R}$-linear map*

$$q : \tilde H^{1,1}_{\mathrm{BC}}(X,\mathbb{R}) \to H^2(X,\mathbb{R})$$

with $p\circ q = \mathrm{id}$. *In particular q is injective.*

Proof. By the uniqueness property of the map Q we have

$$Q(d'\sigma) = d(\sigma + \bar\sigma), \quad \sigma \in \Gamma(X, \mathcal{D}^{1,0}_X).$$

and so Q descends to cohomology. Moreover, $p^{1,1}\circ Q(\omega) = \omega$ by construction so that $p\circ q = \mathrm{id}$. □

Using the existence of such a map Q, we can now complete the proof of the existence of a Kähler metric. For the rest of the proof we use the usual hermitian inner product on the vector space of complex 2-forms:

$$(\alpha,\beta) = \int_X \alpha \wedge \overline{\beta}.$$

For closed real 2-forms this is just the ordinary cup-product on De Rham cohomology.

We arrive at the final step:

(3.16) **Lemma.** *Let T be a weakly positive pluriharmonic $(1,1)$-current which is of the form $T = d'u = \partial\bar{u} + \bar{\partial}u$. Then $T = 0$.*

Proof. Suppose $T = \lim_{k\to\infty} \tau_k$ (in the weak sense, cf. page 40) with τ_k a strictly positive dd^c-closed $(1,1)$-form. Consider the images $[\tau_k] \in \tilde{H}^{1,1}_{\mathrm{BC}}(X,\mathbb{R})$. Since $T = d'u$ this sequence has limit zero. But then also $q[\tau_k] \in H^2(X,\mathbb{R})$ has limit zero. Now $Q\tau_k$ is a d-closed form and so $(Q\tau_k, Q\tau_k)$ is well defined on cohomology and then also tends to zero. Writing $Q\tau_k = \beta_k + \tau_k + \overline{\beta_k}$, by positivity of τ_k, this implies that $(Q\tau_k, Q\tau_k) \geq 2(\beta_k, \beta_k) \geq 0$ and so $\lim(\beta_k,\beta_k) = 0$. By the Schwarz inequality, for any $(0,2)$-form α we have the estimate

$$(\beta_k, \alpha)^2 \leq (\beta_k, \beta_k) \cdot (\alpha, \alpha)$$

and so β_k weakly converges to 0 in $\Gamma(X, \mathcal{D}'^2_X)$. It follows that $T = \lim \tau_k = \lim Q(\tau_k)$ and hence is the weak limit of d-closed forms and so itself d-closed. By Lemma 3.13 we then conclude that $T = 0$. □

4. Deformations of Surfaces

First we formulate a result which is particularly useful for studying deformations of non-minimal surfaces.

(4.1) **Proposition.** *Let $f : X \to S$ be a complex analytic family of compact surfaces.*

(i) *If for some point $0 \in S$ the fibre X_0 contains a (-1)-curve E_0, then there exists an open neighbourhood U of 0 in S and a closed and connected submanifold E of $f^{-1}(U)$ such that $E \cap E_0 = E_0$ and such that $E \cap X_t$ is a (-1)-curve for every $t \in U$.*

(ii) *If there exists a closed submanifold E of X such that $E_s = E \cap X_s$ is a (-1)-curve for all $s \in U$, then there exists a family $g : X' \to S$ and a commutative triangle*

$$\begin{array}{ccc} X & \xrightarrow{h} & X' \\ & {\scriptstyle f}\searrow \quad \swarrow{\scriptstyle g} & \\ & S & \end{array}$$

such that $h|X_s \to X'_s$ is the blowing down of E_s

A proof of the first property – called stability of (-1)-curves – can be found in [Ko63], Theorem 5, whereas the second property (simultaneous blowing down of (-1)-curves in a family) is proved in [Ii70], Appendix I.

Let $f : X \to S$ be a complex-analytic family of compact surfaces, and let $b = b_2(X_s)$, $p_g = p_g(X_s)$ (constant by Proposition 2.7, (iv)). Furthermore, let $\mathcal{H} = f_{*2}\mathbb{C}_X \otimes_{\mathbb{C}_S} \mathcal{O}_S$.

By Theorem I.8.5 the $\mathcal{O}_S$-module $f_{*2}\mathcal{O}_X$ is a locally free sheaf on S of rank p_g with fibres $H^{0,2}(X_s)$.

The canonical injection $\mathbb{C}_X \to \mathcal{O}_X$ gives a mapping of $\mathcal{O}_S$-modules $p : \mathcal{H} \to f_{*2}\mathcal{O}_X$ which is surjective, since it is surjective on each fibre (Proposition 2.12). Each space $H^{0,2}(X_s)$ is Serre-dual to $H^{2,0}(X_s)$. The vector spaces $H^{2,0}(X_s)$ form the fibres of $f_*\Omega^2_{X/S}$, and the compatibility of Serre duality and Poincaré duality implies that p is dual to an injective map of $\mathcal{O}_S$-modules $f_*\Omega^2_{X/S} \to \mathcal{H}$, which on the fibres is the inclusion $H^{2,0}(X_s) \to H^2(X_s, \mathbb{C})$. This proves

(4.2) **Theorem.** *With the above notations*

$$\mathcal{H}^{2,0} = \bigcup_{s \in S} H^{2,0}(X_s)$$

is a holomorphic subbundle of $\mathcal{H}$ of rank $p_g = p_g(X_s) = h^{2,0}(X_s)$, $s \in S$.

Now assume that a euclidean lattice L (see I, Sect. 2) of rank b is fixed with the property that L is isometric to any $H^2(X_s, \mathbb{Z})/\operatorname{Tors} H^2(X_s, \mathbb{Z})$.

We associate to L a period-domain in the following way. We extend the bilinear form (,) on L to $L_{\mathbb{C}} = L \otimes_{\mathbb{Z}} \mathbb{C}$ in a $\mathbb{C}$-bilinear fashion and we set

$$\begin{aligned} D = D(L) = \{P \in \operatorname{Gr}(p_g, L_{\mathbb{C}}) \mid\ & P \text{ is totally isotropic} \\ & \text{and } (p, \bar{p}) > 0 \text{ for all } p \in P,\ p \neq 0\}; \\ G_{\mathbb{C}} = \text{group of } & \text{isometries of } L_{\mathbb{C}}. \end{aligned}$$

It is easy to show that $G_{\mathbb{C}}$ acts transitively on the set of isotropic p_g-dimensional complex subspaces of $L_{\mathbb{C}}$, so in particular D is manifold.

A trivialization

$$\varphi : f_{*2}\mathbb{Z}_X \xrightarrow{\sim} L_S \quad \text{(mod torsion)}$$

with the property that it is an isometry on each fibre is called a marking. It induces the period map

$$\tau : S \to \operatorname{Gr}(p_g, L_{\mathbb{C}})$$

by sending $s \in S$ to the image of $H^{2,0}(X_s)$ under the $\mathbb{C}$-linear extension of φ.

(4.3) **Theorem.** *The period map is holomorphic and has its image in the period domain D.*

Proof. The first statement is a consequence of Theorem 4.2. As to the second assertion, for any two classes $[\omega_1]$, $[\omega_2]$ in $H^{2,0}(X_s)$ we have

$$([\omega_1], [\omega_2]) = \int_{X_s} \omega_1 \wedge \omega_2 = 0$$

and similarly for $[\omega] \in H^{2,0}(X_s)$

$$([\omega], [\overline{\omega}]) = \int_{X_s} \omega \wedge \overline{\omega} > 0 \ \text{ if } \ \omega \neq 0. \qquad \square$$

We close this section with a proposition which will be used in Chap. VI, Sect. 8.

(4.4) **Proposition.** *If V is a surface with $\mathcal{K}_V^{\otimes m} = \mathcal{O}_V$ for some $m \geq 1$, $\mathcal{K}_V^{\otimes n} \neq \mathcal{O}_V$ $(0 < n < m)$, then the same holds for every deformation of V.*

Proof. We first prove the proposition for $m = 1$. We have to show two things:

(i) $\mathcal{K}_{X_t} = \mathcal{O}_{X_t}$ for every small deformation $f : X \to S$ of $X_0 = V$, $0 \in S$.

(ii) If $\{X_t\}$ is a complex analytic family over $|t| < \varepsilon$ and if there is a sequence $t_i \to 0$ such that $\mathcal{K}_{X_{t_i}} = \mathcal{O}_{X_{t_i}}$ for $t \neq 0$, then $\mathcal{K}_{X_0} = \mathcal{O}_{X_0}$.

As to (i), we observe that $p_g(X_t)$ being constant (Proposition 2.7, (iv)), the natural map $(f_*\mathcal{K}_X)_t \to H^{2,0}(X_t)$ is surjective by Theorem I.8.5, (iv). Since $\mathcal{K}_{X_0}$ contains a nowhere vanishing section s, we therefore can find a section $\tilde{s}$ of $f_*\mathcal{K}_X$ in a neighbourhood of 0 and $\tilde{s}(t)$ gives a nowhere zero section of $\mathcal{K}_{X_t}$ for t close to 0.

Part (ii) is an immediate consequence of the upper-semi-continuity of both $p_g(X_t)$ and dim $H^0(\mathcal{K}_{X_t}^{-1})$.

Next, we prove the proposition for arbitrary $m \geq 1$. Here we have to prove three things

(i) Every small deformation X_t of $X_0 = V$ occurring in a family $f : X \to S$ has $\mathcal{K}_{X_t}^{\otimes m} = \mathcal{O}_{X_t}$.

(ii) If $\{X_t\}$ is a complex-analytic family over $|t| < \varepsilon$, and if there is a sequence $t_i \to 0$, such that $\mathcal{K}_{X_{t_i}}^{\otimes m} = \mathcal{O}_{X_0}$, then $\mathcal{K}_{X_0}^{\otimes m} = \mathcal{O}_{X_0}$.

(iii) If $f : X \to S$ is a family of surfaces with $\mathcal{K}_{X_t}^{\otimes m} = \mathcal{O}_{X_t}$, $t \in S$, $m > 0$, and $\mathcal{K}_{X_0}^{\otimes n} = \mathcal{O}_{X_0}$ for $0 < n < m$, then $\mathcal{K}_{X_t}^{\otimes n} = \mathcal{O}_{X_t}$ for all $t \in S$.

Now (ii) is proved in the same way as the corresponding statement for $m = 1$, whereas (iii) directly follows from (i) and (ii), so the only thing left to prove is (i). We reduce it to the corresponding statement for $m = 1$ as follows. First, we may assume that f is differentiably trivial with simply-connected base, so that the inclusion induces an isomorphism $\pi_1(X_0) \xrightarrow{\sim} \pi_1(X)$. Then the m-fold unramified covering $X_0' \to X_0$ corresponding to $\mathcal{K}_{X_0}$ extends to an m-fold unramified covering $\rho : X' \to X$ fitting into a commutative diagram

$$\begin{array}{ccc} X' & \xrightarrow{\rho} & X \\ & {\scriptstyle f'}\searrow \quad \swarrow{\scriptstyle f} & \\ & S & \end{array}$$

where $f' : X' \to S$ is a complex-analytic family. By construction $\rho^*(\mathcal{K}_X | X_0')$ is trivial. But this bundle is precisely $\mathcal{K}_{X_0'}$ and so by the corresponding statement for $m = 1$ we see that $\mathcal{K}_{X_t'} = \mathcal{O}_{X_t'}$ for small t. Since $\rho_t = \rho | X_t'$ is an m-fold unramified covering and since for every line bundle $\mathcal{L}$ on X_t with $\rho_t^*(\mathcal{L}) = \mathcal{O}_{X_t'}$ one always has $\mathcal{L}^{\otimes m} = \mathcal{O}_{X_t}$ (by Lemma I. 16.2), it follows that $\mathcal{K}_{X_t}^{\otimes m} = \mathcal{O}_{X_t}$ for those t. This completes the proof of the proposition. □

5. Some Inequalities for Hodge Numbers

Apart from their own interest, the following observations are quite useful, in particular with respect to the question which Chern numbers a surface can have.

(5.1) **Proposition.** *If on the compact surface X there are two linearly independent holomorphic 1-forms ω_1 and ω_2 with $\omega_1 \wedge \omega_2 \equiv 0$, then there exist a smooth curve R of genus ≥ 2, a connected holomorphic map $k : X \to R$ from X onto R and 1-forms α_1, α_2 on R, such that $\omega_1 = k^*(\alpha_1)$ and $\omega_2 = k^*(\alpha_2)$.*

Proof. Let us agree that we shall consider connected coordinate neighbourhoods only. If in local coordinates (z_1, z_2) the 1-form ω_i is given by

$$\omega_i = f_i(z_1, z_2)\, dz_1 + g_i(z_1, z_2)\, dz_2, \quad (i = 1, 2),$$

with $f_2(z_1, z_2) \not\equiv 0$, then the quotient f_1/f_2 is a meromorphic function, which is independent of the coordinate system and thus the restriction of a global meromorphic function h on X. After blowing up the points of indeterminacy of h, we get a surface $\bar{X}$ and a holomorphic map $h : \bar{X} \to \mathbb{P}_1$, such that h is constant along the fibres of h. Using Stein factorization we then obtain connected holomorphic map $k : \bar{X} \to R$, where R is a Riemann surface, and where h is still constant along the fibres of k.

We denote by B the union of all curves on $\bar{X}$ which are mapped onto points by the canonical projection from $\bar{X}$ onto X. Let $p \in \bar{X} \setminus B$, such that there exist local coordinates (z_1, z_2), $|z_1| < A$, $|z_2| < A$, covering a neighbourhood U of p, with

a) $U \cap B = \varnothing$;
b) the restriction $k|U$ is given by $t = z$;
c) if for $i = 1, 2$ the restriction $\omega_i|U$ is given by $\omega_i = f_i\, dz_1 + g_i\, dz_2$, then f_1 and f_2 do not vanish on U.

On all but a finite number of fibres of k (namely, the fibres for which each component has multiplicity ≥ 2) there are points p, satisfying this condition.

Now let the point p and the coordinates (z_1, z_2) be as just described. Then using the identities $f_1 = hf_2$, $g_1 = hg_2$ and the fact that ω_1 and ω_2 are closed (Lemma 2.1), we obtain

$$\begin{aligned} 0 &= f_2 \frac{\partial h}{\partial z_2} = \frac{\partial f_1}{\partial z_2} - h \frac{\partial f_2}{\partial z_2} \\ &= \frac{\partial g_1}{\partial z_1} - h \frac{\partial g_2}{\partial z_1} = g_2 \frac{\partial h}{\partial z_1} \end{aligned}$$

i.e., g_2 vanishes on U. Since $\dfrac{\partial f_2}{\partial z_2} = \dfrac{\partial g_2}{\partial z_1}$, we find that $\omega_2|U = f_2(z_1)\, dz_1$, that is, there exists a holomorphic 1-form α_2 on $k(U)$, such that $k^*(\alpha_2) = \omega_2$

on $k^{-1}(kU)$). Hence there are finitely many points on R: $a_1, \ldots, a_n$, such that there is a holomorphic 1-form α on $R \setminus \bigcup_{i=1}^{n} a_i$, with $k^*(\alpha_2) = \omega_2$ on $k^{-1}\Big(R \setminus \bigcup_{i=1}^{n} a_i\Big)$. What we have to do is to show that α_2 can be extended to a holomorphic 1-form on all of R.

Given any i, $i = 1, \ldots, n$, there is a point $p \in k^{-1}(a_i)$ with the property that in a neighbourhood of p there is a connected non-singular curve C, such that with respect to suitable local coordinates on X and R the restriction $k|C$ is given by

$$t = u^n,$$

with p being $(0, v)$, say, and $t = 0$ being the point a_i. Outside of a_i we have that

$$\alpha_2 = f(t)\ dt$$

with f holomorphic outside a_i. We know that $(k|C \setminus p)^*(\alpha_2)$ can be extended to a holomorphic form on all of C, i.e.,

$$nf(u^n)u^{n-1}\ du$$

is holomorphic around $u = 0$. But this implies that also for $t = 0$ the function f is holomorphic.

In the same way we find a holomorphic 1-form α_1 on R with $k^*(\alpha_1) = \omega_1$. Finally, k factors through X because of Lüroth's theorem. □

(5.2) **Proposition.** *If the compact surface X does not admit a holomorphic map onto a curve of genus ≥ 2, then $h^{2,0}(X) \geq 2h^{1,0}(X) - 3$.*

Proof. The preceding proposition implies that the kernel of the natural homomorphism from $\bigwedge^2 H^{1,0}(X)$ into $H^{2,0}(X)$ meets the cone of decomposable elements only in 0. This cone has dimension $2h^{1,0} - 3$, so $2h^{1,0} - 3 \leq h^{2,0}$. □

Remark. If X is kählerian, then we can rewrite this inequality as $p_g(X) \geq 2q(X) - 3$.

(5.3) **Proposition.** *If the compact surface X with $h^{1,0}(X) \geq 2$ does not admit any holomorphic map onto a curve of genus ≥ 2, then $h^{1,1}(X) \geq 2h^{1,0}(X) - 1$.*

Proof. By Theorem 2.10 we can identify $H^{1,1}(X)$ with the space of d-closed $(1,1)$-forms modulo d-boundaries.

Let $h : H^{1,0}(X) \times \overline{H^{1,0}(X)} \to H^{1,1}(X)$ be defined by $h(\omega, \rho) = \omega \wedge \rho$.

We claim that for fixed $\omega \neq 0$ and also for fixed $\rho \neq 0$ the restrictions $h|\omega \times \overline{H^{1,0}(X)}$ and $h|H^{1,0}(X) \times \rho$ are injective. To see this, let $h(\omega, \bar{\theta}) = 0$, with θ a holomorphic 1-form. Then we have $\omega \wedge \bar{\theta} = d\alpha$, with α a complex 1-form on X. It follows that

$$\omega \wedge \theta \wedge \bar{\omega} \wedge \bar{\theta} = d\beta,$$

with β a 3-form on X. Therefore, by Stokes' theorem we find

$$\int_X \omega \wedge \theta \wedge \bar{\omega} \wedge \bar{\theta} = 0.$$

But this implies that the holomorphic 2-form $\omega \wedge \theta$ vanishes identically on X (compare the proof of Lemma 2.1). Because of Proposition 5.1 we thus find that θ is linearly dependent on ω. However, this being the case, we conclude from $\omega \wedge \bar{\omega} = d\alpha'$ that $\omega \wedge \eta \wedge \bar{\omega} \wedge \bar{\eta} = d\beta'$, with η independent of ω (such an η exists by assumption). Using Stokes' theorem again we are led to a contradiction.

The injectivity of $h|H^{1,0}(X) \times \rho$ is proved in the same way. Thus h induces a regular map

$$\mathbb{P}(H^{1,0}(X)) \times \mathbb{P}(\overline{H^{1,0}(X)}) \to \mathbb{P}(H^{1,1}(X))$$

which is injective on the fibres in both directions. Since each holomorphic map from a product of projective spaces onto any complex space of strictly lower dimension factors through one of the projections ([R-V63], p. 155) we see that the image of $\mathbb{P}(H^{1,0}(X)) \times \mathbb{P}(\overline{H^{1,0}(X)})$ in $\mathbb{P}(H^{1,1}(X))$ has the same dimension as the first of these manifolds. Consequently $h^{1,1}(X) - 1 \geq 2(h^{1,0}(X) - 1)$, that is $h^{1,1}(X) \geq 2h^{1,0}(X) - 1$. □

(5.4) **Corollary.** *If the compact Kähler surface X does not admit any connected fibration with base genus ≥ 2, then $h^{1,1}(X) \geq 2h^{1,0}(X) - 1$.*

6. Projectivity of Surfaces

Theorem 6.2 below and its corollaries play a very crucial role in this book, for they are essential in handling non-projective surfaces. We obtain the results as simple consequences of Grauert's criterion I.19.3, together with the Riemann-Roch theorem for surfaces. Firstly, we observe that as a special case of Theorem I.19.3 we have

(6.1) **Theorem** (Grauert's ampleness criterion for surfaces). *A line bundle $\mathcal{L}$ on the compact surface X is ample if and only if it has the following two properties:*
(i) *for some $n \geq 1$ there is an effective divisor D with $\mathcal{L}^{\otimes n} \cong \mathcal{O}_X(D)$;*
(ii) *given any irreducible curve C on X, then for some $n \geq 1$ (depending on C) there is a section in $\mathcal{L}^{\otimes n}|C$, which vanishes somewhere on C, but not everywhere on C.*

Using the Riemann-Roch theorem formula of II, Sect. 3 for an irreducible curve we see that to prove the ampleness of a line bundle $\mathcal{L}$ on a compact surface X it is sufficient to exhibit on X an effective divisor B with $\mathcal{L} = \mathcal{O}_X(B)$ such that $BD > 0$ for every irreducible curve D. This is exactly what we shall do in the proof of Theorem 6.2.

(6.2) Theorem. *A compact surface X is projective if an only if there exists on X a line bundle $\mathcal{L}$ with $c_1^2(\mathcal{L}) > 0$.*

Proof. In one direction the theorem is trivial: if $\mathcal{L}$ is very ample then $c_1^2(\mathcal{L}) > 0$. So let X be a compact surface and $\mathcal{L}$ a line bundle on X with $c_1^2(\mathcal{L}) > 0$. We start by applying Riemann-Roch to the line bundle $\mathcal{L}^{\otimes n}$:

$$h^0(\mathcal{L}^{\otimes n}) - h^1(\mathcal{L}^{\otimes n}) + h^2(\mathcal{L}^{\otimes n}) = \frac{n}{2}\big(nc_1(\mathcal{L}) + c_1(X)\big)c_1(\mathcal{L}) + \chi(X).$$

By Serre duality $h^2(\mathcal{L}^{\otimes n}) = h^0(\mathcal{K}_X \otimes \mathcal{L}^{\otimes(-n)})$, so for n large we have that either $h^0(\mathcal{L}^{\otimes n}) \geq 2$ or $h^0(\mathcal{K}_X \otimes \mathcal{L}^{\otimes(-n)}) \geq 2$. Since not only $c_1^2(\mathcal{L}^{\otimes n}) > 0$, but also $c_1^2(\mathcal{K}_X \otimes \mathcal{L}^{\otimes(-n)}) > 0$ for large n there always exists on X an effective divisor D with $D^2 > 0$ and $\dim |D| \geq 1$.

By Sect. 1 there always is a rational or irrational pencil $\mathcal{P}$ on X with the following properties :

(i) a general member C of $\mathcal{P}$ is an irreducible curve;
(ii) there exists a positive integer k_0 and a non-negative divisor F (the fixed part of $|D|$) with $k_0C + F$ homologous to D for all $C \in \mathcal{P}$.

From here on we distinguish between two cases:

a) $\mathcal{P}$ has at least one base point, i.e., $C^2 > 0$. Let $E_1, \ldots, E_m$ be the (possibly empty) set of all those components of members of $\mathcal{P}$ which do not pass through any base point. By Zariski's Lemma III.8.2 and Corollary I.2.11 there exist non-negative integers $a_1, \ldots, a_m$ such that $\Big(\sum_{i=1}^{m} a_iE_i\Big)E_j < 0$ for all $j = 1 \leq m$. We observe: if $n \geq n_0$, then $\big(nC - \sum a_iE_i\big)G > 0$ for every irreducible component G of any member of $\mathcal{P}$. On the other hand if $n \geq n_1$, then $\big|nC - \sum a_iE_i\big|$ can be represented by an effective divisor, consisting only of fibre components of elements of $\mathcal{P}$, and containing with some positive multiplicity an irreducible C, hence $\big(nC - \sum a_iE_i\big)G > 0$ for any irreducible curve G not contained in a member of $\mathcal{P}$. Application of Grauert's criterion 6.1 to $\mathcal{O}_X\big(nC - \sum a_iE_i\big)$ with $n \geq \max(n_0, n_1)$ now completes the proof.

b) $\mathcal{P}$ has no base points. If $F = \sum b_kF_k$, then there must be at least one curve F_j, which is not contained in a fibre of $\mathcal{P}$; otherwise we would have $D^2 = (k_0C + F)^2 \leq 0$ by Zariski's lemma. We may assume that F_1 is such a curve. Let $E_1, \ldots, E_m$ be all those components of members of $\mathcal{P}$ which do not meet F_1. As above there are integers $a_1, \ldots, a_m$ such that $\Big(\sum_{i=1}^{m} a_iE_i\Big)E_j < 0$ for $j = 1, \ldots, m$. And as before we find that we can apply Grauert's criterion to $\mathcal{O}_X\big(nC + mF_1 - \sum a_iE_i\big)$ with n, m large and $n > -\frac{F_1^2}{CF_1}m$, to obtain the desired result. □

(6.3) Corollary. *Every compact surface X with $c_1^2(X) = \mathcal{K}_X^2 > 0$ is projective.*

(6.4) **Corollary** (Nakai's criterion). *A line bundle $\mathcal{L}$ on a compact surface X is ample if and only if $c_1^2(\mathcal{L}) > 0$ and $(\mathcal{L}, D) > 0$ for every effective divisor D on X.*

Proof. By Theorem 6.2 we know already that X is projective. We take a very ample divisor H on X and fix an n_0, such that $(H, \mathcal{K}_X \otimes \mathcal{L}^{\otimes(-n)}) < 0$ for $n \geq n_0$. This is possible by assumption. On the other hand, by the Riemann-Roch theorem for surfaces we know that either $\mathcal{L}^{\otimes n}$ or $\mathcal{K}_X \otimes \mathcal{L}^{\otimes(-n)}$ has sections for n large enough. But the second possibility is excluded, since $(H, \mathcal{K}_X \otimes \mathcal{L}^{\otimes(-n)}) < 0$ for $n \geq n_0$, whereas $HD > 0$ for every effective D. So $\mathcal{L}^{\otimes n}$ has a section and we can apply the remark following Theorem 6.1. □

(6.5) **Corollary.** *A compact surface X is projective if and only if it has algebraic dimension 2.*

Proof. Let X be a surface with $\mathrm{a}(X) = 2$. Take a meromorphic function on X and consider the associated pencil $\mathcal{P}$ with general member C. If $C^2 > 0$ we can conclude by Theorem 6.2. If $C^2 = 0$, then $\mathcal{P}$ defines a regular map $f : X \to \mathbb{P}_1$. Let $f = hg$ be its Stein factorization. If every meromorphic function on X would be constant on the fibres of g, then $\mathrm{a}(X)$ would be 1. Hence there is an irreducible curve D on X, which is not contained in any fibre of g. If G denotes a general fibre of g, then $(nG + D)^2 > 0$ for n sufficiently large. Hence X is algebraic by Theorem 6.2. □

Remark. The preceding result is no longer true if X is allowed to have singularities. The following example of Hironaka (see [Gr62]) shows much more: it is possible to obtain from a smooth projective surface a normal non-projective complex space by blowing down a smooth (non-rational!) curve.

Let $C \subset \mathbb{P}_2$ be a smooth cubic, and $p_1, \ldots, p_{10}$ ten points on C, such that there is no curve $D \subset \mathbb{P}_2$ with the property that set-theoretically $C \cap D \subset p_1 \cup \cdots \cup p_{10}$. (The possibility of such a choice of points on C follows from the classical geometrical interpretation of the addition on C, once this curve is made into an abelian variety by choosing a base point, compare [Wal], p. 192.) Then let X be obtained from $\mathbb{P}_2$ by blowing up $p_1, \ldots, p_{10}$. If $\overline{C}$ denotes the proper transform of C on X, we have $\overline{C}^2 = -1$, and by Theorem III.2.1, $\overline{C}$ can be blown down in X, such that the result is a normal complex space Y. If $q \in Y$ is the image of $\overline{C}$, then there are two independent meromorphic functions on $Y \setminus q = X \setminus \overline{C}$, hence on Y by Levi's theorem (Theorem I.8.7). But Y is not projective. For if Y were projective, then there would be a curve on Y, not containing q; in other words there would be a curve D on $\mathbb{P}_2$, such that $C \cap D \subset p_1 \cup \cdots \cup p_{10}$, which is impossible by construction.

An abstract algebraic variety is defined by the usual patching procedure, using affine open sets and rational morphisms. An abstract algebraic variety need not be projective. However, Corollary 6.5 immediately yields

(6.6) **Corollary.** *Every (smooth) compact abstract algebraic surface is projective.*

Because of this result it is not dangerous to call a smooth projective-algebraic surface simply an algebraic surface, as we shall often do (and did!).

(6.7) **Corollary.** *Let X be a compact surface and Y be obtained from X by blowing up a point. Then X is projective if and only if Y is projective.*

(6.8) **Theorem.** *Let X, Y be compact surfaces and $f : X \to Y$ a finite map. Then X is projective if and only if Y is projective.*

Proof. If Y is projective, we can take a line bundle $\mathfrak{L}$ on Y with $c_1^2(\mathfrak{L}) > 0$. Then $c_1^2(f^*(\mathfrak{L})) = (\deg)c_1^2(\mathfrak{L}) > 0$, and X is projective by Theorem 6.2. Conversely, if X is projective, we can take a divisor D on X with $DC > 0$ for every effective divisor C on X. Then, for every effective divisor E on Y we have $f_*(D)E = Df^*(E) > 0$, so Y is algebraic by Nakai's criterion 6.4. □

7. The Nef Cone

We start with some properties of convex cones in general. Let V a real finite dimensional vector space and let $C \subset V$ be some convex cone. Its dual is the closed convex cone

$$C^\vee = \{f \in V^\vee \mid f(v) \geq 0, \text{ for } v \in V\}.$$

If V has a non-degenerate bilinear form $(-,-)$ we can use it to identify V and $V^\vee$, which we shall do in the sequel.

(7.1) **Lemma.** *Suppose that V has a non-degenerate bilinear form $(-,-)$ with signature $(1, \dim V - 1)$. Let C be one of the two components of the set $\{v \in V \mid (v,v) > 0\}$. The closure of the cone C is self-dual.*

Proof. The dual cone of the half-line spanned by an isotropic $v \neq 0$ in the closure of C is the half-space passing through v tangent to the isotropic cone and containing C. If v varies over such elements, the intersection of these half-spaces is exactly the closure of C. It follows that $w \in \bar{C}$ if and only if $(v,w) \geq 0$ for all $v \in \bar{C}$ that are isotropic. Since any element in $\bar{C}$ is a convex linear combination of boundary points, it follows that $w \in \bar{C}$ if and only if $(v,w) \geq 0$ for all $v \in \bar{C}$. □

As a corollary of the proof we have

(7.2) **Corollary.** *Under the same hypothesis we have*

$$(5) \quad \begin{cases} v, w \in \bar{C} \text{ implies } (v,w) \geq 0 \text{ and if } v, w \neq 0 \text{ strict inequality holds} \\ \text{if either } v \text{ or } w \text{ is contained in } C. \end{cases}$$

Let X be a surface with b_1 even. As we have seen (Theorem 2.14), the intersection form on $H^{1,1}_{\mathbb{R}}(X)$ is non-degenerate with signature $(1, h^{1,1} - 1)$ and so the previous discussion applies here. In particular, the cone of classes a with $(a, a) > 0$ consists of two disjoint connected components and only one of these contains the Kähler classes. This component, $\mathcal{C}^+_X$, will be called the positive half-cone.

For the rest of this section, we suppose that X is projective.

Consider the Néron-Severi group

$$\mathrm{NS}(X) = H^{1,1}(X) \cap i^* \left(H^2(X, \mathbb{Z})\right).$$

This is a lattice inside the real vector space

$$\mathrm{NS}_{\mathbb{R}}(X) := \mathrm{NS}(X) \otimes \mathbb{R} \subset H^{1,1}_{\mathbb{R}}(X)$$

of dimension $\rho(X)$. Since X contains at least one ample divisor, the intersection pairing has signature $(1, \rho(X) - 1)$ and the positive half-cone intersects $\mathrm{NS}_{\mathbb{R}}(X)$ properly in the convex cone:

$$\mathrm{NS}_+(X) = \mathcal{C}^+_X \cap \mathrm{NS}_{\mathbb{R}}(X).$$

We claim:

(7.3) **Lemma.** *A rational point in* $\mathrm{NS}_+(X)$ *is a rational multiple of the class of an effective divisor.*

This follows immediately from:

(7.4) **Proposition.** *If for a divisor D on a surface X one has $D^2 > 0$, then $DH \neq 0$ for any ample divisor H. If $DH > 0$ some positive multiple of D is effective, in fact, $h^0(mD)$ grows quadratically in m. If $DH < 0$, some negative multiple of D is effective (and then $h^0(-mD)$ grows quadratically in m).*

Proof. The first assertion follows from the Algebraic Index Theorem.

The Riemann-Roch inequality shows that $h^0(mD) + h^0(-mD + K_X) \geq \frac{1}{2}m^2D^2$+linear term in m. If $DH > 0$, there can be no divisor in $|-mD+K_X|$ for m large and so $|mD|$ must contain effective divisors for m large enough. The proof of the second assertion is similar. □

The non-negative linear combinations of classes of nef divisors span a convex sub-cone of $\mathrm{NS}_{\mathbb{R}}(X)$, which we call the nef cone and denote $\mathrm{Nef}(X)$. The nef cone is the dual of the cone generated by all effective divisors:

$\mathrm{Ef}(X)$ =Cone in $\mathrm{NS}_{\mathbb{R}}(X)$ generated by the classes of the effective divisors

$\mathrm{Nef}(X) = \mathrm{Ef}^{\vee}(X)$.

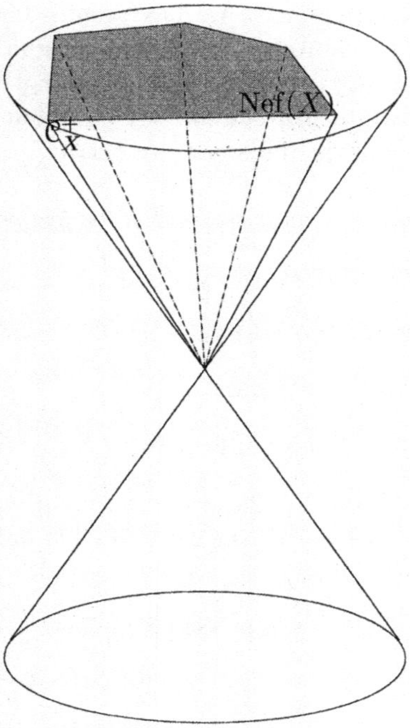

Figure 3. The Nef Cone

Using Nakai's Criterion (Corollary 6.4) we find the following simple criterion for ampleness in terms of this cone.

(7.5) **Proposition.** *L is ample if and only if $Lc > 0$ for all $c \neq 0$ in the closure of* $\mathrm{Ef}(X)$.

Proof. If L is ample, $LD > 0$ for all effective divisors D and so $Lc \geq 0$ for all c in the closure of the effective cone. If $Lc = 0$ for some c, the Hodge index theorem (2.16) tells us that $c^2 \leq 0$ with equality if and only if $c = 0$. Since c is in the closure of the effective cone, if $c \neq 0$, we can find an effective divisor C' with $C'c < 0$. Then $(nL + C')c < 0$. On the other hand $nL + C'$ will be ample for n large enough by Nakai's criterion (for at worst finitely many of components D of C' you will have $DC' < 0$ and these can be taken care of by making n large enough). This is a contradiction and so $Lc > 0$.

Conversely, by Nakai's criterion it suffices to show that $L^2 > 0$. For some ample line bundle H we consider the function $c \mapsto f(c) = Lc/Hc$. It is constant under homotheties. So to study its values, one can restrict to the (compact) closure of $\mathrm{Ef}\, X$ in the unit ball with respect to some metric on the real vector space $N_{\mathbb{R}}(X)$. It has a positive (rational) minimum ϵ and so $(L - \frac{1}{2}\epsilon H)c > 0$ for all $c \in \mathrm{Ef}\, X$ and in particular $L - \frac{1}{2}\epsilon H$ is nef and so has non-negative self-intersection. But then $L^2 = (L - \frac{1}{2}\epsilon H)(L - \frac{1}{2}\epsilon H) + \epsilon H(L - \frac{1}{2}\epsilon H) + \frac{1}{4}\epsilon^2 H^2 > 0$. □

Since by Lemma 7.3 $\mathrm{NS}_+(X) \subset \mathrm{Ef}(X)$, we have $\mathrm{Nef}(X) = \mathrm{Ef}(X)^\vee \subset \mathrm{NS}_+(X)^\vee = \overline{\mathrm{NS}_+(X)}$, where the last equality follows from Lemma 7.1. We thus arrive at the

(7.6) **Observation.** *If for a divisor D one has $DC \geq 0$ for all irreducible curves C then $D^2 \geq 0$. In other words, any nef divisor has non-negative self intersection.*

(7.7) **Corollary.** *The cone consisting of positive rational multiples of ample divisors forms an open convex cone in* $\mathrm{NS}(X) \otimes \mathbb{Q}$ *and its closure is the (rational) nef cone.*

Proof. The Nakai criterion implies that the ample cone is convex and contained in $\mathrm{Nef}(X)$. To prove that it is open in $\mathrm{NS}(X) \otimes \mathbb{Q}$ we apply Proposition 7.5. Let H be an element of the ample cone, and S the unit sphere in some metric on $\mathrm{NS}(X) \otimes \mathbb{R}$. Then by Proposition 7.5, the continuous function $f_H : S \cap \overline{\mathrm{Ef}(X)} \to \mathbb{R}$ defined by $f_H(x) = H \cdot x$ has a strictly positive minimum on the compact set $S \cap \overline{\mathrm{Ef}(X)}$. Consequently, there is an open ball around H such that for G in this ball the function f_G still has a strictly positive minimum on $S \cap \overline{\mathrm{Ef}(X)}$. This implies that $G \cdot x > 0$ for all non-zero $x \in \overline{\mathrm{Ef}(X)}$ and so, again by Proposition 7.5, G is a rational multiple of an ample divisor and so the ample cone is indeed open in $\mathrm{Nef}(X) \otimes \mathbb{Q}$

Finally, to show that the ample cone is the interior of the (closed) cone $\mathrm{Nef}(X)$, since it is clearly entirely contained in the interior of the nef-cone, it suffices to show that it is dense therein. Let D be a rational nef-divisor and let H be an ample class. Then for all $t > 0$ the divisor $D + tH$ is ample by Nakai's criterion. So indeed, the (rational) ample cone is dense in the rational nef-cone. □

8. Surfaces of Algebraic Dimension Zero

Let X be a compact surface. The property $\mathrm{a}(X) = 0$ readily implies several noteworthy properties, some of which we shall use later on in this book.

(8.1) **Proposition.** *Let X be a compact surface with* $\mathrm{a}(X) = 0$. *Then*
(i) *$h^0(\mathcal{L}) \leq 1$ for every line bundle $\mathcal{L}$ on X, and in particular $p_g(X) = h^0(\Omega^2_X) \leq 1$;*
(ii) *$h^{1,0}(X) \leq 2$.*

Proof. (i) Trivial.
(ii) Let ω_1, ω_2 and ω_3 be three linearly independent holomorphic 1-forms. Then $\omega_1 \wedge \omega_2$ and $\omega_1 \wedge \omega_3$ do not vanish identically, otherwise $\mathrm{a}(X)$ would be at least 1 (Proposition 5.1). But (i) implies that for some λ, $\mu \neq 0$ we have $\lambda\omega_1 \wedge \omega_2 + \mu\omega_1 \wedge \omega_3 \equiv 0$, i.e., $\omega_1 \wedge (\lambda\omega_2 + \mu\omega_3) \equiv 0$ and we would find again that $\mathrm{a}(X) \geq 1$, contrary to our assumption. So $h^{1,0}(X) \leq 2$. □

Remark. All of these bounds are sharp, as is clear from the fact that there exist 2-tori of algebraic dimension 0.

(8.2) **Theorem.** *If X is a compact surface with* $\mathrm{a}(X) = 0$, *then the number of irreducible curves on X is finite and at most equal to* $h^{1,1}(X) + 2$.

Proof. If $\mathcal{M}_X^1$ denotes the sheaf of germs of meromorphic 1-forms on X, then there is a natural injection $\Omega_X^1 \hookrightarrow \mathcal{M}_X^1$. We denote the quotient sheaf by $\mathcal{Q}_X$. To each curve C on X we can attach an element of $\Gamma(\mathcal{Q}_X)$ by writing C locally as $f = 0$ and then consider $\frac{df}{f}$. It is easily verified that the images in $\Gamma(\mathcal{Q}_X)$ of a finite number of different irreducible curves are linearly independent. So the exact cohomology sequence

$$\cdots \longrightarrow \Gamma(\mathcal{M}_X^1) \longrightarrow \Gamma(\mathcal{Q}_X) \longrightarrow H^1(\Omega_X^1) \longrightarrow \cdots$$

shows that it is sufficient to prove that $\dim \Gamma(\mathcal{M}_X^1) < 3$, i.e., that every three sections of $\mathcal{M}_X^1$ are linearly dependent. But if ω_1 and ω_2 are meromorphic 1-forms on X, independent over $\mathbb{C}$, then ω_1 and ω_2 must be independent over the field of meromorphic functions $\mathcal{M}(X)$, since $\mathrm{a}(X) = 0$. It follows that every third meromorphic form must be dependent on ω_1 and ω_2 over $\mathcal{M}(X)$, and since $\mathrm{a}(X) = 0$, it must be dependent on ω_1 and ω_2 over $\mathbb{C}$. □

Remark. Again this inequality is sharp, as follows for example from the existence of Hopf surfaces, without meromorphic functions, with two curves (see Proposition V.18.2).

9. Almost-Complex Surfaces without any Complex Structure

In this section we shall show how the preceding results easily yield compact, oriented 4-dimensional differentiable manifolds, which admit almost-complex structures, but no complex structure.

We shall use the concept of a connected sum of two differentiable manifolds. We refer for a formal treatment to [B-J] §10, and just say the following.

Let X and Y be two oriented, connected n-dimensional differentiable manifolds. Let $D \subset \mathbb{R}^n$ be the unit disc, and let $f_1 : \mathbb{R}^n \to X$, $f_2 : \mathbb{R}^n \to Y$ be orientation-preserving embeddings. The connected sum $X\#Y$ (with respect to f_1 and f_2) is the naturally oriented n-dimensional differentiable manifold, obtained from the disjoint union

$$\left(X - f_1\left(\frac{1}{3}D\right)\right) \cup \left(Y - f_2\left(\frac{1}{3}D\right)\right)$$

by identifying $f_1(tx)$ with $f_2((1-t)x)$ for all $\frac{1}{3} < t < \frac{2}{3}$, $x \in S^{n-1}$.

If X and Y are compact, then so is $X\#Y$. Since the special choice of f_1 and f_2 plays no role here, we simply denote by $X\#Y$ some connected sum of X and Y (they are all diffeomorphic anyway).

What we need about the topology of connected sums is contained in the following simple result.

(9.1) **Proposition.** *Let G be either $\mathbb{Z}$ or a field. Then for all i, $1 \leq i \leq n-1$, there are natural isomorphisms*

$$\lambda_i : H^i(X, G) \oplus H^i(Y, G) \xrightarrow{\sim} H^i(X \# Y, G)$$

such that if $a \in H^k(X, G)$, $b \in H^{n-k}(X, G)$, then $\lambda_k(a)\lambda_{n-k}(b) = \lambda_n(ab)$.

Now we are ready for the examples.

(9.2) **Theorem.** *A connected sum $W = S^1 \times S^3 \# S^1 \times S^3 \# \mathbb{P}_2$ admits almost-complex structures, but no complex structure.*

The existence of almost-complex structures on W is a consequence of the following criterion ([W-R], p. 74, together with Theorem I.3.1).

(9.3) **Proposition.** *Let W be an oriented compact connected 4-dimensional differentiable manifold, and let $h \in H^2(W, \mathbb{Z})$. Then there exist almost-complex structures $\mathcal{A}$ on W with $c_1(\mathcal{A}) = h$ (and of course $c_2(\mathcal{A}) = e(W)$) if and only if the following two conditions are satisfied:*
(i) $h \equiv w_2(W) \bmod 2$,
(ii) $h^2 = 3\tau(W) + 2e(W)$.

Proof of Theorem 9.2. From Proposition 9.1 we find $H^2(W, \mathbb{Z}) \cong \mathbb{Z}$, such that if g is a generator of $H^2(W, \mathbb{Z})$, then $g^2 = 1$. Proposition 9.1 also yields that $H^2(W, \mathbb{Z}/2\mathbb{Z}) \cong \mathbb{Z}/2\mathbb{Z}$. Since $w_2 g \equiv g^2 \pmod 2$ by [M-S], p. 732, we see that $w_2 \neq 0$, so $w_2 \equiv g \pmod 2$. Applying Proposition 9.1 once more we find $e(W) = -1$, and since $\tau(W) = 1$, the conditions of Proposition 9.3 are satisfied (with $h = g$). Consequently there exist almost-complex structures $\mathcal{A}$ on W with $c_1(\mathcal{A}) = g$. Conversely, every almost-complex structure on W has g as its first Chern class.

It remains to be seen that there is no complex structure on W. If there were one, it would be a projective one, for, as already observed, it would have Chern class g, hence it would be projective by Theorem 6.2. Furthermore, $b_1(W) = 2$ by Proposition 9.1, so $\mathrm{Alb}(W)$ would be an elliptic curve, and there would exist a surjective map $f : W \to \mathrm{Alb}(W)$. If F is any fibre of f, then $F^2 = 0$, but F is not homologous to 0 (Lemma I.13.1). This, however, is impossible since $H^2(W, \mathbb{Z}) \cong \mathbb{Z}$. Hence W has no complex structure. □

There are many examples of this type, like $S^1 \times S^3 \# S^1 \times S^3 \# S^2 \times S^2$, or the connected sum of $2k+1$, $k \geq 1$, copies of a smooth surface of degree 4 in $\mathbb{P}_3$. This last example is simply-connected, since $\pi_1(X \# Y)$ is the free product of $\pi_1(X)$ and $\pi_1(Y)$, whereas a smooth surface in $\mathbb{P}_3$ is simply-connected by Lefschetz' Theorem I.20.4.

If X is a compact complex surface, then it follows from Noether's formula that $c_1^2(X) + c_2(X) \equiv 0$ (12). This remains true if X is any almost-complex surface ([Hir3], p. 125). Conversely, given any ordered pair (p, q) of integers, with $p + q \equiv 0$ (12), there always exists a compact differentiable 4-manifold with an almost-complex structure $\mathcal{A}$, such that $c_1^2(\mathcal{A}) = p$, $c_2(\mathcal{A}) = q$. To see this, it is sufficient to apply Proposition 9.1 and 9.3 to a suitable connected sum

$$\mathbb{P}_2 \# \cdots \# \mathbb{P}_2 \# \overline{\mathbb{P}}_2 \cdots \# \overline{\mathbb{P}}_2 \# \mathbb{P}_1 \times R \# \cdots \# \mathbb{P}_1 \times R,$$

where $\overline{\mathbb{P}}_2$ is $\mathbb{P}_2$ with the orientation reversed, and R a curve of genus 2 ([Ve66], p. 1625).

On the other hand, as we shall see in Chap. VII, there are many pairs (p, q), with $p + q \equiv 0$ (12), such that there is no compact complex surface with these Chern numbers. In this way many more examples of the type above can be produced. A deeper study yields still more examples, but up to now these examples, however abundant, reveal no pattern as to which almost-complex structures can be deformed into an integrable one. And there is not a single example of a (compact) higher-dimensional differentiable manifold with almost-complex structures, but no complex structure. Few doubt that these exist, but there is no method to handle the higher-dimensional case. In particular, a simple projectivity criterion like Theorem 6.2 does not exist in higher dimensions.

10. Bogomolov's Theorem

In this section we shall prove Bogomolov's theorem on the instability of rank 2 vector bundles over an algebraic surface X. We shall follow Reid's exposition [Rei77]. Recall first that a 0-dimensional subscheme of a complex projective variety X is nothing but an analytic subspace of X supported in points. In other words, it is locally given by the vanishing of a finite number of holomorphic functions which do not vanish along a common divisor.

(10.1) **Theorem** (Bogomolov). *Let $\mathcal{E}$ be a rank 2 vector bundle over a projective surface X. Then the following statements are equivalent:*

(i) $c_1^2(\mathcal{E}) - 4c_2(\mathcal{E}) > 0$.

(ii) *There exist line bundles $\mathcal{L}, \mathcal{M} \in \operatorname{Pic} X$, a 0-dimensional subscheme $Z \subset X$ and an exact sequence*

$$0 \to \mathcal{L} \to \mathcal{E} \to \mathcal{J}_Z \cdot \mathcal{M} \to 0 \tag{6}$$

such that

$$(\mathcal{L} \otimes \mathcal{M}^{-1})^2 > 4 \deg Z, \tag{7}$$

and

$$(\mathcal{L} \otimes \mathcal{M}^{-1}, H) > 0 \quad \text{for every ample divisor } H \text{ on } X. \tag{8}$$

(10.2) *Remark.* Using Prop. I, 5.2 the exact sequence (6) implies

$$\begin{aligned} c_1(\mathcal{E}) &= c_1(\mathcal{L}) + c_1(\mathcal{M}) \\ c_2(\mathcal{E}) &= (\mathcal{L}, \mathcal{M}) + \deg Z. \end{aligned}$$

Before we give the proof of the above theorem we shall make some comments and preparations. First note that the preceding remark and (7) implies that $(\mathcal{L}, H) > (c_1(\mathcal{E}), H)/2$ for every ample H. In particular, $\mathcal{E}$ is H-unstable in the sense of Maruyama (by definition). We also remark that the line bundle $\mathcal{L}$ is uniquely determined. Assume that $\mathcal{L}'$ were another line subbundle of $\mathcal{E}$ fulfilling (7), (8) for some $\mathcal{M}'$ such that $\mathcal{L}' \otimes \mathcal{M}' = \mathcal{L} \otimes \mathcal{M}$. Then there exists a non-zero homomorphism $\mathcal{L}' \to \mathcal{M}$ and hence $\mathcal{L}' = \mathcal{M}(-D)$ for some effective divisor $D \geq 0$. We then have $c_1(\mathcal{L}' \otimes \mathcal{M}'^{-1}) = c_1(\mathcal{M} \otimes \mathcal{L}^{-1}(-2D))$ and hence $(\mathcal{L}' \otimes \mathcal{M}'^{-1}, H) < 0$ contradicting (8).

The proof of the above theorem is based on the auxiliary vector bundles

$$\mathcal{H}_n := S^{2n}\mathcal{E} \otimes (\det \mathcal{E})^{-n}.$$

(10.3) **Lemma.** *The following holds:*

(i) $\operatorname{rank} \mathcal{H}_n = 2n + 1$,

(ii) $\mathcal{H}_n^\vee \cong \mathcal{H}_n$ *and* $\det \mathcal{H}_n = \mathcal{O}_X$,

(iii) $c_1(\mathcal{H}_n) = 0,\ c_2(\mathcal{H}_n) = (c_1^2(\mathcal{E}) - 4c_2(\mathcal{E}))\left(\sum_{i=0}^{n} i^2\right)$,

(iv) $\chi(X, \mathcal{H}_n) = (2n+1)\chi(X) + (c_1^2(\mathcal{E}) - 4c_2(\mathcal{E}))\left(\sum_{i=0}^{n} i^2\right)$.

Proof. Assertion (i) is clear and (ii) follows from standard canonical isomorphisms in linear algebra. Assertion (iii) is a straightforward Chern class calculation. Assertion (iv) is the Riemann-Roch theorem for the bundle $\mathcal{H}_n$ on X. See I. (9). □

(10.4) **Proposition.** *Let $\mathcal{E}$ be a rank 2 vector bundle on X with $c_1^2(\mathcal{E}) - 4c_2(\mathcal{E}) > 0$. Then there exists a constant $c > 0$ such that $h^0(\mathcal{H}_n) > cn^3$ for $n \gg 0$.*

Proof. It follows from the above lemma (iv) that $h^0(\mathcal{H}_n) + h^2(\mathcal{H}_n) > cn^3$. By Serre duality $h^2(\mathcal{H}_n) = h^0(\mathcal{H}_n^* \otimes \mathcal{K}_X) = h^0(\mathcal{H}_n \otimes \mathcal{K}_X)$. Hence $h^0(\mathcal{H}_n) + h^0(\mathcal{H}_n \otimes \mathcal{K}_X) > cn^3$. The assertion of the proposition then follows from the

Claim. Let $\mathcal{L} \in \operatorname{Pic}(X)$. Then there exists a constant $c' > 0$ such that $h^0(\mathcal{H}_n \otimes \mathcal{L}) - h^0(\mathcal{H}_n) < c'n^2$.

Proof of the Claim. We write $c_1(\mathcal{L}) = C - D$ where C and D are smooth curves on X. Then we have the exact sequence

$$0 \to \mathcal{H}_n \otimes \mathcal{O}_X(-D) \to \mathcal{H}_n \otimes \mathcal{L} \to \mathcal{H}_n \otimes \mathcal{L}|_C \to 0.$$

Hence

$$h^0(\mathcal{H}_n \otimes \mathcal{L}) - h^0(\mathcal{H}_n \otimes \mathcal{O}_X(-D)) \leq h^0(\mathcal{H}_n \otimes \mathcal{L}|_C).$$

Since $h^0(\mathcal{H}_n \otimes \mathcal{O}_X(-D)) \le h^0(\mathcal{H}_n)$ it suffices to show that $h^0(\mathcal{H}_n \otimes \mathcal{L}|_C) < c'n^2$. Recall that every rank 2 vector bundle on a smooth curve is the extension of two line bundles. (Just tensor the vector bundle with a sufficiently ample line bundle and take a section.) Hence we can write

$$0 \to \mathcal{L}_1 \to \mathcal{E}|_C \to \mathcal{L}_2 \to 0$$

for line bundles $\mathcal{L}_1, \mathcal{L}_2 \in \operatorname{Pic} C$. Again by standard linear algebra arguments (cf. also [Ha77], Exercise II.5.16.c) there exists a filtration

$$S^{2n}\mathcal{E}|_C = F^0 \supseteq F^1 \supseteq \ldots \supseteq F^{2n+1} = 0$$

with

$$F^p/F^{p+1} \cong \mathcal{L}_1^p \otimes \mathcal{L}_2^{2n-p}.$$

It follows that $\mathcal{H}_n \otimes \mathcal{L}|_C$ has a filtration

$$\mathcal{H}_n \otimes \mathcal{L}|_C = G^0 \supseteq G^1 \supseteq \ldots \supseteq G^{2n+1} = 0$$

with

$$G^p/G^{p+1} \cong \mathcal{L}_1^{p-n} \otimes \mathcal{L}_2^{n-p} \otimes \mathcal{L}.$$

It follows from this filtration that

$$h^0(\mathcal{H}_n \otimes \mathcal{L}|_C) \le \sum_{p=0}^{2n} h^0(\mathcal{L}_1^{p-n} \otimes \mathcal{L}_2^{n-p} \otimes \mathcal{L}) < c'n^2.$$

The latter inequality follows since the Riemann-Roch formula for a line bundle on a curve is linear in the degree of the line bundle. $\square$

Let $Y = \mathbb{P}(\mathcal{E}^\vee)$ and $\mathcal{O}_Y(1)$ the dual of the tautological line subbundle, i.e., $\pi_*\mathcal{O}_Y(1) = \mathcal{E}$ where $\pi : Y \to X$ is the natural projection. On Y we consider the line bundle

$$\mathcal{R}_n = \mathcal{O}_Y(2n) \otimes (\pi^* \det \mathcal{E})^{-n}, \quad Y = \mathbb{P}(\mathcal{E}^\vee).$$

By the projection formula $\pi_*\mathcal{R}_n = \mathcal{H}_n$. In particular $H^0(Y, \mathcal{R}_n) = H^0(X, \mathcal{H}_n)$.

Definition. A quasi-section of the $\mathbb{P}_1$-bundle $\pi : \mathbb{P}(\mathcal{E}^\vee) \to X$ is an irreducible subvariety $S \subset \mathbb{P}(\mathcal{E}^\vee)$ such that $\pi|_S : S \to X$ is a birational morphism.

If S is a quasi-section, then clearly the intersection number $S\pi^{-1}(x) = 1$ for $x \in X$. Either S intersects the fibre $\pi^{-1}(x)$ transversally in one point or it contains it. The latter can happen only for a finite number of points in X, since otherwise S would have a component lying over a curve in X.

(10.5) **Lemma.** *There exists a bijection between quasi-sections of* $\mathbb{P}(\mathcal{E}^\vee)$ *and exact sequences of the form*

$$0 \to \mathcal{L} \to \mathcal{E} \to \mathcal{J}_Z \cdot \mathcal{M} \to 0$$

where $\mathcal{L}, \mathcal{M} \in \operatorname{Pic}(X)$ *and* $Z \subset X$ *is* 2*-codimensional.*

Proof. Let $S \subset \mathbb{P}(\mathcal{E}^\vee)$ be a quasi-section. Then we have an exact sequence

$$0 \to \mathcal{O}_{\mathbb{P}(\mathcal{E}^\vee)}(1) \otimes \mathcal{O}_{\mathbb{P}(\mathcal{E}^\vee)}(-S) \to \mathcal{O}_{\mathbb{P}(\mathcal{E}^\vee)}(1) \to \mathcal{N} \to 0 \tag{9}$$

where $\mathcal{N} = \mathcal{O}_{\mathbb{P}(\mathcal{E}^\vee)}(1)|_S$. The line bundle $\mathcal{O}_{\mathbb{P}(\mathcal{E}^\vee)}(1) \otimes \mathcal{O}_{\mathbb{P}(\mathcal{E}^\vee)}(-S)$ is trivial on all fibres of $\mathbb{P}(\mathcal{E}^\vee)$ and hence of the form $\pi^*\mathcal{L}$ for some line bundle $\mathcal{L}$ on X. Pushing the above exact sequence (9) down to X we find an exact sequence

$$0 \to \mathcal{L} \to \mathcal{E} \to \pi_*\mathcal{N} \to 0. \tag{10}$$

Locally over an open set U the homomorphism $\mathcal{L} \to \mathcal{E}$ is given by two functions $f, g \in \mathcal{O}_X(U)$ and the quasi-section S is given in suitable local coordinates by

$$S \cap \pi^{-1}(U) = \{(x, y) \mid fx + gy = 0\}.$$

Since S contains only finitely many fibres of $\mathbb{P}(\mathcal{E}^\vee)$ the functions f, g do not vanish on a common divisor. Hence over U sequence (10) reads

$$0 \to \mathcal{O}_U \xrightarrow{(f,g)} \mathcal{O}_U \oplus \mathcal{O}_U \xrightarrow{\begin{pmatrix} f \\ -g \end{pmatrix}} \mathcal{J}_{Z \cap U} \cdot \mathcal{O}_U \to 0 \tag{11}$$

where $Z \cap U = \{f = g = 0\}$. This shows that $\pi_*\mathcal{N} = \mathcal{J}_Z \cdot \mathcal{M}$ for some line bundle $\mathcal{M}$ on X and a codimension 2 subscheme Z of X. Reading the above argument backwards shows that every exact sequence of the form given in the statement of the proposition gives rise to a quasi-section $S \subset \mathbb{P}(\mathcal{E}^\vee)$. □

The following proposition is crucial for the proof of Bogomolov's theorem.

(10.6) **Proposition.** *Assume that* $c_1^2(\mathcal{E}) - 4c_2(\mathcal{E}) > 0$. *Then for* $n \gg 0$ *there exists a section* $0 \neq s \in H^0(\mathbb{P}(\mathcal{E}^\vee), \mathcal{R}_n)$ *whose divisor vanishes of order at least* $n+1$ *on a quasi-section* S *of* $\mathbb{P}(\mathcal{E}^\vee)$.

Before we can give the proof of this proposition we need some elementary facts from geometric invariant theory. Let W be the standard representation of $\mathrm{GL}(2, \mathbb{C})$. The centre of $\mathrm{GL}(2, \mathbb{C})$ acts trivially on the induced representation $W_n = S^{2n}W \otimes (\det W)^{-n}$, i.e., W_n is a $\mathrm{PGL}(2, \mathbb{C})$-module. Recall that $\mathrm{PGL}(2, \mathbb{C})$ is a reductive group. An element $f \in W_n$ is called unstable with respect to the group $\mathrm{PGL}(2, \mathbb{C})$ if the following two equivalent conditions are fulfilled

(i) The closure of the $\mathrm{PGL}(2, \mathbb{C})$-orbit of f contains 0,

(ii) If P is a $\mathrm{PGL}(2, \mathbb{C})$-invariant function on W_n, then $P(f) = 0$.

By a result of Hilbert these two conditions are equivalent to

(iii) f has a root of order at least $n+1$, i.e., there exists a basis e_1, e_2 of W such that $f = e_1^{n+1} f' / (e_1 \wedge e_2)^n$.

For references for these statements see [Mu7], Prop. 4.1. It is the equivalence of (ii) and (iii) which is used to prove Proposition 10.6. For $m \geq 1$ we denote by $I_{n,m}$ the space of PGL(2, $\mathbb{C}$)-invariant polynomials of degree m on W_n. We then have a canonical decomposition

$$S^m(W_n^\vee) = I_{n,m} \oplus S'$$

where S' is the sum over all non-trivial summands of $S^m(W_n^\vee)$. The vector bundle $\mathcal{H}_n$ is the vector bundle induced from $\mathcal{E}$ via the representation W_n. The above decomposition of $S^m(W_n^\vee)$ induces a decomposition

$$S^m(\mathcal{H}_n^\vee) = \mathcal{I}_{n,m} \oplus \mathcal{S}'$$

where $\mathcal{I}_{n,m}$ is the trivial bundle of rank equal to the dimension of $I_{n,m}$. Evaluation defines a canonical pairing

$$S^m(\mathcal{H}_n^\vee) \otimes \mathcal{H}_n \to \mathcal{O}_X \tag{12}$$

Proof of Proposition 10.6. Let $x \in X$ and set

$$U(n,x) := \{s \in H^0(X, \mathcal{H}_n) \mid s(x) = 0\}.$$

Since rank $\mathcal{H}_n = 2n+1$ and $h^0(X, \mathcal{H}_n) > cn^3$ for $n \gg 0$ by Proposition 10.4 it follows that $U(n,x) \neq \emptyset$ for $n \gg 0$. Choose some section $0 \neq s \in U(n,x)$. Every PGL(2, $\mathbb{C}$)-invariant polynomial $P \in I_{n,m}$ gives rise to a section of $\mathcal{I}_{n,m}$ and hence of $S^m(\mathcal{H}_n^\vee)$. We denote this section again by P. Since $s(x) = 0$ it follows that under the evaluation pairing (12) one has $P(s) = 0$. Under the isomorphism $H^0(X, \mathcal{H}_n) = H^0(Y, \mathcal{R}_n)$ we can consider s as a section of $\mathcal{R}_n$. Since $\mathcal{R}_n$ restricts to $\mathcal{O}_{\mathbb{P}_{x'}}(2n)$ on the fibre $\mathbb{P}_{x'}$ of $Y = \mathbb{P}(\mathcal{E}^\vee)$ over $x' \in X$, we can view $s|\mathbb{P}_{x'}$ as an element of W_n. By the equivalence of (ii) and (iii) it follows that $s|\mathbb{P}_{x'}$ either vanishes or has a zero of multiplicity at least $n+1$. Since the components of the divisor $\{s = 0\} \subset Y$ are not all fibres, there must exist a quasi-section S of $Y \to X$ such that $\{s = 0\}$ contains S with multiplicity $(n+1)$. □

We are now in a position to give the
Proof of Bogomolov's Theorem 10.1. We shall first show the easy direction (ii)⇒(i). Using the expressions for the Chern classes of $\mathcal{E}$ as given in Remark 10.2 we find

$$\begin{aligned} c_1^2(\mathcal{E}) - 4c_2(\mathcal{E}) &= (c_1(\mathcal{L}) + c_1(\mathcal{M}))^2 - 4(\mathcal{L}, \mathcal{M}) - 4 \deg Z \\ &= (\mathcal{L} \otimes \mathcal{M}^{-1})^2 - 4 \deg Z > 0, \end{aligned}$$

where the last inequality follows from (7).

Now assume (i). Let $S \subset \mathbb{P}(\mathcal{E}^\vee)$ be a quasi-section as in Proposition 10.6. By Lemma 10.5 this corresponds to an exact sequence

$$0 \to \mathcal{L} \to \mathcal{E} \to \mathcal{I}_Z \cdot \mathcal{M} \to 0$$

and we want to prove that this sequence fulfils (7) and (8). The assertion (7) is easy: By the above computation

$$(\mathcal{L} \otimes \mathcal{M}^{-1})^2 > 4 \deg Z.$$

We shall prove (8) by showing that for some $i > 0$ we have $h^0((\mathcal{L} \otimes \mathcal{M}^{-1})^i) > 0$. This clearly suffices, since a non-zero section of $(\mathcal{L} \otimes \mathcal{M}^{-1})^i$ vanishes on an effective curve C. By (7), which we have already proved, it follows that C is a non-zero effective curve and hence $HC > 0$ for every ample H.

We shall now prove that $h^0((\mathcal{L} \otimes \mathcal{M}^{-1})^i) > 0$ for some $i > 0$. The quasi-section S is associated to some section $s \in H^0(\mathbb{P}(\mathcal{E}^\vee), \mathcal{R}_{n_0})$ for a suitable integer $n_0 > 0$. Since the line bundle $\mathcal{R}_{n_0}$ has degree $2n_0$ on each fibre of $\mathbb{P}(\mathcal{E}^\vee)$ it follows that $h^0(\mathcal{R}_{n_0}(-(2n_0+1)S)) = 0$. For each k in the range $n_0 + 1 \leq k \leq 2n_0$ we consider the exact sequence

$$0 \to \mathcal{R}_{n_0}(-(k+1)S_0) \to \mathcal{R}_{n_0}(-kS_0) \to \mathcal{R}_{n_0}(-kS_0)|_{S_0} \to 0.$$

This sequence implies

$$h^0(\mathcal{R}_{n_0}(-kS_0)|_{S_0}) \geq h^0(\mathcal{R}_{n_0}(-kS_0)) - h^0(\mathcal{R}_{n_0}(-(k+1)S_0)).$$

Summation over k gives

$$\sum_{k=n_0+1}^{2n_0} h^0(\mathcal{R}_{n_0}(-kS_0)|_{S_0}) \geq h^0(\mathcal{R}_{n_0}(-(n_0+1)S_0)) > 0.$$

In particular there is some $i > 0$ such that

$$h^0(\mathcal{R}_{n_0}(-(n_0+1)S_0)|_{S_0}) > 0.$$

Now recall sequence (9) from the proof of Lemma 10.5 where $\pi^*\mathcal{L} = \mathcal{O}_{\mathbb{P}(\mathcal{E}^\vee)}(1) \otimes \mathcal{O}_{\mathbb{P}(\mathcal{E}^\vee)}(-S)$ and $\mathcal{N} = \mathcal{O}_{\mathbb{P}(\mathcal{E}^\vee)}(1)|_S$. Then

$$\begin{aligned}
\mathcal{R}_{n_0}((-(n_0+i)S))|_S &= \mathcal{O}_{\mathbb{P}(\mathcal{E}^\vee)}(2n_0) \otimes (\pi^* \det \mathcal{E})^{-n_0} \otimes \mathcal{O}_{\mathbb{P}(\mathcal{E}^\vee)}(-(n_0+i)S)|_S \\
&= (\mathcal{N} \otimes \pi^*\mathcal{L} \otimes \pi^*(\det \mathcal{E})^{-1})^{n_0} \otimes (\pi^*\mathcal{L} \otimes \mathcal{N}^{-1})^i \\
&= (\mathcal{N} \otimes \pi^*\mathcal{M}^{-1})^{n_0} \otimes (\pi^*\mathcal{L} \otimes \mathcal{N}^{-1})^i \\
&= (\pi^*\mathcal{M})^{-n_0} \otimes \mathcal{N}^{(n_0-i)} \otimes (\pi^*\mathcal{L})^i.
\end{aligned}$$

Moreover $\pi_*\mathcal{N} = \mathcal{I}_Z \cdot \mathcal{M} \subset \mathcal{M}$. Hence $\mathcal{N}|_{S\backslash\pi^{-1}(Z)} \subset \pi^*\mathcal{M}|_{S\backslash\pi^{-1}(Z)}$. Together with the above this shows that

$$\mathcal{R}_{n_0}(-(n_0+i)S)|_{S\backslash\pi^{-1}(Z)} \subset \pi^*(\mathcal{L} \otimes \mathcal{M}^{-1})^i|_{S\backslash\pi^{-1}(Z)}.$$

Hence $H^0(S\backslash\pi^{-1}(Z), \pi^*(\mathcal{L} \otimes \mathcal{M}^{-1})^i|_{S\backslash\pi^{-1}(Z)}) \neq 0$ and pushing down with π_* we find that $H^0(X\backslash Z, (\mathcal{L} \otimes \mathcal{M}^{-1})^i) \neq 0$. Since Z is 2-codimensional we can extend sections over Z and find that $H^0(X, (\mathcal{L} \otimes \mathcal{M}^{-1})^i) \neq 0$ as desired.
□

11. Reider's Method

We start this section with a brief account of the Serre construction which gives a connection between codimension 2 subvarieties and rank 2 vector bundles.

(11.1) **Theorem.** (Serre). *Let X be a surface and $\mathcal{L} \in \operatorname{Pic} X$ be a line bundle on X with $H^2(X, \mathcal{L}^{-1}) = 0$. Assume $Y \subset X$ is a 0-dimensional subscheme of X which is a locally complete intersection. Then there exists a rank 2 vector bundle $\mathcal{E}$ on X with* $\det \mathcal{E} = \mathcal{L}$ *and a section $s \in \Gamma(X, \mathcal{E})$ with $\{s = 0\} = Y$.*

Proof. Since $Y \subset X$ is a locally complete intersection the conormal bundle $\mathcal{N}_{Y/X} = \mathcal{J}_Y/\mathcal{J}_Y^2$ is a locally free rank 2 $\mathcal{O}_Y$-module supported on a 0-dimensional scheme. Hence we can choose an isomorphism

$$e : \det \mathcal{N}_{Y/X}^\vee \to \mathcal{L}^{-1}|_Y.$$

By the local fundamental isomorphism ([A-K], Theorem 4.5) we have an isomorphism

$$\mathcal{H}om_{\mathcal{O}_Y}(\det \mathcal{N}_{Y/X}^\vee, \mathcal{L}^{-1}|_Y) \cong \mathcal{E}xt^2_{\mathcal{O}_X}(\mathcal{O}_Y, \mathcal{L}^{-1}).$$

On the other hand, by applying $\mathcal{H}om_{\mathcal{O}_X}(-, \mathcal{L}^{-1})$

$$0 \to \mathcal{J}_Y \to \mathcal{O}_X \to \mathcal{O}_Y \to 0$$

we find that

$$\mathcal{E}xt^1_{\mathcal{O}_X}(\mathcal{J}_Y, \mathcal{L}^{-1}) \cong \mathcal{E}xt^2_{\mathcal{O}_X}(\mathcal{O}_Y, \mathcal{L}^{-1}).$$

Hence the isomorphism e we chose above induces a global section

$$e \in H^0(X, \mathcal{H}om_{\mathcal{O}_Y}(\det \mathcal{N}_{Y/X}^\vee, \mathcal{L}^{-1}|_Y)) = H^0(X, \mathcal{E}xt^1_{\mathcal{O}_X}(\mathcal{J}_Y, \mathcal{L}^{-1})).$$

The lower term sequence of the spectral sequence

$$E_2^{pq} = H^p(X, \mathcal{E}xt^q_{\mathcal{O}_X}(\mathcal{J}_Y, \mathcal{L}^{-1})) \Rightarrow E^{p+q} = \operatorname{Ext}^{p+q}(\mathcal{J}_Y, \mathcal{L}^{-1})$$

(see [G-H78a], p. 706) gives the following local-global sequence

$$\begin{aligned} 0 \to H^1(X, \mathcal{L}^{-1}) \to \operatorname{Ext}^1(\mathcal{J}_Y, \mathcal{L}^{-1}) \to H^0(X, \mathcal{E}xt^1_{\mathcal{O}_X}(\mathcal{J}_Y, \mathcal{L}^{-1})) \\ \to H^2(X, \mathcal{H}om_{\mathcal{O}_X}(\mathcal{J}_Y, \mathcal{L}^{-1})). \end{aligned}$$

Since

$$H^2(X, \mathcal{H}om_{\mathcal{O}_X}(\mathcal{J}_Y, \mathcal{L}^{-1})) = H^2(X, \mathcal{L}^{-1}) = 0$$

it follows that we can lift $e \in H^0(X, \mathcal{E}xt^1_{\mathcal{O}_X}(\mathcal{J}_Y, \mathcal{L}^{-1}))$ to an element in $\operatorname{Ext}^1(\mathcal{J}_Y, \mathcal{L}^{-1})$ which corresponds to an extension

$$0 \to \mathcal{L}^{-1} \to \mathcal{E}^\vee \to \mathcal{J}_Y \to 0.$$

The sheaf $\mathcal{E}^\vee$ is locally free. This follows by a local algebra result due to Serre (cf. [O-S-S], p. 98) from the fact that $0 \neq e_y \in \mathcal{E}xt^1_{\mathcal{O}_X}(\mathcal{J}_Y, \mathcal{L}^{-1})_y$ is a local generator for each point $y \in Y$.

Tensoring the above sequence with $\mathcal{L}$ gives the desired extension.
□

Remark. The above argument goes through for higher dimensional varieties X and locally complete intersections $Y \subset X$, provided there is an isomorphism $\det \mathcal{N}_{Y/X}^{\vee} \cong \mathcal{L}^{-1}|_Y$, i.e., provided the determinant of the conormal bundle of Y can be extended to a line bundle on X with vanishing second cohomology. In higher dimension this is a non-trivial condition. The theorem then says in geometric language that, if the determinant of the normal bundle can be extended to a line bundle, then the normal bundle itself can be extended to a rank 2 vector bundle on X. This result was rediscovered independently by several authors.

Here we shall need a version of this construction which is due to Griffiths and Harris ([G-H78b], [G-H78a], p. 731).

Definition. Let Y be a 0-dimensional subscheme of X which consists of simple points and let $\mathcal{L} \in \operatorname{Pic} X$. The set Y fulfils the Cayley-Bacharach property with respect to $\mathcal{L} \otimes \mathcal{K}_X$ if for every point $P \in Y$ we have

$$H^0(X, \mathcal{L} \otimes \mathcal{K}_X \otimes \mathcal{I}_Y) = H^0(X, \mathcal{L} \otimes \mathcal{K}_X \otimes \mathcal{I}_{Y-\{P\}}).$$

(11.2) **Theorem** (Griffiths-Harris). *Let Y be a reduced 0-dimensional subscheme of X and $\mathcal{L} \in \operatorname{Pic} X$. Then there exists a rank 2 vector bundle $\mathcal{E}$ on X with $\det \mathcal{E} = \mathcal{L}$ and a section $s \in H^0(X, \mathcal{E})$ with $Y = \{s = 0\}$ if and only if Y fulfils the Cayley-Bacharach property with respect to $\mathcal{K}_X \otimes \mathcal{L}$.*

Proof. We look at part of the local-global sequence

$$(13) \qquad \to \operatorname{Ext}^1(\mathcal{I}_Y, \mathcal{L}) \to H^0(X, \mathcal{E}xt^1_{\mathcal{O}_X}(\mathcal{I}_Y, \mathcal{L}^{-1})) \xrightarrow{\phi} H^2(X, \mathcal{L}^{-1}).$$

We have the following isomorphisms of $\mathcal{O}_X$-modules:

$$\begin{aligned} &\mathcal{E}xt^1_{\mathcal{O}_X}(\mathcal{I}_Y, \mathcal{L}^{-1}) \cong \mathcal{E}xt^2_{\mathcal{O}_X}(\mathcal{O}_Y, \mathcal{L}^{-1}) \cong \\ &\mathcal{H}om_{\mathcal{O}_X}(\det \mathcal{N}_{Y/X}^{\vee}, \mathcal{L}^{-1}|_Y) \cong (\mathcal{K}_X \otimes \mathcal{L})^{-1} \otimes \mathcal{O}_Y \end{aligned}$$

where the last isomorphism comes from the adjunction formula. By duality theory, as explained in detail in [G-H78a], p. 729–731, the homomorphism induced by the map ϕ from (13),

$$\bigoplus_{y \in Y} (\mathcal{K}_X \otimes \mathcal{L})_y^{-1} \cong H^0(X, \mathcal{E}xt^1_{\mathcal{O}_X}(\mathcal{I}_Y, \mathcal{L}^{-1})) \to H^0(X, \mathcal{K}_X \otimes \mathcal{L})^{\vee},$$

is dual to the evaluation map

$$H^0(X, \mathcal{K}_X \otimes \mathcal{L}) \longrightarrow \bigoplus_{y \in Y} (\mathcal{K}_X \otimes \mathcal{L})_y.$$

If $Y = \{P_1 \ldots, P_n\}$, then

$$(\mathcal{K}_X \otimes \mathcal{L})^{-1} \otimes \mathcal{O}_Y = \bigoplus_{i=1}^{n} (\mathcal{K}_X \otimes \mathcal{L})_{P_i}^{-1} \cong \mathbb{C}^n.$$

The existence of a locally free $\mathcal{E}$ is now equivalent to the statement that the kernel of ϕ is not contained in a coordinate hyperplane of $\bigoplus_{i=1}^{n}(\mathcal{K}_X \otimes \mathcal{L})_{P_i}^{-1} \cong \mathbb{C}^n$. Dually this is exactly the Cayley-Bacharach property. □

For Reider's theorem we shall also need a version of this result in case where Y is non-reduced. We shall limit ourselves here to the simple case where length $Y = 2$, since this is sufficient for our application. For a discussion of the general case see [Cat90].

(11.3) **Lemma.** *Let $Y \subset X$ be a non-reduced subscheme of length 2 supported at a point P. Assume that P is not a base point of $\mathcal{K}_X \otimes \mathcal{L}$, but that $H^0(X, \mathcal{K}_X \otimes \mathcal{L}) \to \mathcal{K}_X \otimes \mathcal{L} \otimes \mathcal{O}_Y$ is not surjective. Then there exists a rank 2 vector bundle $\mathcal{E}$ with $\det \mathcal{E} = \mathcal{L}$ and a section $s \in H^0(X, \mathcal{E})$ with $\{s = 0\} = Y$.*

Proof. By our assumption the image of the evaluation map

$$H^0(X, \mathcal{K}_X \otimes \mathcal{L}) \to \mathcal{K}_X \otimes \mathcal{L} \otimes \mathcal{O}_Y$$

does not contain the maximal ideal. Dually this means that $\ker \phi$ is not contained in the annihilator of the maximal ideal. It follows that we can find an element in the kernel of ϕ which is a generator at P. This gives us the desired vector bundle. □

(11.4) **Theorem.** (Reider). *Let X be a projective surface and $\mathcal{L} \in \operatorname{Pic} X$ be a nef line bundle.*

(i) *Assume $\mathcal{L}^2 \geq 5$. If P is a base point of $|\mathcal{K}_X \otimes \mathcal{L}|$, then there exists a curve D with $P \in D$ and such that*

(a) $(D, \mathcal{L}) = 0$ *and* $D^2 = -1$ *or*

(b) $(D, \mathcal{L}) = 1$ *and* $D^2 = 0$.

(ii) *Assume $\mathcal{L}^2 \geq 9$ and assume that P and Q are not base points of $|\mathcal{K}_X \otimes \mathcal{L}|$. If $|\mathcal{K}_X \otimes \mathcal{L}|$ fails to separate two points P and Q (possibly infinitely near), then there exists a curve D with $P, Q \in D$ and such that*

(a) $(D, \mathcal{L}) = 0$ *and* $D^2 = -2$ *or* -1 *or*

(b) $(D, \mathcal{L}) = 1$ *and* $D^2 = -1$ *or* 0 *or*

(c) $(D, \mathcal{L}) = 2$ *and* $D^2 = 0$ *or*

(d) $\mathcal{L}^2 = 9$, *$c_1(\mathcal{L})$ cohomologous to $3D$ modulo torsion, and hence* $D^2 = 1$.

Proof. We shall first prove (i). Assume that P is a base point of $|\mathcal{K}_X \otimes \mathcal{L}|$. This means that $Y = \{P\}$ fulfils the Cayley-Bacharach property with respect to $\mathcal{K}_X \otimes \mathcal{L}$ and hence we have an extension

$$0 \to \mathcal{O}_X \xrightarrow{s} \mathcal{E} \to \mathcal{J}_P \otimes \mathcal{L} \to 0.$$

Since $c_1(\mathcal{E}) = c_1(\mathcal{L})$ and $c_2(\mathcal{E}) = 1$ we have

$$c_1^2(\mathcal{E}) - 4c_2(\mathcal{E}) = \mathcal{L}^2 - 4 > 0$$

which implies that $\mathcal{E}$ is Bogomolov unstable. Hence by Bogomolov's theorem we obtain a diagram

$$\begin{array}{ccccccccc} & & & & 0 & & & & \\ & & & & \downarrow & & & & \\ & & & & \mathcal{O}_X(A) & \searrow & & & \\ & & & & \downarrow & t' & & & \\ 0 & \to & \mathcal{O} & \xrightarrow{s} & \mathcal{E} & \to & \mathcal{J}_P \otimes \mathcal{L} & \to & 0 \\ & & & \searrow^{t} & \downarrow & & & & \\ & & & & \mathcal{J}_Z \cdot \mathcal{O}_X(B) & & & & \\ & & & & \downarrow & & & & \\ & & & & 0 & & & & \end{array}$$

with divisors A and B such that

(14) $$A + B = c_1(\mathcal{E}) = c_1(\mathcal{L}),$$

(15) $$(A - B)^2 > 0, (A - B)H > 0 \quad \text{for every ample } H.$$

We first claim that $t \not\equiv 0$ (and hence $t' \not\equiv 0$). If $t \equiv 0$ then there is a non-zero homomorphism $\mathcal{L} \to \mathcal{O}_S(B)$ which implies that $-A \geq 0$. On the other hand by (15) and (14)

$$(A + A)H > (A + A)H + (B - A)H = (\mathcal{L}, H) \geq 0$$

which implies $AH > 0$, a contradiction to $-A \geq 0$. Let $D := \{t = 0\}$. Since $s(P) = 0$ it follows that $P \in D$. It remains to prove that D has the right intersection numbers. By construction $\mathcal{O}_X(B) \cong \mathcal{O}_X(D)$ and hence also $\mathcal{O}_X(A) \cong \mathcal{L}(-D)$. We first claim that

(16) $$0 \leq (D, c_1(\mathcal{L}) - D) \leq 1.$$

Indeed, let $D = \Sigma n_i D_i$ be the decomposition of D into irreducible components. Since $t|_{D_i} \equiv 0$ we have a non-zero homomorphism $\mathcal{O}_{D_i} \to \mathcal{O}_X(A)|_{D_i} = \mathcal{L}(-D)|_{D_i}$. This implies $(c_1(\mathcal{L}) - D)D_i \geq 0$ and summation over i gives $(c_1(\mathcal{L}) - D, D) \geq 0$. On the other hand $1 = c_2(\mathcal{E}) = (D, c_1(\mathcal{L}) - D) + \deg Z$. Since $\deg Z \geq 0$ this implies that $(D, c_1(\mathcal{L}) - D) \leq 1$. Our next claim is

(17) $$D^2 \leq 0.$$

Indeed, from the assumption $\mathcal{L}^2 \geq 5$ it follows that

$$\mathcal{L}^2 = ((c_1(\mathcal{L}) - D) + D)^2 = (c_1(\mathcal{L}) - D)^2 + 2(c_1(\mathcal{L}) - D, D) + D^2 \geq 5.$$

Using (16) this shows that

$$(c_1(\mathcal{L}) - D)^2 + D^2 \geq 3. \tag{18}$$

On the other hand $A - B = c_1(\mathcal{L}) - 2D$ and we know from the proof of Bogomolov's theorem that a positive multiple of $A - B$ is effective. Hence, since $\mathcal{L}$ is nef, it follows that

$$(\mathcal{L}, c_1(\mathcal{L}) - 2D) = ((c_1(\mathcal{L}) - D) + D, (c_1(\mathcal{L}) - D) - D) \geq 0$$

and this shows $(c_1(\mathcal{L}) - D)^2 \geq D^2$, which by (18) gives $(c_1(\mathcal{L}) - D)^2 \geq 2$. Now by the Hodge index theorem

$$(c_1(\mathcal{L}) - D)^2 D^2 \leq (c_1(\mathcal{L}) - D, D)^2 \leq 1$$

where the second inequality comes from (16). Hence $D^2 \leq 0$.

Since $D \geq 0$ and $\mathcal{L}$ is nef it follows that $(\mathcal{L}, D) \geq 0$. Hence

$$-D^2 \leq (\mathcal{L}, D) - D^2 = (c_1(\mathcal{L}) - D, D) \leq 1$$

where the last inequality follows from (16). Together with (17) this shows $-1 \leq D^2 \leq 0$. First assume $D^2 = 0$. By (16) this implies $0 \leq (\mathcal{L}, D) \leq 1$. We cannot have $(\mathcal{L}, D) = 0$ since $D^2 = (\mathcal{L}, D) = 0$ and $\mathcal{L}^2 > 0$ would contradict the Hodge index theorem. Hence in this case $(\mathcal{L}, D) = 1$ and we are in case (b). Now assume $D^2 = -1$. Then by (16) we have $(\mathcal{L}, D) \leq 0$, i.e., since $D \geq 0$ and $\mathcal{L}$ is nef $(\mathcal{L}, D) = 0$ which is case (a).

The proof of (ii) goes along similar lines. Let Y be the length 2 subscheme given by P and Q. Under the assumptions of (ii) we can either apply Theorem 11.2 or Lemma 11.3 and construct a rank 2 vector bundle $\mathcal{E}$ with $\det \mathcal{E} = \mathcal{L}$ and a section $s \in H^0(X, \mathcal{E})$ whose zero locus is Y. Again the bundle $\mathcal{E}$ is Bogomolov unstable since

$$c_1^2(\mathcal{E}) - 4c_2(\mathcal{E}) = \mathcal{L}^2 - 4 \deg Y > 0$$

by our assumption $\mathcal{L}^2 \geq 9$. As a consequence we obtain a diagram exactly as in the proof of part (i). As before we set $D = \{t = 0\}$ and by construction $Y \subset D$. Arguing as in the proof of formula (16) we find that

$$0 \leq (D, c_1(\mathcal{L}) - D) \leq 2. \tag{19}$$

As in the proof of formula (18) we can conclude that

$$(c_1(\mathcal{L}) - D)^2 + D^2 \geq 5 \quad (\text{resp. } \geq 6 \text{ if } \mathcal{L}^2 \geq 10) \tag{20}$$

and also

$$(c_1(\mathcal{L}) - D)^2 \geq D^2. \tag{21}$$

By the Hodge index theorem

$$\det \begin{pmatrix} (c_1(\mathcal{L}) - D)^2 & (c_1(\mathcal{L}) - D, D) \\ (c_1(\mathcal{L}) - D, D) & D^2 \end{pmatrix} \leq 0.$$

Using formulae (19)–(21) we see that this implies $D^2 \leq 0$ with the only possible exception $c_1^2(\mathcal{L}) = 9$ and $D^2 = 1$. Since $D \geq 0$ and $\mathcal{L}$ is nef it follows that $(\mathcal{L}, D) \geq 0$ and hence

$$-D^2 \leq (\mathcal{L}, D) - D^2 = (c_1(\mathcal{L}) - D, D) \leq 2,$$

and therefore $D^2 \geq -2$. First assume $D^2 = 0$. Then by (19) we know that $0 \leq (\mathcal{L}, D) \leq 2$. Since $(\mathcal{L}, D) = 0$ is excluded by the Hodge index theorem we are in case (b) or (c). Similarly $D^2 = -1$ implies $0 \leq (\mathcal{L}, D) \leq 1$ and hence we are in case (b) or (a). In the same way $D^2 = -2$ implies $(\mathcal{L}, D) = 0$, i.e., case (a). It remains to treat the case $D^2 = 1$ in which case $c_1^2(\mathcal{L}) = 9$. By (19) we have $(\mathcal{L}, D) \leq 3$. The Hodge index theorem gives $9 = \mathcal{L}^2 D^2 \leq (\mathcal{L}, D)^2 \leq 9$ and this implies that $c_1(\mathcal{L}) = 3[D]$ in rational cohomology. □

Here we gave the original version of Reider's theorem. Meanwhile many refinements have been proven of which we only want to give one example. It concerns k-very ample line bundles on a projective surface X, i.e., line bundles $\mathcal{L}$ having the property that for every length $(k+1)$ subscheme Z of X the natural restriction map $H^0(X, \mathcal{L}) \to H^0(Z, \mathcal{L} \otimes \mathcal{O}_Z)$ is surjective. Clearly, 0-very ample line bundles are those without base points and 1-very ample line bundles are exactly the very ample ones. We have [B-So]:

Theorem. *Let $\mathcal{L}$ be a nef line bundle on a smooth projective surface X and suppose that $\mathcal{L}^2 \geq 4k + 5$. Then either $\mathcal{K}_X \otimes \mathcal{L}$ is k-very ample or there exists an effective divisor D containing some 0-dimensional scheme of length $\leq k + 1$ along which k-ampleness fails, such that a power of the line bundle $\mathcal{L}(-2D)$ has sections and such that*

$$(D, \mathcal{L}) - k - 1 \leq D^2 < \frac{1}{2}(D, \mathcal{L}) < k + 1.$$

12. Vanishing Theorems on Surfaces

We start with a vanishing theorem due to Mumford, who pointed out that it can be deduced from Bogomolov's theorem.

(12.1) **Theorem** (Mumford). *Let $\mathcal{L}$ be a nef line bundle on a smooth projective surface X with $c_1^2(\mathcal{L}) > 0$. Then $H^1(X, \mathcal{L}^{-1}) = 0$.*

Proof. Since $H^1(X, \mathcal{L}^{-1}) \cong \mathrm{Ext}^1(\mathcal{L}, \mathcal{O}_X)$ every element of $H^1(X, \mathcal{L}^{-1})$ determines an extension

$$0 \to \mathcal{O}_X \to \mathcal{E} \to \mathcal{L} \to 0.$$

We have to show that under the given assumptions on $\mathcal{L}$ this sequence splits. First notice that we are in a position to apply Bogomolov's theorem, since $c_1^2(\mathcal{E}) = c_1^2(\mathcal{L}) > 0$ and $c_2(\mathcal{E}) = 0$. Hence Bogomolov's theorem gives rise to a diagram

$$
\begin{array}{ccccccc}
 & & & 0 & & & \\
 & & & \downarrow & & & \\
 & & & \mathcal{O}_X & & & \\
 & & & \downarrow & & & \\
0 \to & \mathcal{L}' & \to & \mathcal{E} & \to & \mathcal{I}_Z \cdot \mathcal{M}' & \to 0 \\
 & & \overset{\varphi}{\searrow} & \downarrow & & & \\
 & & & \mathcal{L} & & & \\
 & & & \downarrow & & & \\
 & & & 0 & & &
\end{array}
$$

where the horizontal sequence comes from Bogomolov's theorem. In particular $c_1^2(\mathcal{L}' \otimes \mathcal{M}'^{-1}) > 0$ and $(\mathcal{L}' \otimes \mathcal{M}'^{-1}, H) > 0$ for every ample H. We first show that $\varphi \neq 0$. If φ were the zero map we would have $\mathcal{L}' = \mathcal{O}_X(-D)$ for some $D \geq 0$. This implies $\mathcal{M}' = \mathcal{L}(D)$ and $\mathcal{L}' \otimes \mathcal{M}'^{-1} = \mathcal{L}^{-1}(-2D)$. But then $(\mathcal{L}' \otimes \mathcal{M}'^{-1}, H) = -(\mathcal{L}, H) - 2(D, H) \leq 0$, a contradiction. Hence we can conclude that $\mathcal{L}' = \mathcal{L}(-D)$ with φ being the natural inclusion. It suffices to show that $D = 0$, since we have then constructed a lifting of the quotient $\mathcal{E} \to \mathcal{L}$ and thus shown that the extension splits.

It follows from $\mathcal{L}' = \mathcal{L}(-D)$ that $\mathcal{M}' = \det \mathcal{E} \otimes \mathcal{L}'^{-1} = \mathcal{L} \otimes \mathcal{L}^{-1}(D) = \mathcal{O}_X(D)$. On the other hand

$$
\begin{aligned}
0 = c_2(\mathcal{E}) &= (\mathcal{L}', \mathcal{M}') + \deg Z \\
&= (c_1(\mathcal{L}) - D, D) + \deg Z
\end{aligned}
$$

and this implies that $D^2 \geq (\mathcal{L}, D)$. We also know from the proof of Bogomolov's theorem that a positive multiple of $\mathcal{L}' \otimes \mathcal{M}'^{-1} = \mathcal{L}(-2D)$ is effective. Hence by our assumption on $c_1(\mathcal{L})$ it follows that $(\mathcal{L}, c_1(\mathcal{L}) - 2D) \geq 0$ or in other words $\mathcal{L}^2 \geq 2(\mathcal{L}, D)$. The two inequalities which we have just proved show that either

$$
\det \begin{pmatrix} \mathcal{L}^2 & (\mathcal{L}, D) \\ (\mathcal{L}, D) & D^2 \end{pmatrix} = \mathcal{L}^2 D^2 - (\mathcal{L}, D)^2 > 0
$$

or $D^2 = (\mathcal{L}, D) = 0$. The first possibility contradicts the index theorem. Since D is effective, the latter implies by once more using the index theorem that $D = 0$. □

Remark. For any line bundle $\mathcal{L}$ as in the theorem $h^0(\mathcal{L}^m)$ grows quadratically in m. This follows from Prop. 7.4. In current parlance one calls such a line bundle "big". Conversely, for a nef and big line bundle one can show that $c_1(\mathcal{L})^2 > 0$. Indeed, $\mathcal{L}$ being nef implies (Observation 7.6) that $c_1(\mathcal{L})^2 \geq 0$ and we have to exclude $c_1(\mathcal{L})^2 = 0$. Replacing $\mathcal{L} = \mathcal{O}_X(D)$ by a suitable power we may assume that $|D|$ is not composed with a pencil (for this terminology see Sect. 1). So, if we write $D = E + V$, where V is the fixed part, we have $E^2 > 0$. Suppose now that $D^2 = 0$. Using that E is nef (since it moves in a linear system) and D is effective and nef, from $0 = D^2 = (E+V)D = ED + VD$, we infer that both terms vanish and so $E^2 = E \cdot (D - V) = 0$, contradicting $E^2 > 0$.

The previous theorem can thus be paraphrased by saying that if $\mathcal{L}$ is nef and big, then $H^1(X, \mathcal{L}^{-1}) = 0$. In this formulation, the theorem is a special case of the vanishing theorem of Kawamata [Kaw], and Viehweg [Vie82] which we now state for surfaces:

(12.2) **Theorem** (Kawamata-Viehweg). *Let $\mathcal{L}$ be a line bundle on a surface X. If for some positive integer N and some normal crossing divisor D the line bundle $\mathcal{L}^N \otimes \mathcal{O}_X(-D)$ is nef and big, then $H^1(X, \mathcal{L}^{-1}) = 0$.*

The special case $D = 0$ is Mumford's theorem stated above.

In the literature the preceding theorem is usually formulated in terms of fractional divisors. A fractional divisor is a divisor of the form $\sum_{i=1}^{k} a_i D_i$, where the D_i are irreducible divisors and the rational numbers a_i are fractional, i.e., $0 \le a_i < 1$. Then the Kawamata-Viehweg vanishing theorem says the following: let D' be a fractional divisor supported on a normal crossing divisor and let E be a big and nef $\mathbb{Q}$-divisor such that $E + D'$ is a divisor, say $\mathcal{O}_X(E + D') = \mathcal{L}$; then $H^k(X, \mathcal{L}^{-1}) = 0$ for $k < \dim X$.

We now discuss another version of Mumford's theorem:

(12.3) **Corollary** (Mumford's vanishing theorem). *Let X be a compact Kähler surface and $\mathcal{L}$ a line bundle on X. If for $N \gg 0$ the bundle $\mathcal{L}^N$ is globally generated and has three algebraically independent sections, then*

$$H^1(\mathcal{L}^{-1}) = 0.$$

Proof. Since $\mathcal{L}^N$ is globally generated the line bundle $\mathcal{L}^N$ and hence also $\mathcal{L}$ must be nef. The fact that $\mathcal{L}^N$ has three algebraically independent sections implies that $h^0(\mathcal{L}^N)$ grows quadratically ([Ue75], p. 86). So $\mathcal{L}$ is nef and big. By the previous remark $c_1(\mathcal{L})^2 > 0$ and so X must be projective (Theorem 6.2) and the result follows from Theorem 12.1. □

Mumford's vanishing theorem contains as a special case a theorem, which is the algebraic formulation of Kodaira's vanishing theorem (for surfaces).

(12.4) **Theorem** (Kodaira's vanishing theorem). *Let X be a compact surface and $\mathcal{L}$ an ample line bundle on X. Then $h^1(\mathcal{L}^\vee) = 0$.*

We close this section with Ramanujam's theorem. It is a criterion for the vanishing of $H^1(\mathcal{O}_X(-D))$, where D is an effective divisor on a compact Kähler surface X.

(12.5) **Theorem** (Ramanujam's vanishing theorem). *Let X be a compact Kähler surface and D an effective divisor on X, such that*

(i) *D is 1-connected,*

(ii) *$h^0(\mathcal{O}_X(nD)) \ge 2$ for some $n \ge 1$,*

(iii) *the linear system* $|nD|$ *is not composed with an irrational pencil.*
Then $h^1(\mathcal{O}_X(-D)) = 0$.

For the concept of "being composed with an irrational pencil" we again refer to Sect. 1.

The following corollary is a version of theorem 12.1 where the hypothesis $\mathcal{L}$ nef is replaced by $\mathcal{L} = \mathcal{O}_X(D)$, D effective and 1-connected:

(12.6) **Corollary.** *Let* X *be a compact Kähler surface and* D *an effective, 1-connected divisor on* X *with* $D^2 > 0$. *Then* $h^1(\mathcal{O}_X(-D)) = 0$.

Proof of the Corollary. By Prop. 7.4 we have that $h^0(\mathcal{O}_X(nD))$ grows quadratically in n and so linear system $|nD|$ cannot be composed with a pencil, for n large. □

Before we come to the proof of Theorem 12.5, we have to deal with a few auxiliary results.

(12.7) **Lemma.** *Let* $C \subset X$ *be an effective divisor and* $\mathrm{restr}_C : H^1(\mathcal{O}_X) \to H^1(\mathcal{O}_C)$ *the restriction map. If* η *is an element in* $H^1(\mathcal{O}_X)$ *with* $\mathrm{restr}_C(\eta) = 0$, *then already* $\eta|U = 0$ *for an open neighbourhood* U *of* C *in* X. *In particular, the kernel of* $\mathrm{restr}_{nC} : H^1(\mathcal{O}_X) \to H^1(\mathcal{O}_{nC})$ *is independent of* n.

Proof. By Proposition II.2.1 the subgroup $H^1(C, \mathbb{Z}) \subset H^1(\mathcal{O}_C)$ is discrete. Since X is Kähler, $H^1(X, \mathbb{Z})$ spans $H^1(\mathcal{O}_X)$ over $\mathbb{R}$ (I. Sect. 13). It follows that the kernel of restr_C is spanned over $\mathbb{R}$ by the kernel of $H^1(X, \mathbb{Z}) \to H^1(C, \mathbb{Z})$. This is the same as the kernel of $H^1(X, \mathbb{Z}) \to H^1(U, \mathbb{Z})$, where U is a suitable open neighbourhood of C in X (cf. Theorem I.8.8). □

(12.8) **Lemma.** *Let* D *be an effective divisor on* X *with linear system* $|D|$ *free of base points. Then*

$$\mathrm{restr}_D : H^1(\mathcal{O}_X) \to H^1(\mathcal{O}_D)$$

is injective if and only if the system $|D|$ *is not composed with an irrational pencil.*

Proof. Let $\varphi_D = \varphi : C \to \mathbb{P}_N$, $N = \dim |D|$, be the map determined by $|D|$. Like in Sect. 1, we distinguish between two cases:

The case $\dim \varphi(X) = 1$. Let $\varphi = \rho\psi$, with $\psi : X \to S$ and S a non-singular Riemann surface, be the Stein factorization of φ. Then $\psi_* \mathcal{O}_X = \mathcal{O}_S$ ([Rem57], p. 363), and Leray's spectral sequence gives an exact sequence

$$0 \longrightarrow H^1(\mathcal{O}_S) \xrightarrow{\psi^*} H^1(\mathcal{O}_X) \longrightarrow H^0(\psi_{*1}\mathcal{O}_X) \longrightarrow 0.$$

If S has genus ≥ 1, then $\psi^*(H^1(\mathcal{O}_S)) \subset H^1(\mathcal{O}_X)$ is a non-trivial subspace in the kernel of restr_D. If S is rational, we consider a class $\eta \in H^1(\mathcal{O}_X)$ with $\mathrm{restr}_D(\eta) = 0$. Let X_s, $s \in S$, be a connected component of D. Then η vanishes on X_s too, and by Lemma 12.7, η even vanishes in a neighbourhood of X_s. By Corollary III.11.2 the direct image sheaf $\psi_{*1}\mathcal{O}_X$ is locally free. The image of η in $H^0(\psi_{*1}\mathcal{O}_X)$ vanishes on a neighbourhood of s, i.e., it is the zero section. But now $H^1(\mathcal{O}_X) \to H^0(\psi_{*1}\mathcal{O}_X)$ is injective and we find $\eta = 0$.

The case $\dim \varphi(X) = 2$. We use Bertini's theorem I.20.2 saying in particular that each $D' \in |D|$ is connected. We pick D' such that $\mathrm{supp}(D) \cap \mathrm{supp}(D')$ is finite, and blow up this intersection by way of $\tau : \bar{X} \to X$ such that the meromorphic map $X \to \mathbb{P}^1$ defined by the 1-dimensional system $|\lambda D + \mu D'|$, $\lambda, \mu \in \mathbb{C}$, induces a holomorphic map $\psi : \bar{X} \to \mathbb{P}_1$. This ψ is connected, and one fibre $\bar{X}_s$, $s \in \mathbb{P}_1$, is contained in the total transform $\tau^*(D)$ of D. By the commutativity of

$$\begin{array}{ccccc} H^1(\mathcal{O}_X) & \xrightarrow{\text{restr}} & H^1(\mathcal{O}_D) & & \\ \| & & \downarrow{\scriptstyle \tau^*} & & \\ H^1(\mathcal{O}_{\bar{X}}) & \xrightarrow{\text{restr}} & H^1(\mathcal{O}_{\tau^* D}) & \xrightarrow{\text{restr}} & H^1(\mathcal{O}_{\bar{X}_s}) \end{array}$$

it suffices to show that $\mathrm{restr} : H^1(\mathcal{O}_{\bar{X}}) \to H^1(\mathcal{O}_{\bar{X}_s})$ is injective, which can be done as in the case $\dim \varphi(X) = 1$ above. □

Proof of Theorem 12.5. We consider the exact sequence

$$0 \longrightarrow \Gamma(\mathcal{O}_X) \longrightarrow \Gamma(\mathcal{O}_D) \longrightarrow H^1(\mathcal{O}_X(-D)) \longrightarrow H^1(\mathcal{O}_X) \longrightarrow H^1(\mathcal{O}_D).$$

Since D is 1-connected, $\Gamma(\mathcal{O}_D) \cong \mathbb{C}$ by Lemma II.12.2, and it suffices to prove injectivity of $\mathrm{restr}_D : H^1(\mathcal{O}_X) \to H^1(\mathcal{O}_D)$. By Lemma 12.7 this is equivalent to the injectivity of restr_{nD} for sufficiently large $n \in \mathbb{N}$. We choose n as in the theorem and put $nD = F + C$, with F the fixed part of nD. Since restr_C factors through restr_{nD}, it suffices to prove injectivity of restr_C. Next we blow up the fixed points of $|C|$ via $\sigma : \bar{X} \to X$ such that $\sigma^*(C) = \overline{F} + B$ with $\overline{F}$ the fixed part of $|\sigma^*(C)|$ and $|B|$ without base points. Just as in the proof of Lemma 12.8 above it suffices to show the injectivity of $\mathrm{restr}_B : H^1(\mathcal{O}_X) \to H^1(\mathcal{O}_B)$, which is again obtained by the argument used to prove Lemma 12.8. □

Historical Remarks

Practically all results in Sects. 2, 6 and 8 are due to Kodaira, though he never mentions explicitly the Fröhlicher spectral sequence. Most of his results in this direction appear in [Ko66] part I.

The proof given in Sect. 3 closely follows [Lama] but incorporates simplifications based on [Dema-P]. There is another proof due to Buchdahl ([Buch]), likewise based on Demailly's regularity theorem

Propositions 5.1 and 5.2 go back to Castelnuovo and de Franchis ([Cas05]), whereas Proposition 5.3 can be found in [Ve66] or [Ve76].

Corollary 6.5 is due to Chow and Kodaira (see [C-K]).

A first bound for the number of curves on a surface X with $a(X) = 0$ was given in [Ko60] (Theorem 5.0); for the bound given in Sect. 8 see [Kr] and [F-F].

The first examples of almost-complex surfaces without complex structure appeared in [Ve66], and other examples can e.g. be found in [Y76], [Bro], [Bra].

The exposition in Sect. 10 follows very closely Reid's exposition of Bogomolov's theorem [Rei77]. As explained there, the inequality $c_1^2 \leq 4c_2$ can be deduced from it.

The proof of Reider's theorem in Sect. 11 can be found in [Reider]. There are refinements in [Cat90].

For the results of Sect. 12 we refer to [Mu62], [Ram], [Bom73]. Del Busto [Bus] has shown that one can also deduce Bogomolov's theorem from the Kawamata-Viehweg vanishing theorem on surfaces.

Chapter V. Examples

In this chapter a surface will again mean a connected 2-dimensional complex manifold.

Some Classical Examples

1. The Projective Plane $\mathbb{P}_2$

A universally known and in many ways most basic compact complex surface is the projective plane. Some obvious questions concerning $\mathbb{P}_2$ have fascinated many a geometer, in particular the question (raised by Severi in [Sev]) whether a surface which is homeomorphic to $\mathbb{P}_2$, is also isomorphic to $\mathbb{P}_2$. The last and very difficult step towards the affirmative answer was done only in the late 1970s by S. - T. Yau. The striking point is that the only known proof uses analysis (hidden in Riemann-Roch for non-algebraic surfaces) and differential geometry as well as the methods of analytic and algebraic geometry.

(1.1) **Theorem.** *Let X be a compact surface with $b_1(X) = 0, b_2(X) = 1$. Then X is algebraic. If furthermore either* (i) $P_2(X) = 0$ *or* (ii) $\pi_1(X)$ *is finite, then X is (algebraically) isomorphic to $\mathbb{P}_2$.*

Proof. We either have $b^+(X) = 0$ or $b^+(X) = 1$. The first possibility is excluded since the index formula would give $c_1^2(X) = 3$, leading to a violation of $c_1^2(X) + e(X) \equiv 0\,(\mathrm{mod} 12)$. If $b^+(X) = 1$, the index formula yields $c_1^2(X) = 9$, and hence X is algebraic by Corollary IV. 6.3.

Now let $P_2(X) = 0$, and let g be a generator for $H_2(X, \mathbb{Z})/(\text{torsion})$, such that for some $n > 0$ the class ng is the class of a hyperplane section H. Then there is no effective divisor homologous (mod torsion) to ag with $a \leq 0$. We have (again by the index formula) that $c_1(X) = \pm 3g$ (mod torsion). The case $c_1(X) = -3g$ (mod torsion) is excluded since here Riemann-Roch would imply $P_2(X) > 0$. In the case that $c_1(X) = 3g$ (mod torsion), we take a line bundle $\mathcal{L}$ with $c_1(\mathcal{L}) = g$ (this is possible by the exponential cohomology sequence, since $h^{2,0}(X) = 0$), and we put $\mathcal{L} = \mathcal{O}_X(L)$. Then Riemann-Roch yields that dim $|L| \geq 2$. No element of $|L|$ can be reducible, otherwise there would be an effective divisor on X, homologous (mod torsion) to ag, with $a \leq 0$. So $|L|$ has no fixed components. Since $L^2 = 1$ and dim $|L| \geq 2$ there cannot be any fixed points either. Hence $f_{\mathcal{L}}$ is everywhere defined and maps X birationally onto a surface of degree 1, i.e., onto the projective plane (it follows that dim $|L| = 2$). If $f_{\mathcal{L}}$ would map any curve onto a point, we would have two curves on X which do not intersect, which is impossible since $b_2(X) = 1$. So $f_{\mathcal{L}}$ is an isomorphism by Lemma III. 4.3. This proves (i). To prove (ii) it is sufficient to show that if $P_2(X) > 0$, then $\pi_1(X)$ is infinite. In this case $c_1(X) = -3g$ (mod torsion), and there exists a positive integer m, such that $-mc_1(X)$ is a hyperplane class H' (with torsion taken into account). Therefore by Theorem I. 15.2 and Corollary I. 15.5 the universal covering of X is the unit ball in $\mathbb{C}^2$, so $\pi_1(X)$ is infinite. □

(1.2) *Remark.* As was shown by Mumford in [Mu79], there exists at least one other algebraic surface X with $b_1(X) = 0, b_2(X) = 1$. It follows as before that X is a quotient of the unit ball $E \subset \mathbb{C}^2$. But Mumford has obtained his example in a totally different way and it is not yet known how to obtain it as a quotient of E. Ishida [Is88] discusses an elliptic surface Y covered by a (blow-up) of Mumford's surface. The surface Y, although its properties are stated explicitly, could up to now not be constructed in a direct way. So this does not yet yield an alternative approach to Mumford's surface. As we shall prove later on (Theorem VII. 5.1) $f_{\mathcal{K}^{\otimes 5}}$ provides an embedding of X as a surface of degree $25c_1^2(X) = 225$ in $\mathbb{P}_{90}$. Since (by a theorem of Calabi and Vesentini, see [C-V]) any small deformation of a quotient of E is trivial, and since the subset on a Chow scheme representing smooth varieties consists of finitely many connected components, we see that in any case there exists only a finite number of surfaces X with $b_1(X) = 0, b_2(X) = 1$. But there might be many "fake projective planes"!

(1.3) **Example.** *Every deformation of* $\mathbb{P}_2$ *is isomorphic to* $\mathbb{P}_2$.

This fact is of course an immediate consequence of Theorem1.1, (ii) but it can also be proved without making use of Yau's results. In fact, by upper-semi-continuity all plurigenera of a small deformation of $\mathbb{P}_2$ vanish, hence every small deformation of $\mathbb{P}_2$ is isomorphic to $\mathbb{P}_2$ by the elementary part (i) of Theorem 1.1. To complete the proof it is sufficient to show that if we have a 1-dimensional deformation of a surface X, say X_t, with $X_0 = X$, such that X_{t_i} is isomorphic to $\mathbb{P}_2$ for a sequence $t_i \to 0$, then also X_0 is isomorphic to $\mathbb{P}_2$. This follows again from semi-continuity, but this time applied to $h^0(\mathcal{K}_{X_t}^{\vee})$, which does not vanish for $t = t_i$, hence does not vanish for $t = 0$. If also $h^0(\mathcal{K}_{X_0}) \neq 0$ it would follow that $\mathcal{K}_{X_0}$ were trivial, contradicting Prop. IV.5.4. So $h^0(\mathcal{K}_{X_0}) = 0$. So, if now $h^0(\mathcal{K}_{X_0}^2) \neq 0$, the fact that $h^0(\mathcal{K}_{X_0}^{\vee}) \neq 0$ would lead to a section of $\mathcal{K}_{X_0}$, contradicting what we have already proved. So indeed $P_2(X_0) = 0$ and X_0 must be isomorphic to $\mathbb{P}_2$.

(1.4) **Example.** *Let* X *be a compact surface, and* $f : \mathbb{P}_2 \to X$ *a non-constant holomorphic map. Then* f *is finite and* X *is again* $\mathbb{P}_2$.

Proof. The image $f(\mathbb{P}_2) \subset X$, which is a connected analytic subset of X by Remmert's theorem (Theorem I. 8.4) cannot be a curve, otherwise we would obtain two curves in $\mathbb{P}_2$ (namely two fibres) which do not intersect. So $f(\mathbb{P}_2) = X$. By Corollary I. 1.2 we know that $b_1(X) = 0$ and $b_2(X) = 0$ or 1. The case $b_2(X) = 0$ is not possible: we would have $c_1^2(X) = 0, c_2(X) = 2$ and $c_1^2(X)+c_2(X) \not\equiv 0(12)$. In the second case we find that X is $\mathbb{P}_2$ by Theorem1.1, (i) since $P_2(X) \leq P_2(\mathbb{P}_2) = 0$. Finally, f must be finite, otherwise we could again find a couple of non-intersecting curves on $\mathbb{P}_2$. □

The fact that every surface X with $b_2(X) = 0, b_2(X) = 1$ is algebraic has been known since Kodaira proved Corollary IV. 6.3 in [Ko66]. Theorem 1.1, (i) was proved for Kähler surfaces already in [H-K], whereas Theorem 1.1, (ii) was proved in [Y77]. Example 1.3 is due to Kodaira and Spencer and Example 1.4 was given in [R-V60].

2. Complete Intersections

Let $d_i \in \mathbb{Z}, d_i \geq 2$ for $i = 1, \ldots, n-2$. A smooth complete intersection of type $(d_1, \ldots, d_{n-2})$ in $\mathbb{P}_n$ is a surface X which is the transversal intersection of $n-2$ hypersurfaces of degree $d_1, \ldots, d_{n-2}$ respectively. (So the hypersurfaces may have singularities, but not along X.) Repeated application of Bertini's theorem I. 20.2 yields that $n-2$ "general hypersurfaces" of degree $d_1, \ldots, d_{n-2}$ meet in such a smooth complete intersection.

(2.1) **Proposition.** *If X is a smooth complete intersection of type $(d_1, \ldots, d_{n-2})$, then*

(i) $\pi_1(X) = 0$

(ii) $\mathcal{K}_X = \mathcal{O}_X\left(\sum_{i=1}^{n-2} d_i - (n+1)\right)$

(iii) $c_1^2(X) = \left(\sum_{i=1}^{n-2} d_i - (n+1)\right)^2 \prod_{i=1}^{n-2} d_i$

(iv) $e(X) = \left[\binom{n+1}{2} - (n+1)\sum_{i=1}^{n-2} d_i + \sum_{i=1}^{n-2} d_i^2 + \sum_{i\neq j} d_i d_j\right] \prod_{i=1}^{n-2} d_i.$

Proof. Property (i) is an immediate consequence of Corollary I. 20.5. The properties (ii), (iii) and (iv) follow from the adjunction formula since $\mathcal{N}_{X/\mathbb{P}_n} \cong \mathcal{O}_X(d_1) \oplus \ldots \oplus \mathcal{O}_X(d_{n-2})$. □

Since $h^{0,1} = h^{1,0} = \frac{1}{2}b_1(X) = 0$ by (i), and since we know $\chi(X) = \frac{1}{12}(c_1^2(X) + e(X))$ by (iii) and (iv), we can use Noether's formula to obtain $h^{2,0}(X) = h^{0,2}(X)$. From this, together with (i) and (iv) we then can find $h^{1,1}(X)$.

From (ii) we see that

$$\operatorname{kod}(X) = \begin{cases} -\infty & \text{if } X \text{ is of type (2), (3) or (2,2)} \\ 0 & \text{if } X \text{ is of type (4), (2,3) or (2,2,2)} \\ 2 & \text{otherwise} \end{cases}$$

A smooth complete intersection of type (2) is a quadric in $\mathbb{P}_3$, hence isomorphic to $\mathbb{P}_1 \times \mathbb{P}_1$ and rational. As is well known, all smooth intersections of type (3) in $\mathbb{P}_3$ and (2,2) in $\mathbb{P}_4$ are also rational, and they can be obtained from $\mathbb{P}_2$ by blowing up six and five points respectively.

Reference: [Hir66], Sect. 22.1, [Ha77], p. 395 and [G-H78a], p. 480 and 550.

3. Tori of Dimension 2

Tori in general were already discussed in Chap. I. Here we shall only prove a special result for the 2-dimensional case, which will be needed later.

Let L be a free $\mathbb{Z}$-module of rank 4. An orientation for L is an isomorphism $\det : \bigwedge^4 L \xrightarrow{\sim} \mathbb{Z}$. it gives rise to a (symmetric) integral bilinear form on $\bigwedge^2 L$ by $(u, v) = \det(u \wedge v)(u, v \in \bigwedge^2 L)$. This form is unimodular.

(3.1) **Proposition.** *Let L and L' be two oriented free $\mathbb{Z}$-modules of rank 4, and let $\varphi : \bigwedge^2 L \to \bigwedge^2 L'$ be an isometry. If the* mod-2 *reduction of φ is of the form $\Psi_2 \wedge \Psi_2$ for some isomorphism $\Psi_2 : L \otimes \mathbb{F}_2 \to L' \otimes \mathbb{F}_2$, then $\varphi = \pm\Psi \wedge \Psi$ where $\Psi : L \to L'$ is an isomorphism.*

Proof. For any field k we denote $L \otimes_{\mathbb{Z}} k$ by $L_k, \varphi \otimes \mathrm{id}_k = \varphi_k$, etc.

The isotropic lines in $\bigwedge^2 L_{\mathbb{Q}}$ are exactly the lines in $\bigwedge^2 L_{\mathbb{Q}}$ which as points of $\mathbb{P}(\bigwedge^2 L_{\mathbb{Q}})$ are the points of the Plücker quadric $\mathrm{Gr}(1, \mathbb{P}(L_{\mathbb{Q}}))$. So φ induces a projective linear map $\mathbb{P}(\varphi_{\mathbb{Q}})$ from $\mathbb{P}(\bigwedge^2 L_{\mathbb{Q}})$ onto $\mathbb{P}(\bigwedge^2 L'_{\mathbb{Q}})$ which maps $\mathrm{Gr}(1, \mathbb{P}(L_Q))$ onto $\mathrm{Gr}(1, \mathbb{P}(L'_{\mathbb{Q}}))$. Now on these Grassmann varieties there are two systems of 2-dimensional linear spaces: those given by the lines through a point of $\mathbb{P}(L_{\mathbb{Q}})$ or $\mathbb{P}(L'_{\mathbb{Q}})$, and those given by the lines in a plane of $\mathbb{P}(L_{\mathbb{Q}})$ or $\mathbb{P}(L'_{\mathbb{Q}})$. The map $\mathbb{P}(\varphi_{\mathbb{Q}})$ either maps the planes of the first system onto the planes of the first system or it maps the planes of the first system onto the planes of the second system. The "main theorem of projective geometry" tells us that in the first case $\mathbb{P}(\varphi_{\mathbb{Q}})$ is induced by an isomorphism from $L_{\mathbb{Q}}$ onto $L'_{\mathbb{Q}}$, whereas in the second case $\mathbb{P}(\varphi_{\mathbb{Q}})$ is induced by an isomorphism from $L_{\mathbb{Q}}$ onto $(L'_{\mathbb{Q}})^{\vee}$. Since by assumption we are in the first case when we reduce modulo 2, we must also be in the first case after tensoring with $\mathbb{Q}$. In other words: there exists an isomorphism $\tilde{\Psi} : L_{\mathbb{Q}} \to L'_{\mathbb{Q}}$ such that $\tilde{\Psi} \wedge \tilde{\Psi} = \mu\varphi_{\mathbb{Q}}$, with $\mu \in \mathbb{Q}$. Replacing $\tilde{\Psi}$ by $n\tilde{\Psi}(n \in \mathbb{N})$ we can achieve that $\tilde{\Psi}(L) \subset L'$. Now given a sublattice M' of a lattice M, there is always a sublattice M'' of M, with $M' = mM''(m \in \mathbb{N})$ such that M'' contains a primitive vector. So multiplying $\tilde{\Psi}$ again with a suitable rational number, we obtain an isomorphism from $L_{\mathbb{Q}}$ onto $L'_{\mathbb{Q}}$ which we denote again by $\tilde{\Psi}$, such that

(i) $\tilde{\Psi} \wedge \tilde{\Psi} = \lambda\varphi$ with $\lambda \in \mathbb{Q}$,
(ii) $\tilde{\Psi}(L) \subset L'$,
(iii) L' contains a primitive vector $\tilde{\Psi}(e_1)$.

Then e_1 is primitive itself, and therefore can be complemented to a basis $e_1, \ldots, e_4$ of L. Since $\tilde{\Psi}(e_1) \wedge \tilde{\Psi}(e_1) = \lambda(e_1 \wedge e_i)$ and since $\varphi(e_i \wedge e_j)$ is primitive, we have that $\lambda \in \mathbb{Z}$. Now $\tilde{\Psi}(e_1) \wedge \tilde{\Psi}(e_i) \in \lambda(\bigwedge^2 L')$, so after replacing e_i by a suitable sum $e_i + p_i e_1$ ($p_i \in \mathbb{Z}$) we have $\tilde{\Psi}(e_i) \in \lambda L'$. Hence $\tilde{\Psi}(e_2) \wedge \tilde{\Psi}(e_3) = \lambda\varphi(e_2 \wedge e_3) \in \lambda^2 \bigwedge^2 L'$. But $\varphi(e_2 \wedge e_3)$ is primitive, therefore $\lambda = \pm 1$. To conclude that $\tilde{\Psi} = \Psi_{\mathbb{Q}}$, with Ψ as described in the proposition, it remains to be shown that $\tilde{\Psi}(L) = L'$. In other words we still have to show that all elementary divisors of $\tilde{\Psi}$, considered as a homomorphism from L into L', are equal to 1. This, however, is an immediate consequence of the fact that all elementary divisors of φ are equal to 1. □

For a 2-torus $T = V/\Gamma$, the lattice $H^1(T, \mathbb{Z})$ caries a natural orientation, induced by the complex structure, for $\bigwedge^4 H^1(T, \mathbb{Z}) \cong H^4(T, \mathbb{Z})$. In fact, the abstract form on $\bigwedge^2 H^1(T, \mathbb{Z})$ given by the orientation is nothing but the usual quadratic form on $H^2(T, \mathbb{Z})$.

(3.2) **Theorem.** *Let T, T' be 2-tori, and let $\varphi : H^2(T', \mathbb{Z}) \to H^2(T, \mathbb{Z})$ be an isometry which preserves the Hodge decomposition. If there exists an isomorphism $\Psi_2 : H^1(T', \mathbb{Z}/2\mathbb{Z}) \to H^1(T, \mathbb{Z}/2\mathbb{Z})$, such that $\Psi_2 \wedge \Psi_2$ equals the* mod-2 *reduction of φ, then $\pm\varphi$ is induced by an isomorphism from T onto T'.*

Proof. From the remarks above and from Proposition 3.1 it follows that $\varphi = \pm\Psi \wedge \Psi$, where $\Psi : H^1(T', \mathbb{Z}) \to H^1(T, \mathbb{Z})$ is an isomorphism. We claim that Ψ induces an isomorphism $\Psi_{\mathbb{C}} : H^1(T', \mathbb{C}) \to H^1(T, \mathbb{C})$, which preserves the Hodge decomposition. This immediately follows from the fact that $\Psi_{\mathbb{C}} \wedge \Psi_{\mathbb{C}}$ maps the point on $\mathrm{Gr}(1, \mathbb{P}(H^1(T', \mathbb{C})))$ corresponding to $H^{1,0}(T')$, to the point on $\mathrm{Gr}(1, \mathbb{P}(H^1(T, \mathbb{C})))$ corresponding to $H^{1,0}(T)$. Indeed, these points correspond to $H^{2,0}(T')$, $H^{2,0}(T)$ respectively , and $\Psi_{\mathbb{C}} \wedge \Psi_{\mathbb{C}} = \varphi_{\mathbb{C}}$ preserves the Hodge decomposition. So we can apply the Torelli theorem I. 14.2 to obtain the desired result. □

The results of this section have been known for some time and they appear in [L-P], Sect. 3.

Fibre Bundles

4. Ruled Surfaces

A ruled surface is a compact surface which admits a ruling, more precisely, a compact surface which (in at least one way) is the total space of an analytic fibre bundle with fibre $\mathbb{P}_1$ and structure group $\mathrm{PGL}(2, \mathbb{C})$ over a smooth, connected curve B. Actually, as we shall presently see, such an analytic fibre bundle is always equivalent to an algebraic one.

Remark. Some authors use "ruled surface" for ruled surfaces embedded in a projective space, such that all fibres are lines. We shall call these geometrically ruled surfaces.

It will be proved later that, with the single exception of $\mathbb{P}_1 \times \mathbb{P}_1$, a surface admits at most one ruling.

Obvious examples are provided by the projective bundles $\mathbb{P}(\mathcal{V})$ of algebraic 2-vector bundles over a smooth, connected, compact curve. In fact there are no other examples, as follows from

(4.1) **Proposition.** *Any analytic fibre bundle with fibre* $\mathbb{P}_n$ *and structural group* $\mathrm{PGL}(n+1,\mathbb{C})$ *over a smooth, compact curve* B *is isomorphic to* $\mathbb{P}(\mathcal{V})$, *where* $\mathcal{V}$ *is an algebraic* $(n+1)$*-vector bundle over* B.

Proof. The general theory of fibre bundles yields an " exact sequence" of cohomology sets

$$\to H^1(B,\mathcal{GL}(n+1,\mathbb{C})) \to H^1(B,\mathcal{PGL}(n+1,\mathbb{C})) \to H^2(B,\mathcal{O}_B^*),$$

where $\mathcal{GL}(n+1,\mathbb{C})$, $\mathcal{PGL}(n+1,\mathbb{C})$ and $\mathcal{O}_B^*$ are the sheaves of germs of analytic maps from B into $\mathrm{GL}(n+1,\mathbb{C}), \mathrm{PGL}(n+1,\mathbb{C})$, $\mathbb{C}^*$ respectively, and where $H^1(B,\mathcal{GL}(n+1,\mathbb{C}))$, $H^1(B,\mathcal{PGL}(n+1,\mathbb{C}))$ classify analytic $(n+1)$-vector bundles, respectively $\mathbb{P}_n$-bundles over B. Since B is a curve, we immediately find from the exponential cohomology sequence that $H^2(B,\mathcal{O}_B^*) = 0$. The proposition follows from this and from GAGA (every analytic vector bundle over B is algebraic). □

As a consequence, every ruled surface with base B is birationally equivalent to $B \times \mathbb{P}_1$.

Remark. The preceding proof also works if $\dim B \geq 2$, provided that $H^2(B,\mathcal{O}_B) = H^3(B,\mathbb{Z}) = 0$ (the proposition is however not true for any smooth base).

Actually, a ruling can be characterised as a fibration over a smooth connected, compact curve B, such that all fibres are isomorphic to $\mathbb{P}_1$. In that case the projection is everywhere of maximal rank, and the statement is an immediate consequence of the Grauert-Fischer theorem (Theorem I. 10.1). A direct proof can be given in the following way.

Let $f : X \to B$ be a fibration such that all fibres are isomorphic to $\mathbb{P}_1$. The adjunction formula implies that all fibres are smooth, and from the differentiable point of view, f is locally trivial fibration by Ehresmann's theorem (compare [M-K] p. 19). So, by elementary obstruction theory there is a section. This implies that there is a complex (topological) line bundle $\mathcal{L}$, such that $\mathcal{L}|F \cong \mathcal{O}_F(1)$ for every fibre F. A canonical divisor has negative intersection number with all fibres and so $p_g(X) = h^2(\mathcal{O}_X) = 0$. Applying the exponential cohomology sequence (I, Sect. 6), we deduce that the dual cohomology class of this section is the Chern class of a *holomorphic* line bundle.

Let $\mathcal{M}$ be some very ample line bundle on B and let $\mathcal{L}_n = f^*(\mathcal{M}^{\otimes n}) \otimes \mathcal{L}$. We claim that for n large enough $f_{\mathcal{L}_n} : X \to \mathbb{P}_N$ is everywhere defined, mapping every F isomorphically onto a line in $\mathbb{P}_n$. This will follow as soon as we know that $H^1(X, \mathcal{L}_n \otimes \mathcal{O}_X(-F)) = 0$. By Leray's spectral sequence this last group is equal to $H^1(B, f_*\mathcal{L} \otimes \mathcal{O}_X(-F) \otimes \mathcal{M}^{\otimes n})$. It vanishes for n large enough by "Theorem B" [Se55b]. This means that we obtain a regular map from B into the Grassmann variety $\mathrm{Gr}(1, N)$, such that X as a fibre space is the pull-back of the universal subbundle on $\mathrm{Gr}(1, N)$. So f is locally trivial as claimed.

If $B = \mathbb{P}_1$, then by a theorem of Grothendieck ([Gk57a]) every algebraic vector bundle over B is isomorphic to a direct sum of line bundles. So in this case every ruled surface over $\mathbb{P}_1$ is of the form $\mathbb{P}(\mathcal{O}_{\mathbb{P}_1} \oplus \mathcal{O}_{\mathbb{P}_1}(n))$ for some $n \geq 0$ (since $\mathrm{Pic}(\mathbb{P}_1) \cong \mathbb{Z}$ and $\mathbb{P}(\mathcal{V} \otimes \mathcal{L}) \cong \mathbb{P}(\mathcal{V})$ for any algebraic line bundle $\mathcal{L}$ on B). As is customary, we denote the surface $\mathbb{P}(\mathcal{O}_{\mathbb{P}_1} \oplus \mathcal{O}_{\mathbb{P}_1}(n))$ by Σ_n and call it the n-th Hirzebruch surface. The surfaces Σ_n are birationally equivalent to $\mathbb{P}_1 \times \mathbb{P}_1$, hence to $\mathbb{P}_2$, so they are all rational. The surfaces Σ_n can be characterised in many ways. For example, Σ_0 is $\mathbb{P}_1 \times \mathbb{P}_1, \Sigma_1$ is $\mathbb{P}_2$ blown up in one point, and $\Sigma_n, n \geq 2$, is obtained by desingularizing the cone in $\mathbb{P}_{n+1}$ over a rational normal curve spanning $\mathbb{P}_n$ (see [G-H78a] p. 523).

Let $n \geq 1$ and $C_n = \mathbb{P}(\mathcal{O}_{\mathbb{P}_1}(n)) \subset \Sigma_n$. We claim that $C_n^2 = -n$. To see this, it is sufficient to prove that $\mathcal{N}_{C_n/\Sigma_n} \cong \mathcal{O}_{C_n}(-n)$ (Proposition I.6.2). This follows by restricting the standard sequence I, (3) to C_n, since the tautological bundle restricted to C_n is $\mathcal{O}_{C_n}(n)$, whereas the bundle along the fibres restricted to C_n is $\mathcal{N}_{C_n/\Sigma_n}$. We furthermore claim that any irreducible curve D on Σ_n with $D^2 \leq 0$ is either C_n or a fibre F. Indeed, such a curve D is homologous to $rC_n + sF$, with $r, s \in \mathbb{Z}$. Now if D is neither C_n nor a fibre, then $(rC_n + sF)F > 0$ and $(rC_n + sF)C_n \geq 0$, but these inequalities are incompatible with $(rC_n + sF)^2 \leq 0$.

In particular we have

(4.2) **Proposition.** *The Hirzebruch surfaces $\Sigma_n, n \geq 0$ are all biregularly distinct. Except for $\Sigma_0 = \mathbb{P}_1 \times \mathbb{P}_1$, each such surface has only one ruling.*

Remark. Since any surface with two different rulings is a $\mathbb{P}_1$-bundle over $\mathbb{P}_1$, by Lüroth's theorem, we have shown at the same time that $\mathbb{P}_1 \times \mathbb{P}_1$ is the only surface with more than one ruling.

If B is elliptic, then there are 2-bundles on B which do not split (that is, which are not direct sums of two 1-subbundles). To see this, we start from an *extension*

$$0 \to \mathcal{O}_B \xrightarrow{i} \mathcal{V} \to \mathcal{O}_B \to 0$$

which does not split (this is possible since these extensions are classified by $H^1(B, \mathcal{O}_B \otimes \mathcal{O}_B^\vee) = H^1(B, \mathcal{O}_B) \cong \mathbb{C}$ – compare [Gk57a]) and show that the *bundle* $\mathcal{V}$ does not split either. Indeed, let us assume that $\mathcal{V} \cong \mathcal{L}_1 \oplus \mathcal{L}_2$, with $\mathcal{L}_1$ and $\mathcal{L}_2$ 1-subbundles of $\mathcal{V}$. If, say $\mathcal{L}_1$, were $i(\mathcal{O}_B)$, then $\mathcal{L}_2$ would also be isomorphic to $\mathcal{O}_B$ (for $\mathcal{L}_1 \otimes \mathcal{L}_2 \cong \mathcal{O}_B$) and the extension would split. If neither $\mathcal{L}_1$ nor $\mathcal{L}_2$ were $i(\mathcal{O}_B)$, then, since $\mathcal{L}_1 \otimes \mathcal{L}_2 \cong \mathcal{O}_B$, and both line bundles would admit non-trivial homomorphisms onto $\mathcal{O}_B$, we would have $\mathcal{L}_1 \cong \mathcal{L}_2 \cong \mathcal{O}_B$. It would follow that, say $\mathcal{L}_1$ and $i(\mathcal{O}_B)$ span $\mathcal{V}$ everywhere, which is impossible.

Similarly there exists an extension

$$0 \to \mathcal{O}_B \to \mathcal{W} \to \mathcal{L} \to 0,$$

such that the bundle $\mathcal{W}$ does not split, where now $\mathcal{L}$ is some line bundle of degree 1. Atiyah has shown in [At57] that $\mathbb{P}(\mathcal{V})$ and $\mathbb{P}(\mathcal{W})$ are the only ruled surfaces over an elliptic curve B which are not the projective bundle of a splitting vector bundle of rank 2.

For the classification of ruled surfaces over a base of genus ≥ 2 we refer to [Tj64], [Tj65].

(4.3) **Proposition.** *Let X be a compact surface and C a smooth rational curve on X.*

(i) *If $C^2 = 0$, then there exists a modification $\varphi : X \to Y$, where Y is ruled, such that C meets no exceptional curve of φ, and $\varphi(C)$ is a fibre of Y;*

(ii) *if $C^2 > 0$, then X is either $\mathbb{P}_2$, a Hirzebruch surface, or a blow-up of one of these surfaces.*

Proof. (i) We start by observing that X is algebraic by Theorem IV. 6.2, for $c_1^2(\mathcal{O}_X(nC) \otimes \mathcal{K}_X^\vee) > 0$ for n large enough.

This said, we distinguish between two cases: $q(X) = 0$ and $q(X) > 0$.

In the first of these cases, considering the exact cohomology sequence of

$$0 \to \mathcal{O}_X \to \mathcal{O}_X(C) \to \mathcal{O}_C(C) \to 0,$$

we see immediately that $h^0(\mathcal{O}_X(C)) = 2$. Since the irreducible curve C is a member of $|C|$, also the general member of $|C|$ is irreducible, and $f = f_{\mathcal{O}_X(C)}$ provides a regular map onto $\mathbb{P}_1$. The general fibre of f is again rational, since its intersection number with K_X is -2. Now suppose there is a non-smooth fibre, say $\sum_{i=1}^{m} c_i C_i$. Because of the adjunction formula we must have $m \geq 2$. Since $K_X(\Sigma c_i C_i) = -2$, there must be a C_i, say C_1, such that $K_X C_1 < 0$. By Zariski's Lemma III. 8.2 we have $C_1^2 < 0$, and we find from the adjunction formula that $K_X C_1 = C_1^2 = -1$, i.e., C_1 is a (-1)-curve (Proposition III. 2.2). So after blowing down X a finite number of times, we obtain a surface Y admitting a holomorphic map onto $\mathbb{P}_1$, such that all fibres are smooth rational curves, i.e., a Hirzebruch surface.

On the other hand, if $q(X) > 0$ we use the Albanese map $f' : X \to$ Alb(X) (since X is algebraic, dim $\text{Alb}(X) = q(X) > 0$ and f' is not a constant map). Let $f' = h'g'$ be the Stein factorisation of f'. If the general fibre of g' were a point, then C^2 would be strictly negative by Theorem III. 2.1. So f'

maps X onto a curve $D' \subset \mathrm{Alb(X)}$. If D is the desingularization of D', then f' can be lifted to $f : X \to D$. By Zariski's Lemma the fibre of f containing C must be of the form $aC, a \geq 1$, but the adjunction formula immediately yields $a = 1$. Consequently, also the general fibre of f is rational. The proof can now be completed in the same way as in the case $q(X) = 0$.

(ii) If X contains a smooth rational curve A with $A^2 = 0$, then X is a blown-up ruled surface by (i). Since C is mapped onto the base of any ruling, by Lüroth's theorem this base is rational, so in this case we are done.

Now suppose that X does not contain any smooth rational A with $A^2 = 0$. Let $k = \min(B^2), B$ smooth rational, $B^2 \geq 1$, and let D be smooth rational with $D^2 = k$. Without loss of generality it may be assumed that all (-1)-curves on X meet D. We blow up k points $x_1, \ldots, x_k \in D$, obtaining on the resulting surface $\bar{X}$ the corresponding (-1)-curves $E_1, \ldots, E_k$. Since for the proper transform $\bar{D}$ of D we have $\bar{D}^2 = 0$, the surface $\bar{X}$ is a blown-up ruled surface with $\bar{D}$ a fibre of a ruling $p : \bar{X} \to R$. Each E_i is a section of this ruling, so R is rational. If all fibres of p are irreducible, then $\bar{X}$ is a surface Σ_m. Since Σ_m only contains a (-1)-curve if $m = 1$, we must have $m = k = 1$ and X is $\mathbb{P}_2$. On the other hand, if p had a reducible fibre, then this fibre would contain an exceptional curve E which would meet at least one E_i, otherwise we would have on X a (-1)-curve not meeting D. But if E met some E_i, then it would meet this curve transversally in one point, so the image F of E on X would be smooth rational with $0 \leq F^2 \leq k - 1$, which is impossible by assumption. □

5. Elliptic Fibre Bundles

If E is an elliptic curve, we denote by $A(E)$ the group of its biholomorphic automorphisms. After fixing an origin $0 \in E$ this group can be described in the following way: E, acting on itself by translations, forms a normal subgroup of $A(E)$, and the quotient $A(E)/E$ can be identified with the group of automorphisms leaving 0 fixed. So this quotient is the cyclic group $\mathbb{Z}/n\mathbb{Z}$ of order

$$\begin{aligned} n &= 4 \quad \text{if } E = \mathbb{C}/\mathbb{Z} \oplus \mathbb{Z}i, \\ n &= 6 \quad \text{if } E = \mathbb{C}/\mathbb{Z} \oplus \mathbb{Z}\omega, \quad \omega = \mathbf{e}(1/6), \\ n &= 2 \quad \text{in all other cases.} \end{aligned}$$

Then $A(E)$ is the semi-direct product $E \times \mathbb{Z}/n\mathbb{Z}$. We write its elements as $(e, z), e \in E, z \in \mathbb{Z}/n\mathbb{Z}$,

$$(e, z) : x \mapsto e + zx, \quad x \in E.$$

The group operation is given by

$$(e, z)(e', z') = (e + ze', zz').$$

The translation group E is described by the universal covering sequence

$$(1) \qquad 0 \to \Gamma \underset{j}{\longrightarrow} \mathbb{C} \to E \to 0, \qquad \Gamma = \mathbb{Z} \oplus \mathbb{Z}.$$

If B is any smooth, compact, connected curve, then the holomorphic fibre bundles with typical fibre E and base B are classified by the cohomology set $H^1(\mathcal{A}_B)$ and there is an exact sequence of cohomology *sets*

$$(2) \qquad H^1(\mathcal{E}_B) \to H^1(\mathcal{A}_B) \to H^1(B, \mathbb{Z}/n\mathbb{Z}).$$

Here $\mathcal{A}_B$, $\mathcal{E}_B$ are the sheaves of germs of local holomorphic maps from B to $A(E)$, respectively E. We shall call a bundle $X \to B$ a principal bundle if its structure group can be reduced to E. To describe $H^1(\mathcal{E}_B)$ we use the exact cohomology sequence:

$$(3) \qquad H^1(B, \Gamma) \to H^1(\mathcal{O}_B) \to H^1(\mathcal{E}_B) \xrightarrow{c} H^2(B, \Gamma) \to 0$$

which is induced by (1).

(5.1) **Lemma.** a) *The bundle $X \to B$ with typical fibre E is principal if and only if X admits an action of the group E which on all fibres $X_b, b \in B$, induces the translation group.*

b) *Two principal E-bundles defined by cocycles $\zeta = \{\zeta_{ij}\}$ and $\zeta' = \{\zeta'_{ij}\}$ are isomorphic as $A(E)$-bundles if and only if there is some $z \in \mathbb{Z}/n\mathbb{Z}$ such that $\zeta' = z\zeta$ in $H^1(\mathcal{E}_B)$.*

c) *A principal E-bundle with class $\zeta \in H^1(\mathcal{E}_B)$ can be defined by a locally constant cocycle if and only if $c(\zeta) = 0$.*

Proof. a) E being abelian, the "only if" part is trivial. So let $\zeta_{ij} = (e_{ij}, z_{ij})$ be a cocycle with values in $\mathcal{A}_B$ defining X, and assume that the structure group cannot be reduced to E. Because of (2) this means that the cocycle $\{z_{ij}\}$ defines a non-zero class $z \in H^1(B, \mathbb{Z}/n\mathbb{Z})$. Any automorphism t of X, leaving all fibres invariant and acting as a translation on each of them, is given by a collection $t_i : U_i \to E$ such that

$$\begin{array}{rcll} t_i\zeta_{ij} & = & \zeta_{ij}t_j, & \text{i.e.,} \\ (t_i + e_{ij}, z_{ij}) & = & (e_{ij} + z_{ij}t_j, z_{ij}), & \text{i.e.,} \\ t_i & = & z_{ij}t_j. & \end{array}$$

So the collection $\{t_i\}$ is nothing but a section in the bundle $E^{(z)}$ obtained from the trivial bundle $E \times B$ by twisting with the cocycle $\{z_{ij}\}$. Now $\mathbb{Z}/n\mathbb{Z}$ operates equivariantly on the sequence (1). Therefore, this sequence may be operated on by $\{z_{ij}\}$, which yields an exact sequence

$$0 \to \Gamma^{(z)} \to \mathcal{O}_B^{(z)} \to E^{(z)} \to 0.$$

Here $\mathcal{O}_B^{(z)} \in \text{Pic}(B)$ is a non-trivial torsion bundle, because $H^1(B, \mathbb{Z}/n\mathbb{Z}) \to H^1(\mathcal{O}_B^*)$ is injective. So $H^0(\mathcal{O}_B^{(z)}) = 0$ and the group of sections in $E^{(z)}$ must be discrete. This means that X does not admit enough automorphisms to induce all translations on the fibres.

b) ζ and ζ' define the same class in $H^1(\mathcal{A}_B)$ if and only if there are maps $h_i : U_i \to A(E)$ such that $h_i\zeta_{ij} = \zeta'_{ij}h_j$. If we write $h_i = (e_i, z_i)$ as above, then these equations become

$$(e_i + z_i\zeta_{ij}, z_i) = (\zeta'_{ij} + e_j, z_j).$$

This means $z_i = z_j = z \in \mathbb{Z}/n\mathbb{Z}$ and $\zeta'_{ij} - z\zeta_{ij} = e_i - e_j$.

c) Let as usual $\mathbb{C}_B$, respectively E_B, be the sheaf of germs of locally constant maps $B \to \mathbb{C}$, respectively $B \to E$. The diagram of sheaves

$$\begin{array}{ccccccccc} 0 & \longrightarrow & \Gamma & \longrightarrow & \mathbb{C}_B & \longrightarrow & E_B & \longrightarrow & 0 \\ & & \| & & \downarrow & & \downarrow & & \\ 0 & \longrightarrow & \Gamma & \longrightarrow & \mathcal{O}_B & \longrightarrow & \mathcal{E}_B & \longrightarrow & 0 \end{array}$$

induces in cohomology:

$$\begin{array}{ccccccc} H^1(B,\mathbb{C}) & \longrightarrow & H^1(B,E) & \xrightarrow{\gamma} & H^2(B,\Gamma) & \longrightarrow & H^2(B,\mathbb{C}) \\ \downarrow & & \downarrow & & \| & & \\ H^1(\mathcal{O}_B) & \longrightarrow & H^1(\mathcal{E}_B) & \longrightarrow & H^2(B,\Gamma). & & \end{array}$$

Since $H^2(B,\mathbb{Z})$ is free of torsion, the map $H^2(B,\Gamma) \to H^2(B,\mathbb{C})$ is injective and $\gamma = 0$. Because $H^1(B,\mathbb{C}) \to H^1(\mathcal{O}_B)$ is surjective, one finds that $c(\zeta) = 0$ if and only if ζ comes from $H^1(B,E)$. □

Next we show that every principal bundle with typical fibre E admits as unramified covering a holomorphic $\mathbb{C}^*$-bundle. There is a canonical isomorphism $\Gamma \to H^2(B,\Gamma) = H^2(B,\mathbb{Z}) \otimes_{\mathbb{Z}} \Gamma$. For $\zeta \in H^1(\mathcal{E}_B)$ given, we have a primitive embedding $\mathbb{Z} \xrightarrow{i} \Gamma$ such that $c(\zeta)$ is in the image of the induced map $H^2(B,\mathbb{Z}) \to H^2(B,\Gamma)$. Chasing the cohomology diagram of

$$\begin{array}{ccccccccc} 0 & \longrightarrow & \mathbb{Z} & \longrightarrow & \mathbb{C} & \xrightarrow{\mathbf{e}} & \mathbb{C}^* & \longrightarrow & 0 \\ & & {\scriptstyle i}\downarrow & & \| & & \downarrow{\scriptstyle q} & & \\ 0 & \longrightarrow & \Gamma & \longrightarrow & \mathbb{C} & \longrightarrow & E & \longrightarrow & 0 \end{array}$$

i.e.,

$$\begin{array}{ccccccc} H^1(\mathcal{O}_B) & \longrightarrow & H^1(\mathcal{O}_B^*) & \longrightarrow & H^2(B,\mathbb{Z}) & \longrightarrow & 0 \\ \| & & {\scriptstyle H^1(q)}\downarrow & & \downarrow & & \\ H^1(\mathcal{O}_B) & \longrightarrow & H^1(\mathcal{E}_B) & \longrightarrow & H^2(B,\Gamma) & \longrightarrow & 0 \end{array}$$

where q is the quotient with respect to some subgroup $\mathbb{Z}\tau$, $|\tau| > 1$, of $\mathbb{C}^*$, we obtain

(5.2) **Proposition.** *For every* $\zeta \in H^1(\mathcal{E}_B)$ *there is some* $\eta \in H^1(\mathcal{O}_B^*)$ *with* $\zeta = H^1(q)\eta$, i.e., *the bundle space* X *of* ζ *is a* $\mathbb{Z}$*-quotient of the total space of* η *where the* $\mathbb{Z}$*-action is generated by multiplication with* τ.

Since the embedding $i : \mathbb{Z} \to \Gamma$ is primitive, we may complement it to obtain an isomorphism $i \times j : \mathbb{Z} \times \mathbb{Z} \to \Gamma$. Topologically, this induces a splitting $E = S^1 \times S^1$. The topological version of (3) shows that, topologically, principal E-bundles are classified by their class in $H^2(B, \Gamma)$. In our case this implies that topologically $X = C \times S^1$, where C is the S^1-bundle over B determined by η.

The bundle $C \to B$ has a Gysin sequence ([M-S]), p.143)

$$\begin{array}{ccccccccc} 0 & \longrightarrow & H^1(B,\mathbb{Z}) & \longrightarrow & H^1(C,\mathbb{Z}) & \longrightarrow & H^0(B,\mathbb{Z}) & \stackrel{\delta}{\longrightarrow} & \\ & \longrightarrow & H^2(B,\mathbb{Z}) & \longrightarrow & H^2(C,\mathbb{Z}) & \longrightarrow & H^1(B,\mathbb{Z}) & \longrightarrow & 0 \end{array}$$

where δ is multiplication with the Chern class $c(\eta) = c(\zeta)$.

(5.3) **Proposition.** (i) *If* $c(\zeta) = 0$, i.e., *the bundle* $X \to B$ *is topologically trivial, then* $b_1(X) = b_1(B) + 2$ *and* $b_2(X) = 2b_1(B) + 2$,

(ii) *If* $c(\zeta) \neq 0$ *(so* $X \to B$ *is not topologically a product), then* $b_1(X) = b_1(B) + 1$ *and* $b_2(X) = 2b_1(B)$.

Proof. (i) is an immediate consequence of the Künneth formula. To prove (ii) we deduce from the Gysin sequence $b_1(C) = b_1(B)$ and $b_2(C) = b_2(B) + b_1(B) - 1 = b_1(B)$. Applying the Künneth formula to $C \times S^1$ we obtain the result. □

These preliminaries being out of the way, we describe explicitly the situation for $B = \mathbb{P}_1$ or B an elliptic curve.

A) *The Case* $B = \mathbb{P}_1$

Since B is simply connected, by (2) the structure group reduces to the translation group, i.e., every bundle is a principal bundle. Furthermore $H^1(\mathcal{O}_{\mathbb{P}_1}) = 0$, so (3) shows that the topological and analytical classification for these bundles is the same. In particular, in Proposition 5.3 we may replace the term "topologically" by "analytically".

(5.4) **Theorem.** *Any elliptic fibre bundle over* $\mathbb{P}_1$ *is either a product or a Hopf surface (*see Sect. 18).

Proof. Let $X \to \mathbb{P}_1$ be an elliptic fibre bundle with $c(\zeta) \neq 0$. We have to show that X is a Hopf surface. By Proposition 5.2 we may assume that $X \to \mathbb{P}_1$ comes from a $\mathbb{C}^*$-bundle $Y \to \mathbb{P}_1$. The surface Y is an unramified covering of X. Since a Hopf surface is defined as a surface for which the universal covering is isomorphic to $W = \mathbb{C}^2 \setminus \{0\}$, it thus suffices to exhibit an unramified covering $W \to Y$. However, W is nothing but the bundle space of the $\mathbb{C}^*$-bundle of degree -1 over $\mathbb{P}_1$, and this bundle space can be mapped onto Y fibre-wise by $z \mapsto z^{-d}$, where d is the degree of $Y \to \mathbb{P}_1$, which does not vanish since $X \to \mathbb{P}_1$ is not the product bundle. □

B) *The Case of an Elliptic Base B*

We distinguish between the following cases.

B I) Principal bundles:

B Ia) bundles defined by cocycles $\zeta \in H^1(\mathcal{E}_B)$ with $c(\zeta) = 0$,

B Ib) bundles defined by cocycles $\zeta \in H^1(\mathcal{E}_B)$ with $c(\zeta) \neq 0$.

B II) Non-principal bundles.

B Ia) Because of Lemma 5.1.c), the bundle $X \to B$ is defined over some covering $\{U_i\}$ of B by patching the pieces $E \times U_i$ according to a locally constant cocycle $\{\zeta_{ij}\}$ with $\zeta_{ij} \in E$. On each product $E \times U_i$ we consider two kinds of holomorphic tangent vector fields: those coming from E and the ones coming from $B \supset U_i$. Since the cocycle is locally constant, both kinds of vector fields are respected by the patching procedure. This means they extend to all of X and span T_X everywhere. The two one-parameter subgroups of $\mathrm{Aut}(X)$, generated by the two types of vector fields, commute, so they generate an abelian complex Lie group operating transitively on X. This shows that all surfaces of the form considered are quotients of $\mathbb{C}^2$ by a lattice, i.e., are complex tori.

B Ib) *Bundles Defined by Cocycles ζ with $c(\zeta) \neq 0$.*

Such surfaces are called primary Kodaira surfaces. Their invariants are

$$H^1(X,\mathbb{Z}) = \mathbb{Z}^3, \quad H^2(X,\mathbb{Z}) = \mathbb{Z}^4 \quad \text{or} \quad \mathbb{Z}^4 \oplus \mathbb{Z}/m\mathbb{Z},$$

$$e(X) = 0, \quad h^1(\mathcal{O}_X) = 2, \quad h^2(\mathcal{O}_X) = 1, \quad \mathcal{K}_X = \mathcal{O}_X.$$

The topological invariants follow from Proposition 5.3. As to the other invariants, consider the direct image sheaf $f_{*1}\mathcal{O}_X$ on B, where $f : X \to B$ is the projection. Since translations operate trivially on $H^1(\mathcal{O}_E)$, the line bundle $f_{*1}\mathcal{O}_X$ on B is trivial. From the exact sequence

$$(4) \qquad 0 \longrightarrow H^1(\mathcal{O}_B) \xrightarrow{f^*} H^1(\mathcal{O}_X) \to H^0(f_{*1}\mathcal{O}_X) \longrightarrow 0$$

we find $h^1(\mathcal{O}_X) = 2$ and from

$$(5) \qquad 0 \longrightarrow f^*(\omega_B) \longrightarrow \Omega_X \longrightarrow \omega_{X|B} \longrightarrow 0$$

we deduce that $\mathcal{K}_X$ is trivial:

$$\mathcal{K}_X \cong \omega_{X|B} \cong f^*(f_*\omega_{X|B}) \cong f^*(f_{*1}\mathcal{O}_X),$$

the last one being relative duality (Theorem III. 12.3).

Notice that because of $b_1(X) = 3$, Kodaira surfaces are **not kählerian**.

Sometimes a primary Kodaira surface admits a finite, freely-operating group of automorphisms. The smooth quotients thus obtained are called secondary Kodaira surfaces. An example can be constructed in the following way.

Let B be any elliptic curve, and $p \in B$. We consider the line bundle $\mathcal{O}_B(p)$ and denote by L the total space of the associated principal $\mathbb{C}^*$-bundle. Let $a \in \mathbb{C}^*, |a| \neq 1$, and let $g_a : L \to L$ be the automorphism obtained by multiplication with a in each fibre. The quotient $X = L/\langle g_a \rangle$ is an elliptic fibre bundle over B. Since $c_1(\mathcal{O}_B(p)) \neq 0$, the diagram preceding Proposition 5.2 shows that X is a primary Kodaira surface. We want to construct a fixed point-free involution $\rho : X \to X$. To do this, we start from an involution $\iota : B \to B$ which has p as a fixed point. Since $\iota^*(\mathcal{O}_B(p)) \cong \mathcal{O}_B(p)$, there exists a biholomorphic map $\alpha : L \to L$, covering ι. Upon multiplication with a suitable automorphism of L we obtain a biholomorphic map $\beta : L \to L$, covering ι, with $\beta^2 = \mathrm{id}_L$. Now we can simply take $\rho = g_{\sqrt{a}} \circ \beta$.

A secondary Kodaira surface Y satisfies $b_1(Y) = 1$, $b_2(Y) = 0$, $q(Y) = 1$, $p_g(Y) = 0$. This will follow from our considerations in VI, Sect. 4.

B II) *Non-principal Bundles*

Now we consider a bundle $X \to B$ given by a class $\zeta \in H^1(\mathcal{A}_B)$ which has a non-trivial image $\zeta \in H^1(B, \mathbb{Z}/n\mathbb{Z})$ under (1). To study X, the first thing to do is to form the cyclic covering $B' \to B$ of order $m, m|n$, killing ζ. The pull-back X' of X to B' is a principal bundle over B'. Since X' is an unramified cyclic m-fold covering of X, it admits a cyclic group $\mathbb{Z}/m\mathbb{Z}$ operating compatibly with the fibration $X' \to B'$. Let $\sigma \in \mathrm{Aut}(X')$ be a generator of $\mathbb{Z}/m\mathbb{Z}$ and let τ be a translation on B' induced by σ. Then σ induces an isomorphism h between the two principal bundles X' and $\tau^*(X')$ over B'. This h cannot commute with the E-actions on X' and $\tau^*(X')$, because otherwise this E-action would descend to $X \to B$, which is impossible by Lemma 5.1 a) and the assumption that X is not a principal bundle. This means that h does not respect the structure of principal bundles. By Lemma 5.1 b) there is some $z \in \mathbb{Z}/n\mathbb{Z}, z \neq 1$, defining h. For $\zeta' \in H^1(\mathcal{E}_{B'})$ the class defining X', this implies $c(\zeta') = 0$: indeed, τ^* operates trivially on $H^2(B', \Gamma)$, whereas z has no fixed point $\neq 0$ in Γ. So necessarily X' is a torus containing E as closed subgroup with $B' = X'/E$. The translation τ on the base B' is induced by some group element $t \in X'$. Then $\sigma = t \circ h$. Consider the closed subgroup

$$C = \{x \in X' : \sigma(x) = t + x\}$$

of fixed points of h. Since h acts on all the fibres of X' as an automorphism which is not a translation, C intersects each fibre in a finite non-empty set. So C is an elliptic curve projected onto B'. The torus X' is isogeneous to $E \times C$ and if we pull back X' under $C \to B'$, it becomes the product $E \times C$. Altogether this proves:

There is an elliptic curve C such that $B = C/G$, where G is a finite subgroup of the translation group of C. The surface X is the quotient $E \times C/G$ with G acting on C by translations and on E by some representation $G \to A(E)$, which has its image not in the group of translations only.

Such a surface X is called bi-elliptic since it admits an elliptic fibration over an elliptic curve. A more traditional terminology is that of an irregular hyperelliptic surface. This usage stems from the fact that the points of each projective image of X are parametrised by abelian functions in two variables, which were called hyperelliptic functions when these surfaces were discovered. The classification of these surfaces is easy, see [B-M77] p. 36-37. One finds the following classical list for $E = \mathbb{C}/\Gamma$ and $G \subset C$:

Type	Γ	G	Action on E of generators of G
a 1)	arbitrary	$\mathbb{Z}/2\mathbb{Z}$	$e \mapsto -e$
a 2)	arbitrary	$\mathbb{Z}/2\mathbb{Z} \oplus \mathbb{Z}/2\mathbb{Z}$	$e \mapsto -e$,
			$e \mapsto e + e_1$, where $2e_1 = 0$
b 1)	$\mathbb{Z} \oplus \mathbb{Z}\omega$	$\mathbb{Z}/3\mathbb{Z}$	$e \mapsto \omega e$
b 2)	$\mathbb{Z} \oplus \mathbb{Z}\omega$	$\mathbb{Z}/3\mathbb{Z} \oplus \mathbb{Z}/3\mathbb{Z}$	$e \mapsto \omega e$
			$e \mapsto e + e_1$, where $\omega e_1 = e_1$
c 1)	$\mathbb{Z} \oplus \mathbb{Z}\mathrm{i}$	$\mathbb{Z}/4\mathbb{Z}$	$e \mapsto \mathrm{i}e$
c 2)	$\mathbb{Z} \oplus \mathbb{Z}\mathrm{i}$	$\mathbb{Z}/4\mathbb{Z} \oplus \mathbb{Z}/2\mathbb{Z}$	$e \mapsto \mathrm{i}e$
			$e \mapsto e + e_1$, where $\mathrm{i}e_1 = e_1$
d)	$\mathbb{Z} \oplus \mathbb{Z}\omega$	$\mathbb{Z}/6\mathbb{Z}$	$e \mapsto -\omega e$

List of bi-elliptic surfaces from [B-M77] ($\omega = \mathbf{e}(1/6)$)

Since the covering $E \times C$ is projective, by Theorem IV. 6.8 every bielliptic surface X is projective. Its invariants are

$$h^{1,0} = 1, \qquad h^{2,0} = 0, \qquad h^{1,1} = 2,$$

and $\mathcal{K}_X$ is a torsion bundle of order 2, 3, 4 and 6 for the type a, b, c and d respectively.

To see this, we observe that G acts non-trivially (with its first factor) on $H^1(\mathcal{O}_E)$. So $f_{*1}\mathcal{O}_X$ is a torsion line bundle on B of the orders as given above in the different cases. The sequence (4) then shows $h^1(\mathcal{O}_X) = 0$. Similarly, from

$$H^2(\mathcal{O}_B) \to H^2(\mathcal{O}_X) \to H^1(f_{*1}\mathcal{O}_X)$$

one deduces $h^2(\mathcal{O}_X) = 0$. Finally, $\omega_{X/B} = f^*(f_{*1}\mathcal{O}_X)$, again by relative duality. So (5) shows that $\mathcal{K}_X$ is a torsion bundle of the order as stated above.

Furthermore $e(X) = c_2(\Omega_X) = 0$ and

$$h^{1,1} = e(X) + 2b_1(X) - 2 = 2. \quad \square$$

6. Higher Genus Fibre Bundles

If D is a smooth, compact, connected curve of genus ≥ 2, then $\mathrm{Aut}(D)$ is a finite group. Consequently, every D-fibre bundle over a curve C is given by a representation $\pi_1(C) \to \mathrm{Aut}(D)$, and the classification of D-bundles becomes a purely algebraic problem. We shall only need the fact that, given a D-bundle over C, there exists a finite unramified covering $\tilde{C}$ of C such that the lifted bundle on $\tilde{C}$ is isomorphic to $\tilde{C} \times D$.

As to the product case $X = C \times D$, we have (as a special case of a much more general result)

(6.1) **Proposition.** *If C and D are smooth curves and $X = C \times D$, then $q(X) = g(C) + g(D)$ and $P_n(X) = P_n(C) \cdot P_n(D)$. In particular we have*

$$\mathrm{kod}(X) = \begin{cases} -\infty & \textit{if } C \textit{ or } D \textit{ rational} \\ 0 & \textit{if } C \textit{ and } D \textit{ are elliptic} \\ 1 & \textit{if } g(C) = 1,\ g(D) \geq 2 \\ & \textit{or conversely} \\ 2 & \textit{if } g(C) \geq 2,\ g(D) \geq 2. \end{cases}$$

Proof. Let $f_1 : X \to C$ and $f_2 : X \to D$ be the projections. Then

$$\mathcal{K}_X = f_1^*(\mathcal{K}_C) \otimes f_2^*(\mathcal{K}_D),$$

hence (by the Künneth formula)

$$\Gamma(\mathcal{K}_X^{\otimes n}) \cong \Gamma(\mathcal{K}_C^{\otimes n}) \otimes \Gamma(\mathcal{K}_D^{\otimes n}).$$

This gives the formula for $P_n(X)$. And since $H^1(\mathcal{O}_X) \cong H^1(\mathcal{O}_C) \oplus H^1(\mathcal{O}_D)$ again by Künneth, we have $q(X) = g(C) + g(D)$. □

Elliptic Fibrations

We recall that by an elliptic fibration of a surface X we mean a proper, connected holomorphic map $f : X \to S$, such that the general fibre X_s ($s \in S$) is non-singular elliptic (the holomorphic structure may depend on s). Unless otherwise stated we shall always assume that f is (relatively) minimal, i.e., all fibres are free of (-1)-curves. An elliptic surface is a surface admitting an elliptic fibration.

7. Kodaira's Table of Singular Fibres

Let $f : X \to \Delta$ be an elliptic fibration over the unit disk Δ, such that all fibres X_s, $s \neq 0$, are smooth. We shall list all possibilities for the nature of X_0. The result is embodied in Table 3. We use Kodaira's original notation, which is now generally accepted.

We distinguish between three cases:

a) X_0 is irreducible. In this case the adjunction formula immediately yields that X_0 is either smooth elliptic, or rational with a node, or rational with a cusp (types I_0, I_1 and II);

b) X_0 is reducible, but not multiple. We claim that every component C_i of $X_0 = \sum n_i C_i$ is a (-2)-curve. For we have $0 = (\mathcal{K}, X_0) = \sum n_i(\mathcal{K}, C_i) = \sum n_i(-C_i^2 + 2g(C_i) - 2)$ whereas $C_i^2 \le -1$ by Zariski's lemma, and the case $C_i^2 = -1, g(C_i) = 0$ is excluded by our assumption about relative minimality. Applying again Zariski's lemma, we see that for every two different components C_i, C_j we must have that C_iC_j is either 0, 1 or 2. In the last case we must have $X_0 = C_i + C_j$, thus obtaining the possibilities I_1 and III. Otherwise the intersection graph Γ has the property that any two vertices are joined by at most one edge, and we can consider the associated quadratic form $Q(\Gamma)$. Using once more Zariski's lemma we find that this form is positive semi-definite, so we can apply Lemma I. 2.12, obtaining as only possibilities the graphs $\tilde{A}_n, \tilde{D}_{n+4}, \tilde{E}_6, \tilde{E}_7$ and $\tilde{E}_8$. The graph $\tilde{A}_n, n \ge 3$ leads to I_n, whereas for $\tilde{A}_2$ there are two possibilities, namely I_2 and IV. The graphs $\tilde{D}_{n+4}, \tilde{E}_6, \tilde{E}_7$ and $\tilde{E}_8$ give the types I_n^*, II^*, III^* and IV^* respectively.

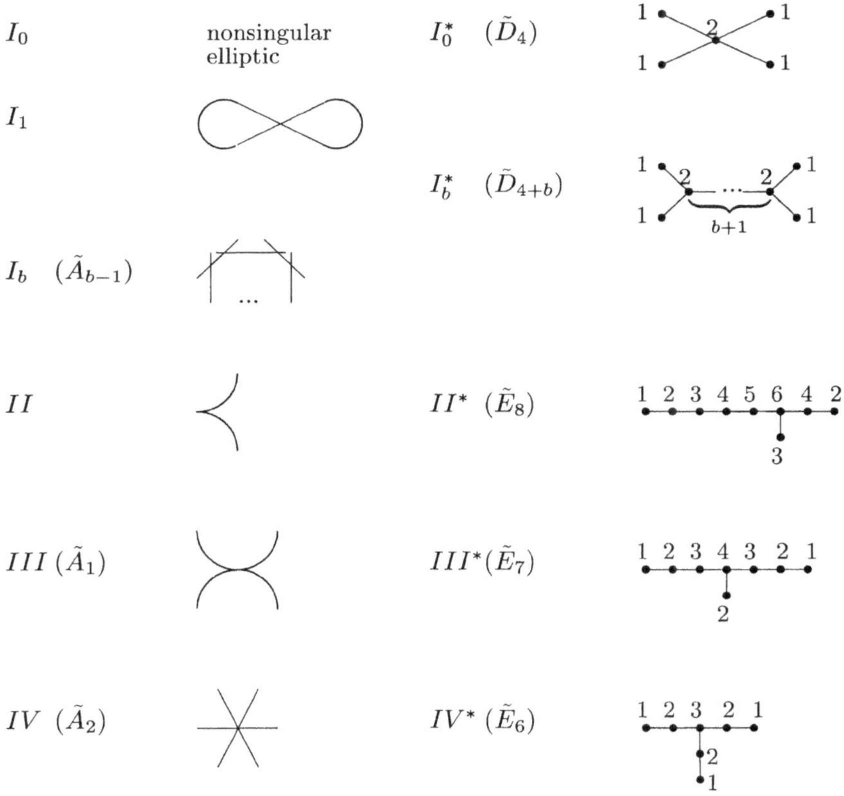

Table 3. Kodaira's table of singular elliptic fibres

c) X_0 is a multiple fibre. Exploiting Zariski's lemma in the same way as we did earlier, we find that $X_0 = mX_0'$, where X_0' is one of the types described before. As a consequence of Lemma III. 8.3, however, X_0' cannot be simply connected, hence we are left with the possibilities $X_0' = \mathrm{I}_0$, I_1 or I_b leading to the types ${}_m\mathrm{I}_0$, ${}_m\mathrm{I}_1$ and ${}_m\mathrm{I}_b$.

Remark 1. The existence of all types will be established later (Sects. 8 and 10).

Remark 2. The table shows that only the fibres of type $\mathrm{I}_b, b \geq 0$, are stable, a fact that was already mentioned before (III, Sect. 10).

Kodaira's table is given in [Ko60], Theorem 6.2.

8. Stable Fibrations

In this section we give the construction of some stable elliptic fibrations $f : X \to \Delta$ over the unit disk and compute their monodromy and period map.

Type I_0. Let $z(s)$ be an arbitrary holomorphic function on Δ with $\mathrm{Im}\, z(s) > 0$. Let $\mathbb{Z} \times \mathbb{Z}$ act on $\mathbb{C}(c) \times \Delta(s)$ by

$$(m, n)(c, s) = (c + m + nz(s), s).$$

The quotient $X = (\mathbb{C} \times \Delta)/(\mathbb{Z} \times \mathbb{Z})$ is a non-singular surface fibred over Δ, such that X_s is an elliptic curve with periods $1, z(s)$. Conversely, every elliptic fibration which has only smooth fibres, can locally be obtained in this way. Obviously, the monodromy is trivial. The period domain is $\mathfrak{H}_1/\Gamma_1 = \mathbb{C}$, where $\mathfrak{H}_1 = \{z \in \mathbb{C}; \mathrm{Im}\, z > 0\}$ is the upper half plane and $\Gamma_1 = \mathrm{SL}(2, \mathbb{Z})/\{\pm 1\}$ is the modular group acting by $z \mapsto \dfrac{az+b}{cz+d}$. The isomorphism $\mathfrak{H}_1/\Gamma_1 \to \mathbb{C}$ is given explicitly by the j-function. Following Kodaira, we shall use the convention in which

$$j(z) = 0 \quad \text{if } z \text{ is equivalent to } \mathbf{e}(\tfrac{1}{6}) \text{ under } \Gamma_1,$$
$$j(z) = 1 \quad \text{if } z \text{ is equivalent to i under } \Gamma_1.$$

So, wherever j vanishes in the upper half-plane, it vanishes there to the third order, and when $j = 1$, then $j - 1$ vanishes to the second order.

The period map $J(s) = j(z(s))$ was called the functional invariant by Kodaira. We see that for a smooth fibration $f : X \to \Delta$, the functional invariant always satisfies

$$(6) \qquad \begin{cases} \text{if } J(0) = 0, \text{ then } \quad J \quad \text{vanishes at 0 to an order } h \equiv 0(3), \\ \text{if } J(0) = 1, \text{ then } J - 1 \text{ vanishes at 0 to an order } h \equiv 0(2). \end{cases}$$

If one has any fibration in Weierstrass normal form

$$X = \{((z_0 : z_1 : z_2), s) \in \mathbb{P}_2 \times \Delta \mid z_0 z_2^2 = 4z_1^3 - g_2(s) z_0^2 z_1 - g_3(s) z_0^3\},$$

then

$$J(s) = \frac{g_2^3}{g_2^3 - 27g_3^2}$$

and X_s is non-singular if and only if $g_2^3(s) \neq 27g_3^2(s)$.

Type I_1. An example in Weierstrass normal form can be constructed using $g_2(s) = 3 - s, g_3(s) = 1 - s$, hence

$$X = \{((z_0 : z_1 : z_2), s) \in \mathbb{P}_2 \times \Delta \mid z_0 z_2^2 = 4z_1^3 + (s-3)z_0^2 z_1 + (s-1)z_0^3\}.$$

It is easily checked that the surface X is non-singular, that X_0 is irreducible rational with node at $(1 : -\frac{1}{4} : 0)$, and that X_s is non-singular elliptic for $s \neq 0$.

The functional invariant is

$$J(s) = \frac{(3-s)^3}{s(27 - 18s - s^2)}$$

and has an ordinary pole at $s = 0$. (In fact, in this case Satake's compactification is $\mathbb{P}_1$.)

We claim that the monodromy in integral homology can be given by

$$T = \begin{pmatrix} 1 & 1 \\ 0 & 1 \end{pmatrix}.$$

Proof. For real $s, 0 < s < 1$, the polynomial $4x^3 + (s-3)x + s - 1$ has three real roots $s_1 < s_2 < s_3$ (as its discriminant is $\neq 0$ for $0 < s < 1$, it suffices to check this for $s = 1$). If $s \to 0$, then both s_1 and s_2 tend to $-\frac{1}{4}$, whereas s_3 tends to 1. The projection $((z_0 : z_1 : z_2), s) \to (z_0 : z_1)$ exhibits the curve X_s as a double covering of $\mathbb{P}_1$, ramified in s_1, s_2, s_3 and ∞. The part of X_s over the real axis consists of four "circles" a, b, a', b' as in Fig. 4b.

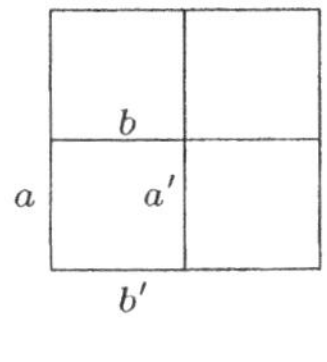

Fig. 4a

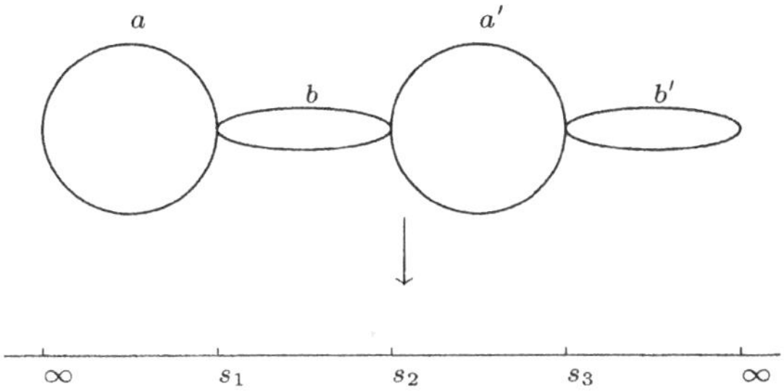

Fig. 4b

Since the intersection number of a and b is -1, we see that a and b span $H_1(X_s, \mathbb{Z})$. Furthermore, b shrinks to a point for $s \to 0$, so it is a vanishing cycle. Again since the intersection number of a and b is -1, we deduce from the Picard-Lefschetz formula (III, Sect. 14) that $T(b) = b$ and $T(a) = a + b$.

Type I_b, $b > 1$. Let $f : X \to \Delta$ be the fibration constructed above with X_0 of type I_1, and let $f^{(b)} : X^{(b)} \to \Delta$ be its b-th root fibration (III, Sect. 9).

There is a diagram

$$\begin{array}{ccccc} X^{(b)} & \xrightarrow{\tau'} & X' & \xrightarrow{\tau} & X \\ {\scriptstyle f^{(b)}}\downarrow & & \downarrow & & \downarrow{\scriptstyle f} \\ \Delta & = & \Delta & \xrightarrow[\delta_b : s \mapsto s^b]{} & \Delta \end{array}$$

where X' is the normalisation of $X \times_\Delta \Delta$ (the fibre product with respect to δ_b). This X' has one singularity of type A_b over the double point of X_0. The map τ is a b-fold covering map ramified along $C_1' = \tau^{-1}(X_0)$. The singularity in X' is resolved under τ' by a string $C_2, \ldots, C_b$ of (-2)-curves such that $C_1, C_2, \ldots, C_b$ form a cycle (C_1 is the proper transform of C_1'). So the singular fibre in $X^{(b)}$ is of type I_b.

Outside of $s = 0$, the fibration $f^{(b)}$ is the pull-back of f under δ_b. This shows that its functional invariant at $s = 0$ has a *pole of order* b and that its monodromy is given by

$$\begin{pmatrix} 1 & 1 \\ 0 & 1 \end{pmatrix}^b = \begin{pmatrix} 1 & b \\ 0 & 1 \end{pmatrix}.$$

The examples in this section are from [Ko60], p. 598–600.

9. The Jacobian Fibration

In this section let $f : X \to S$ be an elliptic fibration without multiple fibres. We describe the construction of $\mathrm{Jac}(f)$, a fibre space over S with fibres $\mathrm{Jac}(f)_s$, the one-dimensional complex Lie groups $\mathrm{Pic}^0(X_s)$. The sheaf of holomorphic sections of $\mathrm{Jac}(f)$ will be the sheaf $f_{*1}\mathcal{O}_X / f_{*1}\mathbb{Z}_X$.

As in III, Sect. 15, let E be the line bundle over S corresponding to the invertible sheaf $f_{*1}\mathcal{O}_X$ and $L \subset E$ the subset corresponding to the subsheaf $f_{*1}\mathbb{Z}_X$. If X_s is non-singular then L_s is a lattice in $E_s \cong \mathbb{C}$, and L is a closed submanifold of E near s. The whole point is to show that $L \subset E$ is closed everywhere. Then it is straight-forward to form $\mathrm{Jac}(f)$ as the quotient E/L. We refer to Kodaira ([Ko60], Sections 8 and 9) for details.

To prove that L is indeed a closed submanifold of E we argue in the following way. The assertion being local with respect to S, we fix some $s_0 \in S$ with X_{s_0} singular, and distinguish between several cases.

Case 1: X_{s_0} is stable. By Proposition III.13.1 and Theorem III.15.2 near s_0 the set L is spanned over $\mathbb{Z}$ by two sections β and $\alpha = z \cdot \beta$, where

$$z = \tilde{z} + \frac{\lambda}{2\pi \mathrm{i}} \log s, \quad 0 < \lambda \in \mathbb{N},$$

with $\tilde{z}$ holomorphic in s_0 and $\tilde{z}(s_0) = 0$.

If we take any sequence $g_1, g_2, \ldots$, with

$$g_n = \mu_n \alpha(s_n) + \nu_n \beta(s_n) = (\mu_n z(s_n) + \nu_n)\beta(s_n) \in L,$$

then convergence of g_n implies $\mu_n = 0$, hence $\nu_n = \nu$ is constant for $n \gg 0$ (this follows since β is holomorphic near s_0, but $\operatorname{Im} z(s_n) \to \infty$). So $g_n \to \nu\beta(s_0) \in L$.

Case 2: X_{s_0} is unstable. We choose a stable reduction

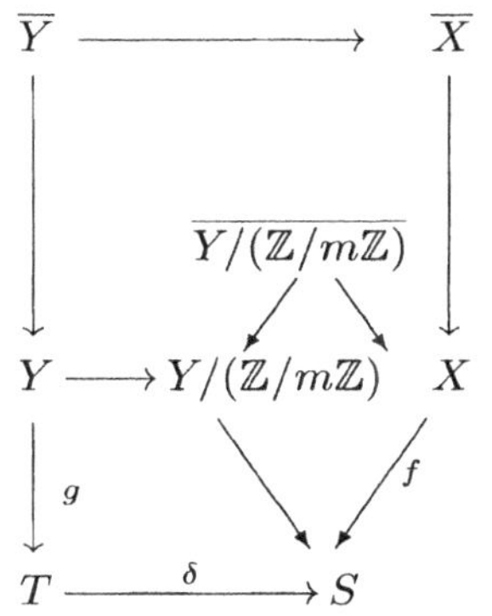

as in III, Sect. 10. It suffices to show that $(\mathbb{Z}/m\mathbb{Z})$ acts non-trivially on $H^1(\mathcal{O}_{Y_{t_0}})$, where $\delta(t_0) = s_0$, because then there is a diagram

$$\begin{array}{ccc} \delta^*(f_{*1}\mathbb{Z}_X) & \longrightarrow & g_{*1}\mathbb{Z}_Y \\ \downarrow & & \downarrow \\ \delta^*(f_{*1}\mathcal{O}_X) & \longrightarrow & g_{*1}\mathcal{O}_Y \end{array}$$

in which the lower horizontal arrow vanishes at $t_0 \in T$, the point over $s_0 \in S$ (Proposition III. 11.3). So if a sequence $g_n \in L = f_{*1}\mathbb{Z}_X$ converges in $E = f_{*1}\mathcal{O}_X$ by case 1, $\delta^*(g_n) = 0$ for $n \gg 0$ for these n. Let us now consider the action of $(\mathbb{Z}/m\mathbb{Z})$ on Y. It must have fixed points on Y_{t_0} (otherwise $X_0 = Y_{t_0}/(Z/m\mathbb{Z})$ would be a multiple fibre), but cannot act trivially on Y_{t_0} (because then $X_0 = Y_{t_0}$ would be stable). This implies already that $(\mathbb{Z}/m\mathbb{Z})$ acts non-trivially on $H^1(\mathcal{O}_{Y_{t_0}})$ if Y_{t_0} is non-singular. And if Y_{t_0} is of type I_b, $b > 0$, it is sufficient to notice that the generator of $(\mathbb{Z}/m\mathbb{Z})$ cannot preserve the orientation of the cycle of rational curves, because then X_0 would not be simply connected. □

(9.1) **Proposition.** *Assume that* $f : X \to S$ *admits a section* $\sigma : S \to X$, *and let* $X^\sigma = X \setminus \{$*irreducible components of fibres* X_s *not meeting* $\sigma(S)\}$. *Then there is a canonical fibre-preserving isomorphism from* $\mathrm{Jac}(f)$ *onto* X^σ, *mapping the zero-section in* $\mathrm{Jac}(f)$ *onto* σ.

Proof. For $x \in X^\sigma$, $f(x) = s$, let $\mathcal{O}_{X_s}(x - \sigma(s)) \in \mathrm{Pic}(X_s)$. From Serre duality and the Riemann-Roch theorem II.3.1 it follows that

(i) $\mathcal{O}_{X_s}(x - \sigma(s)) \in \mathrm{Pic}^0(X_s) = \mathrm{Jac}(f)_s$

(ii) $\mathcal{O}_{X_s}(x - \sigma(s)) = \mathcal{O}_{X_s}(x' - \sigma(s))$ if and only if $x = x'$,

(iii) for $d \in \mathrm{Jac}(f)_s$, there is always some $x \in X_s \cap X^\sigma$ with $d = \mathcal{O}_{X_s}(x - \sigma(s))$.

This shows that there is a bijective map $\mathrm{Jac}(f) \to X^\sigma$ defined by $d \mapsto x$ if $d = \mathcal{O}_{X_s}(x - \sigma(s))$. It is a formality to verify that this map is holomorphic. □

Remark. Proposition 9.1 implies that X is a "properification" of its Jacobian fibration. This can be used to prove uniqueness theorems. For example, consider a fibration $f : X \to \Delta$ with X_0 of type I_b, $b > 0$. By construction, $\mathrm{Jac}(f) = \mathbb{C} \times \Delta / L$ where L is spanned over $\mathbb{Z}$ by 1 and $z = \tilde{z} + \frac{\lambda}{2\pi \mathrm{i}} \log s$, $0 < \lambda \in \mathbb{N}$, with $\tilde{z}$ holomorphic near 0. Replacing s by $se\left(\frac{\tilde{z}}{\lambda}\right)$, we can achieve $z = \frac{\lambda}{2\pi \mathrm{i}} \log s$. This means that all those fibrations which have the same λ are biholomorphically equivalent (cf. Proposition III. 8.5). But λ is determined by the monodromy action, and for the fibration $f^{(b)}$ constructed above we have $\lambda = b$. So every fibration with X_0 of type I_b is (at the point 0) locally equivalent to the fibration $f^{(b)}$, and to the properification of $\mathbb{C} \times \Delta / L$, where $L = \mathbb{Z} + \mathbb{Z}\frac{b}{2\pi \mathrm{i}} \log s$.

Finally, let $X^\#$ be the subset of X where $df \neq 0$, and let $X_s^\#$ be the intersection $X_s \cap X^\#$. Denote by $\mathcal{O}(X^\#)$ the sheaf of local holomorphic sections of $f : X \to S$. For any two sections $\sigma, \tau : S \to X$ there is (by Proposition 9.1 and Proposition III. 8.5) a unique biholomorphic automorphism of X mapping σ to τ and acting as translation on every non-singular fibre X_s. So, upon specifying one section $\sigma : S \to X$ as zero section, the sheaf $\mathcal{O}(X^\#)$ becomes a sheaf of commutative groups. There is an exact sequence

$$0 \to f_{*1}\mathcal{O}_X / f_{*1}\mathbb{Z}_X \to \mathcal{O}(X^\#) \to F \to 0,$$

where F is concentrated at the critical values $s \in S$ and F_s is finite.

Since each fibre X_s which is not a multiple fibre admits local sections (cf. Table 3), the curve $X_s^\#$ carries the structure of a 1-dimensional complex Lie group. If in particular X_s is of type $\mathrm{I}_b, b > 0$, then $F_s = \mathbb{Z}/b\mathbb{Z}$ and $X_s^\# = \mathbb{C}^* \times \mathbb{Z}/b\mathbb{Z}$.

We have closely followed [Ko60], Sect. 9.

10. Stable Reduction

In this section we keep an earlier promise (Sect. 7, Remark 1) and give examples of elliptic fibrations $f : X \to \Delta$ with unstable singular fibres X_0. It turns out that all types in Table 3 actually occur. For the types which are not multiple fibres we list in Table 6 below the behaviour of the functional invariant $J(s)$ and the monodromy in the examples constructed.

Type ${}_m\mathrm{I}_0$. We start with a non-singular fibration $Y = \mathbb{C} \times \Delta/L$, where $L = \mathbb{Z}+\mathbb{Z}\cdot z(s)$ with z holomorphic on Δ and $z(s) = z(0)+\text{const}\cdot s^{mh}$, $h \in \mathbb{N}$. The automorphism of $\mathbb{C} \times \Delta$

$$(c, s) \mapsto \left(c + \frac{1}{m}, \mathbf{e}\left(\frac{1}{m}\right) s\right)$$

generates a group $\mathbb{Z}/m\mathbb{Z}$ acting on Y without fixed points. The quotient has a singular fibre over 0 of type ${}_m\mathrm{I}_0$.

Type ${}_m\mathrm{I}_b$, $b > 0$. Here we start with a fibration $f : Y \to \Delta$ where

$$\mathrm{Jac}(f) = \mathbb{C} \times \Delta/L \text{ with } L = \mathbb{Z} + \mathbb{Z}\frac{mb}{2\pi \mathrm{i}} \log s.$$

So Y_0 is of type I_{mb}, say $Y_0 = C_1 + \ldots + C_{mb}$. The automorphism

$$(c, s) \mapsto \left(c, \mathbf{e}\left(\frac{1}{m}\right) s\right)$$

of $\mathbb{C} \times \Delta$ induces on Y a fibre-preserving automorphism μ of order m, which commutes with all translations from $\mathcal{O}(Y^{\#})_0$. Since $Y_0^{\#} = \mathbb{C}^* \times /(\mathbb{Z}/mb\mathbb{Z})$, there is some section $t \in \mathcal{O}(Y_0^{\#})$ acting as $C_i \to C_{i+b}$ on Y_0 and as translation of order m on the nearby regular fibres. So μt generates a group $\mathbb{Z}/m\mathbb{Z}$ acting on Y without fixed points, such that the quotient has a singular fibre of type ${}_m\mathrm{I}_b$.

Type I_0^*. This time we start with a non-singular fibration $Y = \mathbb{C} \times \Delta/L$, where $L = \mathbb{Z} + \mathbb{Z}z(s)$ with $z(s) = z(0) + s^{2h}, h \in \mathbb{N}$. The map

$$(c, s) \mapsto (-c, -s)$$

defines an involution ι on Y having four isolated fixed points on Y_0. In $Y/\{\mathrm{id}, \iota\}$ these four points give rise to four A_1-singularities, located on the image of Y_0 which is a non-singular rational curve of multiplicity two. Resolution of the four singularities leads to a fibration with singular fibre of type I_0^*.

Type I_b^*. Here we begin with a fibration $f : Y \to S$ of type I_{2b}, where $\mathrm{Jac}(f) = \mathbb{C} \times \Delta/L$ and $L = \mathbb{Z} + \mathbb{Z}\frac{2b}{2\pi \mathrm{i}} \log s$. The map

$$(c, s) \mapsto (-c, -s)$$

induces on $\mathrm{Jac}(f)$ an involution ι. By Proposition III. 8.5, this ι extends to Y, where it interchanges C_i and C_{2b+2-i} and has on both C_1 and C_{b+1} two fixed points. The image of Y_0 in $Y/\{\mathrm{id}, \iota\}$ is a string of b non-singular rational curves of multiplicity two. Resolution of the A_1-singularities on the first and the last of these curves leads to a fibration with singular fibre of type I_b^*.

Table 5.

type	II	II*	III	III*	IV	IV*
h	$2 \bmod 6$	$4 \bmod 6$		$2 \bmod 4$	$2 \bmod 3$	$1 \bmod 3$
$z(s)$	$\dfrac{\mathbf{e}\left(\frac{1}{3}\right) - \mathbf{e}\left(\frac{2}{3}\right) s^h}{1-s^h}$		$\dfrac{\mathrm{i} + \mathrm{i}s^h}{1-s^h}$		$\dfrac{\mathbf{e}\left(\frac{1}{3}\right) - \mathbf{e}\left(\frac{2}{3}\right) s^h}{1-s^h}$	
m	6		4		3	
$z\left(\mathbf{e}(\frac{1}{m})s\right)$	$-\dfrac{1}{z(s)} - 1$	$-\dfrac{1}{z(s)+1}$	$-\dfrac{1}{z(s)}$		$-\dfrac{1}{z(s)+1}$	$-\dfrac{1}{z(s)} - 1$
$\tilde{\mu}(c)$	$-\dfrac{c}{z(s)}$	$\dfrac{c}{z(s)+1}$	$-\dfrac{c}{z(s)}$	$\dfrac{c}{z(s)}$	$-\dfrac{c}{z(s)+1}$	$\dfrac{c}{z(s)}$
$\mu \vert Y_0$	$\mathbf{e}(\frac{1}{6})$	$\mathbf{e}(\frac{5}{6})$	i	$-\mathrm{i}$	$\mathbf{e}(\frac{1}{3})$	$\mathbf{e}(\frac{2}{3})$
fixed points	$\mu:$ $(1,1)$	$(5,1)$	$\mu: 2\times(1,1)$	$2\times(3,1)$		
with	$\mu^2: 2\times(1,1)$	$2\times(2,1)$	$\mu^2: 2\times(1,1)$	$2\times(1,1)$	$3\times(1,1)$	$3\times(2,1)$
weights	$\mu^3: 3\times(1,1)$	$3\times(1,1)$				
quotient singularities	$A_{6,1}\, A_{3,1}\, A_1$	$A_5\, A_2\, A_1$	$2A_{4,1}A_1$	$2A_3\, A_1$	$3A_{3,1}$	$3A_2$

For *all the other types* II-IV* we can start with a non-singular fibration $Y = \mathbb{C}\times\Delta/L$, where $L = \mathbb{Z}+\mathbb{Z}z(s)$, and $z(s)-z(0)$ vanishes at $s=0$ to a certain order $h \in \mathbb{N}$. In Table 5 we give for all types this order h and the form, to which $z(s)$ can be normalised (perhaps after shrinking Δ) by changing the coordinate s. Then a group $\mathbb{Z}/m\mathbb{Z}$ of order $m = 6, 4$ or 3 acting on Y is constructed as follows. On s, the generator μ acts as multiplication with $\mathbf{e}\left(\frac{1}{m}\right)$. So it transforms $z(s)$ into $z\left(\mathbf{e}\left(\frac{1}{m}\right)s\right)$, which is computed in row 5 of Table 5. Then on Y the map μ is induced by the automorphism of $\mathbb{C}\times\Delta$

$$(c, s) \mapsto \left(\tilde{\mu}(c), \mathbf{e}\left(\frac{1}{m}\right)s\right)$$

with $\tilde{\mu}(c), c \in \mathbb{C}$, defined in row 6. So μ acts on Y_0 by multiplication with an easily computed root of unity (row 7).

In $Y/(\mathbb{Z}/m\mathbb{Z})$ the quotient $Y_0/(\mathbb{Z}/m\mathbb{Z})$ becomes a rational curve of multiplicity m. On this curve, there are some singularities of $Y/(\mathbb{Z}/m\mathbb{Z})$ coming from non-trivial isotropy groups. These fixed points, and the weights by which the group acts, as well as the quotient singularities, are easily determined (III, Sect. 5). The minimal resolution of these singularities leads to the fibration X. In the case II*, III*, IV* no further blowing down is necessary, whereas in the other cases the final step is described below (numbers without brackets are multiplicities, and numbers within brackets denote self-intersections):

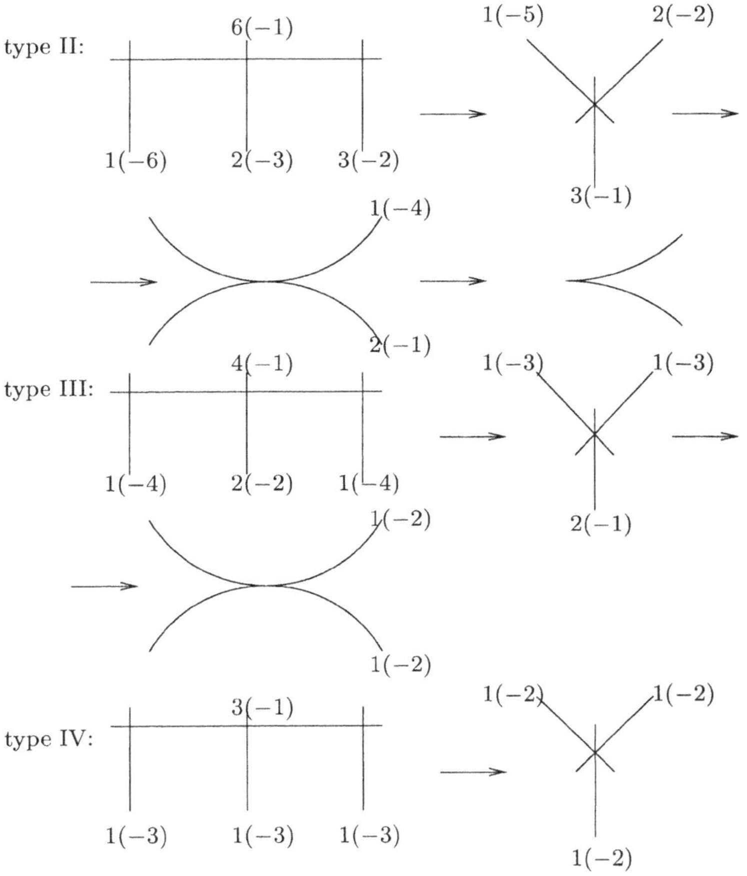

In the examples constructed above, the functional invariant J and the monodromy T are as in the Table below. *Proof.* The behaviour of J is easily deduced from the behaviour of J for the stable reduction. The monodromy of I_0 is trivial and that of I_b was computed in Sect. 8. The monodromy for I_b^* must be a matrix T with

$$T^2 = \begin{pmatrix} 1 & 2b \\ 0 & 1 \end{pmatrix}, \quad T\begin{pmatrix} 1 \\ 0 \end{pmatrix} = \begin{pmatrix} -1 \\ 0 \end{pmatrix}.$$

This leaves us with the possibility $T = -\begin{pmatrix} 1 & b \\ 0 & 1 \end{pmatrix}$.

Table 6.

type	$J(s)$	T
I_0	subject to conditions (6)	$\begin{pmatrix} 1 & 0 \\ 0 & 1 \end{pmatrix}$
I_0^*		$-\begin{pmatrix} 1 & 0 \\ 0 & 1 \end{pmatrix}$
I_b	pole of order b	$\begin{pmatrix} 1 & b \\ 0 & 1 \end{pmatrix}$
I_b^*		$-\begin{pmatrix} 1 & b \\ 0 & 1 \end{pmatrix}$
II	s^h, $h \equiv 1 \bmod 3$	$\begin{pmatrix} 1 & 1 \\ -1 & 0 \end{pmatrix}$
II*	s^h, $h \equiv 2 \bmod 3$	$\begin{pmatrix} 0 & -1 \\ 1 & 1 \end{pmatrix}$
III	$1 + s^h$, h odd	$\begin{pmatrix} 0 & 1 \\ -1 & 0 \end{pmatrix}$
III*		$\begin{pmatrix} 0 & -1 \\ 1 & 0 \end{pmatrix}$
IV	s^h, $h \equiv 2 \bmod 3$	$\begin{pmatrix} 0 & 1 \\ -1 & -1 \end{pmatrix}$
IV*	s^h, $h \equiv 1 \bmod 3$	$\begin{pmatrix} -1 & -1 \\ 1 & 0 \end{pmatrix}$

For all other types, the stable reduction Y_0 is non-singular. So T is just the action on $H_1(Y_0, \mathbb{Z}) = H_1(Y, \mathbb{Z})$ by the generator μ^{-1} of the ramification group $\mathbb{Z}/m\mathbb{Z}, m = 2, 3, 4$ or 6. □

The results in this section are due to Kodaira ([Ko60], Sects 8,9, [Ko66], Sect. 4).

11. Classification

Let $f : X \to S$ be an elliptic fibration and let $s_1, \ldots, s_k \in S$ be the critical values and $S^* = S \backslash \{s_1, \ldots, s_k\}$. Then $L = f_{*1}\mathbb{Z}_X$ is a locally constant sheaf on S^* with fibre $\mathbb{Z} \oplus \mathbb{Z}$. This sheaf, or equivalently its global monodromy which is an equivalence class of representations $R : \pi_1(S^*) \to SL(2, \mathbb{Z})$, was called by Kodaira the homological invariant of the fibration.

In this section it is essential to distinguish carefully between $SL(2, \mathbb{Z})$ and the modular group $\Gamma_1 = SL(2, \mathbb{Z})/\{\pm 1\}$. Any representation R as above induces by composition a representation $r : \pi_1(S^*) \to \Gamma_1$. The equivalence class of the representation r is already uniquely determined by the functional invariant $J(s)$. For let $\mathfrak{H}^*$ be the upper half plane $\{z \in \mathbb{C}; \operatorname{Im} z > 0\}$ with all points in the orbits $\Gamma_1 \cdot \mathbf{e}(\frac{1}{3})$ and $\Gamma_1 \cdot \mathrm{i}$ removed. Then the covering $j : \mathfrak{H}^* \to \mathbb{C}\backslash\{0, 1\}$ makes $\mathfrak{H}^*$ into a Γ_1-principal bundle over $\mathbb{C}\backslash\{0, 1\}$, thus fixing a class of representations $\pi_1(\mathbb{C}\backslash\{0, 1\}) \to \Gamma_1$. Any meromorphic function J on S defines a map of $S^* = S\backslash J^{-1}(\{0, 1, \infty\}) \to \mathbb{C}\backslash\{0, 1\}$, hence uniquely determines a class of representations $r : \pi(S^*) \to \Gamma_1$ by pull-back. In terms of the holomorphic multivalued period function $z(s)$, this r is the monodromy imposed on z by the condition $j(z(s)) = J(s)$. Kodaira said it this way: a homological invariant $R : \pi_1(S^*) \to SL(2, \mathbb{Z})$ belongs to J, if it induces the representation class of $r : \pi_1(S^*) \to \Gamma_1$ defined by J.

(11.1) **Theorem** (Classification theorem for elliptic fibrations without multiple fibres). *Let S be a smooth compact connected curve, $S^* = S\backslash\{s_1, \ldots, s_k\}$ and J a meromorphic function on S with $J(s) \neq 0, 1, \infty$ for $s \in S^*$.*

(a) *If $k \geq 1$, then there are exactly $2^{g(S)+k-1}$ inequivalent homological invariants L belonging to J.*

(b) *Given J and a homological invariant L belonging to J, there is exactly one elliptic fibration $f : X \to S$ with these invariants, admitting a section.*

(c) *All elliptic fibrations, without multiple fibres, with given invariants J and L form a set $\mathfrak{F}(J, L)$ parametrised by the abelian group $H^1(\mathfrak{Jac}(f))$. Here $\mathfrak{Jac}(f)$ is the jacobian fibration for one, hence by* (b), *for all members of $\mathfrak{F}(J, L)$ and $\mathfrak{Jac}(f)$ is the sheaf (of abelian groups) of local holomorphic sections in $\mathfrak{Jac}(f)$.*

Proof. (a) Let $l_1, \ldots, l_{2g}$ be a canonical set of generators for $\pi_1(S)$ and $l_{2g+1}, \ldots, l_{2g+k}$ be loops around the points $s_1, \ldots, s_k$. Then $\pi_1(S^*)$ is the group with generators $l_1, \ldots, l_{2g+k}$ and one relation

$$l_1 l_2 l_1^{-1} l_2^{-1} \ldots l_{2g-1} l_{2g} l_{2g-1}^{-1} l_{2g}^{-1} l_{2g+1} \ldots l_{2g+k} = 1.$$

So for any given $r(l_i) \in \Gamma_1, i = 1, \ldots, 2g + k - 1$, a representative $R(l_i) \in SL(2, \mathbb{Z})$ can be chosen arbitrarily, but for $R(l_{2g+k})$ there is no choice.

(b)*Local existence.* By inspection of Table 6, it is possible to choose for each point $s \in S$ a neighbourhood $N \subset S$ and a holomorphic injection

$(N, s) \to (\Delta, 0)$ such that the function $J|N$ is transformed into one which is given in column 2 of this table. Even the two different possibilities for T, differing by sign, are realised by a row of this table.

Local uniqueness. It was observed already (Remark after Proposition 9.1), that a stable fibration with section is locally determined uniquely by its functional invariant $J(s)$. Now let $f : X \to \Delta$ be a fibration with section σ and functional invariant $J \neq 0, 1, \infty$ for $s \in \Delta^*$, and let $g : Y \to \Delta$ be its stable reduction. Then X is bimeromorphically equivalent to $Y/(\mathbb{Z}/m\mathbb{Z})$, where the generator μ of $(\mathbb{Z}/m\mathbb{Z})$ acts on Y by fibre-preserving automorphisms, and on Δ by $s \to \mathbf{e}\left(\frac{1}{m}\right) \cdot s$. Now g is determined uniquely by the pull-back of J. The map μ is determined by the monodromy up to a translation on the fibres. But, since μ respects the pull-back of the section σ, it is totally determined by the monodromy of f.

Global existence. Choose a covering $S = \bigcup \Delta_i$ by unit disks such that *either* Δ_i does not contain any one of the points $s_1, \ldots, s_k$ *or* Δ_i contains exactly one of the points $s_1, \ldots, s_k$, this point being its centre.

The $\Delta_i \subset S$ may be chosen so small that by Table 6, there is an elliptic fibration $f_i : X_i \to \Delta_i$ admitting a section σ_i, having the functional invariant $J|\Delta_i$, and $f_{*i}\mathbb{Z}_{X_i} = L|\Delta_i$. Over an intersection $\Delta_i \cap \Delta_j$, both fibrations f_i and f_j are smooth with the same functional invariant and trivial monodromy. So there is an isomorphism $c_{ij} : X_j|\Delta_i \cap \Delta_j \to X_i|\Delta_i \cap \Delta_j$ preserving the fibration, transforming $\sigma_j|\Delta_i \cap \Delta_j$ into $\sigma_i|\Delta_i \cap \Delta_j$, and inducing the identity on $f_{*1}\mathbb{Z}_{X_i} = L|\Delta_i \cap \Delta_j = f_{*1}\mathbb{Z}_{X_j}$. For any three indices i, j, k the composition $c_{ij}c_{jk}c_{ki}$ is an automorphism of $X_i|\Delta_i \cap \Delta_j \cap \Delta_k$ preserving the section σ and acting trivially on the homology of the fibres. So $c_{ij}c_{jk}c_{ki} = \mathrm{id}$, and the c_{ij} can be used to define a fibration $f : X \to S$ by patching the X_i together.

Global uniqueness. This follows from the construction as given before, since the only freedom in the construction of X is the choice of the Δ_i.

(c) If X is any element of $\mathcal{F}(J, L)$ there is a covering $S = \bigcup \Delta_i$ as above with isomorphisms $c_i : X_i \to X|\Delta_i$ inducing the identity on $L|\Delta_i$. It follows that $c_j^{-1}c_i$ is a cocycle with values in $\mathrm{Jac}(f)$ which glues the collection of X_i together to build X. Visibly, cohomologous $c_j^{-1}c_i$ give isomorphic X. □

For this section, we refer to [Ko60], Sect. 10. We observe that Theorem 11.1 does not describe moduli. Seiler [Sei87a,87b] made an essential step in solving the moduli problem, a complete solution of which can be found in the book [F-M94].

12. Invariants

We start with Kodaira's formula for the canonical bundle of an elliptic surface (with or without multiple fibres).

(12.1) **Theorem** (Canonical bundle formula for elliptic fibrations). *Let $f : X \to S$ be a relatively minimal elliptic fibration, such that its multiple fibres are $X_{s_1} = m_1F_1, \dots, X_{s_k} = m_kF_k$. Then*

$$\mathcal{K}_X = f^*(\mathcal{K}_S \otimes (f_{*1}\mathcal{O}_X)^\vee) \otimes \mathcal{O}_X(\sum(m_i - 1)F_i).$$

Proof. We have seen (Proposition III. 12.1) that $f_*\mathcal{K}_X$ is locally free of rank one. By relatively duality

$$f_*\mathcal{K}_X = f_*\omega_{X/S} \otimes \mathcal{K}_S = \mathcal{K}_s \otimes (f_{*1}\mathcal{O}_X)^\vee.$$

This proves that

$$\mathcal{K}_X = f^*(\mathcal{K}_S \otimes (f_{*1}\mathcal{O}_X)^\vee) \otimes \mathcal{O}_X(D)$$

where D is the zero divisor of the canonical morphism

$$\lambda : f^*(f_*\mathcal{K}_X) \to \mathcal{K}_X.$$

This λ is an isomorphism on each non-singular fibre X_s. On a singular fibre the map λ cannot vanish identically, i.e., to the same order as the multiplicity of the fibre, for the following reason. Locally with respect to S, any 2-form on X is of the type $f^*(\omega)$, with ω a section in $f_*\mathcal{K}_X$. If now $\lambda|X_s$ were zero, there would be on X a 2-form $(f^*(\omega))/g$, with g an equation for X_s, which is not of the type $f^*(\omega)$.

Let D_s be the part of the divisor D over s. It is a canonical divisor on a neighbourhood of X_s. By Kodaira's table we have $D_s \cdot C = 0$ for all irreducible components $C \subset X_s$. In particular $D_s^2 = 0$, and by Zariski's lemma $D_s = rX_s$ with $r \in \mathbb{Q}, r < 1$. So if $D_s \neq 0$, then X_s must be a multiple fibre.

We see therefore that $D = \sum n_iF_i$, where the summation is taken over all multiple fibres and $n_i < m_i$. To determine the precise value of n_i, we use the adjunction formula

$$\omega_{F_i} = \mathcal{K}_X \otimes \mathcal{O}_{F_i}(F_i) = \mathcal{O}_{F_i}(F_i)^{\otimes(n_i+1)}.$$

Since F_i is a curve of type I_b, $b \geq 0$ there is an isomorphism $\omega_{F_i} = \mathcal{O}_{F_i}$. But $\mathcal{O}_{F_i}(F_i)$ is a torsion bundle of order exactly m_i (Lemma III. 8.3). So m_i divides $n_i + 1$, and $n_i = m_i - 1$. □

Remark. By Theorem III. 18.2 we have

$$\deg\,(f_{*1}\mathcal{O}_X)^\vee = \deg\,(f_*\omega_{X/S}) > 0$$

unless all the smooth fibres of f are isomorphic and the singular fibres are of type ${}_m\mathrm{I}_0$ only.

(12.2) **Proposition.** $\deg (f_{*1}\mathcal{O}_X)^\vee = \chi(\mathcal{O}_X)$.

Proof. By Leray's spectral sequence for f and $\mathcal{O}_X$ there are isomorphisms

$$H^0(\mathcal{O}_X) \cong H^0(f_*\mathcal{O}_X), \quad H^2(\mathcal{O}_X) \cong H^1(f_{*1}\mathcal{O}_X)$$

and an exact sequence

$$0 \to H^1(f_*\mathcal{O}_X) \to H^1(\mathcal{O}_X) \to H^0(f_{*1}\mathcal{O}_X) \to 0.$$

So by Riemann-Roch on S

$$\chi(\mathcal{O}_X) = \chi(f_*\mathcal{O}_X) - \chi(f_{*1}\mathcal{O}_X) = -\deg (f_{*1}\mathcal{O}_X). \qquad \square$$

In many cases a weak consequence of 12.1 and 12.2 is sufficient:

(12.3) **Corollary.** *Let $f : X \to S$ be an elliptic fibration, such that its multiple fibres are $X_{s_1} = m_1F_1, \dots, X_{s_k} = m_kF_k$. Then*

$$\mathcal{K}_X = f^*(\mathcal{L}) \otimes \mathcal{O}_X(\textstyle\sum(m_i - 1)F_i)$$

where $\mathcal{L}$ is a line bundle of degree $\chi(\mathcal{O}_X) - 2\chi(\mathcal{O}_S)$ on S.

(12.4) **Corollary.** *If a compact surface X admits a relatively minimal elliptic fibration over $\mathbb{P}_1$, then*
(i) $\mathrm{kod}(X) = 0$ *if and only if* $\mathcal{K}_X^{\otimes 12} \cong \mathcal{O}_X$,
(ii) $\mathrm{kod}(X) = -\infty$ *if and only if* $P_{12}(X) = 0$.

Proof. (i) If $\mathcal{K}_X^{\otimes 12} \cong \mathcal{O}_X$, then $\mathrm{kod}(X) = 0$. To prove the converse, we consider a minimal elliptic fibration $X \to \mathbb{P}_1$ with k multiple fibres of multiplicity $m_1, \dots, m_k$ respectively. From Corollary 12.3 we infer that $\mathrm{kod}(X) = 0$ implies

$$\chi(\mathcal{O}_X) - 2 + k - \sum_{i=1}^{k} 1/m_i = 0.$$

(If this number were strictly positive, then $P_n(X) \geq 2$ for n large enough and hence $\mathrm{kod}(X) \geq 1$, whereas if this number were strictly negative, K_X would be rationally homologous to a strictly negative multiple of a fibre and then $\mathrm{kod}(X) = -\infty$.) Since $\chi(\mathcal{O}_X) \geq 0$ by Proposition 12.2 and the remark preceding it, we see that there remain only the following possibilities:

$$\begin{aligned}
&\chi(\mathcal{O}_X) = 1,\ k = 2,\ m_1 = m_2 = 2\\
&\chi(\mathcal{O}_X) = 2,\ k = 4,\ m_1 = m_2 = m_3 = m_4 = 4\\
&\chi(\mathcal{O}_X) = 2,\ k = 3,\ m_1 = m_2 = m_3 = 3\\
&\qquad\qquad\qquad\qquad m_1 = 2,\ m_2 = m_3 = 4\\
&\qquad\qquad\qquad\qquad m_1 = 2,\ m_2 = 3, m_3 = 6.
\end{aligned}$$

In all cases we have $\mathcal{K}_X^{\otimes 12} \cong \mathcal{O}_X$.

(ii) If $\mathrm{kod}(X) = -\infty$, then $P_{12}(X) = 0$. For the other direction, let $X \to \mathbb{P}_1$ be a relatively minimal elliptic fibration, with general fibre F and multiple fibres $F_1, \ldots, F_k$ of multiplicity $m_1, \ldots, m_k$ respectively, such that $P_{12}(X) = 0$. There exist non-negative integers l_i, r_i with $0 \le r_i \le m_i$, such that

$$m_i l_i = 12 + r_i, \quad i = 1, \ldots, k.$$

(Hence $l_i \le 6$ and $l_i \ne 5$.) By Corollary12.3 we have

$$\mathcal{K}_X^{\otimes 12} = \mathcal{O}_X \left\{ \left(12\chi(\mathcal{O}_X) - 24 + \sum_{i=1}^{k} (12 - l_i) \right) F + \sum_{i=1}^{k} r_i F_i \right\}.$$

So $\sum_{i=1}^{k} (12 - l_i) < 24 - 12\chi(\mathcal{O}_X)$. Since $\chi(\mathcal{O}_X) \ge 0$, this implies already that $k \le 3$. The case $k = 2$ being trivial, it is sufficient to consider the case $k = 3$. Assuming $l_1 \ge l_2 \ge l_3$, we find two possibilities: $l_1 = l_2 = 6$ and $l_1 = 6, l_2 = 4, l_3 \le 3$. In the first case we obtain $m_1 = m_2 = 6$ and in the second case $m_1 = 2, m_2 = 3, m_3 \le 5$. In both cases $\mathrm{kod}(X) = -\infty$. □

Following [Wall86] we show that the following invariant of an elliptic fibration f:

$$\delta(f) := \chi(\mathcal{O}_X) + \left(2g(S) - 2 + \sum_{i=1}^{k} (1 - m_i^{-1}) \right)$$

determines the Kodaira dimension of X:

(12.5) **Proposition.** *Let $f : X \to S$ be a relatively minimal elliptic fibration with X compact. Then always* $\mathrm{kod}(X) \le 1$. *More precisely,* $\mathrm{kod}(X) = -\infty$ *if and only if* $\delta(f) < 0$, $\mathrm{kod}(X) = 0$ *if and only if* $\delta(f) = 0$, *and* $\mathrm{kod}(X) = 1$ *if and only if* $\delta(f) > 0$. *In more geometric terms,* $\mathrm{kod}(X) = 1$ *if one of the following holds:*

(i) *for some $n \ge 0$ there is an effective n-canonical divisor,*

(ii) $g(S) \ge 2$,

(iii) $g(S) = 1$ *and f is not locally trivial.*

Proof. Let m be the l.c.m. of all the multiplicities m_i appearing in Theorem 12.1. Then for $\mu \in \mathbb{N}$

$$\begin{aligned} \mathcal{K}_X^{\otimes \mu m} &= f^*(\mathcal{K}_S \otimes (f_{*1}\mathcal{O}_X)^\vee)^{\otimes \mu m} \otimes \mathcal{O}_X(\sum (m_i - 1) m F_i)^{\otimes \mu} \\ &= f^* \left(\mathcal{K}_S \otimes (f_{*1}\mathcal{O}_X)^\vee)^{\otimes m} \otimes \mathcal{O}_X(\sum (m_i - 1)(m/m_i) F \right)^{\otimes \mu} \\ &= f^*(D^{\otimes \mu}) \end{aligned}$$

where $D = \mathcal{K}_S^{\otimes m} \otimes (f_{*1}\mathcal{O}_X)^{\otimes -m} \otimes \mathcal{O}_S(\Sigma(m_i - 1)(m/m_i)s_i)$ is a line bundle on S of degree $m\delta(f)$. So $h^0(\mathcal{K}_X^{\otimes \mu m})$ cannot grow faster than linearly in μ. And if $\delta(f) > 0$, then by Riemann-Roch on S this growth is linear and so $\text{kod}(X) = 1$. If there is an effective n-canonical divisor, then there is also a μmn-canonical one and so in this case $\delta(f) > 0$, proving Case i).

If $g(S) \geq 2$, then $\deg(\mathcal{K}_S) > 0$ and $\deg((f_{*1}\mathcal{O}_X)^\vee) \geq 0$ enforce $\deg(D) > 0$, proving Case ii).

If $g(S) = 1$, then $\deg(D) = 0$ is only possible if both $\deg(f_{*1}\mathcal{O}_X)$ and $\sum(m_i - 1)(m/m_i)$ vanish. The vanishing of $\deg(f_{*1}\mathcal{O}_X)$ ensures that f is locally trivial outside of the multiple fibres (Theorem III. 18.2) whereas $\sum(m_i-1)(m/m_i) = 0$ implies that no multiple fibres are present. This proves Case iii).

There remains the case of a locally trivial bundle over an elliptic curve and the case of a fibration over $\mathbb{P}_1$. For the first case we go back to the classification of locally trivial elliptic bundles over an elliptic curve (Section 5): both the primary Kodaira surfaces and the bi-elliptic surfaces have $\text{kod}(X) = 0$ in accord with $\delta(f) = 0$. For the second case we apply Corollary 12.4. □

For the results in Sect. 12 we refer to [Ko60], Theorem 12.1 and [Ko66] Theorem 12. Corollary 12.4 is more or less equivalent to [Ko66], Theorems 28, 35. Our treatment of the canonical bundle formula is inspired by [B-Hu]. It has the advantage that it can be generalised immediately to algebraic elliptic fibrations in positive characteristic. The numerical criterion for the Kodaira dimension has been formulated by Wall ([Wall86], Lemma 7.1).

13. Logarithmic Transformations

A logarithmic transformation is a means of replacing any given fibre of type I_b in an elliptic fibration by a multiple fibre of type ${}_mI_b, m \geq 2$. To be more precise, given an elliptic fibration $X \to S$ and a point $0 \in S$, over which the fibre is of type I_b, a new elliptic fibration $X' \to S$ is constructed, such that X_0' is of type ${}_mI_b$, whereas $X \backslash X_0$ and $X' \backslash X_0'$ are isomorphic as fibre spaces over $S\backslash\{0\}$. This process, still depending on the choice of a suitable $k \in \mathbb{N}$, is called a logarithmic transformation of order m at 0 or with centre 0. The process can be reversed in the sense that given $X' \to S$ with X_0' of type ${}_mI_b$, then a fibration $X \to S$ can be constructed with X_0 of type I_b, such that $X' \to S$ can be derived from $X \to S$ by applying a logarithmic transformation at 0.

i) The case $b = 0$.

We start with $X \to S$ and $0 \in S$ as above. Since X_0 is smooth, over a coordinate disk $\Delta(t)$, centred at 0, this fibration is isomorphic to a fibration

$$Y = \mathbb{C} \times \Delta / L,$$

where $L = \mathbb{Z} + \mathbb{Z}\omega(t)$, with ω holomorphic on Δ and $\operatorname{Im}\omega(t) \neq 0$. Now we imitate the procedure of Sect. 10 to construct an elliptic fibre space over Δ whose only singular fibre lies over 0 and is of type ${}_m\mathrm{I}_0$. Let $t = s^m$. We take a $k \in \mathbb{N}$, with $(k, m) = 1$ and form the smooth quotient Y' of $\mathbb{C} \times \Delta(s)/(\mathbb{Z} + \mathbb{Z}\omega(s^m))$ by the cyclic group of order m, generated by

$$(c, s) \mapsto \left(c + k/m, \mathbf{e}\left(\frac{1}{m}\right) s\right).$$

If we define $Y' \to \Delta(t)$ by $(c, s) \mapsto s^m = t$, then $Y' \to \Delta(t)$ has indeed only one singular fibre, namely Y_0', which is of type ${}_m\mathrm{I}_0$. The map given by

$$(c, s) \mapsto (c - (k/2\pi\mathrm{i}) \log s, s^m)$$

induces a biholomorphic fibre-preserving map $f : Y' \setminus Y_0' \xrightarrow{\sim} Y \setminus Y_0$. If we identify $Y' \setminus Y_0'$ with the part of X lying over $\Delta \setminus \{0\}$, by way of f, then the union $(X \setminus X_0) \cup Y'$ becomes a Hausdorff space, which in fact is an elliptic fibration $X' \to S$ with the two required properties: $X' \setminus X_0' \cong X \setminus X_0$ and $X_0' \cong Y_0'$ is of type ${}_m\mathrm{I}_0$.

Conversely, to show that it is always possible to replace a fibre of type ${}_m\mathrm{I}_0$ by a smooth fibre, it is clearly sufficient to prove that a neighbourhood of it can be obtained as the quotient of an elliptic fibre space

$$\mathbb{C} \times \Delta / (\mathbb{Z} + \mathbb{Z}\omega(s^m))$$

by a cyclic group

$$(c, s) \mapsto \left(c + k/m, \mathbf{e}\left(\frac{1}{m}\right) s\right)$$

for a suitable k with $(k, m) = 1$. So let $Y' \to \Delta$ be an elliptic fibration with a single singular fibre Y_0 of type ${}_m\mathrm{I}_0$. Let $Y'' \to \Delta$ be its m-th root fibration (III, Sect. 9). Then Y'' is smooth and $Y'' \to Y'$ is an unramified cyclic covering, whose covering group G is generated by an automorphism g of Y of order m, such that $gY_t'' = Y_{\rho t}''$, $\rho = \mathbf{e}\left(\frac{1}{m}\right)$. Now $Y'' = \mathbb{C} \times \Delta / (\mathbb{Z} + \mathbb{Z}\omega'(s^m))$, with ω' holomorphic, and g can be given as

$$(c, s) \mapsto \left(c + \beta(s), \mathbf{e}\left(\frac{1}{m}\right) s\right)$$

for some holomorphic function β on Δ. Since the order of g is m we have

$$\sum_{\nu=0}^{m-1} \beta(g^\nu s) = p + q\omega'(s^m), \quad p, q \in \mathbb{Z}.$$

If $\beta = \sum_{n=0}^{\infty} \beta_n s^n$, we define $\gamma(s) = \sum_{n=0}^{\infty} \sum_{\nu=1}^{m-1} \frac{\beta_{nm+\nu}}{1 - \rho^\nu} s^{mn+\nu}$. Then γ is holomorphic, and the preceding relation implies that

$$\gamma(\rho s) - \gamma(s) = -\beta(s) + \frac{1}{m}(p + q\omega'(s^m)).$$

Let $(p, q) = k$, and let $a, b \in \mathbb{Z}$ satisfy $ap - bq = k$. If we put $p = kp', q = kq'$, then

$$\begin{pmatrix} a & b \\ p' & q' \end{pmatrix} \in \mathrm{SL}(2, \mathbb{Z}),$$

so introducing the new fibre coordinate $c' = (c - \gamma(s))/(q'\omega'(s^n) + p')$ the lattice $\mathbb{Z} + \mathbb{Z}\omega'(s^m)$ becomes $\mathbb{Z} + \mathbb{Z}\omega(s^m)$, where

$$\omega(s^m) = (a\omega'(s^m) + b)/(q'\omega'(s^m) + p').$$

Since

$$(\beta(s) - \gamma(\rho s) - \gamma(s))/(p' + q'\omega'(s^m)) = k/m,$$

we find that g can be given by

$$(c', s) \mapsto (c' + k/m, \rho s),$$

as required.

ii) The case $b \neq 0$.

We start by describing a (small) neighbourhood of a fibre of type I_n, $n \geq 2$. Since such neighbourhoods are unique up to fibre preserving automorphisms (cf. the proof of Theorem 11.1) it is sufficient to exhibit one such neighbourhood. We take

$$U_k = \{(u_k, v_k) \in \mathbb{C}^2; |u_k v_k| < 1\}, \quad k \in \mathbb{Z},$$

and then glue U_k and U_{k+1} by identifying (u_{k+1}, v_{k+1}) with $(v_k^{-1}, u_k v_k^2)$ in as far as this makes sense. The infinite cyclic group operates on the resulting manifold by $(u_k, v_k) \mapsto (u_{k+n}, v_{k+n})$ and the quotient Z_n admits a surjective holomorphic map

$$\begin{aligned} Z_n &\to \Delta(s) \\ (u_k, v_k) &\mapsto s = u_k v_k. \end{aligned}$$

For the fibre over $s \neq 0$ we clearly have

$$(Z_n)_s \cong \mathbb{C}^* / (s^n)^{\mathbb{Z}} \cong \mathbb{C}/L_s(n), \quad L_s(n) = \mathbb{Z} \oplus \frac{n}{2\pi \mathrm{i}} \log(s)\mathbb{Z}$$

so it is a smooth elliptic curve, whereas $(Z_n)_0$ is of type I_n.

Suppose now that $n = mb$, then $(\mathbb{Z}/m\mathbb{Z})$ acts on Z_n by

$$(u_k, v_k) \mapsto (\varepsilon\rho^{1-k} u_{k+b}, \varepsilon^{-1}\rho^k v_{k+b}), \quad \varepsilon = \mathbf{e}\left(\frac{b(m-1)}{2m}\right),$$

where we recall that $\rho = \mathbf{e}(1/m)$. Indeed, one verifies that this formula is compatible with the glueing between U_k, U_{k-1} and between U_{k+b} and U_{k+b-1}. The resulting action is however not compatible with the projection onto the

s-disk. Instead, it is compatible with the map $(u_k, v_k) \mapsto t = s^m = (u_k v_k)^m$ and so the quotient Y' is an elliptic fibre space over the t-disk with $t = s^m$. It follows that Y_0', the fibre over 0, is of type ${}_m\mathrm{I}_b$. Now $Y' \setminus Y_0' \to \Delta^*$ has monodromy (conjugate to) $\begin{pmatrix} 1 & b \\ 0 & 1 \end{pmatrix}$ and the J-invariant has a pole of order b at 0. So, by the local classification, $Y' \setminus Y_0'$ is isomorphic to the family $Z_b \setminus (Z_b)_0$ and we can glue Y' and Z_b together over Δ^*. This is the desired family. As in the case of $b = 0$, one can give an explicit identification

$$Y' \setminus Y_0' \xrightarrow{\sim} Z_b \setminus (Z_b)_0$$

using logarithms and using these formulas, the preceding process can be reversed. For details we refer to Kodaira ([Ko66] part I, p.768–771).

A logarithmic transformation should by no means be seen as something close to a birational transformation. It can completely change the topological as well as the analytic nature of an elliptic surface, as may be illustrated by the following examples.

(13.1) **Example.** Let E be a smooth elliptic curve and consider $X = \mathbb{P}_1 \times E$ as an elliptic fibre space $X \to \mathbb{P}_1$. It follows from the canonical bundle formula that if we apply logarithmic transformations to sufficiently many fibres $p \times E$, then we obtain a surface Y with $\mathrm{kod}(Y) = 1$. Since $\mathrm{kod}(X) = -\infty$, we see that *application of logarithmic transformations can change the Kodaira dimension.*

(13.2) **Example.** Let X be as in the preceding example and let Y be obtained from X by applying a single logarithmic transformation. We claim: *Y is a non algebraic surface.* To see this, let $p : Y \to \mathbb{P}_1$ be the projection and F the multiple fibre of p. Since $c_1^2(Y) = p_g(Y) = 0$ by Corollary 12.3 and $c_2(Y) = c_2(X) = 0$, we find by the Todd-Hirzebruch formula that $q(Y) = 1$. Hence Theorem IV.2.6 yields that either $b_1(Y) = 2, h^{1,0}(Y) = 1$ or $b_1(Y) = 1, h^{1,0}(Y) = 0$. So it is sufficient to show that $h^{1,0}(Y) = 0$. We prove this by deriving a contradiction from the assumption that $h^{1,0}(Y) = 1$. If $h^{1,0}(Y) = 1$, then there is a surjective Albanese map $f : Y \to \mathrm{Alb}(Y)$, where $\mathrm{Alb}(Y)$ is a 1-dimensional torus. Let C be a smooth component of a general fibre of f. Then C is not a fibre of p, otherwise f would factor through p, which is impossible by Lüroth's theorem. So $p|C : C \to \mathbb{P}_1$ is a ramified covering, which has degree ≥ 2, since $p : Y \to \mathbb{P}_1$ has a multiple fibre by construction. We claim that $p|C$ can only be ramified in the points of $F \cap C$. In other words, we assert that in all other points $y \in C$ we have that C intersects $p^{-1}(p(y))$ transversally. Since outside of F the map p is everywhere of maximal rank, this is however an immediate consequence of the fact that the restriction of f to a fibre $p^{-1}(p(y))$ is an *unramified* covering map onto E. So, $\mathbb{P}_1 \setminus p(F)$ being simply-connected, we have arrived at our contradiction. Thus we conclude: *application of a logarithmic transformation may change an algebraic surface into a non-algebraic one.*

Logarithmic transformations have been introduced by Kodaira in [Ko66, part I]. We have closely followed him for the case $b = 0$, whereas our approach for $b \neq 0$ has been motivated by toroidal embeddings.

For a thorough discussion of the moduli problem for elliptic surfaces, possibly containing multiple fibres, and the resulting classification of all deformation types of elliptic surfaces, we refer to the book [F-M94]. Logarithmic transformations play a crucial role in it.

Kodaira Fibrations

14. Kodaira Fibrations

A connected fibration $f : X \to C$ of the compact surface X over the smooth curve C is called a **Kodaira fibration** if f is everywhere of maximal rank but *not* a complex analytic fibre bundle map. In view of the Grauert-Fischer theorem I. 10.1 this means that, though all fibres are smooth curves, their complex structure varies.

It follows immediately from the uniqueness of $\mathbb{P}_1$ as a curve of genus 0 and the existence of the J-fibration (Sect. 9) that the fibre genus of a Kodaira fibration is at least 2. Theorem III. 15.4 implies that this also holds for the base genus.

Every surface X admitting a Kodaira fibration is algebraic: if F is a fibre, then $\mathcal{K}_X F \geq 2$, hence $c_1^2(\mathcal{K}_X \otimes \mathcal{O}_X(nF)) > 0$ for n large enough, and X is algebraic by Theorem IV. 6.2.

A surface admitting at least one Kodaira fibration is sometimes called a Kodaira surface. However, as we have seen in Sect. 5, this name is also used for a totally different type of surface. In this book we shall reserve the name Kodaira surface for the surfaces described in Sect. 5.

It follows readily from the projectivity of the Satake compactification for the moduli space of curves of genus g that there exist Kodaira fibrations with any given genus $g, g \geq 3$. We want to give here, however, a more direct construction.

In fact, for certain values of g we shall construct connected fibrations $f : X \to C$ of fibre genus g which are everywhere of maximal rank such that the index $\tau(X)$ is strictly positive. Then f is automatically a Kodaira fibration. For if f were a locally trivial bundle map, then by Sect. 7 there would exist an unramified covering Y of X, which is the product of two curves, and $\tau(X)$ would vanish.

Let C and D be smooth, compact, connected curves of genus ≥ 2, $h : C \to D$ a regular map and G a finite group operating analytically and without fixed points on D. If B_g denotes the graph of $g \circ h, g \in G$, then $B = \bigcup_{g \in G} B_g$ is a smooth curve on $C \times D$, consisting of $|G|$ disjoint components. Suppose that for some $r \geq 2$ there exists an r-fold cyclic covering X of $C \times D$ which is ramified exactly over B. The surface X admits a natural projection $f : X \to C$ and we claim that it is everywhere of maximal rank and connected, whereas $\tau(X) > 0$. The first claim is clear since f admits local sections everywhere. As to the connectedness of f, if there were a non-connected fibre, then over $c \times D (c \in C)$ there would lie on X a non-connected smooth curve, which is impossible since there are points on $C \times D$, over which there is only one point of X. So we only have to calculate $\tau(X)$.

Let $p = g(C), q = g(D)$ and d the degree of h. Then we have first of all

$$\begin{aligned} e(X) = & \quad e(C)(re(D) - (r-1)|G|) \\ = & \quad 4r(p-1)(q-1) + 2(p-1)(r-1)|G|. \end{aligned}$$

Furthermore, by Lemma I. 17.1 there is a line bundle $\mathcal{L}$ on $C \times D$ with $\mathcal{O}(B) = \mathcal{L}^{\otimes r}$. So we have by Lemma I. 17.1,(iii) that

$$\mathcal{K}_X = k^*(\mathcal{K}_{C \times D} \otimes \mathcal{L}^{\otimes (r-1)})$$

where $k : X \to C \times D$ denotes the projection. Hence if $C_0 = c_0 \times D$ and $D_0 = C \times d_0$, we find

$$K_X^2 = r \left[(2q-2)C_0 + (2p-2)D_0 + \frac{r-1}{r} B \right]^2 .$$

Since $B^2 = \sum_{g \in G} B_g^2 = -2d|G|(q-1)$ by the adjunction formula, it follows that

$$K_X^2 = 8r(p-1)(q-1) + 4(r-1)(p-1)|G| + 2(r-1) \cdot \frac{r+1}{r} \cdot d|G|(q-1).$$

So $\tau(X) = \frac{1}{3}(K_X^2 - 2e(X)) > 0$.

Since $K_X^2 > 0$ and $\Gamma((\mathcal{K}_X^\vee)^{\otimes n}) = 0$ for all $n \geq 1$ we find, using Riemann-Roch, that $\mathrm{kod}(X) = 2$ for our example (it can be proved that this holds for any surface X admitting a Kodaira fibration).

If we put $g(D/G) = s$, we find

$$\frac{K_X^2}{e(X)} = \frac{c_1^2(X)}{c_2(X)} = 2 + \frac{1 - \frac{1}{r^2}}{(2s-1) - \frac{1}{r}}$$

and when we fix s, which has to be at least 2, we find $\lim_{r \to \infty} \dfrac{c_1^2(X)}{c_2(X)} = 2 + \dfrac{1}{2s-1}$.

In particular we see that we always have $2 < \dfrac{c_1^2(X)}{c_2(X)} < 7/3$.

We still have to show that there really exist sets (C, D, G, h, r) as described above. By I, Sect. 18 it is sufficient to produce sets (C, D, G, h, r) with C, D curves, with G a finite analytic fixed-point-free transformation group of D, and with $h : C \to D$ a morphism, such that $\mathcal{O}(B)$ is divisible by r in $\mathrm{Pic}(C \times D)$. This can be accomplished, even if r and s are given in advance (but not C and D), by the following method.

Let $r, s \in \mathbb{N}, r, s \geq 2$. We start with a smooth curve D_0 with $g(D_0) = s$ and take an unramified Galois covering $D \to D_0$ of degree $kr (k \geq 1)$. Then our G will of course be the group of this covering. Let us now assume for a moment that we have another curve C and a regular map $h : C \to D$ with the property that

(*) all elements of $h^*(H^1(D, \mathbb{Z}))$ are r-divisible in $H^1(C, \mathbb{Z})$.

We claim that then $\mathcal{O}(B)$ is r-divisible in $\mathrm{Pic}(C \times D)$. Indeed, since $\mathrm{Pic}^0(C \times D)$ is r-divisible anyhow, it is sufficient to show that $c_1(\mathcal{O}(B))$ is r-divisible in $H^2(C \times D, \mathbb{Z})$. The cupproduct induces on this torsion-free group a perfect pairing with itself and it is therefore sufficient to prove that

$$(c_1(\mathcal{O}(B)), \alpha) \equiv 0 \pmod r \text{ for all } \alpha \in H^2(C \times D, \mathbb{Z}).$$

By Künneth's formula this will be true if we prove it in three cases: if $\alpha \in p_1^*(H^2(C, \mathbb{Z}))$, if $\alpha \in p_2^*(H^2(D, \mathbb{Z}))$ and if $\alpha = p_1^*(\beta) \cdot p_2^*(\gamma)$, where $p_1 : C \times D \to C$ and $p_2 : C \times D \to D$ are the projections. The first two cases follow from the fact that the intersection number of B with both "horizontal" and "vertical" fibres is divisible by $|G|$, hence by r. And in the last case we have by the projection formula

$$\begin{aligned}(c_1(\mathcal{O}_{C\times D}(B)), \alpha) &= \sum_{g\in G} (c_1(\mathcal{O}_{C\times D}(B_g)), p_1^*(\beta) \cdot p_2^*(\gamma)) \\ &= \sum_{g\in G} (\beta, p_{1!}(c_1(\mathcal{O}_X(B_g)) \cdot p_2^*(\gamma))) \\ &= \sum_{g\in G} (\beta, g^* h^*(\gamma)) \equiv 0 \,(\mathrm{mod} r)\end{aligned}$$

by assumption (*).

It remains to convince the reader that assumption (*) can be satisfied. In other words, given a curve D and $r \in \mathbb{N}$, we have to find a curve C and a morphism $h : C \to D$, such that (*) holds. But that is rather easy. For if we consider $H_1(D, \mathbb{Z}/r\mathbb{Z})$ as a quotient of $\pi_1(D)$ by way of a canonical surjections

$$\pi_1(D) \to H_1(D, \mathbb{Z}) \quad \text{and} \quad H_1(D, \mathbb{Z}) \to H_1(D, \mathbb{Z}/r\mathbb{Z}),$$

then we can take the unbranched covering $h : C \to D$ with covering group $H_1(D, \mathbb{Z}/r\mathbb{Z})$, obtaining an exact sequence

$$0 \to \pi_1(C) \to \pi_1(D) \to \mathbb{Z}/r\mathbb{Z} \to 0.$$

It follows that $h_*(H^1(D,\mathbb{Z}))$ is r-divisible in $H_1(D,\mathbb{Z})$, so by transposition $h^*(H^1(D,\mathbb{Z}))$ is r-divisible in $H^1(C,\mathbb{Z})$.

Remark. If, as in the case just described, the map $h : C \to D$ is an unramified covering, then X admits a natural fibration in two ways: over C *and* over D.

References: [Ko67], [At69], [Ks]. In this last paper it is proved that the fibre genus of any Kodaira fibration is at least 3. Kas also proved that every small deformation of a Kodaira surface is again a Kodaira surface. His result was extended to the case of any deformation by Jost and Yau (see [J-Y]).

Finite Quotients

15. The Godeaux Surface

Let $(\zeta_0 : \ldots : \zeta_3)$ be a system of homogeneous coordinates in $\mathbb{P}_3$. We define an action of the cyclic group $G = \mathbb{Z}/5\mathbb{Z}$ by putting

$$(1_G)(\zeta_0 : \ldots : \zeta_3) = (\zeta_0 : \rho\zeta_1 : \ldots : \rho^3\zeta_3),$$

where $\rho = \mathbf{e}(\frac{1}{5})$. The Fermat surface Y, given by $\sum_{i=0}^{3} \xi_i^5 = 0$, is G-invariant and does not pass through any of the four fixed points. Consequently, the quotient Y/G is a smooth surface X, which is algebraic by Theorem IV. 6.8. By Proposition 2.1 we have $\pi_1(Y) = 0$, so $\pi_1(X) \cong \mathbb{Z}/5\mathbb{Z}$, and $q(X) = \frac{1}{2}b_1(X) = 0$. Since $\chi(X) = \frac{1}{5}\chi(Y) = 1$ (Proposition 2.1) we have $p_g(X) = 0$. Furthermore, if $p : Y \to X$ denotes the projection, then

$$H^i(\mathcal{K}_X^{\otimes n}) = H^i(p_*\mathcal{K}_X^{\otimes n}) \supset H^i(\mathcal{K}_X^{\otimes n}) \quad \text{(Lemma I. 17.2)},$$

so $H^i(\mathcal{K}_X^{\otimes n}) = 0$ for $i = 1, 2$ as soon as $H^i(\mathcal{K}_Y^{\otimes n}) = 0$ for $i = 1, 2$. Since $\mathcal{K}_Y = \mathcal{O}_Y(1)$ we find by Serre duality

$$H^i(\mathcal{K}_Y^{\otimes n}) = H^{2-i}(\mathcal{O}_Y(-n)),$$

so using Theorem IV. 12.4 we obtain

$$H^i(\mathcal{K}_X^{\otimes n}) = 0 \text{ for } i = 1, 2 \text{ and } n \geq 2.$$

Riemann-Roch then yields $P_n(X) = 1 + \frac{1}{2}n(n-1)$ for $n \geq 2$. In particular we have that $\mathrm{kod}(X) = 2$.

The Godeaux surface provides a historically important example of a surface X with $q(X) = p_g(X) = 0$ which is not rational. (In VII, Sect. 11 we shall give some more information about these surfaces.)

The Godeaux surface appears in [Gx31]. In [Gx49] Godeaux constructed a similar example, with Y an intersection of four quadrics in $\mathbb{P}_6$ and $G \cong \mathbb{Z}/8\mathbb{Z}$.

16. Kummer Surfaces

Let T be a 2-torus on which a base point has been chosen. The involution $\iota : T \to T$, defined by $\iota(x) = -x$ has exactly sixteen fixed points, namely the points of order 2 on T. At each of these the exponents are $(-1,-1)$, so $T/\langle 1, \iota\rangle$ has sixteen ordinary double points (of type A_1). Resolving the double points we obtain a smooth surface X, the Kummer surface $\mathrm{Km}(T)$ of T, or associated to T. We let $g : X \to T/\langle 1, \iota\rangle$ be the blowing down map.

Let $\tilde{T}$ be obtained from T by blowing up the sixteen fixed points of ι. Then ι can be lifted to an involution $\tilde{\iota}$ of $\tilde{T}$, which leaves the sixteen exceptional curves point-wise invariant but has no fixed points otherwise. So the quotient is a smooth surface Y, admitting a map onto $T/\{1, \iota\}$ which is 1–1, except that over each singularity there lies a (-2)-curve. It follows that Y is isomorphic to X in such a way, that the obvious diagram

$$\begin{array}{ccc} \tilde{T} & \longrightarrow & X \\ {\scriptstyle f}\downarrow & & \downarrow{\scriptstyle g} \\ T & \longrightarrow & T/\{1,\iota\} \end{array}$$

is commutative. Since $H^1(\tilde{T}, \mathbb{Q}) = f^*(H^1(T, \mathbb{Q}))$ and since the only element in $H^1(T, \mathbb{Q})$, invariant under ι, is the zero element, we have $H^1(X, \mathbb{Q}) = 0$. Similarly we see that $p_g(X) = 1$. Now a 2-form on X could only vanish on components of the ramification divisor, but then we would have on $\tilde{T}$ a 2-form vanishing to a higher order on (-1)-curves, which is impossible. Hence $\mathcal{K}_X \cong \mathcal{O}_X$.

Kummer surfaces are special K 3-surfaces and as such will play an important role in Chapter VIII. They are all kählerian since $b_1 = 0$ (Theorem IV, 3.1). For a simpler proof see [Fu].

17. Quotients of Products of Curves

We shall give here an example in which one of the curves is elliptic.

Let E be an elliptic curve, C some other smooth compact, connected curve, G a finite subgroup of $\mathrm{Aut}(C)$, which also acts as translation group of E. So G acts without fixed points on $C \times E$, and if $C \times E/G$, then the natural projection $f : X \to D = C/G$ is an elliptic fibration. If $g : C \to D$ is the projection, then all fibres, except those lying over branch points of g, are smooth elliptic curves, isomorphic to E/G, and if $g^{-1}(d)$ contains $m < |G|$ points, then $g^{-1}(d)$ is a fibre of type ${}_n\mathrm{I}_0$, with $n = \dfrac{|G|}{m}$.

As to the invariants of X, we have for example

$$H_1(X,\mathbb{Q}) \cong H_1(E\times C,\mathbb{Q})^G \cong H_1(E,\mathbb{Q})\oplus H_1(C,\mathbb{Q})^G \cong H_1(E,\mathbb{Q})\oplus H_1(D,\mathbb{Q}),$$

and

$$\Gamma(\mathcal{K}_X^{\otimes n}) \cong \Gamma(\mathcal{K}_E^{\otimes n} \otimes \mathcal{K}_C^{\otimes n})^G \cong \Gamma(\mathcal{K}_C^{\otimes n})^G \cong \Gamma(\mathcal{K}_D^{\otimes n}).$$

Also $c_1^2(X) = \dfrac{1}{|G|}c_1^2(C \times E) = 0,\ c_2(X) = \dfrac{1}{|G|}\ c_2(C \times E) = 0.$

There exist many examples of this type. Given a smooth compact, connected curve D and integers, a, b, it is always possible to construct a surface as above with this D and $G \cong \mathbb{Z}/a\mathbb{Z} \oplus \mathbb{Z}/b\mathbb{Z}$. For if x, y and z are any points on D, there exists a covering C of D of degree ab, with group G (which is ramified exactly over x, y and z), whereas G can always be taken as a translation group on an elliptic curve.

The existence of the covering C can be proved in the following way. If we set $g(D) = p$, then $\pi_1(D\backslash\{x, y, z\})$ is generated by elements $\alpha_1, \dots, \alpha_{2p}, \beta_1, \beta_2$ and β_3, with one relation, namely

$$\alpha_1\alpha_2\alpha_1^{-1}\alpha_2^{-1}\dots\alpha_{2p-1}^{-1}\alpha_{2p}^{-1}\beta_1^{-1}\beta_2^{-1}\beta_3^{-1} = 1,$$

where β_1, β_2 and β_3 are represented by suitable loops around x, y and z. There is a homomorphism h from $\pi_1(D\backslash\{x, y, z\})$ onto $\mathbb{Z}/a\mathbb{Z} \oplus \mathbb{Z}/b\mathbb{Z}$ with $h(\alpha_j) = 0, j = 1, \dots, 2p, h(\beta_1) = (1, 0), h(\beta_2) = (0, 1)$ and $h(\beta_3) = (-1, -1)$. This homomorphism h determines a covering over $D\backslash\{x, y, z\}$ as described before. Restricting the covering to punctured disks around x, y and z it becomes readily clear that it can be extended to a ramified covering over all of D.

Infinite Quotients

18. Hopf Surfaces

Let W be the complex surface obtained from $\mathbb{C}^2$ by deleting the origin. A compact complex surface is called a Hopf surface if its universal covering is (analytically) isomorphic to W.

The original Hopf surface, as defined in [Hop48] by Hopf in 1948, is the quotient of W by the infinite cyclic group, generated by the homothety

$$(z_1, z_2) \mapsto \left(\frac{1}{2}z_1, \frac{1}{2}z_2\right).$$

Historically, this surface H has played an important role for several reasons. For example, it was the first example of a compact surface that does not admit any Kähler metric. This follows from Corollary I. 13.5 for H is diffeomorphic to $S^1 \times S^3$, hence $b_1(H) = 1$. The surface H is an elliptic fibre bundle over $\mathbb{P}_1$ (compare Sect. 5), and it is homogeneous.

Hopf's construction can be generalised immediately to the case where W is divided out by the infinite cyclic group G which is generated by the automorphism

$$(z_1, z_2) \mapsto (\alpha_1 z_1, \alpha_2 z_2),$$

with $0 < |\alpha_1| \leq |\alpha_2| < 1$. Putting $\alpha = (\alpha_1, \alpha_2)$, we shall denote W/G by H_α.

(18.1) **Proposition.** *The surfaces H_α have the following properties:*
(i) *H_α is diffeomorphic to $S^1 \times S^3$, and therefore does not admit any Kähler metric;*
(ii) $h^{1,0}(H_\alpha) = h^{2,0}(H_\alpha) = h^{0,2}(H_\alpha) = h^{1,1}(H_\alpha) = 0$ *and* $h^{0,1}(H_\alpha) = 1$;
(iii) $\mathrm{Pic}(H_\alpha) \cong \mathbb{C}^*$.

Proof. (i) This is a consequence of the fact that H_α can be deformed differentiably into H.

(ii) Since $b_2(H_\alpha) = 0$, we find from Theorem IV. 2.10, that

$$h^{2,0}(H_\alpha) = h^{0,2}(H_\alpha) = h^{1,1}(H_\alpha) = 0.$$

Combining this with $c_1^2(H_\alpha) = e(H_\alpha) = 0$, we find from the Todd-Hirzebruch formula that $h^{0,1}(H_\alpha) = 1$. Finally, Theorem IV. 2.10 yields $h^{1,0}(H_\alpha) = 0$.

(iii) It is obvious from the exponential cohomology sequence that the map $H^1(H_\alpha, \mathbb{Z}) \to H^1(H_\alpha, \mathcal{O}_{H_\alpha})$ is injective, so

$$\mathrm{Pic}(H_\alpha) = H^{0,1}(H_\alpha)/H^1(H_\alpha, \mathbb{Z}) \cong \mathbb{C}^*. \qquad \square$$

Though a surface H_α always contains two elliptic curves (namely the images C and D of the punctured z_1- and z_2-axes), it is not always elliptic.

In fact we have

(18.2) **Proposition.** *H_α is an elliptic fibre space over $\mathbb{P}_1$ if and only if $\alpha_1^k = \alpha_2^l$ for some $k, l \in \mathbb{Z}$. Otherwise, H_α contains exactly two irreducible curves.*

Proof. If the condition is satisfied, then $z_1^k z_2^{-l}$ is a non-constant meromorphic function on H_α, defined in every point. This function gives a surjective holomorphic map $f' : H_\alpha \to \mathbb{P}_1$. Using Stein factorisation, we obtain from f' a connected map $f : H_\alpha \to B$, where B is some smooth curve. This curve is again rational, since otherwise there would be a holomorphic 1-form on H_α which is impossible by Proposition 18.1,(ii). Finally, the adjunction formula shows immediately that the fibration is an elliptic one.

Conversely, if H_α is in some way a fibre space, then mC is a fibre for some $m \in \mathbb{Z}, m \geq 1$; otherwise C would intersect some other curve in a strictly positive number of points, which is impossible, since every curve is homologous to 0. So there exists a meromorphic function g' on W, which has the z_1-axis, taken with multiplicity m as polar divisor, whereas g' is left invariant under the automorphism $(z_1, z_2) \mapsto (\alpha_1 z_1, \alpha_2 z_2)$. Then the function $g = z_1^m g'$ is a holomorphic function on W, hence on $\mathbb{C}^2$ (Theorem I.8.7), satisfying $g(\alpha_1 z_1, \alpha_2 z_2) = \alpha_1^m g(z_1, z_2)$. But straightforward calculation shows that this is only possible if $\alpha_1^k = \alpha_2^l$ for suitable $k, l \in \mathbb{N}$.

So if $\alpha_1^k \neq \alpha_2^l$ for all $k, l \in \mathbb{N}$, there is no meromorphic function at all on H_α, i.e., H_α has algebraic dimension 0. Then, by Theorem IV. 8.2 the number of irreducible curves on H_α is finite. Now given any two points in $H_\alpha \backslash (C \cup D)$, there exists an automorphism of H_α (induced by a linear automorphism of $\mathbb{C}^2$), carrying the first of these points into the second. So if there were any (closed) irreducible curves on H_α, different from C and D, there would be an infinity of such curves. Consequently, in the case that H_α has no elliptic fibration, there are no irreducible curves on H_α but C and D. □

(18.3) *Remark.* This example shows that the bound given in Theorem IV. 8.2 is sharp, for $h^{1,1}(H_\alpha) = 0$ by Proposition 18.1,(ii).

There are many Hopf surfaces which are not of the type H_α, or not even diffeomorphic to $S^1 \times S^3$. For example, one can take the quotient of H by a suitable fibre preserving cyclic group of order n to obtain a Hopf surface X, which is still an elliptic fibre bundle over $\mathbb{P}_1$, with $\pi_1(X) \cong \mathbb{Z} \oplus \mathbb{Z}/n\mathbb{Z}$.

Kodaira has treated Hopf surfaces extensively in [Ko66], part II and III. We mention some of his results.

Every Hopf surface has Kodaira dimension $-\infty$.

A Hopf surface X is called primary if $\pi_1(X) \cong \mathbb{Z}$. Any Hopf surface admits a finite, unramified covering which is a primary Hopf surface.

(18.4) **Theorem.** *A compact surface X is a primary Hopf surface if and only if either one of the following (equivalent) conditions is satisfied:*
(i) *X is homeomorphic to $S^1 \times S^3$,*
(ii) *$b_2(X) = 0$ and $\pi_1(X) \cong \mathbb{Z}$.*

Kodaira has furthermore shown

(18.5) **Theorem.** *A minimal compact surface X with $a\,(X) = 1$ is a Hopf surface if and only if $P_{12}(X) = 0$.*

We shall see in the next chapter that every surface X with $a(X) = 1$ is elliptic. Combining this fact with Corollary 12.4,(ii), Proposition 12.5 and the results of Sect. 5 we obtain

(18.6) **Theorem.** *The minimal compact surfaces X with* $\mathrm{a}(X) = 1, \mathrm{kod}(X) = -\infty$ *are exactly the Hopf surfaces.*

As to the case $\mathrm{a}(X) = 0$, we have

(18.7) **Theorem.** *A compact surface X with $a(X) = 0$ is a Hopf surface if and only if $b_1(X) = 1, b_2(X) = 0$ and there is a curve on X.*

Additional reference: [Ka78a], [Weh], [Dl84], [Dl89] and [Da]. The last paper deals with the existence of global moduli spaces.

19. Inoue Surfaces

Let $M = (m_{i,j}) \in \mathrm{SL}(3, \mathbb{Z})$, and suppose that the eigenvalues of M satisfy the following conditions: one of them, say α, is real with $\alpha > 1$, and the other two, β and $\bar{\beta}$, are not real, i.e., $\beta \neq \bar{\beta}$. For example, we can take for M a matrix

$$\begin{pmatrix} 0 & 1 & 0 \\ n & 0 & 1 \\ 1 & 1-n & 0 \end{pmatrix}$$

with $n \in \mathbb{Z}$. Starting from such a matrix M, a compact surface S_M can be constructed in the following way. Let (a_1, a_2, a_3) be a real eigenvector of M corresponding to α, and (b_1, b_2, b_3) an eigenvector of M corresponding to β. Since $(a_1, a_2, a_3), (b_1, b_2, b_3)$ and $(\bar{b}_1, \bar{b}_2, \bar{b}_3)$ are independent over $\mathbb{C}$, the vectors $\begin{pmatrix} a_1 \\ b_1 \end{pmatrix}, \begin{pmatrix} a_2 \\ b_2 \end{pmatrix}$, and $\begin{pmatrix} a_3 \\ b_3 \end{pmatrix}$ are independent over $\mathbb{R}$.

Now let as usual $\mathfrak{H}$ be the upper half plane, and G_M the group of analytic automorphisms of $\mathfrak{H} \times \mathbb{C}$, generated by g_0, g_1, g_2, g_3, where

$$g_0(w, z) = (\alpha w, \beta z)$$
$$g_i(w, z) = (w + a_i, z + b_i) \quad i = 1, 2, 3.$$

Furthermore, let $G \subset G_M$ be the subgroup generated by g_1, g_2 and g_3. The independence of $\begin{pmatrix} a_1 \\ b_1 \end{pmatrix}, \begin{pmatrix} a_2 \\ b_2 \end{pmatrix}$, and $\begin{pmatrix} a_3 \\ b_3 \end{pmatrix}$ over $\mathbb{R}$ implies that the group G, which is isomorphic to the free abelian group $\mathbb{Z}^3$, acts properly and discontinuously on $\mathfrak{H} \times \mathbb{C}$, without fixed points. In fact, G transforms any real 3-dimensional affine linear variety

$$(w_0, z_0) + (\mathbb{R}(a_1, b_1) \oplus \mathbb{R}(a_2, b_2) \oplus \mathbb{R}(a_3, b_3))$$

into itself. Since these spaces are parametrised by $\mathrm{Im}(w_0)$, we see that $\mathfrak{H} \times \mathbb{C}/G$ is a real 3-torus bundle over $\mathbb{R}^+$ (in fact the product bundle). Because of

$$\begin{pmatrix} \alpha a_j \\ \beta b_j \end{pmatrix} = \sum_{k=1}^{3} m_{j,k} \begin{pmatrix} a_k \\ b_k \end{pmatrix}, \quad j = 1, 2, 3$$

the transformation g_0 acts in a fibre-preserving way on $\mathfrak{H} \times \mathbb{C}/G$, and since $\alpha > 1$ by assumption, the quotient is a 3-torus bundle over S^1, at least from the real point of view. Thus G_M acts properly and discontinuously without fixed points on $\mathfrak{H} \times \mathbb{C}$, so by [Car] the quotient is a compact complex surface. This is the desired surface S_M.

From the relations

$$g_i g_j = g_j g_i \quad (i = 1, 2, 3)$$
$$g_0 g_i g_0^{-1} = g_1^{m_{i,1}} g_2^{m_{i,2}} g_3^{m_{i,3}}$$

we see that

$$H_1(S_M, \mathbb{Z}) \cong \pi_1(S_M)/[\pi_1(S_M), \pi_1(S_M)] \cong G_M/[G_M, G_M]$$
$$\cong \mathbb{Z} \oplus \mathbb{Z}/e_1\mathbb{Z} \oplus \mathbb{Z}/e_2\mathbb{Z} \oplus \mathbb{Z}/e_3\mathbb{Z},$$

where e_1, e_2 and e_3 are the elementary divisors of $M - \mathbb{1}_3$.

Since $e(S_M) = 0$, we find that also $b_2(S_M) = 0$.

(19.1) **Proposition.** *The surface S_M does not contain any (closed) curve.*

Proof. Since $b_2(S_M) = 0$, the adjunction formula immediately yields that any irreducible curve on S_M is either smooth elliptic, or rational with one ordinary double point or one cusp.

Let D be any irreducible compact curve on S_M, and let $p : \mathfrak{H} \times \mathbb{C} \to S_M$ be the projection. Then $\mathfrak{H} \times \mathbb{C}$ contains at least one irreducible curve $\tilde{D}$, such that $p : \tilde{D} \to D$ exhibits $\tilde{D}$ as an unramified covering of D. This fact already excludes the rational cases, for otherwise $\mathfrak{H} \times \mathbb{C}$ would contain a compact curve. For the same reason, if D is elliptic, then $\tilde{D}$ has to be non-compact, so isomorphic to $\mathbb{C}$ or $\mathbb{C}^*$. But every holomorphic map from $\mathbb{C}$ or $\mathbb{C}^*$ into $\mathfrak{H}$ is constant, so $\tilde{D}$ would have to be contained in some fibre $w_0 \times C, w_0 \in \mathfrak{H}$. Then $\tilde{D}$ would be dense in $w_0 \times \mathbb{C}$, so $p(w_0 \times \mathbb{C})$ would be D again.

Now we observe that $w_0 \times \mathbb{C}$ is contained in the real affine space

$$L = (w_0, 0) + \sum_{i=1}^{3} \mathbb{R}(a_i, b_i).$$

Indeed, given any point (w_0, z), the system of equations

$$\sum_{i=1}^{3} \lambda_i a_i = 0, \quad \sum_{i=1}^{3} \lambda_i b_i = z$$

has a real solution, as follows from the independence of $\begin{pmatrix} a_1 \\ b_1 \end{pmatrix}, \begin{pmatrix} a_2 \\ b_2 \end{pmatrix}$ and $\begin{pmatrix} a_3 \\ b_3 \end{pmatrix}$ over $\mathbb{R}$.

The description of S_M as a real 3-torus bundle over S^1 tells us that $p|w_0 \times \mathbb{C}$ is the restriction to $w_0 \times \mathbb{C}$ of the projection from L onto the torus L/G. So as soon as we can prove that the rank over $\mathbb{Q}$ of

$$(w_0 \times \mathbb{C}) \cap \left((w_0, 0) + \sum_{i=1}^{3} \mathbb{Z}(a_i, b_i) \right)$$

is at most one, we are done, for then it would follow that D is not compact. This rank statement follows once we prove that the solution space $V \subset$

$\mathbb{R}^3(x) = \mathbb{R}^3(x_1, x_2, x_3)$ of the equation $\sum_{i=1}^{3} a_i x_i = 0$ has rank at most 1 over $\mathbb{Q}$. Suppose that this rank were 2. Since $({}^tMx, a) = (x, Ma) = \alpha(x, a)$, we see that ${}^tM$ leaves this V invariant, i.e., if $v = 0$ is an equation for V, then $Mv = \lambda v$, with $\lambda \in \mathbb{Q}$. This however would imply that M has a rational eigenvalue, which is actually not the case (by assumption, M has only one real eigenvalue $\alpha > 1$, which is clearly irrational). □

(19.2) **Proposition.** *For any surface S_M we have $p_g(S_M) = 0$, $q(S_M) = 1$, $h^{1,0}(S_M) = 0$.*

Proof. Since $b_2(S_M) = 0$, Theorem IV. 2.10 gives that

$$h^{1,1}(S_M) = h^{2,0}(S_M) = h^{0,2}(S_M) = p_g(S_M) = 0.$$

Furthermore $c_1^2(S_M) = 0$ (since $b_2(S_M) = 0$) and $e(S_M) = 0$ (since S_M is a torus bundle), so Noether's formula yields $q(S_M) = 1$. Applying IV.2.9 again, we find that $h^{1,0}(S_M) = 0$. □

For a long time the only compact surfaces X known with $\mathrm{kod}(X) = -\infty, b_1(X) = 1$ were the Hopf surfaces (Sect. 18). In 1972 Inoue and Bombieri independently found the example described above. For more details and more examples we refer to [In74] and [In77]. All these surfaces have $b_2 = 0$. In [Bog76] Bogomolov claims to have completely classified those which in addition do not contain curves. His proof raised doubts; finally Teleman [Te] clarified the situation. In 1974 (see [In77]) Inoue and later Hirzebruch constructed examples with $b_2 > 0$ (the so-called Inoue-Hirzebruch surfaces). For an elementary construction see [Dl88] and for a characterisation see [O-T-Z].

Later Kato ([Ka78b]) gave a construction method of many more surfaces all of which contain a global spherical shell. By definition this is a neighbourhood of the three sphere in $\mathbb{C}^2 \setminus \{0\}$ embedded in such way in the surface that the complement is connected. In fact all known examples of class VII surfaces with $b_2 > 0$ have global spherical shells and Kato conjectured that these exist as soon as the surface has b_2 rational curves. This conjecture has recently been proven. See [D-O-T] and the references given there.

See also [En82], [Na84], [Na90], [Bog83], [Dl84].

20. Quotients of Bounded Domains in $\mathbb{C}^2$

Whereas in most cases the examples in this chapter are either elementary or based on general theorems, we shall use in this and the next section some specific deep results which can only be quoted.

Let D be a bounded symmetric domain in $\mathbb{C}^2$. Then D is isomorphic to either the unit ball E (with isometry group $\mathrm{SU}(2,1)$) or the product P of two disks (with isometry group $\mathrm{PSL}(2,\mathbb{R}) \times \mathrm{PSL}(2,\mathbb{R})$).

Suppose that $G \subset G_D$ operates freely and properly discontinuously on D, such that the quotient is a compact surface X.

According to Hirzebruch's proportionality theorem [(Hir58a)] the Chern numbers of X are proportional, with a strictly positive proportionality factor, to the Chern numbers of the dual homogeneous complex manifold of D. This dual manifold is $\mathbb{P}_2$, if $D = E$, and it is $\mathbb{P}_1 \times \mathbb{P}_1$ if $D = P$.

A theorem of Borel (valid for all bounded symmetric domains) ([Bor]) yields for $D = E$ the existence of many arithmetical subgroups of G_D satisfying the condition above. (Mostow ([Mo78]), ([Mo80]) has found non-arithmetical groups with compact quotient.)

As a consequence we have

(20.1) **Theorem.** *For infinitely many $a \in \mathbb{N}$ there exists a surface X which has the unit ball in $\mathbb{C}^2$ as its universal covering, such that $c_1^2(X) = 3a, c_2(X) = a$.*

Borel's theorem does not give any specific value of a which actually occurs.

These surfaces with $c_1^2(X) = 3c_2(X)$ (all of them algebraic by Cor. IV. 6.3) play an important role as extreme cases in the geography of Chern numbers. We shall return to this point in Chapter VII.

As to the case of P, every product of two smooth compact curves, both of genus ≥ 2, is a quotient of P, but there are infinitely many others. In particular, there is a "fake quadric" X due to Kuga ([Shav]), with $c_1^2(X) = 8, c_2(X) = 4$ and $b_1(X) = 0$.

21. Hilbert Modular Surfaces

For the many facts, some of them highly non-trivial, which are only indicated in this section we mention [Hir73], [H-V74], [H-Z] as general references.

Let p be a square-free natural number and let $\mathfrak{o}$ be the ring of algebraic integers in the field $\mathbb{Q}(\sqrt{p})$. To avoid some technical points – we are only giving examples anyway – we shall restrict ourselves to the case p prime and $p \equiv 1 \,(\mathrm{mod} 4)$.

The group $\mathrm{SL}(2, \mathfrak{o})$ operates on the product $\mathfrak{H} \times \mathfrak{H}$ of the upper half plane by itself:

$$\begin{pmatrix} a & b \\ c & d \end{pmatrix} (z_1, z_2) = \left(\frac{az_1 + bz_2}{cz_1 + dz_2}, \frac{a'z_1 + b'z_2}{c'z_1 + d'z_2} \right),$$

where $a', \ldots$ are the conjugates of $a, \ldots$ in $\mathbb{Q}(\sqrt{p})$. This action becomes effective if we divide out by $\pm$ (identity). It is properly discontinuous, so by [Car] the quotient is a normal complex space $X(p)$. This non-compact space has only a finite number of singular points, called the quotient singularities. They arise from the points on $\mathfrak{H} \times \mathfrak{H}$ with non-trivial isotropy group in $\mathrm{PSL}_2(\mathfrak{o})$. But for the case $p = 5$, where the order 5 occurs, all these points have order 2 or 3. The corresponding quotient singularities are of type A_1 if the order is 2 and either of type $A_{3,1}$ or $A_{3,2} = A_2$ if the order is 3. So their minimal resolution consist of a (-2)-curve, a (-3)-curve, and two (-2)-curves intersecting transversally in one point, respectively. The number $a_2(p)$ of quotient singularities of type A_1 equals $h(-p)$, whereas of each of the types $A_{3,1}$ and $A_{3,2}$ there are $\frac{1}{2}a_3(p) = \frac{1}{2}h(-3p)$ quotient singularities. Here $h(q)$ denotes the class number of $\mathbb{Q}(\sqrt{q})$.

The space $X(p)$ can be compactified to an analytic space $\bar{X}(p)$ by adding a finite number of points, the cusps, in a specific way. In fact, there is a natural 1-1-correspondence between these cusps and the ideal classes in the ring $\mathfrak{o}$. If c is any of the cusps, then its minimal resolution can be described in the following way. The cusp c determines a $\mathbb{Z}$-module $M = \mathbb{Z}\cdot 1 + \mathbb{Z}\cdot w$ in $\mathbb{Q}(\sqrt{p})$ with $0 < w' < 1 < w$. The number w admits an expansion as a continued fraction:

$$w = [[b_0 b_1 b_2 \ldots]] = b_0 - 1\big\rfloor b_1 - 1\big\rfloor b_2 \ldots, \quad b_i \in \mathbb{Z}, b_i \geq 2$$

for $i \geq 1$ with $b_{i+r} = b_i$ for all $i \geq 0$, i.e., w is purely periodic with period r. We denote this by $w = [[\overline{b_0 b_1 b_2 \ldots b_r}]]$. If $p > 5$ the minimal resolution of c consists of an r-gon of smooth rational curves, with self-intersection $-b_0, -b_1, \ldots, -b_{r-1}$ in this order (For $p = 5$ there is one cusp, which is resolved by a rational curve with a node.) In particular, if $M = \mathbb{Z}\cdot 1 + \mathbb{Z}\cdot w_0$, with $0 < w_0' < 1 < w_0$, is a principal ideal, then w_0 has a continued fractional expansion of the form

$$w_0 = [[\overline{b_0 b_1 \ldots b_t b_t b_{t-1} \ldots b_1}]],$$

where $b_0 = \{\sqrt{p}\}$ (the smallest odd integer $\geq \sqrt{p}$). The corresponding curves in the cusp resolution are denoted by $S_0, S_1, \ldots, S_t, S_{-t}, \ldots, S_{-1}$ so $S_{\pm i}^2 = -b_i$ (compare [H-V74], p. 14).

Let $Y(p)$ be obtained from $\bar{X}(p)$ by resolving its singularities (both the quotient singularities and the cusps). The surface $Y(p)$ is called the Hilbert modular surface associated to p.

By Levi's extension Theorem I. 8.7 the field of meromorphic functions on $Y(p)$ is isomorphic to that of $\bar{X}(p)$ which in its turn is isomorphic to that of $X(p)$. This last field is isomorphic to the field of meromorphic functions on $\mathfrak{H}\times\mathfrak{H}$ which are automorphic with respect to the action of $\mathrm{SL}(2,\mathfrak{o})$. It is known that this field is an algebraic function field of transcendency degree 2. In fact, in [B-B] it is proved that suitable automorphic forms yield an embedding of $\bar{X}(p)$ in some projective space.

On $Y(p)$ we have several special types of curves:

1) the curves resolving the quotient singularities;

2) the curves resolving the cusps;

3) the curves F_N. They are obtained as follows. Let N be a strictly positive integer. Consider the set of points (z_1, z_2) satisfying any equation of the form

$$a\sqrt{p}z_1z_2 + \lambda z_2 - \lambda' z_1 + b\sqrt{p} = 0,$$

with $a, b \in \mathbb{Z}$, $\lambda \in \mathfrak{o}, a, b, \lambda$ primitive (i.e., not divisible by a common integral factor) and $\lambda\lambda' + abp = N$. This point set is left invariant by $\mathrm{SL}(2, \mathfrak{o})$ and its projection on $X(p)$ is a curve which is either compact or becomes a closed curve on $\bar{X}(p)$ by adding some of the cusps. Its proper transform on $Y(p)$ is the curve F_N; it may be reducible and it may be empty. The curves F_1, F_4 and, if present, F_2, F_3 are irreducible, smooth and rational. In many cases it is possible to determine the nature of F_N (number of components, singularities, genus). It is always possible to calculate F_N^2 and the intersection behaviour of F_N with respect to the curves of type 1) and 2) and with respect to the other curves F_N (see [H-Z]). For example, F_1 is always a (-1)-curve, and F_p is smooth and irreducible, with $F_p^2 = -(p+1)/6 + g(F_p)$, where the genus $g(F_p)$ is given by a classical formula from the theory of modular curves. It turns out that F_p is rational if $p = 5, 13, 17, 29$ and 41.

The configuration formed by the curves 1), the curves 2), some of the curves F_N (namely F_1, F_4, F_p and, if present, F_2 and F_3), is called the basic configuration. In Fig. 7 the basic configuration is shown for the case that $p > 5$ and there is only one cusp. The curves $B_1, B_2, E, L, C_1, C_1', \ldots, C_k, C_k'$, $D_1, \ldots, D_l, U_1, U_1', \ldots, U_m, U_m', V_1, V_1', \ldots, V_n, V_n'$ are the resolutions of the quotient singularities.

The numbers δ and ε are defined by

$$\delta = \begin{cases} 0 & \text{if } p \equiv 5 \pmod 8 \\ 1 & \text{if } p \equiv 1 \pmod 8 \end{cases}$$

$$\varepsilon = \begin{cases} 0 & \text{if } p \equiv 2 \pmod 3 \\ 1 & \text{if } p \equiv 1 \pmod 3 \end{cases}$$

and the numbers $k, \ldots, n$ are given by

$$\begin{aligned} k &= \frac{1}{2}a_3(p) \\ l &= \frac{1}{2}a_2(p) \\ m &= \frac{1}{4}a_2(p) - \frac{1}{2}(1+\delta) \\ n &= \frac{1}{4}a_3(p) - \frac{1}{2}(1+\varepsilon). \end{aligned}$$

All curves are smooth, all except possibly F_p, are rational, all intersections are transversal, self-intersection numbers as indicated, and the only intersections not shown are some intersections of the curves F_N with the curves $C_1, C_1', \ldots, V_n, V_n'$. Intersection points which are different in the figure are different on $Y(p)$. In several cases the information contained in the basic configuration is already more than sufficient to determine the nature of the surface $Y(p)$. For example, if $p = 17$, we obtain the configuration of Fig. 8.

So we can blow down F_1, then (the image of) E, then B_1, then F_{17} and finally C_1. After all this blowing down, the curve F has become a smooth rational curve with self-intersection +1. Hence $Y(17)$ is a rational surface by Proposition 4.3.

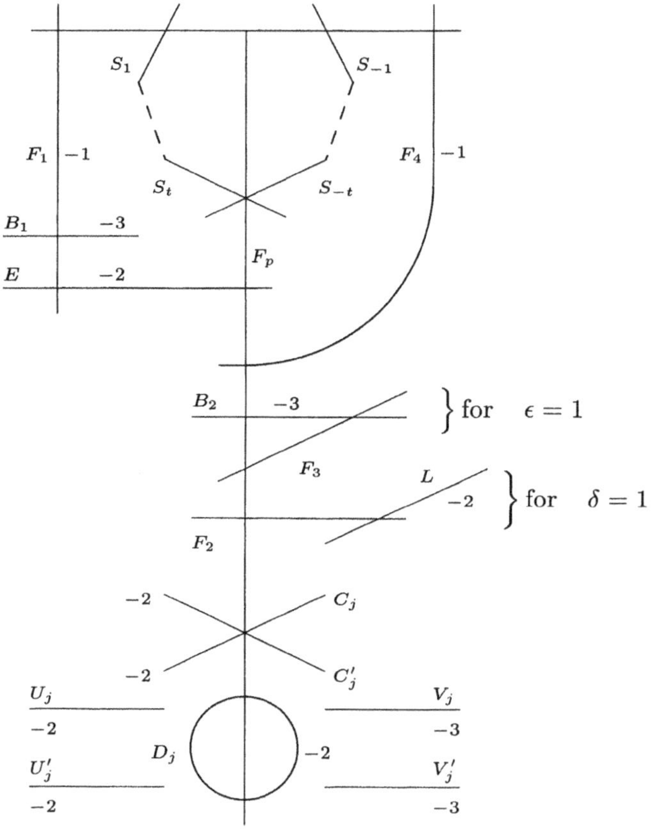

Fig. 7

As a corollary we obtain the fact that $Y(17)$ is simply connected. Actually, this fact holds for any surface $Y(p)$ ([Sv]).

Given p, we can always blow down the smooth rational curves F_1, E, B_1, F_4 and then F_2, L, F_3 if they are present. The smooth surface thus obtained is denoted by $Y^0(p)$. It is conjectured that if $Y^0(p)$ is not rational, i.e., if $p \neq 5, 13, 17$, then this surface is minimal. The conjecture has been verified for many $p's$ ([G-V], [Hir78]), but a general proof seems difficult.

Independent of the conjecture it is possible to give all surfaces $Y(p)$ their place in the Enriques-Kodaira classification of surfaces ([H-V74], Theorem

III.1). Apart from the basic configuration and the fact that $b_1(Y(p)) = 0$ one needs also the values of $e(Y(p))$ and $\chi(Y(p))$. They are given by

$$\begin{aligned} e(Y(p)) &= 2\zeta_{\mathbb{Q}(\sqrt{p})}(-1) + \frac{3}{2}h(-p) + \frac{13}{6}h(-3p) + l(p) \\ \chi(Y(p)) &= \frac{1}{4}(2\zeta_{\mathbb{Q}(\sqrt{p})}(-1) + \frac{3}{2}h(-p) + \frac{13}{6}h(-3p)), \end{aligned}$$

where $l(p)$ denotes the total number of all curves occurring in the cusp resolutions. (Compare [H-V74], p.12 and 20.)

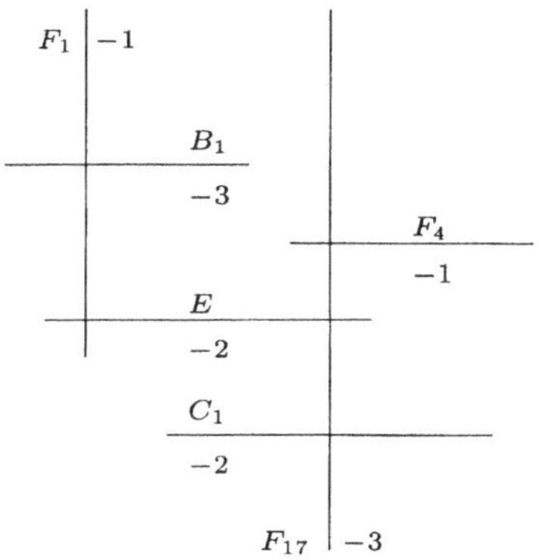

Fig. 8

We would like to illustrate this by considering the case $p = 29$. In this case one finds $\chi(Y^0(29)) = 2$ and $c_1^2(Y^0(29)) = 0$. It is easily verified that on $Y^0(29)$ the image of S_0 satisfies $KS_0 = S_0^2 = 0$, whereas S_1 has remained a (-2)-curve intersecting S_0 transversally in one point (thus we find a confirmation of the fact that $Y^0(29)$ is projective, for $(2S_0 + S_1)^2 > 0$, and this is enough by Theorem IV. 6.2). We want to show that $Y^0(29)$ is an *elliptic* K 3-*surface*. Applying Riemann-Roch we find

$$\dim |S_0| + \dim |K - S_0| \geq 1.$$

If $\dim |K - S_0|$ were strictly positive, then also $\dim |K|$ would be strictly positive, whereas

$$\chi(Y^0(29)) = 2 \text{ and } q(Y^0(29)) = \frac{1}{2}b_1(Y^0(29)) = \frac{1}{2}b_1(Y(29)) = 0$$

yield $\dim |K| = p_g(Y^0(29)) - 1 = 0$. So $\dim |S_0| \geq 1$. Since S_0 is irreducible this linear system cannot have any fixed components or base points and since $S_0^2 = 0$ it provides a map onto $\mathbb{P}_1$, the general fibre of which is elliptic by the adjunction formula. This elliptic fibration is relatively minimal since otherwise $c_1^2(Y^0(29))$ would be strictly negative by Theorem 12.3, and it cannot have any multiple fibres because of the existence of the section S_1. So Theorem 12.3 tells us that $\mathcal{K}_{Y^0(29)}$ is trivial. Combined with $b_1(Y^0(29)) = 0$ this yields the desired result.

In the situations above (for $p = 17$ and 29) we have used only a small part of the available information. This will make it clear, on the one hand, that there are numerous other ways to arrive at the same conclusion, and, on the other hand, that much more can be said about the surfaces in question.

Coverings

22. Invariants of Double Coverings

We recall the following facts from I, Sect. 17 and III, Sect. 7. A reduced divisor B on the compact surface Y, such that $\mathcal{O}_Y(B) = \mathcal{L}^{\otimes 2}$ for some $\mathcal{L} \in \mathrm{Pic}(Y)$, determines a double covering $\pi : X \to Y$ which is ramified exactly over B. The surface X is normal, and if B has a simple singularity at $y_1 \in Y$, then X has a rational singularity of the same type at $\pi^{-1}(y_1)$.

If $\sigma : \bar{X} \to X$ is the canonical resolution of singularities, and $p = \pi \circ \sigma$, we always have

(7) $$p_* \mathcal{O}_{\bar{X}} = \mathcal{O}_Y \oplus \mathcal{L}^{-1}, \; p_{*i}(\mathcal{O}_{\bar{X}}) = 0 \text{ for } i \geq 1 \,.$$

And if moreover B has at most simple singularities, then

(8) $$\mathcal{K}_{\bar{X}} = p^*(\mathcal{K}_Y \otimes \mathcal{L}).$$

Formula (8) has been established in III, Sect. 7. As to (7), the case of a smooth B has already been treated in I, Sect. 17. The general case can be dealt with in the following way. If

$$\begin{array}{ccc} \bar{X} & \xrightarrow{\sigma} & X \\ \downarrow{\scriptstyle \bar{\pi}} & \searrow^{p} & \downarrow{\scriptstyle \pi} \\ \bar{Y} & \longrightarrow & Y \end{array}$$

is the canonical resolution diagram (Theorem III.7.2), then the involution of $\bar{X}$, corresponding to $\bar{\pi}$ induces a canonical splitting

$$p_* \mathcal{O}_{\bar{X}} = \mathcal{O}_Y \oplus \mathcal{M},$$

where $\mathcal{M}$ is a line bundle on Y. Since outside of the finitely many singular points of B we are in the smooth case and hence there $\mathcal{M} \cong \mathcal{L}^{-1}$, we must have $M \cong \mathcal{L}^{-1}$ everywhere on Y. This proves the first part of (7). As to the second part, this follows from the corresponding statement of σ and the very definition of (higher) direct image sheaves.

Remark. It is an immediate consequence of Corollary IV. 6.5 that $\bar{X}$ is algebraic as soon as Y is.

Now let B have at most simple singularities. From (7), the Leray spectral sequence, Riemann-Roch and Serre duality, we obtain

$$\begin{aligned} \chi(\bar{X}) &= 2\chi(Y) + \frac{1}{2}(\mathcal{L}, \mathcal{K}_Y) + \frac{1}{2}(\mathcal{L}, \mathcal{L}) \\ p_g(\bar{X}) &= p_g(Y) + h^0(Y, \mathcal{K}_Y \otimes \mathcal{L}) \\ c_1^2(\bar{X}) &= 2c_1^2(Y) + 4(\mathcal{L}, \mathcal{K}_Y) + 2(\mathcal{L}, \mathcal{L}) \\ c_2(\bar{X}) &= 2c_2(Y) + 2(\mathcal{L}, \mathcal{K}_Y) + 4(\mathcal{L}, \mathcal{L}). \end{aligned} \tag{9}$$

For example, if we take $Y = \mathbb{P}_2$, and for B a curve of degree $2m$ with at most simple singularities, we obtain:

$$\begin{aligned} p_g(\bar{X}) &= 1 + \frac{1}{2}m(m-3) \\ q(\bar{X}) &= 0 \\ c_1^2(\bar{X}) &= 2(m-3)^2 \\ c_2(\bar{X}) &= 4m^2 - 6m + 6. \end{aligned}$$

If $m \geq 3$, then the surface $\bar{X}$ is always minimal, since $(\mathcal{K}_{\bar{X}}, \mathcal{O}_{\bar{X}}(C)) \geq 0$ for every curve C on $\bar{X}$, whereas $(\mathcal{K}_{\bar{X}}, \mathcal{O}_{\bar{X}}(E))$ would be -1 for a (-1)-curve on $\bar{X}$. (If $m = 1$, then $\bar{X}$ is either $\mathbb{P}_1 \times \mathbb{P}_1$ or Σ_2 and for $m = 2$ one obtains a non-minimal rational surface). Formula (8) also immediately yields that

$$\operatorname{kod}(\bar{X}) = \begin{cases} -\infty & \text{if } m = 1, 2 \\ 0 & \text{if } m = 3 \\ 2 & \text{if } m \geq 4. \end{cases}$$

Next we investigate the effect on the invariants of the minimal resolution $\bar{X}$ when B acquires certain non-simple singularities. Suppose that the point $y_1 \in Y$ is an ordinary d-fold point of B. Let $\sigma : \bar{Y} \to Y$ be the blowing-up of Y at y_1. The proper transform $\bar{B}$ of B on $\bar{Y}$ is smooth, and there exists a 2-fold covering $p : \bar{X} \to \bar{Y}$, branched over B_1, where $B_1 = \bar{B}$ if $d = 2m$ and $B_1 = \bar{B} + E$, if $d = 2m + 1 (E = \sigma^{-1}(y_1))$. In both cases the branch curve has at most simple singularities, and the formulae above apply.

More generally, we may assume that y_1 is a singular point of B, such that $\bar{B}$ has only simple singularities. If the curve B has r of these points, say $y_1, \ldots, y_r$ and if $\sigma : \bar{Y} \to Y$ is the blowing-up of Y at $y_1 \cup \ldots \cup y_r$, then there is again a 2-fold covering of $\bar{Y}$ with branch curve the union of $\bar{B}$ and some of the exceptional curves. If y_j is a point of multiplicity d_j on B, with $d_j = 2m_j$ or $d_j = 2m_j + 1$, then application of (9) yields:

$$(10)\quad \begin{cases} \chi(\bar{X}) &= 2\chi(Y) + \frac{1}{2}(\mathcal{L}, \mathcal{K}_Y \otimes \mathcal{L}) - \sum_{j=1}^{r} \frac{1}{2} m_j(m_j - 1) \\ c_1^2(\bar{X}) &= 2c_1^2(Y) + 4(\mathcal{L}, \mathcal{K}_Y) + 2(\mathcal{L}, \mathcal{L}) - 2\sum_{j=1}^{r}(m_j - 1)^2. \\ \text{And if } p_g(Y) = 0: & \\ p_g(\bar{X}) = \begin{cases} \text{dimension of the subspace of } \Gamma(\mathcal{K}_Y \otimes \mathcal{L}), \\ \text{consisting of those sections,} \\ \text{vanishing of order at least } m_j - 1 \text{ in } y_j, j = 1, \ldots, r. \end{cases} \end{cases}$$

It will be clear that taking double coverings, as described in this section, provides a rich source for examples of all possible sorts of surfaces. This method has been applied very frequently to obtain surfaces with given invariants (compare VII, Sects. 9 and 10).

23. An Enriques Surface

Preserving the notation introduced in Sect. 22, we now take for Y the quadric $\mathbb{P}_1 \times \mathbb{P}_1$ and for B any curve of bidegree (4,4) with at most simple singularities. Using (8) and (9) we find that the resulting surface $\bar{X}$ satisfies:
(i) $\mathcal{K}_{\bar{X}} = \mathcal{O}_{\bar{X}}$
(ii) $q(\bar{X}) = 0$
So $\bar{X}$ is a K 3-surface.

Let $(x_0 : x_1, y_0 : y_1)$ be bihomogeneous coordinates on $\mathbb{P}_1 \times \mathbb{P}_1$, and let i be the involution on $\mathbb{P}_1 \times \mathbb{P}_1$ given by $i(x_0 : x_1, y_0 : y_1) = (x_0 : -x_1, y_0 : -y_1)$. It has four isolated fixed points, namely $p_1 = (0 : 1, 0 : 1), p_2 = (0 : 1, 1 : 0), p_3 = (1 : 0, 0 : 1)$ and $p_4 = (1 : 0, 1 : 0)$. The i-invariant polynomials of bidegree (4,4) form a 13-dimensional vector space with a basis consisting of the nine elements $x_0^{2k} x_1^{4-2k} y_0^{2j} y_1^{4-2j} (0 \leq k, j \leq 2)$ and the four elements $x_0^3 x_1 y_0^3 y_1, x_0 x_1^3 y_0^3 y_1, x_0^3 x_1 y_0 y_1^3, x_0 x_1^3 y_0 y_1^3$. The corresponding linear system of curves with bidegree (4,4) has no base points, hence by Bertini's theorem I.20.2 its general member is smooth and irreducible. So there exist plenty of i-invariant curves B of bidegree (4,4) not passing through any of the points p_i and having at most simple singularities. We fix such a curve B and form $\bar{X}$ as before. We claim that i lifts to a fixed-point-free involution j of $\bar{X}$. To see this, we note that i operates on the total space F of the line bundle $\mathcal{F} = \mathcal{K}^{\vee}_{\mathbb{P}_1 \times \mathbb{P}_1}$ in which X is embedded as the pull-back under the squaring map $\mathcal{F} \to \mathcal{F}^{\otimes 2}$ of an i-invariant section. On F the fibres over p_j are pointwise fixed under i, so $i|X$ has eight fixed points and the involution which is the composition of $i|X$ and the covering involution of $X \to \mathbb{P}_1 \times \mathbb{P}_1$ has no fixed points on X. It lifts to a fixed-point-free involution j of the minimal resolution $\bar{X}$ of X. We let $\pi : \bar{X} \to Y$ be the quotient map. Then Y is a compact surface with the following properties
(i) $\mathcal{K}_Y \neq \mathcal{O}_Y$
(ii) $\mathcal{K}_Y^{\otimes 2} = \mathcal{O}_Y$

(iii) $q(Y) = 0$

To prove this, we first observe that $\chi(\mathcal{O}_Y) = \frac{1}{2}\chi(\mathcal{O}_X) = 1$, so - since $q(Y) \leq q(\bar{X}) = 0$ - it follows that $p_g(Y) = q(Y) = 0$. In particular $\mathcal{K}_Y$ is not trivial. Since however $\pi^*(\mathcal{K}_Y) = \mathcal{K}_{\bar{X}} = \mathcal{O}_{\bar{X}}$, we see that $\mathcal{K}_Y$ is the 2-torsion bundle defining π, so $\mathcal{K}_Y^{\otimes 2} = \mathcal{O}_Y$.

A compact surface with properties (i)-(iii) is called an **Enriques surface.** In Chap. VIII we shall deal intensively with these surfaces. Here we only mention two properties:

(i) An Enriques surface is projective. In fact, since $p_g(Y) = 0$ it follows from the exponential cohomology sequence that for every $c \in H^2(Y, \mathbb{Z})$ there is a line bundle $\mathcal{L}$ with $c_1(\mathcal{L}) = c$. But by Lemma IV. 2.6 we have $b_1(Y) = 0$, so Theorem IV. 2.7,(iii) implies $b^+(Y) = 1$ and there is a holomorphic line bundle $\mathcal{L}$ on Y with $c_1^2(\mathcal{L}) > 0$. Therefore by Theorem IV. 6.2 the surface Y is projective.

(ii) An Enriques surface is a minimal surface, since $\mathcal{K}_Y$ restricted to a curve always has degree zero.

In Chap. VIII we need an Enriques surface having an E_8-configuration of curves. We close this section by constructing an example having this property.

In the preceding construction we take for B the reducible curve consisting of the lines $L_1 = \{y_0 = y_1\}, L_2 = \{y_0 = -y_1\}$ and the curve C with the equation

$$(11) \qquad a x_1^4 (y_0^2 - y_1^2) + b x_0^2 x_1^2 (y_0^2 - y_1^2) + x_0^4 (c y_0^2 + d y_1^2) = 0$$

where a, b, c and d have to be chosen suitably, namely such that C is smooth and does not pass through the points p_j. This last condition is satisfied if $a \neq 0, c \neq 0$ and $d \neq 0$. Then also C is smooth at the points $q_1 = (0:1, 1:-1)$ and $q_2 = (0:1, 1:1)$. Since (11) defines a linear system having q_1 and q_2 as base points, the general member C will be smooth except possibly at q_1 and q_2, but if $a \neq 0, c \neq 0$ and $d \neq 0$ this is automatically the case, as we have seen. The curve $B = C \cup L_1 \cup L_2$ has two A_7-singularities, namely at q_1 and q_2. If we apply the results of Chap. III, Table 1 we obtain the configuration of (-2)-curves on X shown in Fig.9. The curve $\bar{L}_i$ is the pre-image on X of the proper transform of the curve $L_i (i = 1, 2)$ and $M_1 \cup M_2$ is the pre-image of $\{x_0 = 0\}$. The E_i give the A_7-singularity at q_1 and the F_j the A_7-singularity at q_2. The involution interchanges $\bar{L}_1$ and $\bar{L}_2, M_1$ and M_2, E_i and F_i. If we omit E_{-3} and F_{-3} we obtain two E_8-configurations interchanged by j, so their image on Y is an E_8-configuration.

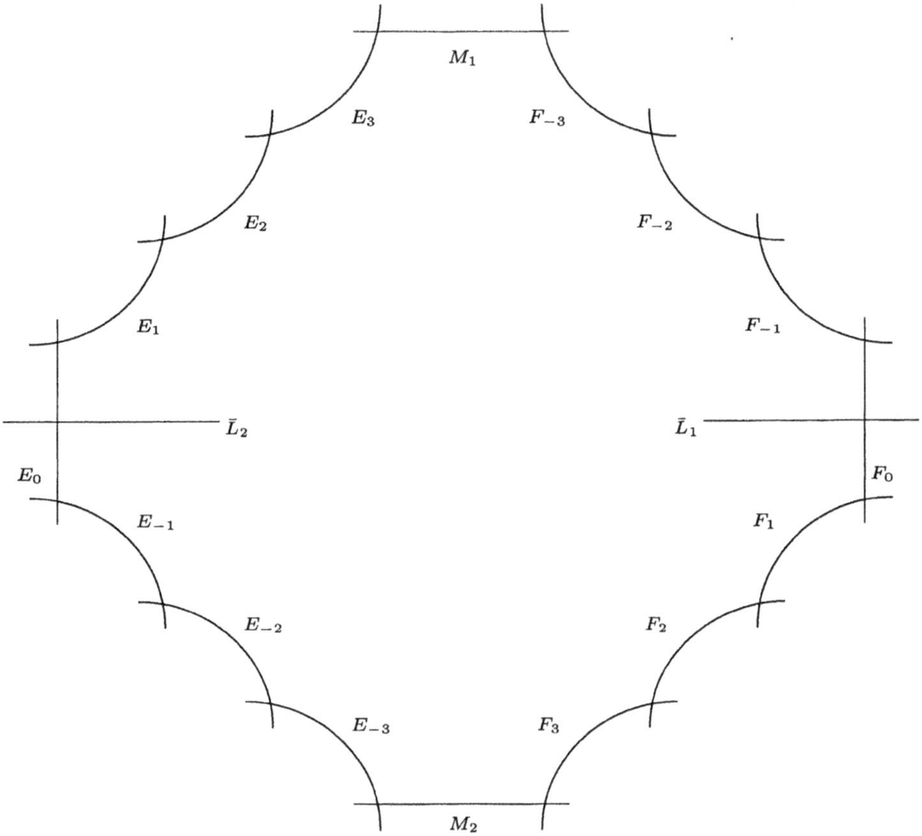

Fig. 9

24. Kummer Coverings

We refer to [B-H-H, Kap. 3] for details on what follows.

Start with k homogeneous linear forms $L_1, \ldots, L_k$ in three variables z_0, z_1, z_2 defining k lines in $\mathbb{P}_2$ and consider the abelian field extension (with group $G := (\mathbb{Z}/n\mathbb{Z})^{k-1}$)

$$K\left(\sqrt[n]{(L_2/L_1)}, \ldots, \sqrt[n]{(L_k/L_1)}\right) \supset K := \mathbb{C}(z_1/z_0, z_2/z_0).$$

This extension corresponds to a covering $X' \to \mathbb{P}_2$ ramified in the k given lines. If at most two lines pass through any given point the configuration of lines is called regular. In this case the surface X' is smooth. Otherwise it is always singular above each point where three or more of these lines meet.

Blowing up $\mathbb{P}_2$ at these points we get $\sigma : \tilde{\mathbb{P}}_2 \to \mathbb{P}_2$ and we can form the G-covering

$$f : X \to \tilde{\mathbb{P}}_2$$

ramified over the exceptional curves and the proper transform of these curves. Then X is smooth. If p is a point where $r = r_p$ lines meet, there are n^{k-1-r} isomorphic curves C_p over this point in X. Each of these is mapped to the corresponding exceptional curve E_p in $\tilde{\mathbb{P}}_2$ with degree n^{r-1}. There are r branch points on E_p over each of which one has n^{r-2} points. The Riemann-Hurwitz formula gives the Euler-number:

$$e(C_p) = n^{r-1}(2-r) + rn^{r-2}. \tag{12}$$

Above each point of a line where no other lines intersect there are $n^{k-1}/n = n^{k-2}$ points. Above the points where only two of the lines meet there are exactly n^{k-3} points. So, if we let L be the union of the k lines, sing L be the singular locus of L and

$$t_r := \text{ number of points where } r \text{ lines meet}$$
$$f_1 := \sum rt_r, \quad f_0 := \sum t_r$$

we get

$$\begin{aligned}
&e(X'\backslash \operatorname{sing} X') \\
&\quad = n^{k-1}e(\mathbb{P}_2 \setminus L) + n^{k-2}e(L \setminus \operatorname{sing} L) + n^{k-3}(t_2) \\
&\quad = n^{k-1}(3 - 2k + \sum (r-1)t_r) + n^{k-2}(2k - \sum rt_r) + n^{k-3}t_2, \\
e(X) &= e(X' \setminus \operatorname{sing} X') + \sum_{r \geq 3} n^{k-1-r} t_r \left(n^{r-1}(2-r) + rn^{r-2}\right) \\
&= n^{k-1}(3 - 2k + f_1 - f_0) + 2n^{k-2}(k - f_1 + f_0) + n^{k-3}(f_1 - t_2).
\end{aligned}$$

The canonical divisor on X can be calculated using the Hurwitz formula (Lemma I. 16.1). Denoting by H a line on $\mathbb{P}_2$ and r_p the number of lines meeting in a point p we get

$$K_X = f^* \left(\sigma^* K_{\mathbb{P}_2} + \sum E_p + \frac{n-1}{n} \left[\sum E_p + \sigma^*(kH) - \sum_p r_p E_p \right] \right).$$

Using this one can calculate the Kodaira dimension of X. One finds almost always surfaces of general type. If the configuration contains six lines such that at most three of them pass through a common point the calculations on loc. cit. p. 120–121 show that X is minimal.

We want to investigate a particular configuration which leads to a surface with $c_1^2 = 3c_2$. Take three distinct non-concurrent lines L, M, N and add the three lines through the three points $L \cap M$, $L \cap N$, $M \cap N$ and a fourth point P not on $L \cup M \cup N$. So we have $k = 6$, $t_2 = 3$, $t_3 = 4$ and the other

t_r vanish. Observe that the arrangement is essentially unique. We let $n = 5$ and we abuse notation by letting X be the surface obtained by the preceding construction. We find

$$c_2(X) = 15 \cdot 5^3.$$

To calculate $c_1^2(X)$ we use the above formula for the canonical divisor on X:

$$\begin{aligned} K &= f^* \left(\sigma^* K_{\mathbb{P}_2} + \sum E_p + 4/5 \left[\sum E_p + \sigma^*(6H) - 3\sum E_p\right]\right) \\ &= f^* \left(\sigma^*(9/5H) - 3/5 \sum E_p\right), \end{aligned}$$

since $r_p = 3$ for the four points p where the lines meet. So we get

$$c_1^2(X) = 5^5(81/5^2 - 36/5^2) = 45 \cdot 5^3.$$

Comparing the two numbers we see indeed that X is on the line $c_1^2 = 3c_2$.

Remark

i) Ishida ([Is88]) calculated the irregularity: $q = 30$.

ii) He also found 4 groups of order 5^2 which operate freely on X giving quotients with

$$c_1^2 = 3c_2 = 225; \quad q = 10, 6, 4, 0$$

and a group of order 5^3 operating freely with quotient having invariants

$$c_1^2 = 3c_2 = 45; \quad q = 2.$$

Using the unbranched covering trick (I. 18.1) we find surfaces with $c_1^2 = 3c_2 = 45k$ for $k = 1, 2, \ldots$.

iii) Using other arrangements one can find at least two more examples on the line $c_1^2 = 3c_2$:

$L=$ the twelve lines in a Hesse pencil with $c_1^2 = 3c_2 = 48 \cdot 3^{10}$,

$L=$ dual of the configuration of the 9 inflection points on a cubic and the connecting lines with $c_1^2 = 3c_2 = 333 \cdot 5^6$.

iv) Let p be one of the points where three lines meet ($r = 3$). The lines through p define a fibration $X \to C_p$. The curve C_p has Euler number -10 (use (12)) and hence genus 6. This will be used to construct surfaces with given slopes $c_1^2/3c_2$ in Chap. VII, Sect. 8.

Chapter VI. The Enriques Kodaira Classification

In this chapter a surface will be a compact, connected 2-dimensional complex manifold. As defined in II, Sect. 1, a curve on a surface is always a closed 1-dimensional subvariety, locally given by one equation (essentially, an effective divisor).

In this chapter we deal with the Enriques-Kodaira classification. In the first edition of this book the case (2, 1) of Iitaka's conjecture (see the Introduction) was used in the proof of Theorem 1.1. In the meantime it has become customary to give a slightly different proof for the classification theorem. The new proof, a form of which is presented below, rests on a systematic use of nef divisors (IV. Sect. 7). A central role is played by the Rationality Theorem which we prove first. We then show how the full classification of surfaces with K_X not nef follows in an astonishingly simple way, and deduce Castelnuovo's criterion as a corollary. After all this the classification of minimal algebraic surfaces becomes quite easy. The classification of non-algebraic surfaces which follows next, is the same as in the first edition of the book. In Sect. 8 we prove Iitaka's results concerning the stability of the ten classes under deformations.

1. Statement of the Main Result

In I, Sect. 7 it has been explained that for given n the n-dimensional compact, connected complex manifolds X can be classified according to their Kodaira dimension $\mathrm{kod}(X)$, which can assume the values $-\infty, 0, 1, \ldots, n$. In the case $n = 2$ the surfaces in the classes $\mathrm{kod}(X) = -\infty$ or $\mathrm{kod}(X) = 0$, and to a lesser extent those with $\mathrm{kod}(X) = 1$ can be classified in much more detail. Thus, starting from the rough classification by Kodaira dimension, surfaces are divided into ten classes. This classification is called the Enriques-Kodaira classification and is embodied in the following central result.

(1.1) **Theorem.** *Every surface has a minimal model in exactly one of the classes* 1) *to* 10) *of Table* 10. *This model is unique (up to isomorphisms) except for the surfaces with minimal models in the classes* 1) *and* 3).

Contrary to the Kodaira dimension, the algebraic dimension $\mathrm{a}(X)$ is not the same for all surfaces in one class; the values that occur are shown in the table. The table also gives some more details about the plurigenera $P_n(X)$ for the case $\mathrm{kod}(X) = 0$, and about the first Betti number $b_1(X)$ whenever this is possible. This information is sufficient to characterise the different classes by the plurigenera and the first Betti number. (Blowing up changes neither of these, compare Theorem I.9.1 (iv) and (viii).)

For convenience we give below the definitions of all these classes, though practically all of them have appeared earlier. These definitions are the standard ones, except perhaps for the classes 5) and 6) and in particular class 2). They vary widely in explicitness: sometimes (e.g. for tori) they are as explicit as anybody can ask for; in other cases (e.g. for K 3-surfaces) they are very formal.

The surfaces in several classes are minimal by definition. The minimality of the surfaces in class 3) is a consequence of Lüroth's theorem for curves (the image of a rational curve is again rational), whereas the minimality in the classes 4)-8) is due to the fact that $(\mathcal{K}, E) = -1$ for a (-1)-curve E.

Table 10.

Class of X	kod(X)	smallest $n > 0$ with $\mathcal{K}_X^{\otimes n} = \mathcal{O}_X$	$b_1(X)$	possible value of a(X)	c_1^2	c_2
1) minimal rational surfaces			0	2	8 or 9	4 or 3
2) minimal surfaces of class VII	$-\infty$		1	$0, 1$	≤ 0	≥ 0
3) ruled surfaces of genus $g \geq 1$			$2g$	2	$8(1-g)$	$4(1-g)$
4) Enriques surfaces		2	0	2	0	12
5) bi-elliptic surfaces		$2, 3, 4, 6$	2	2	0	0
6) Kodaira surfaces	0					
a) primary		1	3	1	0	0
b) secondary		$2, 3, 4, 6$	1	1	0	0
6) K 3-surfaces		1	0	$0, 1, 2$	0	24
8) tori		1	4	$0, 1, 2$	0	0
9) minimal properly elliptic surfaces	1			$1, 2$	0	≥ 0
10) minimal surfaces of general type	2		$\equiv 0(2)$	2	> 0	> 0

A rational surface is a surface that is birationally equivalent to $\mathbb{P}_2$. Apart from $\mathbb{P}_2$ we have described in V, Sect. 4 an infinite sequence of other minimal rational surfaces, namely the Hirzebruch surfaces Σ_n, $n = 0, 2, 3, \ldots$ It will be shown later that there are no others.

A surface of class VII is a surface X with $\text{kod}(X) = -\infty$ and $b_1(X) = 1$. (Minimal surfaces in this class are often called of class VII_0.) We have met two types of examples, namely Hopf surfaces (V, Sect 18) and Inoue surfaces (V, Sect. 19). We have already mentioned that the minimal surfaces X with

$a(X) = 1$ which are contained in class VII are exactly the Hopf surfaces of algebraic dimension 1. For surfaces X with $a(X) = 0$ there are, apart from the Hopf surfaces and the Inoue surfaces, other examples known (compare the references at the end of V, Sect. 19), but a complete classification is still lacking.

It follows from Theorem IV.2.7 that $q(X) = 1, h^{1,0} = 0$ for every surface of class VII.

The name "class VII" comes from Kodaira's presentation of the Enriques classification ([Ko66], part I and part IV). However, our "class VII" is not Kodaira's class VII, (which contains surfaces of different Kodaira dimension) but a subclass of it, namely Kodaira's class 7). We have chosen this name since on the other hand we like to keep the traditional name "class VII" for Hopf surfaces, Inoue surfaces,..., whereas on the other hand it is not possible to have a class containing surfaces of different Kodaira dimension.

A ruled surface of genus g is a surface X admitting a ruling, i.e. an analytically locally trivial fibration with fibre $\mathbb{P}_1$ and a structure group $PGL(2, \mathbb{C})$ over a smooth curve of genus g. Ruled surfaces, which are all algebraic, have been discussed in V, Sect. 4. At that place we also saw that the fibration is always equivalent to an *algebraically* locally trivial one.

An Enriques surface X is a surface with $q(X)$ (or equivalently $b_1(X) = 0$), for which $\mathcal{K}_X^{\otimes 2} \cong \mathcal{O}_X$ but $\mathcal{K}_X \neq \mathcal{O}_X$. Such a surface appeared in V, Sect. 23, where it was also proved that any Enriques surface is projective. The second part of Chap. VIII will deal with the classification of Enriques surfaces.

A bi-elliptic surface is a surface X with $b_1(X) = 2$, admitting a holomorphic, locally trivial fibration over an elliptic curve with an elliptic curve as a typical fibre. Their classification has been described in V, Sect. 5. In particular, every bi-elliptic surface is algebraic.

A primary Kodaira surface (V, Sect. 5) is a surface with $b_1(X) = 3$, admitting a holomorphic locally trivial fibration over an elliptic curve with an elliptic curve as typical fibre.

A secondary Kodaira surface (V, Sect. 5) is a surface, which itself is not a primary Kodaira surface but which admits a primary Kodaira surface as unramified covering. They are elliptic fibre spaces over rational curves, with first Betti number equal to 1.

A K 3-surface is a surface X with $q(X) = 0$ and $\mathcal{K}_X = \mathcal{O}_X$. Examples of such surfaces appeared several times in Chap. V, namely as complete intersections of type (4), (2, 3) and (2, 2, 2) in V, Sect. 2, as Kummer surfaces in V, Sect. 16 and as double coverings of $\mathbb{P}_2$, ramified over a curve of degree 6 with simple singularities in V, Sect. 22. A large part of Chap. VIII will be devoted to the classification theory of these surfaces.

A **torus** is a surface isomorphic to the quotient of $\mathbb{C}^2$ by a lattice of real rank 4.

A **properly elliptic surface** is an elliptic surface X (V, Sect. 7) with $\mathrm{kod}(X) = 1$. A very simple example is provided by the product of two curves, one elliptic and the other of genus ≥ 2. (V, Sect. (6)).

A **surface of general type** is a surface X with $\mathrm{kod}(X) = 2$. Examples of such surfaces are manifold: complete intersections of sufficiently high degree (V, Sect. 2), products of curves of genus ≥ 2 (V, Sect. (6)), Kodaira fibrations (V, Sect. 14), quotients of symmetric domains (V, Sect. 20) and "practically all" ramified double coverings of $\mathbb{P}_2$ (V, Sect. 22). These surfaces are general in the same sense as are curves of genus ≥ 2. Since always $\mathrm{a}(X) \geq \mathrm{kod}(X)$, every surface of general type is algebraic by Corollary IV.6.5. Chapter VII deals with the classification of surfaces of general type.

Remarks.

1) The "size" of the various classes is quite different. The surfaces in the classes 4), 7) and 8) "form one irreducible family", so they would form one main class in any classification. But other classes consist of an infinite number of families.
2) It might seem more natural to take the algebraic dimension $\mathrm{a}(X)$ as the primary invariant instead of $\mathrm{kod}(X)$. But in doing so you tear apart the "families" of tori and K 3-surfaces as well as certain "families" of properly elliptic surfaces, which is a rather cruel thing to do.

As mentioned in the introduction, the proof of Theorem 1.1 will be based on

(i) the results of Chap. IV, in particular Sects. 2, 6, 8;
(ii) the basic facts about fibrations and the canonical bundle formula for elliptic fibrations (Theorem V.12.1);
(iii) the rationality theorem and its consequences, such as the Castelnuovo criterion. It uses Chap. IV. Sect. 7 in an essential way.

The Enriques classification of algebraic surfaces seems to appear for the first time in [Enr14], but the classical reference is [Enr49]. Also [Ge] should be mentioned. The extension to non-algebraic surfaces is due to Kodaira who at the same time gave a modern (and much extended) treatment of the complex-algebraic case ([Ko66], part IV, p. 1062). The case of characteristic 0 has also been the subject of seminars by Zariski and by Shafarevich ([Sh]).

The Enriques classification for characteristic $p > 0$ is due to Mumford and Bombieri ([Mu69], [B-M76], [B-M77]). Other treatments for the case $p = 0$ can be found for example in [Be78], [G-H78a], [Bad] and [Kur].

2. Characterising Minimal Surfaces whose Canonical Bundle is Nef

In Chapter III we saw that a for a minimal surface X with $\mathrm{kod}(X) \geq 0$ the canonical bundle is nef (Corollary III, 2.4). As a first step in the proof of the classification theorem we prove the converse, thereby arriving at the following criterion.

(2.1) **Theorem.** *Let X be a minimal algebraic surface. Then K_X is nef if and only if* $\mathrm{kod}(X) \geq 0$.

Before embarking on the proof we formulate a result, which we shall use a few times later on as well.

(2.2) **Claim.** *Suppose that X is a minimal algebraic surface such that K_X is a nef non-torsion divisor. Then*

$$h^0(mK_X) \geq \frac{1}{2}m(m-1)K_X^2 + \chi(\mathcal{O}_X), \quad m \geq 2. \tag{1}$$

This follows immediately from the Riemann-Roch formula and the fact that $h^0(-kK_X) = 0$ for $k \geq 1$ by Lemma I.6.2.

Proof of Theorem 2.1. As we observed above, it suffices to assume that K_X is nef and to prove that $\mathrm{kod}(X) \geq 0$. If K_X is torsion, then $\mathrm{kod}(X) = 0$. If $K_X^2 > 0$, this is not the case and by the preceding claim, formula (1) implies that $\mathrm{kod}(X) = 2$. Since K_X being nef implies that $K_X^2 \geq 0$ (Observation IV, 7.6), the case $K_X^2 = 0$ remains to be considered.

We distinguish between two cases:

(i) X admits a connected morphism $f : X \to S$ onto a smooth curve of genus ≥ 2. If the genus g of the general fibre of f is also at least 2, we find from Corollary III.11.6 and the Noether formula that $\chi(\mathcal{O}_X) = \frac{1}{12}c_2(X) > 0$ and so by (1) $\mathrm{kod}(X) \geq 0$. The general fibre F cannot be rational, otherwise K_X is not nef. If F is elliptic, we infer from Prop. III.11.4 that $c_2(X) \geq 0$. Since $p_g > 0$ implies $\mathrm{kod}(X) \geq 0$, we see from $1-q = \frac{1}{12}c_2(X)$ that we only have to take into account the case $q(X) = 1$, but this is excluded since the existence of f implies that $q(X) \geq 2$.

(ii) X does not admit a connected morphism $f : X \to S$ onto a smooth curve of genus ≥ 2. If $q(X) \geq 2$ by Prop. IV.5.2, we have $p_g(X) \geq 1$ implying $\mathrm{kod}(X) \geq 0$. If $q(X) = 0$, $\chi(\mathcal{O}_X) \geq 1$ and so $\mathrm{kod}(X) \geq 0$ by formula (1). We are left with the case $q(X) = 1$, $p_g = 0$, hence $c_2(X) = 0$. So let $\alpha_X : X \to C = \mathrm{Alb}(X)$ be the Albanese map. Since $c_2(X) = 0$, by Remark III.11.5 the resulting fibration $X \to C$ is either a smooth fibration of curves of genus ≥ 2 or an elliptic fibration whose only singular fibres are multiples of smooth fibres. In the first case, by Theorem I, 10.1 this is a locally trivial fibration and since the automorphism group of the fibre is finite, the fibration becomes a product after a finite unramified base change. Since the Kodaira dimension does not change under such base changes (I.7.4), we find $\mathrm{kod}(X) = 1$ in this case. In the second case, the canonical bundle formula for elliptic fibrations (Corollary V.12.3) implies that $\mathrm{kod}(X) \geq 0$ as soon as we have one multiple fibre. There remains the case of an elliptic fibre bundle over an elliptic curve. Such bundles have been discussed in Chap. V, Sect. 5. The only possibility for $q(X) = 1$ is when X is a bi-elliptic surface. Since then $\mathrm{kod}(X) = 0$, we are done. □

Remark. We explain briefly what is known to be true in higher dimensions. We refer to [Cl-K-M] for more details. The generalization of the notion of a ruled surface to higher dimensional algebraic geometry is the concept of a **ruled variety**: a variety birational to a product $Y \times \mathbb{P}^1$. Clearly the Kodaira dimension of such a variety equals $-\infty$. The converse is false since there exist so called uniruled varieties which are not ruled (see the discussion after the proof of the Castelnuovo Rationality Criterion (3.4) in the next section). A variety is called **uniruled** if it is dominated by a ruled variety, i.e. if there exists a ruled variety mapping onto the given variety by means of a generically finite surjective morphism. Uniruled varieties are covered by rational curves in the sense that through a general point passes at least one rational curve. Conversely, a variety covered by rational curves can be shown to be uniruled. It is conjectured that $\kappa = -\infty$ forces the variety to be covered by rational curves. This has been proven in dimensions ≤ 3. So one half of Theorem 2.1 holds also in dimension 3 and conjecturally holds in higher dimension as well. The converse also holds in dimension 3 provided we use the correct definition of "minimal". A minimal threefold in this sense may have singularities but they have been classified. In higher dimensions this is not known and constitutes the main stumbling block in showing that a minimal X with K_X not nef is covered by rational curves.

3. The Rationality Theorem and Castelnuovo's Criterion

In this section we prove, essentially following [Wi78], the rationality theorem for the Néron-Severi group of algebraic surfaces and deduce some consequences, in particular Castelnuovo's Criterion.

If X is an algebraic surface with K_X not nef (everything in the real sense, i.e. in $\mathrm{NS}(X) \otimes \mathbb{Q}$) and H_0 an ample divisor class on X, then $H_t = H_0 + tK_X$ is nef for $t = 0$ and not nef for t big enough. Writing $H_t = t(K_X + (1/t)H_0)$ whenever $t > 0$, it is obvious that H_t is nef for $0 \leq t \leq t_0$ if H_{t_0} is nef. If we set $b = \sup\{t \in \mathbb{R} \mid H_t \text{ is nef }\}$, then a priori this number b need not be rational. The Rationality theorem says that it always is. This meas that $H_b \in \mathrm{NS}(X) \otimes \mathbb{Q}$, in other words that H_b is a rational divisor class. The Rationality theorem, on which the classification of algebraic surfaces will be based, should really be seen in th e light of general Mori theory, but in accordance with our policy we restrict ourselves to what is needed for the surface case.

(3.1) **Rationality theorem.** *Let X be an algebraic surface and let H be an ample divisor on X. Assume that K_X is not nef. Then the non-negative number*

$$b := \sup\{t \in \mathbb{R} \mid H_t = H + tK_X \text{ is nef}\}. \tag{2}$$

is rational.

Before giving the proof of this theorem we make some remarks.

(3.2) *Remarks.*
1) In fact $b > 0$ since the ample cone is open by IV, 7.7. This also implies that $H_b = H + bK_X$ does not belong to the ample cone, although H_b is still nef (the nef-cone is closed).
2) Since for ample H and rational nef D the class $(1-u)H + uD$ is ample if $0 \leq u < 1$ by the Nakai criterion (IV. 6.4), we have that $H_t = (1-(t/b)H + (t/b)(H + bK_X)$ is ample for $t \in \mathbb{Q}$, $0 \leq t < b$.

Proof. (See [Wi]) Let

$$P(v,u) := \chi(vH + uK_X), \quad v,u \in \mathbb{Z}.$$

By Riemann-Roch this is a (in general not homogeneous) quadratic polynomial in v, u. If u and v are strictly positive with $u - 1 < vb$, then by the remark above, the divisor $vH + (u-1)K_X$ is ample and so by Kodaira Vanishing (IV.12.4) $H^i(vH + uK_X) = 0$ for $i = 1, 2$. It follows that $P(v,u) \geq 0$ for $u - 1 < bv$. Since for n large $nH + K_X$ is very ample, by a similar argument $P(n, 1) > 0$ and so P does not vanish identically. Assuming that b is irrational we shall arrive at a contradiction. A result from number theory ([H-W], Theorem 167) implies that b can be approximated as close as we like by rational numbers of the form p/q, where p and q are arbitrarily large integers, in such a way that

$$p/q - 1/(3q) < b < p/q.$$

Consider the line L given by $vp = uq$ in the (v,u)-plane. The line L meets the conic $P(v,u) = 0$ in at most two points. For $k = 1, 2, 3$ the numbers $v = kq$ and $u = kp$ satisfy $(u-1)/v < b$ and hence $P(kq, kp) \geq 0$ for these three values of k. So for at least one of these values P must be strictly positive.

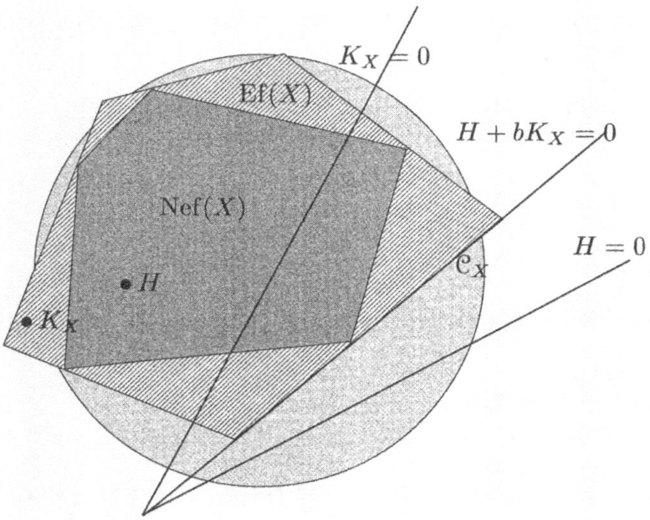

Figure 11. The Rationality Theorem

For the corresponding points (v, u) we have dim $H^0(vH + uK_S) > 0$. Then the effective divisor (with rational coefficients)

$$L := H + (u/v)K_X$$

is not nef since $u/v > b$. Let us write $L = \sum_j a_j \Gamma_j$ with $a_j > 0$ and Γ_j irreducible. Any irreducible curve C with $(L, C) < 0$ must be one of the Γ_j. Suppose that $C = \Gamma_j$, $j = 1, \ldots, N$ are the curves that occur in this way. Of course $(K_X, C_j) < 0$ and since $(H + tK_X, C_j)$ is a strictly decreasing linear function of t, positive for $t = 0$ and negative for $t = u/v$, it has a unique zero t_j which is a rational number. We claim that $b = \min\{t_j, j = 1 \ldots, N\}$, a rational number, contrary to our assumption that b be irrational. Indeed, if $t > \min\{t_j, j = 1 \ldots, N\}$ for at least one j we have $(H + tK_X, C_j) < 0$ and so H_t is not nef. But if $t \leq \min\{t_j, j = 1 \ldots, N\}$ we have $(H + tK_X, C_j) \geq 0$, $j = 1, \ldots, N$ and so H_t is nef, completing the proof that $b = \min\{t_j, j = 1 \ldots, N\}$. □

As a first application we have:

(3.3) **Proposition.** *A minimal algebraic surface X with K_X not nef is either $\mathbb{P}_2$ or a ruled surface. In particular, if $q = 0$, we have $X = \mathbb{P}_2$ or $X = \Sigma_n$ with $n \neq 1$.*

Proof. Let us first look at the positive half ray in NS$(X) \otimes \mathbb{Q}$ spanned by $-K_X$. There are two possibilities. The first possibility is that all ample classes of X are on this line and hence $-K_X$ is ample. Since, if D is any divisor, $D - mK_X$, $m \gg 0$ being ample must also be on this half ray, it follows that Pic(X) has rank 1. Kodaira vanishing (IV.12.4) implies that $h^0(K_X) = h^1(K_X) = 0$ and so $p_g = q = 0$. This implies that all classes in $H^2(X, \mathbb{Z})$ are algebraic and so, since $\mathrm{Pic}(X)$ has rank 1, we have $b_2 = 1$. Moreover, since $-K_X$ is ample, $P_2(X) = 0$. Theorem V.1.1 then implies that $X = \mathbb{P}_2$.

So one may assume that there exists an ample H such that its class in $\mathrm{NS}(X) \otimes \mathbb{Q}$ does not belong to the positive half-ray spanned by $-K_X$. Now apply the Rationality theorem to H and K_X. With $b = u/v$, $u, v > 0$ as in this theorem, we have a nef but not ample divisor

$$L = vH + uK_X.$$

In particular (IV.7.6) $L^2 \geq 0$.

We exclude $L^2 > 0$ as follows. Since L is nef, for any effective divisor D, one has $LD \geq 0$. We first exclude the case of equality. Indeed, any irreducible curve D for which $LD = 0$ must be an exceptional curve of the first kind: from the definition of L one sees that $K_X D < 0$, while the Algebraic Index Theorem applied to L and D shows that $D^2 < 0$. In combination with the adjunction formula this shows that D has to be an exceptional curve of the first kind. By assumption these do not exist on X. It follows that $LD > 0$ for all curves D and so, by the Nakai criterion (IV. 6.4) L is ample. But this is impossible by construction.

Hence we must have $L^2 = 0$. We shall show that for sufficiently large m the system mL gives a ruling. Since L is nef one has $LH \geq 0$, and if $LH = 0$ an application of the Algebraic Index Theorem shows that L is numerically trivial. In this last case, the class of H in $\mathrm{NS}(X) \otimes \mathbb{Q}$ would be on the positive half-ray spanned by the class of $-K_X$, which has been excluded. So $LH > 0$. From $0 = (1/v)L^2 = L(H + (u/v)K_X)$ one infers that $LK_X < 0$. We want to apply Riemann-Roch in this situation. Since $u/v = b$, hence $(mu-1)/mv < b$, Remark 3.2 2) implies that

$$mL - K_X = mv\left[H + \frac{mu-1}{mv}K_X\right]$$

is ample for all $m \geq 1$. Serre duality implies that $h^2(mL) = h^0(-(mL - K_X)) = 0$ and so by Riemann-Roch

$$\dim H^0(mL) \geq \chi(mL) = \chi(\mathcal{O}_X) - \frac{m}{2}LK_X.$$

It follows that for large enough m, $\dim |mL| \geq 1$. Replace L by mL and write $L = L' + L_{\mathrm{fixed}}$, where L_{fixed} is the fixed part of $|L|$. Since L' moves in a linear system, it is nef and so $L'L \geq 0$ and $L'L_{\mathrm{fixed}} \geq 0$. Since L is nef each of the summands in $0 = L^2 = L'L + L_{\mathrm{fixed}}L$ is non-negative and so $L'L = L_{\mathrm{fixed}}L = 0$. Similarly, each of the summands in $0 = L'L = (L')^2 + L'L_{\mathrm{fixed}}$ is non-negative and so $(L')^2 = 0$. Let now $nD + R$ be a member of $|L'|$, where R does not contain D. The equality $LL' = 0$ implies that $LD = 0$ and since $L'L_{\mathrm{fixed}} = 0$ one also has $DL_{\mathrm{fixed}} = 0$. So $L'D = 0$ and hence $nD^2 + DR = 0$ so that $D^2 \leq 0$. Since $LD = 0$, it follows from the definition of L that $DK_X < 0$. By the Adjunction Formula D then is a smooth rational curve with $D^2 = 0$. Now apply Prop. V.4.3 to conclude that X is a ruled surface. If moreover $q(X) = 0$, by the classification done in V. Sect.4, the surface X must actually be a Hirzebruch surface. □

As a corollary we have:

(3.4) **Castelnuovo's Rationality Criterion.** *An algebraic surface X is rational if and only if $P_2(X) = q(X) = 0$.*

Proof. Since q and P_2 are birational invariants (Corollary III.6.4) it is sufficient to consider the case of a minimal surface X.

A surface is rational if it is birationally equivalent to $\mathbb{P}_2$ and so the "only if"-part follows from the birational invariance of q and P_2.

To prove the converse, we may assume X to be minimal. If $P_2(X) = 0$, then automatically $p_g(X) = 0$, so $\chi(\mathcal{O}_X) = 1 - q(X) + p_g(X) = 1$. By Riemann-Roch

$$h^0(-K_X) = h^0(2K_X) + h^0(-K_X) \geq K_X^2 + 1.$$

If $K_X^2 \geq 0$ it follows that $h^0(-K_X) \geq 1$. Now K_X cannot be trivial, since then $p_g(X) > 0$, so K_X cannot be nef (I.6.2). But if $K_X^2 < 0$ it can neither be nef (Observation IV, 7.6). So the preceding result implies that $X = \mathbb{P}_2$ or one of the Hirzebruch surfaces Σ_n, $n \neq 1$, i.e. X is rational. □

As an application of Castelnuovo's criterion we shall prove that every unirational surface is rational. An algebraic variety of dimension n is called rational if it is birationally equivalent to $\mathbb{P}_n$.

Definition. An algebraic variety V is called unirational if there exists a generically finite surjective morphism $f : W \to V$ where W is rational.

There exist unirational varieties of all dimensions ≥ 3 that are not rational, whereas every unirational curve is rational. This is also true in dimension 2:

(3.5) **Theorem.** *Every unirational surface is rational.*

Proof. Using resolution of singularities (III, Sect. 6) we may assume that V and W are smooth surfaces, that $f : W \to V$ is generically finite and surjective. Then $q(V) \leq q(W) = 0$, $P_2(V) \leq P_2(W) = 0$. Hence V is rational by Castelnuovo's criterion. □

Castelnuovo's criterion remains one of the corner stones of surface theory. It was proved by Castelnuovo in 1896 (Sulle superficie di genero 0, Mem. Soc. Ital. Sci., ser. III, X).

We shall now show that every minimal surface belongs to one of the classes 1) to 10), leaving it to the reader to verify that two different classes never contain birationally equivalent surfaces.

4. The Case a$(X) = 2$

The case $\mathrm{a}(X) = 2, \mathrm{kod}(X) = -\infty$

Since $\mathrm{kod}(X) = -\infty$ the canonical bundle cannot be nef (Theorem 2.1) and so by Prop. 3.3 $X = \mathbb{P}_2$ or a ruled surface. □

The case $\mathrm{a}(X) = 2, \mathrm{kod}(X) = 0$

It follows from Corollary III, 2.4 that K_X is nef, hence $K_X^2 \geq 0$. If $K_X^2 > 0$, Claim 2.2 would imply that $\mathrm{kod}(X) = 2$. So $K_X^2 = 0$. Since $\mathrm{kod}(X) = 0$ implies that $p_g(X) \leq 1$, we see by writing out Noether $12(1-q(X)+p_g(X)) = 2 - 4q(X) + b_2(X)$ and using $b_2(X) \geq 0$ that

$$8q - 10 \leq 12p_g \leq 12.$$

This gives five possibilities:

(i) $p_g(X) = q(X) = 0$;
(ii) $p_g(X) = 0,\ q(X) = 1$;
(iii) $p_g(X) = 1,\ q(X) = 0$;
(iv) $p_g(X) = q(X) = 1$;
(v) $p_g(X) = 1,\ q(X) = 2$.

We treat these cases separately.
Case (i). Here $P_2(X) > 0$ because of Castelnuovo's criterion. Hence $P_2(X) = 1$ since $\mathrm{kod}(X) = 0$. We claim that $P_3(X) = 0$. If not, then $P_3(X) = 1$. Let $D_2 \in |2K_X|$ and $D_3 \in |3K_X|$. So $3D_2$ and $2D_3$ are both divisors in $|6K_X|$. Since $P_6(X) \leq 1$ we must have $3D_2 = 2D_3$ and there must be an effective divisor D with $D_2 = 2D$ and $D_3 = 3D$. Then $D_3 - D_2 = D$, an effective divisor in $|K_X|$, but $p_g = 0$. So indeed $P_3(X) = 0$.

Applying the Riemann-Roch inequality to $3K_X$ we find

$$h^0(3K_X) + h^0(-2K_X) \geq 1$$

and hence $h^0(-2K_X) \geq 1$. Since $P_2(X) = h^0(2K_X) = 1$ this is only possible if $2K_X$ is trivial. It follows that X is an Enriques surface by definition.

Case (ii). We are going to show that X is bi-elliptic in this case.

Since $q(X) = 1$, the Albanese variety of X is an elliptic curve C and by Lemma I.13.9 the Albanese mapping $\alpha : X \to C = \mathrm{Alb}(X)$ has connected fibres. By the Noether formula $c_2(X) = e(X) = 0$. From Proposition III.11.4 and the remark following this proposition we conclude that $\alpha : X \to C = \mathrm{Alb}(X)$ is either a genus g fibration with $g \geq 2$ and α everywhere of maximal rank, or an elliptic fibration with no other singular fibres but multiples of smooth elliptic curves. In the first case, we can apply Theorem III.15.4 to conclude that $\alpha : X \to C = \mathrm{Alb}(X)$ is locally trivial. But then, since the fibres are curves of genus ≥ 2, a finite unramified covering $\hat{X} \to X$ becomes a product. For a finite unramified covering the Kodaira-dimension does not change (Theorem I.7.4) and so $\mathrm{kod}(X) = 1$ in this case.

If $\alpha : X \to C = \mathrm{Alb}(X)$ is elliptic and has multiple fibres, an application of the canonical bundle formula for elliptic fibrations (Theorem V.12.1) shows that $\mathrm{kod}(X) = 1$ in this case too.

There remains the possibility that $\alpha : X \to C = \mathrm{Alb}(X)$ is an elliptic fibration over an elliptic curve. By Chap.V, Sect.5 the fact that $p_g = 0$, $q = 1$ then implies that X is bi-elliptic.

Case (iii). In this case one has a K 3-surface. Indeed, Riemann-Roch applied to $2K_X$ yields $h^0(2K_X) + h^0(-K_X) \geq 2$. Since $\mathrm{kod}(X) = 0$ this implies that $h^0(-K_X) \geq 1$ and so K_X is trivial.

Case (iv). We are going to exclude this possibility. Since $q(X) > 0$, one can find a non-trivial line bundle $\mathcal{O}_X(E)$ with $\mathcal{O}_X(2E) = \mathcal{O}_X$ (any non-trivial 2-torsion point of the torus $\mathrm{Pic}^0(X)$ gives such a line bundle). Since $p_g = q = 1$, the Riemann-Roch inequality reads

$$h^0(\mathcal{O}_X(E)) + h^0(\mathcal{O}_X(K_X - E)) \geq 1$$

and hence $h^0(\mathcal{O}_X(K_X - E)) \geq 1$. Take $D \in |K_X - E|$ and let K_X be any canonical divisor. One has $2D = 2K_X$ since $P_2 = 1$ and hence $D = K_X$, contradicting the fact that $\mathcal{O}_X(E) \not\cong \mathcal{O}_X$.

Case (v). We are going to show that X is a torus. Let us first look at the possible components of the canonical divisor $K_X = \sum_j m_j C_j$. Since K_X is nef and $K_X^2 = 0$ one finds $K_X C_i = 0$. Writing down

$$0 = K_X C_j = m_j C_j^2 + \sum_{i \neq j} m_i C_i C_j$$

and using the adjunction formula we conclude that every C_j is either rational or smooth elliptic. Let $D = \sum_j n_j C_j$ be a reduced connected component of the canonical divisor. Writing $K_X = mD + R$, with the support of R disjoint from D, the equalities

$$0 = K_X D = mD^2 + DR = mD^2$$

imply $D^2 = 0$.

The image of the Albanese map $X \to \mathrm{Alb}(X)$ is either a curve $C \subset \mathrm{Alb}(X)$ or it is the entire (two-dimensional) Albanese. In the first case, since $q(X) = 2$, Theorem I.13.9 implies that the curve C is a genus 2 curve. Let $f : X \to C$ be the resulting fibration. By the preceding analysis, the irreducible components of the canonical divisor K_X are rational or elliptic. Since no rational or elliptic curve can be mapped onto a curve of genus 2, we see that K_X must be either empty or the sum of effective divisors, each of them set-theoretically contained in some fibre of f. Let F be the divisor class of such a fibre. Since $D^2 = 0$, from Zariski's Lemma (III.8.2), we deduce that $D = a/b \cdot F$ with a, b positive integers. Then $bD = f^*(a[c])$ and hence $h^0(nbD)$ and $h^0(nbK_X)$ become as large as we want when n tends to infinity.

This contradicts $\text{kod}(X) = 0$. The possibility that K_X is trivial is left. In this case, one pulls the fibration back by way of an unramified cover $C' \to C$ of degree ≥ 2 . One gets an unramified cover $X' \to X$ of degree ≥ 2 and $K_{X'}$ is still trivial, $\chi(\mathcal{O}_{X'}) = 0$ and hence $q(X') = 2$ by what we have seen. But $q(X') \geq q(C') \geq 3$, a contradiction.

There remains the case that $\text{Alb}(X)$ is a 2-torus and that the Albanese maps surjectively onto it. It is an elementary fact that in this case $\alpha^* : H^2(\text{Alb}(X), \mathbb{R}) \to H^2(X, \mathbb{R})$ is injective. Since $b_2(X) = 6$ this map α^* then is an isomorphism and so no fundamental cohomology class of a curve maps to zero. In particular, the Albanese map must be a finite morphism. So, if D is a connected component of the canonical divisor it cannot map to a point and hence it must be an elliptic curve E which maps to an elliptic curve $E' \subset \text{Alb}(X)$, which after a suitable translation becomes a subtorus of the Albanese torus. Now form the quotient elliptic curve $E'' = \text{Alb}(X)/E'$ and consider the surjective morphism $X \to E''$. Then Stein factorisation yields a fibration and D is contained in a fibre. By Zariski's lemma it follows that D is a rational multiple of a fibre and as before one concludes that $\text{kod}(X) = 1$. It follows that the only possibility is that K_X is trivial, but then, by the Hurwitz formula (I.Sect. 16) there is no ramification and so X itself is a torus.

The case $\text{a}(X) = 2, \text{kod}(X) = 1$

We shall show that X admits the structure of an elliptic fibration. For n large enough $|nK_X|$ is at least 1-dimensional. Let D_{fix}, $|D|$ be the fixed part, respectively the variable part of this linear system. One has $0 = nK_X^2 = D_{\text{fix}}K_X + DK_X$ and since each term is ≥ 0 (since K_X is nef) these must vanish. Now $0 = nDK_X = D^2 + DD_{\text{fix}}$ and again, each term is non-negative, since D moves and so $D^2 = 0 = DD_{\text{fix}}$. It follows that the rational map $f = \varphi_{nK_X}$ is a morphism and that $f : X \to C$ maps every divisor $D \in |D|$ to a point and so C is a curve. This is true for all n large enough so that $|nK_X|$ is at most 1-dimensional. If D is a smooth fibre of f, the adjunction formula implies that the connected components are elliptic curves and so, taking the Stein factorisation of f, one obtains an elliptic fibration.

The case $\text{a}(X) = 2, \text{kod}(X) = 2$

By definition X is a surface of general type. □

5. The Case $\text{a}(X) = 1$

(5.1) **Proposition.** *Every surface of algebraic dimension* 1 *is elliptic.*

Proof. The fact that $\text{a}(X) = 1$ implies that there is a non-constant meromorphic function m on X. This function gives rise to a surjective holomorphic map $f : X \setminus \bigcup_{i=1}^k p_i \to \mathbb{P}_1$, where $p_1, \ldots, p_k$ are the points of indeterminacy of m. So, according to IV. Sect. 1, after (repeatedly) blowing up X in $p_1, \ldots, p_k$, we obtain a surface $\tilde{X}$ and a surjective holomorphic map $\tilde{f}:\tilde{X}\to\mathbb{P}_1$, such that $\tilde{f}$ agrees with f outside of the exceptional trees. Using Stein factorisation we obtain a connected map $h:\tilde{X}\to Y$, where Y is a smooth curve. We claim that a general fibre F of h is an elliptic curve. For if $g(F)\neq 1$, then $(\mathcal{K}_X,F)\neq 0$ (by the adjunction formula), and $c_1^2(\mathcal{K}_{\tilde{X}}\otimes\mathcal{O}_{\tilde{X}}(nF))>0$ for suitable $n\in\mathbb{Z}$, which is impossible by Theorem IV. 6.2 and Corollary IV.6.7. In the same way we see that every curve C on $\tilde{X}$ is mapped to a point by h, for otherwise (C,F) would be strictly positive and $(C+nF)^2>0$ for large n. This means that h maps exceptional trees above $p_1,\dots,p_k$ onto points of Y. Hence also X admits a holomorphic map onto Y, such that the general fibre is elliptic (from the fact that all curves on X are contained in a fibre of this map it follows that this elliptic fibration is unique). □

Now that we know that every surface X with $\mathrm{a}(X)=1$ admits an elliptic fibration, we also know for the case that X is minimal:
(i) $\mathcal{K}_X^2=0$ by Corollary V.12.3;
(ii) $\chi(X)\geq 0$ by (i), Proposition III.11.4, and the Todd-Hirzebruch formula.

Since always $\mathrm{kod}(X)\leq\mathrm{a}(X)$ we have $\mathrm{kod}(X)\leq 1$. We want to single out those surfaces X for which $\mathrm{kod}(X)$ is either 0 or $-\infty$. In fact, it follows from Table 10 that we have to show: $\mathrm{kod}(X)=0$ only occurs for Kodaira surfaces, K 3-surfaces and tori, whereas $\mathrm{kod}(X)=-\infty$ implies that X is of class VII.

If g denotes the genus of Y, then the canonical bundle formula (Corollary V.12.3) yields immediately that $\mathrm{kod}(X)=1$ as soon as either $g\geq 2$, or $g=1,\chi(X)\geq 1$, or $\chi(X)\geq 3$. So we are left with the following cases:

(i) $g=1,\ \chi=0$
(ii) $g=0,\ \chi=2$
(iii) $g=0,\ \chi=1$
(iv) $g=0,\ \chi=0$,

which we shall treat separately.

In case (i) we find again by Corollary V.12.3 that there cannot be any multiple fibres. Furthermore, $e(X)=\frac{1}{12}\chi(X)-c_1^2(X)=0$, so f has no singular fibres at all. By theorem III.15.4, f is a locally trivial fibre bundle and hence by V, Sect. 5 either a torus or a primary Kodaira surface, since every bi-elliptic surface is projective.

As to case (ii), we see immediately that there cannot be any multiple fibres, hence $\mathcal{K}_X$ is trivial and $p_g(X)=1$. It follows that $q(X)=0$, so that X is a K 3-surface by definition.

Case (iii) is easily excluded. Since $p_g\leq 1$, we have either $q(X)=1$ or $q(X)=0$. In the first case, by the unbranched covering trick there would be an unramified covering X' of X with $\mathrm{a}(X')=\mathrm{a}(X)=1$, $\mathrm{kod}(X')=\mathrm{kod}(X)$, with $\chi(X)\geq 3$, which is impossible by the remarks above. If $q(X)=p_g(X)=0$, we would have $b^+(X)=0$, otherwise X would be algebraic, but this leads via Theorem IV.2.7 to the absurdity $h^{1,0}=-1$.

Finally the case (iv). The restriction $p_g(X) \leq 1$ leads to $p_g(X) = 1, q(X) = 2$ or $p_g(X) = 0, q(X) = 1$. If $p_g = 1$, then it follows from the canonical bundle formula that there cannot be an effective canonical divisor, for a multiple of it would be of the form $f^*(D)$, with D effective on Y, and $\mathrm{kod}(X)$ would be 1. But $\mathcal{K}_X$ trivial is also impossible here, since in that case there would be multiple fibres by Corollary V.12.3, whereas by Lemma III.8.3 and again Corollary V.12.3 the restriction of $\mathcal{K}_X$ to the reduction of a multiple fibre cannot be trivial.

The case $p_g(X) = 0, q(X) = 1$ can actually occur. In this case we have $b^+(X) = 0$ (otherwise the exponential cohomology sequence, together with Theorem V.5.2 and $p_g(X) = 0$ would imply that X is algebraic), hence $b_1(X) = 1$ by Theorem IV.2.7. If there is a positive n_0 such that $\mathcal{K}_X^{\otimes n_0} \cong \mathcal{O}_X$, then there exists a finite unramified covering X' of X with $\mathcal{K}_{X'}$ trivial. This obviously minimal surface must belong to one of the types of surfaces with $\mathrm{a}(X') = 0$ and $\mathcal{K}_{X'}$ trivial which we have encountered before: tori, K 3-surfaces and primary Kodaira surfaces. But X' cannot be a torus, otherwise X would also be a torus, and X' cannot be a K 3-surface for $b_1(X') \geq b_1(X) = 1$. So X' must be a primary Kodaira surface and X is a secondary Kodaira surface by definition. On the other hand, if there is no $n_0 \neq 0$ with $\mathcal{K}_X^{\otimes n_0} \cong \mathcal{O}_X$ then Corollary V.12.4 implies that $P_n(X) = 0$ for all $n \geq 1$. So X is of class VII.

6. The Case $\mathrm{a}(X) = 0$

We know already that $\mathrm{kod}(X) \leq \mathrm{a}(X) = 0$. Furthermore we have

(i) $p_g(X) \leq 1$ and $h^{0,1}(X) \leq 2$ (Proposition IV.8.1)
Combining the second inequality with Theorem IV.2.7 we obtain:
(ii) $q(X) \leq 3$.
Combining (i) and (ii) we find
(iii) $|\chi(X)| \leq 2$.
Using the unbranched covering trick we obtain from this:
(iv) If $q(X) > 0$, then $\chi(X) = 0$.

Taking into account Theorem IV.2.7 we see from (i)-(iv) that we are left with the following possibilities;

a) $q(X), p_g(X) = 0$
b) $q(X) = 0$, $p_g(X) = 1$
c) $q(X) = 1$, $p_g(X) = 0$
d) $q(X) = 2$, $p_g(X) = 1$, $b_1(X) = 3$
e) $q(X) = 2$, $p_g(X) = 1$, $b_1(X) = 4$.

We shall now consider these cases separately, though not in alphabetical order.

Case a) is easily excluded: from $p_g(X) = 0$ we find $b^+(X) = 0$, and this would (because of Theorem IV.2.7) imply that b_1 is negative.

Case b) is equally easy: using Proposition III.2.3 we see that $\mathcal{K}_X^2 \geq 0$. But $\mathcal{K}_X^2 > 0$ would imply that X is projective (Corollary IV.6.3). So $\mathcal{K}_X^2 = 0$ and Riemann-Roch shows that $h^0(\mathcal{K}_X^\vee) \geq 1$ so that $\mathcal{K}_X$ must be trivial. Hence X is a K 3-surface.

In case e) we know from Corollary IV.2.11 that there is a 2-dimensional Albanese torus Alb(X). Let $f : X \to \mathrm{Alb}(X)$ be the standard map. Then $f(X)$ cannot be a curve, otherwise we would have $\mathrm{a}(X) \geq 1$. So f is surjective. If C is any curve on X, then $f(C)$ must be a point, otherwise we would again obtain by translation an infinity of curves on Alb(X) and hence on X, which is excluded by Theorem IV.8.2. So f maps a finite number of curves onto points $p_1, \ldots, p_k$, and is outside of these of maximal rank (ramification would again lead to an infinity of curves on Alb(X) and X). Stein factorisation yields a factorisation $f = h \circ g$, where g is a connected map from X onto some surface X', and $h : X' \to \mathrm{Alb}(X)$ is a finite map. The restriction $h|X' \setminus h^{-1}(p_1 \cup \ldots \cup p_k)$ is an unramified covering of $\mathrm{Alb}(X) \setminus (p_1 \cup \ldots \cup p_k)$. But the embedding of this last subspace into Alb(X) induces an isomorphism of fundamental groups (both injectivity and surjectivity are elementary exercises), hence X' is a smooth unramified covering of Alb(X). Now $g : X \to X'$ is connected; since X is minimal it must be an isomorphism. Hence X is an unramified covering over a torus and therefore is a torus itself.

Next we exclude case d) by proving the rather subtle

(6.1) **Theorem.** *Every minimal surface X with $q(X) = 2$, $p_g = 1$ and $b_1(X) = 3$ is elliptic.*

As a preliminary result we need

(6.2) **Proposition.** *A minimal surface X with* $\mathrm{a}(X) = 0$, $q(X) = 2$, $p_g(X) = 1$ *and* $b_1(X) = 3$ *contains no curves.*

Proof. From the adjunction formula, the minimality of X and Proposition V.4.3 we obtain that $(\mathcal{K}_X, C) \geq 0$ for every curve C, and in particular that $(\mathcal{K}_X, \mathcal{K}_X) = 0$. If $(\mathcal{K}_X, C) > 0$ for some curve C , then $c_1^2(\mathcal{K}_X^{\otimes n} \otimes \mathcal{O}_X(C))$ would be strictly positive for large n, and X would be algebraic (Theorem IV.6.2). So $(\mathcal{K}_X, C) = 0$ for every curve C, and the adjunction formula yields that an irreducible curve C on X is either a (-2)-curve, or a curve of virtual genus $g = 1$ (smooth elliptic curve or rational with a node or a cusp). If X would contain a (-2)-curve, then we would obtain by the unbranched covering trick a minimal surface Y with $\mathrm{a}(Y) = 0$, containing at least five disjoint (-2)-curves, which are independent over $\mathbb{R}$, hence $b_2(Y)$ would be at least 5. The surface Y would have to belong to one of the classes a)-e) above and it is easy to see that it would belong to class d) again. For such a surface however we have $e(X) = 12\chi(X) - K_X^2 = 0$ and $b_2(X) = e(X) - 2 + 2b_1(X) = 4$.

Now suppose that X contains a smooth elliptic curve C. Since $b_1(C) = 2$ and $b_1(X) = 3$ we would then find unbranched coverings Y of X of arbitrarily high degree d, which are trivial over C, i.e. such that the inverse image of C

consists of d disjoint curves isomorphic to C. As before, these surfaces would again be in class d). If D is the union of those d smooth elliptic curves, then, considering the exact sequence

$$H^1(\mathcal{O}_Y) \to H^1(\mathcal{O}_D) \to H^2(\mathcal{O}_Y(-D))$$

we would obtain a contradiction: on the one hand $h^1(\mathcal{O}_Y) = q(Y) = 2$ and $h^2(\mathcal{O}_Y(-D)) = h^0(\mathcal{K}_Y \otimes \mathcal{O}_Y(D)) \leq 1$, but on the other hand $h^1(\mathcal{O}_D) = d$.

The case that C is rational is similar, but simpler. □

In the sequel, if θ is a real closed 1-form, then we shall denote its Hodge decomposition by $\theta = \theta^{1,0} + \bar{\theta}^{1,0}$ with $\theta^{1,0}$ of type $(1,0)$.

Proof of Theorem 6.1

We can assume that $\mathrm{a}(X) = 0$. From Theorem IV.2.7 we know that $h^{1,0}(X) = 1$, in other words there is a holomorphic 1-form $\omega \neq 0$, determined up to a constant. This form cannot vanish on a curve, since there are no curves on X by the previous Proposition. But it also cannot vanish in a number of isolated points, for the number of these (each counted with the proper strictly positive multiplicity) equals $c_2(\mathcal{T}_X^\vee) = e(X) = 12\chi(X) - c_1^2(X) = 0$. There is also a holomorphic 2-form on X which (again by the previous Proposition) vanishes nowhere. From $e(X) = 0$ and $b_1(X) = 3$ we find $b_2(X) = 4$, and from $c_1^2(X) = e(X) = 0$ we find that the index $\tau(X) = 0$, hence $b^+(X) = b^-(X) = 2$.

The real 2-form $\mathrm{i}(\omega \wedge \bar{\omega})$ is exact, for if $c \in H^2(X, \mathbb{R})$ is its de Rham class, then $c^2 = 0$ hence $c = 0$ by Theorem IV.2.14, and there is a real 1-form ρ with $d\rho = \mathrm{i}(\omega \wedge \bar{\omega})$.

Now we come to the first crucial point of the proof, namely the construction of a real closed 1-form σ, which has the following properties:
(i) $(\omega + \bar{\omega})$, $\mathrm{i}(\omega - \bar{\omega})$ and σ form a de Rham basis of $H^1(X, \mathbb{R})$;
(ii) $d\sigma^{1,0} = \omega \wedge \bar{\omega}$.

Since $b_1(X) = 3$, there is a real closed 1-form τ, such that $(\omega + \bar{\omega})$, $\mathrm{i}(\omega - \bar{\omega})$ and τ form a de Rham basis of $H^1(X, \mathbb{R})$ and also of $H^1(X, \mathbb{C})$. From the sequence IV (1)

$$0 \to \Gamma(\Omega_X^1) \to H^1(X, \mathbb{C}) \to H^1(\mathcal{O}_X) \to 0,$$

the fact that $q(X) = h^{0,1}(X) = 2$, and the way the Dolbeault isomorphism is constructed, it follows that $\bar{\omega}$ and $\bar{\tau}^{1,0}$ form a Dolbeault basis for $H^1(\mathcal{O}_X)$. Now $\bar{\partial}\bar{\rho}^{1,0} = 0$, hence $\mathrm{i}\bar{\rho}^{1,0}$ is a $\mathbb{C}$-linear combination of the Dolbeault basis elements modulo $\bar{\partial}$-exact forms:

$$\mathrm{i}\bar{\rho}^{1,0} = \lambda\bar{\omega} + \mu\bar{\tau}^{1,0} + \bar{\partial} f, \quad (\lambda, \mu \in \mathbb{C}).$$

The form

$$-\mathrm{i}\rho + \lambda\bar{\omega} + df = -\mathrm{i}\rho^{1,0} + \mu\tau^{1,0} + \partial f$$

is of type $(1,0)$ and this will be our $\sigma^{1,0}$. Since $d\sigma^{1,0} = -\mathrm{i}d\rho$ it satisfies (ii). As to (i), it is sufficient to show that ω, $\bar{\omega}$ and σ are independent in $H^1(X, \mathbb{C}) = H^1(X, \mathbb{R}) \otimes \mathbb{C}$. If this were not the case, then we would have

$$\sigma^{1,0} = \lambda\omega + \partial g \quad (g \text{ real}),$$

and hence $d\sigma^{1,0} = \bar{\partial}\partial g = \omega \wedge \bar{\omega}$. If g attains a minimum at $x_0 \in X$, we define a holomorphic function F in a neighbourhood of x_0 by requiring $dF = \omega$, with $F(x_0) = 0$. Then we have $\partial\bar{\partial}|F|^2 = dF \wedge \overline{dF} = \omega \wedge \bar{\omega}$ and hence

$$\partial\bar{\partial}\left(|F|^2 + g\right) = 0,$$

i.e. $|F|^2 + g$ is a harmonic function, which has a minimum at x_0. Thus F is constant and $\omega = 0$, a contradiction.

Now let Y be the universal covering space of X. We shall denote a form on X and its pull-back to Y by the same symbol. On Y there is a holomorphic function w, such that $dw = \omega$. Furthermore, we have $d(\sigma^{1,0} + \bar{w}\omega) = 0$, hence there is a second holomorphic function z on Y with $dz = \sigma^{1,0} + \bar{w}\omega$. The holomorphic 2-form $dw \wedge dz = dw \wedge \sigma^{1,0}$ comes from a holomorphic 2-form on X. We claim that it never vanishes. By the remarks made at the very beginning of the present proof it is sufficient to show that it does not vanish identically. So suppose $\omega \wedge \sigma^{1,0} \equiv 0$. Then $\sigma^{1,0} = h\omega$ with h a differentiable function on X, and $\omega \wedge \bar{\omega} = d\sigma^{1,0} = dh \wedge \omega$. It follows that $\bar{\omega} + dh = g\omega$ and $\bar{\omega} = -\bar{\partial}h$, which is impossible since $\bar{\omega} \neq 0$ in Dolbeault cohomology.

Now let $p \in X$ and let γ_1, γ_2 and γ_3 be three closed paths, starting and ending at p, which represent a basis of $H_1(X, \mathbb{Z})$. The vectors

$$\left(\int_{\gamma_i} (\omega + \bar{\omega}), \mathrm{i}\int_{\gamma_i} (\omega - \bar{\omega}), \int_{\gamma_i} \sigma\right), \quad i = 1, 2, 3$$

are independent in $\mathbb{R}^3$, and thus define a 3-torus T. As in the case of the Albanese torus we can define a (differentiable) map $f : X \to T$ by setting

$$f(x) = \left(\int_p^x (\omega + \bar{\omega}), \mathrm{i}\int_p^x (\omega - \bar{\omega}), \int_p^x \sigma\right),$$

where the integral is taken along some path from p to x. The map f has maximal rank everywhere, for

$$\begin{aligned} -\frac{1}{2}(\omega + \bar{\omega}) \wedge (\omega - \bar{\omega}) \wedge \sigma =& \omega \wedge \bar{\omega} \wedge \sigma = \omega \wedge \omega \wedge (\sigma^{1,0} + \bar{\sigma}^{1,0}) \\ =& \omega \wedge \bar{\omega} \wedge \sigma^{1,0} + \omega \wedge \bar{\omega} \wedge \bar{\sigma}^{1,0}. \end{aligned}$$

So (for type reasons) the vanishing of df implies that also $\omega \wedge \bar{\omega} \wedge \sigma^{1,0} = 0$, which cannot happen, since it implies $\omega \wedge \sigma^{1,0} = 0$. Consequently, f maps X onto T. The fibres of f are connected, since by construction $f_*(\pi_1(X)) = \pi_1(T)$. In other words, X is a circle bundle over T. In fact it is even an orientable circle bundle, since X and T are orientable.

As part of the homotopy sequence we obtain a short exact sequence

$$0 \to \pi_1(S^1) \to \pi_1(X) \to \pi_1(T) \to 0.$$

From this it follows that we can replace, if necessary, γ_1, γ_2 and γ_3 by other loops based at p, such that their classes g_1, g_2 and g_3, together with a generator g_0 of $\pi_1(S^1)$ generate $\pi_1(X)$. From here on we assume that this has been done. Since $\pi_1(T)$ is abelian, we must have

$$(3) \qquad g_i g_j g_i^{-1} g_j^{-1} = g_0^{n_{ij}}, \quad n_{ij} \in \mathbb{Z},$$

for $i, j = 1, 2, 3$. We claim that not all n_{ij} vanish. Namely, if they did, then $\pi_1(X)$ would be abelian since $g_i g_0 = g_0 g_i$ in any case, for X is an *oriented* bundle over T. But if this were the case, $b_1(X)$ would be 4. So not all integers n_{ij} vanish (in fact we have three of them, since $n_{ij} = -n_{ji}$).

If we choose an identification of $\pi_1(X)$ with the group of covering transformations of Y over X, then the same relations hold for the corresponding covering transformations.

If we put for $y \in Y$:

$$g_i(w)(y) = w(g_i(y))$$

and similarly for z, then we see from $dz = \omega$ and $dz = \sigma^{1,0} + \bar{w}\omega$ that

$$g_i(w) = w + c_i,$$
$$g_i(z) = z + \bar{c}_i w + \beta_i,$$

with $c_0 = 0$ and $c_i = \int_{\gamma_i} \omega$, $i = 1, 2, 3$. Substituting this into the relation (3) we find $n_{ij}\beta_0 = c_j\bar{c}_i - c_i\bar{c}_j$. Eliminating β_0, a non-zero number, from these relations we find:

$$n_{23}c_1 + n_{31}c_2 + n_{12}c_3 = 0.$$

So c_1, c_2, c_3 generate a lattice L in $\mathbb{C}$. Indeed, two of them must be independent over the reals, since otherwise the three vectors

$$\left(\int_{\gamma_i} (\omega + \bar{\omega}), \mathrm{i}\int_{\gamma_i} (\omega - \bar{\omega}), \int_{\gamma_i} \sigma\right), \quad i = 1, 2, 3$$

cannot be independent in $\mathbb{R}^3$.

The holomorphic map from X onto the elliptic curve $\mathbb{C}/L$ given by ω, shows that there are infinitely many curves on X, contradicting the previous Proposition. So X must be elliptic as claimed. □

Finally, we come to the case c). Since $b^+(X)$ has to vanish, we find from Theorem IV.2.7 that $b_1(X) = 1$. We shall prove that for a surface X of this type $\mathrm{kod}(X) = -\infty$; in other words it is a surface of class VII by definition. Suppose that $\mathrm{kod}(X) = 0$. Then there is an $n_0 \geq 2$ such that $P_{n_0} = 1$ and either $\mathcal{K}_X^{\otimes n_0} \cong \mathcal{O}_X$ or there exists a positive divisor $\sum d_i D_i \in |n_0 K_X|$. In the first case, by the unbranched covering trick, there is an unramified covering X' of X with trivial canonical bundle. Hence X' is minimal and $p_g(X') = 1$. Furthermore $b_1(X') \geq b_1(X) \geq 1$, and $\mathrm{kod}(X') = \mathrm{a}(X') = 0$. Consequently X' is either a surface of class d) or a torus. This last possibility is excluded, since then X itself would be a torus. So X' is a surface of class d) and by Theorem 6.1 we arrive at the contradiction $\mathrm{a}(X') = \mathrm{a}(X) = 1$.

On the other hand, in the second case, as in the proof of the previous Proposition we find $KD_i = D_i^2 = 0$ for all i. Hence D_i is smooth elliptic or rational with a cusp or a node. In any case we have $\mathcal{K}_{D_i} = \mathcal{O}_{D_i}$. Using the adjunction formula and Serre duality on D_i we obtain an exact sequence

$$H^1(\mathcal{O}_X(n(K_X + D_i))) \to H^1(\mathcal{O}_{D_i}) \to H^2(\mathcal{O}_X(n(K_X + D_i) - D_i)).$$

Since $H^2(\mathcal{O}_X(n(K_X + D_i) - D_i)) \cong H^0(\mathcal{O}_X((1-n)K_X + (1-n)D_i)) = 0$ for $n \geq 2$, we conclude that $H^1(\mathcal{O}_X(n(K_X + D_i))) \neq 0$ for $n \geq 2$. By Riemann-Roch it follows that $\dim |n(K_X + D_i)| \geq 0$, for $n \geq 2$. If for all $i, 1 \leq i \leq l$, there is an integer n_i with $n_i(K_X + D_i) = 0$, then we can find an integer $m \neq 0$ with $mK_X = 0$. But this is excluded here since by assumption there is an effective divisor in $|n_0K_X|$. So there must be one effective divisor in $|n(K_X + D_i)|$ for all $n \geq 2$. Taking $n = 2$ and $n = 3$ we see that there is an effective divisor $C \in |K_X + D_i|$. Then $n_0C \in |n_0(K_X + D_i)|$, i.e. n_0C is equal to the divisor $\sum d_jD_j + n_0D_i$. This means that C contains D_i with some strictly positive multiplicity, i.e. $C - D_j \in |K_X|$ is non-negative. Since $p_g = 0$, we have obtained a contradiction. So $\mathrm{kod}(X) = -\infty$.

7. The Final Step

Since $\mathrm{kod}(X), P_n(X)$ and $b_1(X)$ are invariant under blowing up, whereas the classes 1)-10) can be distinguished by these invariants, every surface has indeed a minimal model in exactly one of the classes 1)-10). We know from Proposition III.4.6 that this minimal model is unique (up to isomorphisms) if $\mathrm{kod}(X) \geq 0$. So the only thing left is to show that for surfaces of class VII the minimal model is unique. Otherwise there would exist a surface of class VII with two intersecting exceptional curves; blowing down one of them would yield a curve C on a surface X of class VII, such that $(\mathcal{K}_X, C) \leq -2$. Since $C^2 \leq 0$ by Theorem IV.6.2, the adjunction formula yields that C is a smooth rational curve with $C^2 = 0$. This is impossible by Proposition V.4.3, (i).

Remark. Surfaces in classes 1) and 3) can indeed have several minimal models. A simple example with minimal models in class 1) provides the surface X obtained by blowing up a point p on $\mathbb{P}_1 \times \mathbb{P}_1$. This surface has $\mathbb{P}_1 \times \mathbb{P}_1$ as a minimal model, but also $\mathbb{P}_2$: in X blow down the two disjoint (-1)-curves that are the proper transforms of the "horizontal" and "vertical" fibres through p. The process consisting of blowing up a point x in a ruled surface and then blowing down the proper transform of the ruling through x is known as an elementary transformation.

To complete the picture, we give another example of this type, but now with minimal models in class 3). Let B any elliptic curve and $X = B \times \mathbb{P}_1$. We claim that the surface $\tilde{X}$ obtained by X from blowing up $x_0 \in X$ has two (even topologically) different minimal models. Indeed, in $\tilde{X}$ we can blow down the proper transform of the fibre through x_0, obtaining another ruled surface Y with base B, which is minimal by Lüroth's theorem. It remains to show that X and Y are different. But this is easy: on X all self-intersection numbers are even, whereas on Y there is a section with self-intersection number 1: start from a section on X which does not pass through x_0, take its proper transform and project it onto Y.

8. Deformations

In this section we consider arbitrary analytic deformations of surfaces. The results are summed up in the following theorem.

(8.1) **Theorem.**
(i) *If X is a surface with minimal model in one of the classes* 1) *up to* 10), *then every deformation of X has a minimal model in that same class.*
(ii) *If X is minimal and contained in one of the classes* 3) *to* 10), *then every deformation of X is again minimal.*
(iii) *Every deformation of $\mathbb{P}_2$ is again $\mathbb{P}_2$.*
(iv) *A surface Y is a deformation of Σ_n, if and only if Y is isomorphic to Σ_m with $n \equiv m \,(\mathrm{mod} 2)$.*

Taking into account that every minimal rational surface is either $\mathbb{P}_2$ or Σ_n, $n \neq 1$ (Proposition 3.3), and that in class VII it really happens that a minimal surface *can* be deformed into a non-minimal one ([Ka78b], p. 61), we see that the above result is satisfactory from the point of view of the Enriques-Kodaira classification.

N.B. We do not prove Iitaka's result that all plurigenera are invariant under deformations. In fact this result follows easily from our considerations, here and in Chap. VII, except for one case: properly elliptic surfaces. We remark however that another proof can be given using Seiberg-Witten invariants (see the remarks at the end of the last chapter).

In order to prove Theorem 8.1 we need the following facts.

(I) If X is a minimal surface of general type, then $c_1^2(X) > 0$ (compare Table 10).
(II) If X is a surface of general type, then $h^1(\mathcal{K}_X^{-1}) = 0$ if and only if X is minimal.
(III) If X is a minimal properly elliptic surface X with $p_g(X) = 0, q(X) = 1, b_1(X) = 1$, then every small deformation of X is again properly elliptic.

We shall prove (I) and (II) in a later chapter (Theorem VII.2.2 and Proposition VII.5.3), but (III) will not be proved in this book, and we have to refer to [Ko66], p. 685–694. Kodaira proves this result by first showing that all these surfaces can be obtained by applying suitable logarithmic transformations (V, Sect. 17) to a product $E \times \mathbb{P}_1$, where E is an elliptic curve. By varying both this curve and all but three points on $\mathbb{P}_1$ in which a logarithmic transformation is performed, he obtains an analytic family $\mathcal{F}$ of properly elliptic surfaces. Then Kodaira calculates $h^1(\mathcal{T}_X)$ and finds that it equals the dimension of the parameter space of $\mathcal{F}$. The proof is completed by showing that $\mathcal{F}$ is "effectively parametrized", i.e. that $\mathcal{F}$ induces the whole of $H^1(\mathcal{T}_X)$.

We shall also use the amusing

(8.2) **Example.** *Let X be a ruled surface of genus $g \geq 1$. Then every surface which is homeomorphic to X is also ruled of genus g.*

Proof. Since $\tau(X) = 0$, the integers $b^+(X), b^-(X), p_g(X)$ and c_1^2 are topological invariants of the *non-oriented* underlying manifold of X (compare Theorem IV.2.7). So if $g \geq 2$, the result follows from Table 10, together with the fact that blowing up a point diminishes c_1^2 by 1.

Now let $g = 1$, and Y a surface, homeomorphic to X. Since $p_g(Y) = p_g(X) = 0$, and $b^+(Y) = b^+(X) = 1$, the surface Y is projective by Theorem IV.6.2. If we apply Stein factorisation to the Albanese map of Y, then we obtain a connected holomorphic map $f : Y \to E$ onto a smooth elliptic curve E (E cannot be rational because of Lüroth and $g(E) \geq 2$ is impossible since otherwise there would be too many holomorphic 1-forms on Y); of course $E = \text{Alb}(Y)$. From Remark III.11.5 we conclude that either f is everywhere of maximal rank, or the general fibre of f is elliptic and f has no other singular fibres but multiple elliptic curves. In the first of these cases the general fibre of f must be rational, otherwise the universal covering space of Y would be contractible, whereas the universal cover of X has the homotopy type of S^2. Using again $e(Y) = e(X) = 0$ we conclude that Y is a ruled surface over an elliptic curve.

The second case does not occur, for in this case there exists by stable reduction and Theorem III.15.4 an unramified covering of Y which is a locally trivial fibre bundle over some curve of genus $g \geq 1$, so the universal covering of Y is again contractible. □

In the coming proof we frequently shall use the upper semi-continuity of $h^i(\mathcal{K}_{X_t}^{\otimes n})$ in a family X_t (Theorem I.8.5, (ii)) and the fact that blowing-up changes neither b_1 nor the P_i (Theorem I.9.1, (iv) and (viii)).

Proof of Theorem 8.1

Step A) Any deformation of a surface in class 4)-8) stays in that class.

Since minimal surfaces X with $\text{kod}(X) = 0$ are characterised among all surfaces by the fact that there is some $n \in \mathbb{Z}$ with $\mathcal{K}_X^{\otimes n}$ trivial, it follows from Proposition IV.4.4 that any deformation of such a surface has again Kodaira dimension 0. Now the classes 4)-8) can still be distinguished by topological invariants, so every deformation must preserve the class from which one starts.

Step B) Any deformation of a surface X with minimal model in class 4)-8) has a minimal model in the same class.

Since blow-ups of surfaces in different classes 4) up to 8) can still be distinguished by topological invariants, it is enough to prove that every deformation of X has again Kodaira dimension 0. Thus we have to establish two things: firstly that any small deformation X' of X has again $\mathrm{kod}(X') = 0$, and secondly that if $X_t, 0 \leq t < 1$ is the part of a complex analytic family over the real arc $[0,1)$ with X_t of class VII for $t \neq 0$, then also X_0 is of class VII.

The first point can be reduced to step A) since by Proposition IV.4.1 we may assume X to be minimal.

In the second case we may assume X_0 to be minimal for the same reason. If any of the surfaces X_t, $t \neq 0$, is minimal, we are done. If not, then $c_1^2(X_0) = c_1^2(X_t)$ must be strictly negative. But Table 10 shows that this cannot occur for a minimal surface of Kodaira dimension ≥ 0.

Step C) Any deformation of a minimal surface of general type is a minimal surface of general type.

Let X_t be a small deformation, with X_0 minimal, and of general type. Since $h^0(\mathcal{K}_{X_0}^{\otimes n}) \geq 2$ for n large enough, we have $h^0(\mathcal{K}_{X_0}^{\vee}) = 0$ and there exists by upper semi-continuity a small disc E around 0 such that $h^0(\mathcal{K}_{X_t}^{\vee}) = 0$ for $t \in E$. Now $c_1^2(X_t) = c_1^2(X_0) > 0$ by (I), so X_t is either rational or of general type for all $t \in E$. But if X_t is rational, then by Riemann-Roch $h^0(\mathcal{K}_{X_t}^{\vee}) \neq 0$. So X_t is of general type for all $t \in E$. And if $|t|$ is small enough, then X_t is minimal by (II) and upper semi-continuity.

As in Step B), let $X_t, 0 \leq t < 1$ be the part of a complex analytic family over the real arc $[0,1)$. Assume that X_t is minimal and of general type for $t \neq 0$. Then by Riemann-Roch there exist constants α, β, with $\alpha \geq 1$ by (I), such that for $n \geq 2$ $P_n(X_t) \geq \alpha \frac{n(n-1)}{2} + \beta$. Hence $P_n(X_0) \geq \alpha \frac{n(n-1)}{2} + \beta$ for $n \geq 2$, and X_0 is of general type. And it is minimal because of Proposition IV.4.1.

Step D) Any deformation of a surface of general type is again of general type.

The case of a small deformation follows from Step C) and Proposition IV.4.1, whereas the "limit case" can be proved as in Step C).

Step E) Any deformation of a surface of class VII is again of class VII.

Let X_t be a small deformation with X_0 of class VII. By upper semi-continuity there exists a small disc E around 0 such that $P_{12}(X_t) = 0$ for $t \in E$. Table 10 and Step B) show that any surface $X_t, t \in E$, is either properly elliptic with $\mathrm{a}(X_t) = 1$ or of class VII. But the first possibility is excluded because of Theorem V.18.5.

For the case of a family X_t, $0 \leq t < 1$, with X_t in class VII for $t \neq 0$, Table 10 shows again that X_0 is either in class VII or properly elliptic. But this last case is excluded by (III).

Step F) Any deformation of a ruled surface of genus $g \geq 1$ is again a ruled surface of genus g.

This follows from Example 8.2.

Step G) Any deformation of a blown-up ruled surface of genus g is again a blown-up ruled surface of genus g.

The case of a small deformation follows from Step F) and Proposition IV.4.1.

As to a family $X_t, 0 \leq t < 1$, with X_t a blown-up ruled surface of genus g for $t \neq 0$, we may again assume that X_0 is minimal. So $c_1^2(X_0) = c_1^2(X_t) \leq 0$, with all X_t minimal as well, and $g = 1$ if $c_1^2(X_0) = 0$. So if $c_1^2(X_0) < 0$ we are done by Table 10 and if $c_1^2(X_0) = 0$ we are back in the preceding case.

Step H) Every deformation of a minimal properly elliptic surface is again minimal and properly elliptic.

Combining Table 10 with the preceding steps we see that, as far as small deformations X_t are concerned we only have to show that for sufficiently small $|t|$ none of the surfaces X_t is rational. But $c_1^2(X_0) = 0$ and $h^0(\mathcal{K}_{X_0}^\vee) = 0$, whereas for a rational surface Y with $c_1^2(Y) = 0$ Riemann -Roch yields $h^0(\mathcal{K}_Y)^\vee \neq 0$. So our claim is again a consequence of upper semi-continuity.

As to the case of a limit X_0 of minimal properly elliptic surfaces X_t, $0 < t < 1$, the only case left would be that X_0 is rational. This is impossible by Castelnuovo's criterion and upper semi-continuity.

Step I) Every deformation of a properly elliptic surface is again properly elliptic.

This follows for a small deformation from Step H), combined with Proposition IV.4.1, whereas the "limit case" is again a consequence of the preceding steps and Castelnuovo's criterion.

Step J) Any deformation of a rational surface is rational.

This follows from Table 10 and Steps A)-I).

Step K) Every deformation of $\mathbb{P}_2$ is again $\mathbb{P}_2$.

This has been proven earlier (Example V.1.3).

Step L) Any deformation of a Hirzebruch surface Σ_n is isomorphic to a Σ_m with $m \equiv n \pmod 2$.

From Step J) and the fact that, among rational surfaces, Hirzebruch surfaces are characterised by $b_2 = 2$, we conclude that any deformation of a Σ_n is another Hirzebruch surface. The parity of the intersection pairing on Σ_n is determined by the parity of n and so the parity of n does not change under deformation.

Step M) It remains to be shown that two Hirzebruch surfaces Σ_n and Σ_m with $n \equiv m \,(\mathrm{mod}\, 2)$ can indeed be deformed into each other. This can be done in the following way. Let $(x_0 : x_1)$ and $(y_0 : y_1 : y_2)$ be homogeneous coordinates on $\mathbb{P}_1$ and $\mathbb{P}_2$ respectively. Then, almost by definition, the surface, given in $\mathbb{P}_1 \times \mathbb{P}_2$ by $x_0^n y_1 - x_1^n y_2$ is isomorphic to Σ_n. Now consider in $\mathbb{P}_1 \times \mathbb{P}_2 \times \mathbb{C}$

the hypersurface $x_0^n y_1 - x_1^n y_2 + t x_0^{n-m} x_1^m y_0 = 0$ as a holomorphic family of Hirzebruch surfaces over $\mathbb{C}$. The fibre over 0 is Σ_n, whereas for $t \neq 0$ the surface X_t can be shown to be isomorphic to Σ_{n-2m} (see [M-K], p. 26). □

As mentioned before, Seiberg-Witten theory implies that for surfaces the plurigenera are differentiable invariants so a fortiori invariants under deformation. In [Wi78] Wilson considers the behaviour of the plurigenera when one smooths a possibly singular algebraic surface. Kollár and Mori showed invariance of the plurigenera for threefolds ([K-M]) while Levine ([Lev]) treated several special cases in all dimensions. The most spectacular result is due to Siu ([Siu98]) who shows that the plurigenera remain constant for manifolds of general type of any dimension. An algebro-geometric version of Siu's proof was given by Kawamata ([Kaw99]]).

Chapter VII. Surfaces of General Type

In this chapter the use of the words "surface" and "curve" is the same as in Chapter VI .

Preliminaries

In the first section we state the restrictions on (c_1^2, c_2) for minimal surfaces of general type. In the section following it, the "easy" inequalities $c_1^2 > 0$ and $c_2 > 0$ are proven and we show that there is only a finite number of (-2)-curves.

1. Introduction

The minimal surfaces of general type can be parametrised in a satisfactory way, namely by a countable number of quasi-projective families. More precisely we have Gieseker's theorem ([Gi], p.236):

There exists a quasi-projective coarse moduli scheme for the minimal surfaces of general type X with fixed Chern numbers $c_1^2(X)$ and $c_2(X)$.

Remark. By Theorem III.4.5 and Proposition III.4.6 a surface of general type has a unique minimal model. For that reason classifying surfaces of general type essentially means classifying minimal surfaces of general type.

The proof of Gieseker's theorem is based on the fact that for $n \geq 5$ an n-canonical map f_n (IV, Sect. 1) is a birational morphism from X onto a normal 2-dimensional subvariety of degree $n^2c_1^2(X)$, with only rational double points, of a fixed projective space $\mathbb{P}_N$ (where N only depends on $c_1^2(X)$ and $c_2(X)$). The variety $f_n(X)$ determines X up to isomorphisms, but, given X, the map f_n depends on the choice of a base of $\Gamma(X, \mathcal{K}_X^n)$. For fixed $c_1^2(X)$ and $c_2(X)$, let $\mathcal{S}$ be the set of all varieties in $\mathbb{P}_N$ thus obtained (for all surfaces X and all choices of a basis). Then $\mathrm{PGL}(N, \mathbb{C})$ operates on $\mathcal{S}$ in a natural way, and the isomorphism classes of surfaces X are in one-to-one correspondence with the points of $\mathcal{S}/\mathrm{PGL}(N, \mathbb{C})$. Now Gieseker shows, using a criterion of Hilbert-Mumford, that for n large enough, this quotient is quasi-projective and indeed a coarse moduli space for the surfaces X.

In this chapter we shall prove the result on the n-canonical maps, but Gieseker's work will not be dealt with.

In general, little is known about the structure of the Gieseker scheme. Even the very first question: for which pairs (c_1^2, c_2) is the Gieseker scheme not empty, has not been completely answered. The known restrictions on c_1^2 and c_2 are expressed by

(1.1) **Theorem.** *If X is any minimal surface of general type, then*

(i) $c_1^2(X) + c_2(X) \equiv 0 \bmod 12$,

(ii) $c_1^2(X) > 0$ *and* $c_2(X) > 0$,

(iii) $c_1^2(X) \leq 3c_2(X)$,

(iv) $5c_1^2(X) - c_2(X) + 36 \geq 0$ $(c_1^2(X)$ even)
$5c_1^2(X) - c_2(X) + 30 \geq 0$ $(c_1^2(X)$ odd).

Property (i) is of course an immediate consequence of Noether's formula, and actually holds for any almost-complex surface (compare IV, Sect. 9). The inequalities (ii)-(iv) will be proved below. The inequalities (iv) might appear strange, but they become more natural if you look at them from another side, namely as the Noether inequalities (Theorem 3.1). They are more conveniently stated by using c_1^2 and the Todd genus $\chi = \frac{1}{12}(c_1^2 + c_2)$ instead of c_1^2 and c_2:

$$\text{(iv}')\qquad \begin{cases} \chi(X) \leq \frac{1}{2}c_1^2(X) + 3 & (c_1^2(X) \text{ even}) \\ \chi(X) \leq \frac{1}{2}c_1^2(X) + \frac{5}{2} & (c_1^2(X) \text{ odd}). \end{cases}$$

Many people prefer the use of c_1^2 and χ instead of c_1^2 and c_2. We shall mainly stick to c_1^2 and c_2, leaving the obvious translation to the reader.

Below (Sect. 8) we shall see that "most" pairs of integers, satisfying the conditions of Theorem 1.1, can be realized as the Chern numbers of at least one minimal surface of general type. In particular, most pairs on the boundary of the domain in the plane determined by the theorem, can be realized. However, it is still not known whether all pairs, satisfying the conditions of the theorem, can be represented.

Apart from the existence of at least one surface for most pairs (c_1^2, c_2), satisfying the conditions of Theorem 1.1, some information is available concerning the Gieseker scheme for certain special values of (c_1^2, c_2). For the case that the Noether inequalities are equalities, Horikawa has obtained detailed results. (One or two simple cases were known before.) And, due to the efforts of many, a good deal is known about the Gieseker scheme for some (c_1^2, c_2) with low c_1^2, in particular $c_1^2 = 1$.

The existence of the Hilbert scheme implies that by desingularization of 2-dimensional irreducible varieties of given degree in a fixed $\mathbb{P}_N$ you can obtain only finitely many diffeomorphism types.

So the fact that every minimal surface X of general type can be obtained by desingularizing a 2-dimensional variety (with rational singularities) of degree d in $\mathbb{P}_N$ with d only depending on c_1^2 and c_2 (Theorem 5.1) means in particular that for given c_1^2 and c_2 there is only a *finite number* of diffeomorphism types for such surfaces (because of Theorem 1.1, (ii) and (iii), and the fact that blowing down a (-1)-curve increases c_1^2 by 1 and decreases c_2 by 1, this statement remains true if we consider *all* surfaces of general type, minimal or not). The topological index theorem expresses their index in c_1^2 and c_2, and the Noether inequalities yield an upper bound for $b_1 = 2q$ in terms of c_1^2 and c_2, but apart from such rather obvious remarks not much is known about the diffeomorphism types that actually occur. We refer to Sect. 8 for a more detailed discussion concerning the global properties of the Gieseker scheme.

The further contents of the chapter is as follows. In Sect. 2 we prove a certain number of basic facts, including Theorem 1.1,(ii). Sections 3 and 4 are respectively devoted to the proof of Theorem 1.1,(iv) and (iii). In Sect. 5-7 we study pluricanonical maps and in Sects. 8–10 we briefly describe what is known about surfaces with given c_1^2 and c_2.

2. Some General Theorems

(2.1) **Proposition.** *If there exists on an algebraic surface X an algebraic system of effective divisors, of dimension at least* 1, *such that the general member is a (possibly singular) rational or elliptic curve, then* $\mathrm{kod}(X) \leq 1$.

Proof. Without loss of generality we may assume the system to be irreducible and of dimension 1. Then the assumption implies that there exists a smooth compact curve C with the following property. The product $X \times C$ contains an irreducible 2-dimensional subvariety, projecting onto X, such that its desingularization Y is either a blown-up ruled or an elliptic surface (V, Sect. 4 and Sect. 7). Hence $\mathrm{kod}(Y) \leq 1$ by V, Sect.4 and Proposition V.12.4, and Theorem I.7.4 yields that also $\mathrm{kod}(X) \leq 1$. □

(2.2) **Theorem.** *If X is a minimal surface of general type, then* $K_X^2 = c_1^2(X) > 0$.

Proof. Let H be a smooth hyperplane section of X. We consider the exact sequence

$$0 \to \mathcal{O}_X(nK_X - H) \to \mathcal{O}_X(nK_X) \to \mathcal{O}_H(nK_X) \to 0,$$

and the corresponding cohomology sequence. By Theorem I.7.2. there is a $c > 0$, such that $h^0(\mathcal{O}_X(nK_X)) > cn^2$ for n large, whereas $h^0(\mathcal{O}_H(nK_X))$ grows linearly with n. So there is an $n_0 > 0$, such that $n_0K_X - H$ can be represented by an effective divisor R. Since $K_XR \geq 0$ by Proposition III.2.3, we find:

$$n_0^2K_X^2 = (n_0K_X)(H + R) \geq (n_0K_X)H = H^2 + HR \geq H^2 > 0. \qquad \square$$

(2.3) **Corollary.** *Let X be a minimal surface of general type and C an irreducible curve on X. Then $K_X C \geq 0$ and $K_X C = 0$ if and only if C is a (-2)-curve.*

Proof. The first statement is already included in Proposition III.2.3.

If C is a (-2)-curve, then $KC = 0$ by the adjunction formula. Conversely, if $KC = 0$, then by the Algebraic index theorem IV.2.16 and Theorem 2.2 above we must have $C^2 < 0$, since an effective divisor is never rationally homologous to 0 on an algebraic surface. The adjunction formula now implies that C is a (-2)-curve. □

(2.4) **Proposition.** *If X is any surface of general type, then $c_2(X) > 0$.*

Proof. Since the Euler number goes down when we blow down, we may assume that X is minimal. We distinguish between two cases:

a) if X admits a connected holomorphic map onto a curve of genus ≥ 2, then by Proposition 2.1 the general fibre must have genus ≥ 2, hence $c_2(X) = e(X) \geq 4$ by Proposition III.11.4;
b) if X does not admit such a map, then by Proposition IV.5.2 and Corollary IV.5.4 we find that $e(X) > 0$, unless $q(X) = 1$, $p_g(X) = 0$ or $q(X) = 2$, $p_g(X) = 1$. In the first case, since $q(X) = 1$, there exists a morphism from X onto an elliptic curve, and hence $h^{1,1}(X) \geq 2$. If $h^{1,1}(X) \geq 3$, then $c_2(X) > 0$ and we are ready. If however $c_2(X) = 0$, the Todd-Hirzebruch formula would give $c_1^2(X) = 0$, thus contradicting Theorem 2.2. In the second case there either exists a map from X onto a 2-torus T, and we would have $c_2(X) = 2-8+2+h^{1,1}(X) \geq -4+h^{1,1}(T) = 0$ or a map from X onto a curve of genus at least 2, which is excluded here. So here we are ready unless $c_2(X) = 0$. But as before this leads to the contradiction $c_1^2(X) = 0$. □

(2.5) **Proposition.** *Let X be a minimal surface of general type. Then we have*

(i) *The number of (-2)-curves on X is finite. They are independent over $\mathbb{Q}$ and their number is at most equal to $\rho(X) - 1$, where $\rho(X)$ is the Picard number of X.*
(ii) *On the subspace of $H_2(X, \mathbb{Q})$, spanned by the (-2)-curves, the intersection form is negative definite.*

Proof. To prove the first part of (i) it is sufficient to show the following: if $C_1, \ldots, C_l$ are (-2)-curves on X, such that

$$\sum_{i=1}^{k} \lambda_i C_i = \sum_{j=k+1}^{l} \lambda_j C_j \quad \text{in } H_2(X, \mathbb{Q})$$

for some k, $1 \leq k \leq l$, and with all $\lambda_i, \lambda_j \geq 0$, then $\lambda_i = 0$ for $i = 1, \ldots, l$.

We have

$$\left(\sum_{i=1}^{k} \lambda_i C_i\right)^2 = \left(\sum_{i=1}^{k} \lambda_i C_i\right)\left(\sum_{j=k+1}^{l} \lambda_j C_j\right) \geq 0.$$

If we combine this inequality with $K_X^2 > 0$ (Theorem 2.2), then the index theorem yields $\sum_{i=1}^{k} \lambda_i C_i = 0$ in rational homology, i.e. $\lambda_1 = \ldots = \lambda_l = 0$.

Since each rational homology class contains at most one (-2)-curve, the number of these curves is at most $\rho(X)$. But $K_X H > 0$ for a hyperplane section H, hence H is not homologous to a sum $\sum \lambda_i C_i$, and the number of (-2)-curves is at most $\rho(X) - 1$.

Part (ii) is an immediate consequence of Theorem 2.2 and the index theorem. □

Two Inequalities

In this subchapter the other two inequalities on the Chern numbers are proved: the Noether inequality and the inequality $c_1^2 \leq 3c_2$.

3. Noether's Inequality

(3.1) **Theorem.** (Noether's inequality). *Let X be a minimal surface of general type. Then*

$$p_g(X) \leq \frac{1}{2}c_1^2(X) + 2.$$

Proof. Since $c_1^2(X) \geq 1$ (Theorem 2.2) we may assume that $p_g(X) \geq 3$. Let $|K_X| = |C| + V$, where V is the fixed part of $|K_X|$. The linear system $|C|$ is either composed with a pencil or its general member is irreducible (IV, Sect. 1).

In the first case there is a 1-dimensional algebraic system of effective divisors on X, the general member F of which is irreducible, such that K_X is homologous to $nF+V$ and $p_g = \dim |C|+1 \leq n+1$. We claim that $K_X^2 \geq 2n$. Since $K_X^2 = (nF+V)K_X \geq nFK_X$ by Corollary 2.3, this will be proved as soon as we have shown that $K_X F \geq 2$. If $K_X F \leq 1$ we find from $K_X F = (nF+V)F \leq 1$ that $F^2 \leq 1$. Since also $F^2 \geq 0$, and $K_X F \equiv F^2 \bmod 2$, we only have to consider the cases $K_X F = F^2 = 1$ and $K_X F = F^2 = 0$. The first of these cases leads to $n = 1$ and $p_g(X) \leq 2$, whereas we have assumed $p_g(X) \geq 3$. The second case is excluded by Corollary 2.3.

If $|C|$ is not composed with a pencil, then $C^2 > 0$ (IV, Sect. 1). We take an irreducible member C of $|C|$ and consider the standard exact sequence

$$0 \to \mathcal{O}_X(K_X) \to \mathcal{O}_X(K_X + C) \to \mathcal{O}_C(K_X + C) \to 0.$$

Since $H^1(\mathcal{O}_X(K_X + C)) \cong H^1(\mathcal{O}_X(-C)) = 0$ by Corollary IV.12.6 and also $H^2(\mathcal{O}_X(K_X + C)) \cong H^0(\mathcal{O}_X(-C)) = 0$, we find from the associated exact cohomology sequence and the Riemann-Roch theorem that

$$h^0(\mathcal{O}_C(K_X + C)) = h^0(\mathcal{O}_X(K_X + C)) - p_g(X) + q(X) = \frac{1}{2}(C^2 + K_X C) + 1.$$

Now if $\mathcal{L}$ and $\mathcal{M}$ are two line bundles on an irreducible variety with

$$h^0(\mathcal{L}) \geq 1, \quad h^0(\mathcal{M}) \geq 1, \quad \text{then} \quad h^0(\mathcal{L} \otimes \mathcal{M}) \geq h^0(\mathcal{L}) + h^0(\mathcal{M}) - 1.$$

So we have:

$$\begin{aligned} h^0(\mathcal{O}_C(K_X + C)) &= h^0(\mathcal{O}_C(2C + V)) \geq h^0(\mathcal{O}_C(2C)) \geq 2h^0(\mathcal{O}_C(C)) - 1 \\ &\geq 2h^0(\mathcal{O}_X(C)) - 3 = 2h^0(\mathcal{O}_X(C + V)) - 3 = 2p_g(X) - 3. \end{aligned}$$

But $2K_X^2 - (K_X C + C^2) = 2K_X V + CV \geq 0$, hence $p_g(X) \leq \frac{1}{2}K_X^2 + 2$. □

Remark. In case $|C|$ is not composed with a pencil, the inequality can be sharpened a little bit by refining the argument in the following way. Firstly, letting r_C denote the map induced by restriction to the curve C, we can replace the lines following "So we have" by

$$\begin{aligned} \dim r_C(H^0(\mathcal{O}_X(K_X + C))) &= \dim r_C(H^0(\mathcal{O}_X(2C + V))) \\ &\geq \dim r_C(H^0(\mathcal{O}_X(2C))) \geq \\ &\geq 2 \dim r_C(H^0(\mathcal{O}_X(C))) - 1 \\ &= 2p_g(X) - 3. \end{aligned}$$

Since $\dim H^0(\mathcal{O}_C(K_X + C)) = \dim r_C(H^0(\mathcal{O}_X(K_X + C))) + q(X)$, we thus obtain

$$p_g(X) \leq \frac{1}{2}K_X^2 + 2 - \frac{1}{2}q(X) - \frac{1}{2}K_X V - \frac{1}{4}CV.$$

This generalisation can be quite useful in special situations.

(3.2) **Corollary.** *Let X be a minimal surface of general type. Then*

$$\begin{aligned} 5c_1^2(X) - c_2(X) + 36 \geq 0 \quad & (c_1^2(X) \text{ even}) \\ 5c_1^2(X) - c_2(X) + 30 \geq 0 \quad & (c_1^2(X) \text{ odd}). \end{aligned}$$

Proof. Let $c_1^2(X)$ be even. Then by Noether's formula we have $\chi(X) \leq p_g(X) + 1 \leq \frac{1}{2}c_1^2(X) + 3$, i.e. $5c_1^2(X) - c_2(X) + 36 \geq 0$.

If $c_1^2(X)$ is odd, then we automatically have $p_g(X) \leq \frac{1}{2}c_1^2(X) + \frac{3}{2}$, and we can proceed as before. □

(3.3) **Corollary.** *If X is a minimal surface of general type with*

$$5c_1^2(X) - c_2(X) + 36 = 0 \quad (c_1^2(X) \text{ even}) \quad \text{or}$$
$$5c_1^2(X) - c_2(X) + 30 = 0 \quad (c_1^2(X) \text{ odd}),$$

then $q(X) = 0$.

Remark. In the case that $|K_X|$ is not composed with a pencil there is an equivalent way of phrasing the proof of Theorem 3.1, based on "counting quadrics". This has been pointed out to us by M. Reid. In rough outline he argues as follows.

1) If $S \subset P = \mathbb{P}_n$ is a 2-dimensional irreducible variety spanning P, H a hyperplane of $P, C = S \cap H$, the restriction map

$$Q_S = \Gamma(\mathcal{J}_{S|P}(2H)) \to Q_C = \Gamma(\mathcal{J}_{C|H}(2H))$$

is an isomorphism for generic H (there pass as many quadrics through S as there pass through a generic hyperplane section).

2) Since $\mathbb{P}(Q_C) \subset \mathbb{P}(\Gamma(\mathcal{O}_H(2H)))$ does not meet the $2(n-1)$-dimensional set of reducible quadrics in H, its codimension is $\geq 2(n-1)$.

3) Combining 1) and 2) we find that the codimension of $\mathbb{P}(Q_S)$ in the complete system of quadrics $\mathbb{P}(\Gamma(\mathcal{O}_P(2H)))$ is $\geq 3n-1$.

4) Applying 3) to $S = f_1(X) \subset \mathbb{P}_{p_g-1}$ one finds that

$$\operatorname{rank}(\Gamma(\mathcal{O}_P(2H)) \to \Gamma(\mathcal{O}_X(2K_X))) \geq 3(p_g - 1)$$

and in particular, since $P_2 \geq 1 - q + p_g + K_X^2$ by Riemann-Roch, we find that $K_X^2 \geq 2p_g - 4$.

This proof has some advantages if one wants to obtain more detailed information. For instance, it can be refined to give a proof of Castelnuovo's second inequality (Theorem 7.9 below).

4. The Inequality $c_1^2 \leq 3c_2$

(4.1) **Theorem.** *For every surface of general type X the inequality $c_1^2(X) \leq 3c_2(X)$ holds.*

The proof of this theorem will be preceded by a number of auxiliary results.

(4.2) **Proposition.** *If on the algebraic surface X there is a line bundle $\mathcal{L}$ with $h^0(\mathcal{L}^\vee \otimes \Omega_X^1) \neq 0$, then there is a constant c such that $h^0(\mathcal{L}^k) \leq ck$ for all $k \geq 1$.*

Proof. Since otherwise the result is trivial, we assume that $h^0(\mathcal{L}^{k_0}) \geq 2$ for some $k_0 \geq 1$.

We start with the case $k_0 = 1$. Let $s_1, s_2 \in \Gamma(\mathcal{L})$ be linearly independent, and let h be a homomorphism from $\mathcal{L}$ into $\Omega^1_X, h \neq 0$. Then $h(s_1)$ and $h(s_2)$ are linearly independent 1-forms on X with $h(s_1) \wedge h(s_2) \equiv 0$. So we are in a position to apply Proposition IV.5.1. Consequently, there exists a holomorphic, connected map $f : X \to Y$ from X onto a smooth curve Y, such that both $h(s_1)$ and $h(s_2)$ are the pull-back of 1-forms on Y. It follows that if s_1 vanishes on a curve, then this curve is contained in the sum of some fibres of f. Hence $\mathcal{L} = \mathcal{O}_X(D)$, where every component of the non-negative divisor D is contained in some fibre of f. Since $(D - nF)A < 0$ for n sufficiently large and A ample, there are no non-negative divisors on X, which are homologous to $k(D - nF)$, where F is a fibre, n sufficiently large and k any natural number. Let the divisor F_k consist of ck general (hence smooth) fibres of f.

From the standard exact sequence

$$0 \to \mathcal{O}_X(kD - F_k) \to \mathcal{O}_X(kD) \to \mathcal{O}_{F_k}(kD) \to 0$$

we find $h^0(\mathcal{L}^k) \leq h^0(\mathcal{O}_{F_k}(kD)) \leq ck$ for all $k \geq 1$.

As to the general case, by Theorem I.8.3 and Theorem IV.6.5 there exists an algebraic surface Y and a holomorphic surjective map $g : Y \to X$, such that $g^*(\mathcal{L})$ has two independent sections. Since $h^0(\mathcal{H}om(\mathcal{L}, \Omega^1_X)) \neq 0$ implies $h^0(\mathcal{H}om(g^*(\mathcal{L}), \Omega^1_Y)) \neq 0$, we can apply to Y and $g^*(\mathcal{L})$ the result for $k_0 = 1$. In other words there exists a constant c, such that for all $k \geq 1$ the inequality $h^0((g^*(\mathcal{L}))^k) \leq ck$ holds. But $h^0(\mathcal{L}^k) \leq h^0((g^*(\mathcal{L}))^k)$, and the proposition has been proved. □

(4.3) **Proposition.** *Let X be an algebraic surface, $\mathcal{O}_X(D)$ a line bundle on X, and $\mathcal{F}$ a locally free, rank-two subsheaf of Ω^1_X, such that*
(i) $c_1(\mathcal{F})S \geq 0$ *for every effective divisor S on X,*
(ii) $h^0(\mathcal{H}om(\mathcal{O}_X(D), \mathcal{F})) \neq 0$.
Then $c_1(\mathcal{F})D \leq \max(c_2(\mathcal{F}), 0)$.

Proof. Since $\Gamma(\mathcal{H}om(\mathcal{O}_X(D), \mathcal{F})) \cong \Gamma(\mathcal{F} \otimes \mathcal{O}_X(-D)) \neq 0$, there is a non-negative divisor S on X, such that $\mathcal{F} \otimes \mathcal{O}_X(-D - S)$ admits a section with at most isolated zeros.

Hence

$$c_2(\mathcal{F} \otimes \mathcal{O}_X(-D - S)) = (D + S)^2 - c_1(\mathcal{F})(D + S) + c_2(\mathcal{F}) \geq 0$$
$$c_1(\mathcal{F})D \leq (D + S)^2 - c_1(\mathcal{F})S + c_2(\mathcal{F}).$$

Since by assumption $c_1(\mathcal{F})S \geq 0$, the proposition is already clear if $(D+S)^2 \leq 0$. On the other hand, if $(D+S)^2 > 0$, then application of the Riemann-Roch theorem, together with Serre duality gives

$$h^0(\mathcal{O}_X(n(D + S))) + h^0(\mathcal{O}_X(K_X - n(D + S))) > dn^2$$

for some constant $d > 0$ and n large enough. So we have either

$$h^0(\mathcal{O}_X(n(D+S))) > \frac{1}{2}dn^2$$

or

$$h^0(\mathcal{O}_X(K_X - n(D+S))) > \frac{1}{2}dn^2$$

for an infinite number of positive values of n. The first of these possibilities is excluded by Proposition 4.2 (for $h^0(\mathcal{H}om(\mathcal{O}_X(D+S),\Omega_X^1)) \neq 0$ since $h^0(\mathcal{H}om(\mathcal{O}_X(D+S),\mathcal{F})) \neq 0$). In the second case we have $c_1(\mathcal{F})(D+S) \leq \frac{1}{n}(c_1(\mathcal{F})K_X)$ for an infinite number of n's, i.e. $c_1(\mathcal{F})D \leq -c_1(\mathcal{F})S \leq 0$. □

(4.4) **Proposition.** *Let X be an algebraic surface, $\mathcal{O}_X(D)$ a line bundle on X and $\mathcal{F}$ a locally free rank-two subsheaf of Ω_X^1, such that*
(i) *$c_1(\mathcal{F})S \geq 0$ for every effective divisor S on X,*
(ii) *$h^0(\mathcal{H}om(\mathcal{O}_X(D), S^n\mathcal{F})) \neq 0$.*
Then $c_1(\mathcal{F})D \leq \max(nc_2(\mathcal{F}),0)$.

Proof. Let $Z = \mathbb{P}(\mathcal{F}) = \mathbb{P}(\mathcal{F}^\vee)$, and let $p : Z \to X$ be the projection. Then Theorem I.5.1 implies the existence of a divisor class $H = H_{\mathcal{F}}$ on Z, such that for every divisor E on X there is a canonical isomorphism between $\Gamma(\mathcal{O}_Z(nH + p^*(E)))$ and $\Gamma(\mathcal{O}_X(S^n\mathcal{F} \otimes \mathcal{O}_X(E)))$. So in our case there is an effective divisor G on Z with $\mathcal{O}_Z(G) = \mathcal{O}_Z(nH - p^*(D))$. By the branched covering trick (Theorem I.18.2) there exists an algebraic surface Y, together with a surjective holomorphic map $f : Y \to X$ (of degree k, say) such that under the induced bundle map from $\mathbb{P}(f^*(\mathcal{F}))$ onto $\mathbb{P}(\mathcal{F}) = Z$ the pull-back of G decomposes into a sum of positive divisors, representing $H_{q^*(\mathcal{F})} - q^*(D_i)$, where $q : \mathbb{P}(f^*(\mathcal{F})) \to Y$ denotes the projection. The D_i's need not be effective, but $\sum D_i = D$. Since $f^*(\mathcal{F})$ is a subsheaf of Ω_Y^1, since by construction $h^0(\mathcal{H}om(\mathcal{O}_Y(D_i), f^*(\mathcal{F}))) \neq 0$, and since $c_1(f^*(\mathcal{F}))\cdot T = c_1(\mathcal{F})\cdot f_*(T) \geq 0$ for every effective divisor T on Y, we can conclude from Proposition 4.3:

$$\begin{aligned} c_1(f^*(\mathcal{F}))\cdot D_i &\leq \max(c_2(f^*(\mathcal{F})),0) \\ f^*(c_1(\mathcal{F})\cdot D) &\leq \max(nc_2(f^*(\mathcal{F})),0) \\ kc_1(\mathcal{F})D &\leq k\max(nc_2(\mathcal{F}),0) \\ c_1(\mathcal{F})D &\leq \max(nc_2(\mathcal{F}),0). \quad \square \end{aligned}$$

Now we can prove Theorem 4.1. The idea is quite simple: assuming $c_1^2 > 3c_2$, a contradiction is obtained by showing that for suitable $\lambda \in \mathbb{Q}, n \in \mathbb{Z}$ with $n\lambda \in \mathbb{Z}$, the cohomology groups $H^i(S^n\Omega_X^1 \otimes \mathcal{O}_X(n\lambda K_X))$ vanish for $i = 0,2$ and n large, whereas on the other hand

$$\chi(S^n\Omega_X^1 \otimes \mathcal{O}_X(n\lambda K_X)) > 0 \quad \text{for} \quad n > n_0.$$

Proof of Theorem 4.1. Since blowing up a point decreases c_1^2 by 1 and increases c_2 by 1, we may assume that X is minimal. Then we have $c_1^2(X) > 0$ and $c_2(X) > 0$ by Theorem 2.2 and Proposition 2.4 respectively.

We shall derive a contradiction from the assumption that

$$\alpha = \frac{c_2(X)}{c_1^2(X)} < \frac{1}{3}.$$

Let $\beta = \frac{1}{4}(1 - 3\alpha)$, and let n be a natural number such that $n(\alpha+\beta) \in \mathbb{Z}$. We consider the vector bundle

$$\mathcal{V}_n = S^n\Omega_X^1 \otimes \mathcal{O}_X(-n(\alpha+\beta)K_X)$$

and claim: $h^0(\mathcal{V}_n) = h^2(\mathcal{V}_n) = 0$, provided that n is large enough. The vanishing of $h^0(\mathcal{V}_n)$ is an immediate consequence of Proposition 4.4: you take $\mathcal{F} = \Omega_X^1$ and $D = n(\alpha+\beta)K_X$, then you use Corollary 2.3 and the assumptions concerning α and β. As to $h^2(\mathcal{V}_n)$, we find, using Serre duality and the fact that $\mathcal{F}_X \cong \Omega_X^1 \otimes \mathcal{O}_X(-K_X)$:

$$\begin{aligned} h^2(\mathcal{V}_n) &= h^0(S^n\mathcal{T}_X \otimes \mathcal{O}_X((n(\alpha+\beta)+1)K_X)) \\ &= h^0(S^n\Omega_X^1 \otimes \mathcal{O}_X((n(\alpha+\beta-1)+1)K_X)). \end{aligned}$$

If n is large enough we find as before that this dimension vanishes. We conclude that the Euler characteristic

$$\chi(S^n\Omega_X^1 \otimes \mathcal{O}_X(-n(\alpha+\beta)K_X)) \text{ is } \leq 0$$

if n is large enough. On the other hand, by the Riemann-Roch theorem this Euler characteristic can be written as a polynomial of degree 3 in n with strictly positive leading coefficient.

To see this, let $Y = \mathbb{P}(\mathcal{F}_X)$, let $p : Y \to X$ be the projection and let $\mathcal{L}^\vee$ be the tautological bundle on Y. Then we derive from Theorem I.5.1 and I (10):

$$\begin{aligned} &\chi(S^n\Omega_X^1 \otimes \mathcal{O}_X(-n(\alpha+\beta)K_X)) \\ =&\chi(\mathcal{L}^n \otimes p^*(\mathcal{O}_X(-n(\alpha+\beta)K_X))) \\ =&\frac{c_1^3(\mathcal{L} \otimes p^*(\mathcal{O}_X(-(\alpha+\beta)K_X)))}{6}n^3 + \gamma n^2 + \delta n + \varepsilon. \end{aligned}$$

So what we have to do is to show that $c_1^3(\mathcal{L} \otimes (p^*\mathcal{O}_X(-(\alpha+\beta)K_X))) > 0$. Putting $c_1(\mathcal{L}) = c$, we have by I. Sect.5

$$c^2 + p^*(c_1(X))c + p^*(c_2(X)) = 0.$$

Since $c \cdot p^*$ (natural generator of $H^4(X,\mathbb{Z})$) $=$ natural generator of $H^6(Y,\mathbb{Z})$ we thus find

$$c^3 = c_1^2(X) - c_2(X).$$

We obtain

$$\begin{aligned} c_1^3(\mathcal{L} \otimes p^*(\mathcal{O}_X(-(\alpha+\beta)K_X)))) &= (c - (\alpha+\beta)p^*(-c_1(X)))^3 \\ &= c^3 + 3(\alpha+\beta)c^2p^*(c_1(X)) + 3(\alpha+\beta)^2cp^*(c_1^2(X)) \\ &= \frac{c_1^2}{16}(3\alpha^2 - 22\alpha + 7) > 0. \end{aligned}$$

So the assumption $\alpha < \frac{1}{3}$ leads to a contradiction, and we find $\alpha \geq \frac{1}{3}$, i.e. $c_1^2(X) \leq 3c_2(X)$. $\square$

All the inequalities expressed by Theorem 1.1 are more or less classical, except for (iii). A weak form of this last inequality, namely $c_1^2 \leq 8c_2$, was proved in [Ve66]. In 1975 Bogomolov obtained the inequality $c_1^2 \leq 4c_2$ (see [Bog79] and [Rei77]), and a year later Miyaoka and S.T. Yau independently proved (iii), see [Mi77b] and [Y78]. The proof given here is a simplified version of Miyaoka's (compare [Ve78b]).

In [Sak80] Sakai extends the inequality $c_1^2 \leq 3c_2$ to the case of "logarithmic Chern numbers"; this result is used by Miyaoka ([Mi84]) to prove a.o. the following conjecture of Hirzebruch: *if the minimal surface of general type X contains k disjoint (-2)-curves, then $k \leq \frac{1}{9}(3c_2(X) - c_1^2(X))$.*

The Noether inequality has also been extended to non-compact surfaces ([T-Zh], [Zh]).

Pluricanonical Maps

5. The Main Results

A minimal surface X of general type contains by Proposition 2.5 a finite number of (-2)-curves. Let C be the union of these curves. The connected components $C^{(i)}$ of C are exceptional by Proposition 2.5 ii), combined with Proposition III.2.1. By Proposition III.2.5 the normal singularity obtained by blowing down $C^{(i)}$ is rational and of A-D-E-type. We denote by $Z^{(i)} = \Sigma a_j C_j^{(i)}, a_j > 0$, its fundamental cycle (III. Sect.3). Let X_{can} be the possibly singular surface which one obtains from X after blowing down the curves $C^{(i)}$.

In this section we shall study the canonical maps $f_{\mathcal{K}_X^n} = f_n$ (see Sect. IV.1). If $f_n : X \to \mathbb{P}_N, N = h^0(X, \mathcal{K}_X^n) - 1$, is everywhere defined then it factors through X_{can}, i.e. defines a map $k_n : X_{\text{can}} \to \mathbb{P}_N$. We shall, in particular, prove that $k_n, n \geq 5$ is an embedding. In fact we prove that f_n is everywhere defined and that the induced map k_n defines an embedding of X_{can}. The main theorem of this section is the following

(5.1) **Theorem.** *Let X be a minimal surface of general type. Then:*

(i) *k_n is an embedding for $n \geq 5$;*

(ii) *k_4 is an embedding if $K_X^2 \geq 2$;*

(iii) *k_3 is a morphism if $K_X^2 \geq 2$ and an embedding if $K_X^2 \geq 3$;*

(iv) *k_2 is a morphism if $K_X^2 \geq 5$. If $K_X^2 \geq 10$, then k_2 is birational if and only if X is not a fibre space with a curve of genus 2 as general fibre.*

Before we give the proof of this theorem we shall point out some of its consequences.

(5.2) **Proposition.** *Let X be a surface of general type. Then its canonical ring $R(X)$ is a finitely generated noetherian ring.*

Proof. By Theorem I.9.1(vii) we may assume that X is minimal. We have $R(X) = \bigoplus_{i \geq 0} \Gamma(\mathcal{K}_X^{5i} \otimes \mathcal{S})$ where

$$\mathcal{S} = \mathcal{O}_X \oplus \mathcal{K}_X \oplus \mathcal{K}_X^2 \oplus \mathcal{K}_X^3 \oplus \mathcal{K}_X^4.$$

By Theorem 5.1 the map $f_5 : X \to \mathbb{P}_{k(5)}$, $k(5) = \dim H^0(\mathcal{K}_X^5) - 1$ is a morphism and hence by construction $f_5^*(\mathcal{O}_{\mathbb{P}_{k(5)}}(1)) = \mathcal{K}_X^5$, and it follows that

$$\bigoplus_{i=0}^{\infty} \Gamma(\mathcal{K}_X^{5i} \otimes \mathcal{S}) = \bigoplus_{i=0}^{\infty} \Gamma(\mathcal{T}(i))$$

where $\mathcal{T} = f_{5*}(\mathcal{S})$. If we put $S = \bigoplus_{j=0}^{\infty} \Gamma(\mathcal{O}_{\mathbb{P}_{k(5)}}(j))$, then by [Se2], §3 the module $\bigoplus_{i=0}^{\infty} \Gamma(\mathcal{T}(i))$ is a finitely generated S-module. If we combine this with the fact that S is a finitely generated ring we find that $R(X)$ is also a finitely generated ring. □

The abstract canonical model X_c of X is defined as $\mathrm{Proj}(R(X))$. Since for a surface X of general type $R(X) = \bigoplus_{d \geq 0} R_d(X)$ has transcendence degree 2 over $\mathbb{C}$ (see I, Sect. 7) X_c is an irreducible 2-dimensional projective variety. The subring $R^{(n)}(X) = \bigoplus_{d \geq 0} R_{nd}(X)$ defines a projective variety $X_c^{(n)}$ which is isomorphic to X_c, where the isomorphism is induced by the inclusion $R^{(n)}(X) \subset R(X)$. We define $R^{[n]}(X)$ as the subring of $R^{(n)}(X)$ generated by the sections of $H^0(X, \mathcal{K}_X^n)$. The variety $X_c^{[n]} = \mathrm{Proj}(R^{(n)}(X)) \subset \mathbb{P}_N$, $N = h^0(\mathcal{K}_X^n) - 1$, is the n-th canonical image of X, which is nothing but the image of X under the map f_n. By Theorem 5.1 the (a priori meromorphic) map $f_n : X \to X_c^{[n]}$ factors through $k_n : X_{\mathrm{can}} \to X_c^{[n]}$ and k_n is an isomorphism for $n \geq 5$. In this case, by a theorem of Serre ([Ha77], Theorem III. 5.2) there is some integer d_0 such that $R_d^{[n]}(X) \cong H^0(X_c^{[n]}, \mathcal{O}_{X_c^{[n]}}(d)) \cong H^0(X_{\mathrm{can}}, k_n^* \mathcal{O}_{\mathbb{P}_N}(d)) \cong H^0(X, \mathcal{K}_X^{nd}) = R_d^{(n)}(X)$ for $d \geq d_0$. This implies in fact that $X_c^{[n]} \cong X_c^{(n)} \cong X_c$ for $n \geq 5$ and in particular this shows, if X is minimal, that X_{can} is isomorphic to the abstract canonical model X_c. Under this identification the map $k_n : X_{\mathrm{can}} = X_c \to X_c^{[n]}$ is just the map given by the inclusion $R^{[n]} \subset R(X)$. In other words Theorem 5.1 can be restated by saying that the n-th pluricanonical map $X_c \to X_c^{[n]}$ induced by the inclusion $R^{[n]} \subset R(X)$ is an isomorphism for $n \geq 5$.

Another consequence of Theorem 5.1, which was already used in Chapter VI, is the following.

(5.3) **Proposition.** *Let X be a surface of general type and $n \in \mathbb{Z}, n \neq 0, 1$. Then X is minimal if and only if $H^1(\mathcal{K}_X^n) = 0$.*

Proof. If X is minimal the result follows e.g. from Theorem 5.1, Mumford's vanishing theorem IV.12.1 and Serre duality. Conversely, if $n \geq 2$ and if $\bar{X}$ is obtained from X by blowing up at least once, we have

$$h^1(\mathcal{K}_{\bar{X}}^n) = P_n(\bar{X}) - \frac{n(n-1)}{2} K_{\bar{X}}^2 - \chi(\bar{X}) \text{ (Riemann-Roch)}$$

$$> P_n(X) - \frac{n(n-1)}{2} K_X^2 - \chi(X) = h^1(\mathcal{K}_X^n).$$

For $n \leq -1$ the argument is similar, but with $h^2(\mathcal{K}_X^n)$ instead of $P_n(X)$. □

(5.4) **Corollary.** *If X is a minimal surface of general type, then*

$$P_n(X) = \frac{n(n-1)}{2} K_X^2 + \chi(X)$$

for all $n \geq 2$.

6. Proof of the Main Results

We can now give the proof of Theorem 5.1. The proof falls into two parts. One is to prove that k_n is bijective onto its image and an isomorphism outside the singularities of X_{can}, the other is to show that it is an embedding in a neighbourhood of the double points. The latter is a local consideration and we shall start with this. In order to simplify the notation we shall consider one singularity $P_j = P$ and denote the corresponding fundamental cycle by $Z^{(j)} = Z$. We have natural isomorphisms $\Gamma(\mathcal{O}_{2Z}(mK)) \cong \Gamma(\mathcal{O}_{2Z}) \cong \mathcal{O}_{X_{\text{can},P}}/\mathfrak{m}_P^2$ where the first comes from Proposition III.3.5 stating that K_X is trivial near an A-D-E curve, and the second follows from Proposition III.3.8. By the differential criterion in [G-R71] it is sufficient to prove that the restriction map $\Gamma(\mathcal{O}_X(mK)) \to \Gamma(\mathcal{O}_{2Z}(mK))$ is surjective. In view of the exact sequence

$$0 \to \mathcal{O}_X(mK_X - 2Z) \to \mathcal{O}_X(mK_X) \to \mathcal{O}_{2Z}(mK_X) \to 0$$

it is enough to prove that $H^1(\mathcal{O}_X(mK_X - 2Z)) = 0$. (Note that this is also necessary since $H^1(\mathcal{O}_X(mK_X)) = 0$ for $m \geq 2$ by Proposition 5.3.)

(6.1) **Proposition.** *Let X be a minimal surface of general type. Then for two (not necessarily different) fundamental cycles Z_1 and Z_2 on X one has $H^1(\mathcal{O}_X(mK_X - Z_1 - Z_2)) = 0$ if $m \geq 4$ or $m = 3$ and $K_X^2 \geq 3$.*

Before we can prove this result we need some connectedness properties of pluricanonical divisors. Since we shall use Reider's theorem this will play less of a role than in Bombieri's proof, but it cannot be entirely avoided.

(6.2) Proposition. *Let X be a minimal surface of general type and $D \in |nK_X|, n \geq 1$. Then*
(i) *D is 1-connected;*
(ii) *except for the case $n = 2, K_X^2 = 1$, the divisor D is even 2-connected.*

Proof. Let $D = D_1 + D_2$ be a splitting of D; by definition D_1 and D_2 are effective. If, say $K_X D_1 = 0$, then the index theorem implies that $D_1^2 < 0$. But $D_1^2 \equiv K_X D_1 (\mathrm{mod} 2)$ by the adjunction formula, hence $D_1^2 \leq -2$. So $D_1 D_2 = D_1(nK_X - D_1) \geq 2$ in this case.

So let from here on $K_X D_1 \geq 1$ and $K_X D_2 \geq 1$. We set

$$\frac{K_X D_1}{K_X^2} = \lambda \in \mathbb{Q} \text{ and } D_1 = \lambda K_X + E \in H^2(X, \mathbb{Q}).$$

Then the algebraic index theorem yields $E^2 \leq 0$. Furthermore,

$$\begin{aligned} D_1 D_2 = (\lambda K_X + E)((n - \lambda)K_X - E) &= \lambda(n - \lambda)K_X^2 - E^2 \\ &\geq \lambda(n - \lambda)K_X^2 \geq n - \frac{1}{K_X^2}, \end{aligned}$$

since $1/K_X^2 \leq \lambda \leq n - (1/K_X^2)$. This implies our result for $n \geq 2$. And if $n = 1$, we get $D_1 D_2 \geq 1$ as soon as $K_X^2 \geq 2$. But $D_1 D_2 = 2K_X D_1 - (D_1^2 + K_X D_1) \equiv 0 (\mathrm{mod} 2)$, hence $D_1 D_2 \geq 2$. Finally, the case $K_X^2 = 1, n = 1$ is excluded here since $K_X D_1 + K_X D_2 = 1$ implies that either $K_X D_1$ or $K_X D_2$ vanishes. □

(6.3) Proposition. *Let X be a minimal surface of general type, and $D \in |nK_X|, n \geq 2$. If $D = D_1 + D_2$ is a splitting of D, with $K_X D_1 \geq 1, K_X D_2 \geq 1$, then $D_1 D_2 \geq 3$, except in the cases $n = 2, K_X^2 = 1$ or 2, and $n = 3, K_X^2 = 1$.*

Proof. In the same way as in the proof of Proposition 6.2 we obtain $D_1 D_2 \geq 3$ as soon as $n \geq 3$ and $K_X^2 \geq 2$. So we are left with the case $n = 2$. If $K_X D_1 \geq 2$ and $K_X D_2 \geq 2$, then $D_1 D_2 \geq 4 - 4/K_X^2$, since $2/K_X^2 \leq \lambda \leq n - (2/K_X^2)$, hence $D_1 D_2 \geq 3$ if $K_X^2 \geq 3$. Finally, if, say, $K_X D_1 = 1$, then by the algebraic index theorem $D_1^2 \leq 1/K_X^2$. So (since $K_X^2 \geq 3$ and $D_1^2 \equiv K_X D_1 (\mathrm{mod} 2)$) we find $D_1^2 \leq -1$ and $D_1 D_2 = D_1(2K_X - D_1) \geq 3$. □

We are now in a position to prove the relevant result about 1-connectedness which will be needed in the proof of Proposition 6.1.

(6.4) Proposition. *Let X be a minimal surface of general type and $n \geq 2$. Suppose that neither $K_X^2 = 1$ or $2, n = 2$ nor $K_X^2 = 1, n = 3$. If $Z^{(i)}$ and $Z^{(j)}$ are two not necessarily different fundamental cycles on X, and $D \in |nK_X - Z^{(i)} - Z^{(j)}|$, then D is 1-connected.*

Proof. Firstly we consider the case that $i \neq j$:

Let $D = D_1 + D_2$ with D_1, D_2 effective. We apply Proposition 6.3 to $D_1 + D_2 + Z^{(i)} + Z^{(j)} \in |nK_X|$. Thus if $K_X D_1 \geq 1, K_X D_2 \geq 1$ we have

$$D_1(D_2 + Z^{(i)} + Z^{(j)}) \geq 3$$

and similarly

$$D_2(D_1 + Z^{(i)} + Z^{(j)}) \geq 3.$$

Hence $2D_1D_2 - (Z^{(i)})^2 - (Z^{(j)})^2 \geq 6$, and $D_1D_2 \geq 1$. If $K_X D_1 = 0$, then D_1 consists of (-2)curves only. Let $D_1 = D_1^{(i)} + R$ with $\mathrm{supp}(D_1^{(i)}) \subset \mathrm{supp}(Z^{(i)})$ and $\mathrm{supp}(R) \cap \mathrm{supp}(Z^{(i)}) = \varnothing$. Then $D_1 Z^{(i)} = D_1^{(i)} Z^{(i)} \leq 0$ by Sect. III.3. Similarly $D_1 Z^{(j)} \leq 0$ and $D_1 D_2 \geq 1$ by Proposition 6.2.

If $i = j$, and $K_X D_1 \geq 1, K_X D_2 \geq 1$, we find in the same way $(D_1 + Z)(D_2 + Z) \geq 3$, hence $D_1 D_2 - Z^2 \geq 3$ and $D_1 D_2 \geq 1$. And if, say, $K_X D_1 = 0$, we put again $D_1 = D_1' + R$, with D_1' consisting of (-2)-curves in Z and $RZ = 0$, hence $D_1' Z \leq 0$ and $D_1(D_2 + 2Z) \geq 1$ by Proposition 6.2, so $D_1 D_2 \geq 1$. □

We shall now give the

Proof of Proposition 6.1 (following [Cat90]). If $H^0(\mathcal{O}_X((n-1)K_X - Z^{(i)} - Z^{(j)})) \neq 0$ then it follows from Proposition 6.4 and Corollary IV.12.6 that $H^1(\mathcal{O}_X(-(n-1)K_X + Z^{(i)} + Z^{(j)})) = H^1(\mathcal{O}_X(nK_X - Z^{(i)} - Z^{(j)})) = 0$. Hence we can assume that $H^0(\mathcal{O}_X((n-1)K_X - Z^{(i)} - Z^{(j)})) = 0$.

By Corollary III.3.6 we have $h^2(\mathcal{O}_X((n-1)K_X - Z^{(i)} - Z^{(j)})) = h^0(\mathcal{O}_X((2-n)K_X + Z^{(i)} + Z^{(j)})) = h^0((2-n)K_X)$. So, if $H^0(\mathcal{O}_X((2-n)K_X + Z^{(i)} + Z^{(j)}) \neq 0$ then also $H^0(\mathcal{O}_X((2-n)K_X) \neq 0$. But if $n \geq 3$ then this contradicts the fact that K_X is nef. Consequently, we also have $H^2(\mathcal{O}_X((n-1)K_X - Z^{(i)} - Z^{(j)})) = 0$. Using this, Riemann-Roch for $\mathcal{O}_X((n-1)K_X - Z^{(i)} - Z^{(j)})$ gives us that

$$\frac{1}{2}(n-1)(n-2)K_X^2 + (Z^{(i)} + Z^{(j)})^2 + \chi(\mathcal{O}_X) \leq 0,$$

and so

$$\frac{1}{2}(n-1)(n-2)K_X^2 \leq \begin{cases} 4 - \chi(\mathcal{O}_X) & \text{if } i \neq j \\ 8 - \chi(\mathcal{O}_X) & \text{if } i = j. \end{cases}$$

Since $\chi(\mathcal{O}_X) > 0$ by Theorem 1.1, in view of our conditions on n and K_X^2 this gives immediately a contradiction if $i \neq j$. Hence we can from now on assume that $i = j$ and $Z = Z^{(i)} = Z^{(j)}$. The only possibility which is still open is $n = K_X^2 = 3$ and $\chi(\mathcal{O}_X) = 1$. Now assume that the statement of the proposition is false, i.e., that $h^1(\mathcal{O}_X(3K_X - 2Z)) \neq 0$. Since $H^1(\mathcal{O}_X(3K_X - 2Z)) = H^1(\mathcal{O}_X(2Z - 2K_X)) = \mathrm{Ext}^1_{\mathcal{O}_X}(\mathcal{O}_X(2K_X - 2Z), \mathcal{O}_X)$ this gives rise to an extension

$$(1) \qquad 0 \to \mathcal{O}_X \to \mathcal{E} \xrightarrow{\phi} \mathcal{O}_X(2K_X - 2Z) \to 0.$$

From this sequence we conclude that

$$c_1(\mathcal{E}) = (2K_X - 2Z)^2 = 4, \quad c_2(\mathcal{E}) = 0,$$

and in particular the rank 2 vector bundle $\mathcal{E}$ is unstable in the sense of Bogomolov's theorem (IV.10.1). Then this theorem implies the existence of a sequence

$$0 \to \mathcal{O}_X(M) \to \mathcal{E} \to \mathcal{J}_W \cdot \mathcal{O}_X(D) \to 0 \tag{2}$$

where W is a 0-dimensional subscheme of X and M and D are divisors. If follows from the estimates (7) and (8) in Bogomolov's theorem (IV.10.1) and Proposition IV.7.4 that $h^0(\mathcal{O}_X(M-D))$ grows quadratically in m and this implies that $K_X(M-D) > 0$ since there are divisors in the linear system $|M-D|$ which are not supported on ADE curves. Note also that $M+D = c_1(\mathcal{E}) = 2K_X - 2Z$. This shows that $K_X(M+D) = 2K_X^2 = 6$ and since $K_XM > K_XD$ it follows that $K_XM > 3$ and $K_XD < 3$. Since $K_XM > 3$ we must have $h^0(\mathcal{O}_X(-M)) = 0$. Tensoring (2) by $\mathcal{O}_X(-M)$ we obtain $h^0(\mathcal{E}(-M)) \geq 1$ and tensoring (1) by $\mathcal{O}_X(-M)$ this shows that $h^0(\mathcal{O}_X(2K_X - 2Z - M)) = h^0(\mathcal{O}_X(D)) \geq 1$. In particular we can assume that D is effective or 0. We eventually want to show that $D = 0$.

By our assumption $h^0(\mathcal{O}_X(M+D)) = h^0(\mathcal{O}_X(2K_X - 2Z)) = 0$ and since $D \geq 0$ it follows that also $h^0(\mathcal{O}_X(M)) = 0$. Since $3 = K_X^2 < K_XM$ we also find that $h^0(\mathcal{O}_X(K_X - M)) = 0$ and by Serre duality this implies that $h^2(\mathcal{O}_X(M)) = 0$. It follows from Riemann-Roch for M that

$$\frac{1}{2}M(M-K_X) + \chi(\mathcal{O}_X) \leq 0$$

and hence $M^2 \leq K_XM - 2$. Now $4 = 4(K_X - Z)^2 = (M+D)^2 = M^2 + 2MD + D^2 \leq K_XM - 2 + 2MD + D^2$. Since $K_XM = 6 - K_XD$ we find that

$$K_XD \leq D^2 + 2MD. \tag{3}$$

We want to deduce from this that $D = 0$. By the index theorem $3D^2 = K_X^2D^2 \leq (K_XD)^2 \leq 4$ since we have already seen that $K_XD \leq 2$. Hence $D^2 \leq 1$. On the other hand, by sequence (2)

$$0 = c_2(\mathcal{E}) = \deg W + DM$$

and in particular $DM \leq 0$. But then it follows immediately from (3) and $D^2 \leq 1$ that $DM = 0$ and hence also that $\deg W = 0$. But then again by (3) we find that $K_XD \leq 1$. Hence $3D^2 \leq (K_XD)^2 \leq 1$ and thus $D^2 \leq 0$. Finally using (3) once more this gives $K_XD = D^2 = 0$. Since the self intersection form is strictly negative on the effective divisors orthogonal to K_X this implies $D = 0$.

Now the sequence (2) reads

$$0 \xrightarrow{\psi} \mathcal{O}_X(2K_X - 2Z) \to \mathcal{E} \to \mathcal{O}_X \to 0. \tag{4}$$

We claim that this sequence provides a splitting of (1) (which then implies that $h^1(\mathcal{O}_X(3K_X - 2Z)) = 0$). For this we only have to see that $\phi \circ \psi \not\equiv 0$.

But if this were not the case then $\mathrm{Im}(\psi) = \ker(\phi)$ and hence by (1) this would imply $h^0(\mathcal{O}_X(2K_X - 2Z)) = h^0(\mathcal{O}_X(M)) \neq 0$, a contradiction.
Proof of theorem 5.1: We want to apply Reider's theorem IV.11.4 to the linear system $L = (m-1)K_X$.

(i) Assume $m \geq 3$. Then under the assumptions of the theorem $L^2 \geq 10$ unless $m = 3$ and $K_X^2 = 2$ in which case $L^2 = 8$. Now assume that f_n fails to be a morphism, respectively an embedding. Then by Reider's theorem this implies the existence of a very special curve D through a base point, respectively through two (possibly infinitely near) points which are not separated by the map f_n. We first note that a number of possible cases which are listed in Reider's theorem cannot occur in our situation since necessarily $K_X D \geq 0$ and $K_X D + D^2 \equiv 0 \bmod 2$ for every divisor D on X. Now going through the list we immediately see that our hypothesis implies that f_n is everywhere defined and that the only curves D through two points which are not separated must have the property that $K_X D = 0$ and $D^2 = -2$. In this case D is the union of (-2)-curves. But then it follows immediately from Proposition 6.1 that f_n separates the double points of X_{can} and that it is a local embedding near each of these singularities.

(ii) Let $m = 2$. In this case $L = K_X$ and the cases in (i) of Theorem IV.11.4 are excluded since $K_X D + D^2$ is even for every curve D. Hence f_2 and thus also k_2 is a morphism provided $K_X^2 \geq 5$. Now assume that $K_X^2 \geq 10$ and that k_2 is not birational onto its image. Then by Reider's theorem there must be infinitely many curves D on X, which cover an open part of X, such that $K_X D = 1, D^2 = -1$ or $K_X D = 2, D^2 = 0$. We first remark that there are only finitely many irreducible curves B on X with $K_X B \leq 2, B^2 < 0$. Because of the second condition it is sufficient to show that there is only a finite number of classes $b \in H_2(X, \mathbb{Q})$ which can be represented by such a curve. Suppose there were an infinite number of such classes. Then there would be projections $b - (BK_X/K_X^2)k$ into the orthogonal complement of the class k of K_X with $-\left(b - BK_X/K_X^2 k\right)^2$ large (algebraic index theorem). But this is impossible, since $B^2 \geq -4$ by the adjunction formula. Hence there must be infinitely many curves D on X, covering an open part of X, with the property $K_X D = 2, D^2 = 0$. We next claim that there must be infinitely many *irreducible* curves D with $K_X D = 2$ and $D^2 = 0$. Assume we have a possibly reducible such D. Pick an irreducible component C of D. If $K_X C \leq 1$ then automatically $C^2 < 0$ and there are only finitely many such curves. If however $K_X C = 2$ the index theorem gives $C^2 \leq 0$ and only finitely many of these curves C have negative self intersection. This completes the proof of our claim. The same argument as before shows that there are only finitely many classes of curves D with $K_X D = 2$, $D^2 = 0$. It follows then that from the pairs (x, D) where $x \in D$ and D runs over irreducible such curves, we can construct an algebraic family. This family has no base locus, and no base points (since $D^2 = 0$) and its general member is a smooth curve of genus 2.

Conversely, if X is a fibre space over a curve such that the general fibre F has genus 2, then $\mathcal{K}_X^2|_F = \mathcal{K}_F^2$ by the adjunction formula and Lemma III. 8.1, so $f_2|_F$ cannot be one-to-one, since the bicanonical map of a smooth curve of genus 2 is two-to-one. It follows that f_2 is not birational. □

The classical geometers certainly were aware of the fact that, except for low values of n, the pluricanonical maps f_n are birational ([Enr49]). Mumford proved in [Mu62] that k_n is an embedding for n large enough and that the canonical ring is always finitely generated. In [Saf] Moishezon showed birationality of f_n for $n \geq 8$. After important work of Kodaira (see [Ko68]), Bombieri finally proved Theorem 5.1(i) and (ii) and a slightly weaker form of (iii) and (iv). His proof is based on analysing connectedness properties of pluricanonical divisors. A new approach was opened up by Reider [Reider] and this is the treatment which we follow here. For another approach see also [C-F-H-R].

Pluricanonical rings for singular or non-compact surfaces have been investigated by Sakai ([Sak80, [Sak82], [Sak84]). For non-compact surfaces one has to replace the pluricanonical map by "logarithmic pluricanonical maps".

7. The Exceptional Cases and the 1-Canonical Map

As far as $f_n, n \geq 3$ is concerned, Theorem 5.1 is best possible in the sense that f_4 is not always birational if $K_X^2 = 1$, that f_3 is not always birational if $K_X^2 \leq 2$ and that f_3 is not always a morphism if $K_X^2 = 1$.

Let us first consider the question of the birationality of f_3 and f_4.

(7.1) **Proposition.** *Let X be a minimal surface of general type with $K_X^2 = 1$ and $p_g(X) = 2$. Then f_3 and f_4 are not birational.*

Proof. We shall first deal with the case of f_4. As usual, we put $|K_X| = |C| + V$, where $|C|$ is a linear system of dimension 1 without fixed components, and V the fixed part of $|K_X|$. Then $K_X(C+V) = 1$. Since by Proposition 2.2 we have $K_X D \geq 0$ for every effective D, and $K_X D = 0$ if and only if D is a sum of (-2)-curves, we can conclude that $K_X C = 1, K_X V = 0$. Furthermore, the general C must be irreducible. From $K_X(K_X - C) = 0$ we find by the index theorem that $C^2 \leq 1$, and because of $C^2 \geq 0$ and $K_X C \equiv C^2 (\mathrm{mod}\, 2)$ we find that $C^2 = 1$. So $V^2 = 0$, and since V consists of (-2)-curves, we derive from Theorem III.2.1 that $V = 0$. So K_X is a pencil without fixed part and one base point, in which two general members meet with multiplicity 1. Consequently, the general canonical curve is an irreducible smooth curve, which is of genus 2 by the adjunction formula. The restriction of $\mathcal{K}_X^4$ to such a smooth canonical curve C is $\mathcal{K}_C^2$, and since $\mathcal{K}_C^2$ yields a two-to-one map, we can conclude that f_4 is not birational.

As to f_3, $|3K_X|$ cannot have fixed components, since $|K_X|$ has no fixed components. So if f_3 were birational, then f_4 would also be birational, since $p_g(X) \neq 0$. □

Minimal surfaces X of general type with $K_X^2 = 1$ and $p_g(X) = 2$ do exist; they are extremal with respect to Noether's inequalities, and as such will appear in Sect. 9 below.

Remark. An alternative way of proving Proposition 7.1 is the following. It is not difficult to show that the canonical ring $R(X)$ of X is generated by four elements x_0, x_1, y, z with x_0, x_1 spanning $\Gamma(\mathcal{K}_X)$, and x_0^2, x_0x_1, x_1^2, y spanning $\Gamma(\mathcal{K}_X^2)$, whereas $\Gamma(\mathcal{K}_X^5)$ is spanned by $x_0^5, x_0^4x_1, x_0^3x_1^2, x_0^2x_1^3, x_0x_1^4, x_1^5, x_0^3y, x_0^2x_1y, x_0x_1^2y, x_1^3y, x_0y^2, x_1y^2$ and z. If we give x_0, x_1 degree 1, y degree 2 and z degree 5, then it can be shown that there is exactly one polynomial $f(x_0, x_1, y, z)$ of total degree 10 such that

$$R(X) \cong \mathbb{C}[x_0, x_1, y, z]/f(x_0, x_1, y, z).$$

We must have $R^3(X) \neq R(X)$ and $R^4(X) \neq R(X)$, so in particular f_3 and f_4 cannot be birational.

(7.2) **Proposition.** *Let X be a minimal surface of general type with $K_X^2 = 2$ and $p_g(X) = 3$. Then*
(i) *f_n is a morphism for all $n \geq 1$ and f_1 maps X generically 2-to-1 onto $\mathbb{P}_2$;*
(ii) *f_3 is a morphism of degree 2.*

We have met surfaces X with $K_X^2 = 2, p_g(X) = 3$ before as 2-fold coverings of $\mathbb{P}_2$ ramified over a curve of degree 6 with at most simple singularities. They also form the simplest example of Horikawa surfaces with K_X^2 even (see Sect. 9).

Proof of Proposition 7.2 (i). We start by showing that $|K_X|$ has no fixed component. So let again $|K_X| = |C| + V$, where $|C|$ is of dimension 2, without fixed components and where V is the fixed part of $|K_X|$. Since $K_X(C+V) = K_X^2 = 2$, and $K_XV \geq 0, K_XC \geq 1$, there are two possibilities: $K_XC = K_XV = 1$, or $K_XC = 2, K_XV = 0$.

Suppose that $K_XC = K_XV = 1$. Then $K_X(K_X - 2C) = 0$, hence by the index theorem $(K_X - 2C)^2 \leq 0$. Since on the other hand $C^2 \geq 0$, we must have $C^2 = 0$, which is impossible, for $C^2 \equiv K_XC \pmod 2$.

Now let us assume that $K_XC = 2, K_XV = 0$. From $K_X(K_X - C) = 0$ and the index theorem we deduce $C^2 \leq 2$. Since $C^2 \geq 0$ and $C^2 \equiv K_XC \pmod 2$, we must have $C^2 = 2$ or $C^2 = 0$. In the first of these cases we have that $K_X - C$ represents $0 \in H_2(X, \mathbb{Q})$, so V is indeed the zero divisor. It remains to exclude the second possibility, i.e. $C^2 = 0$. To do this, we start by observing that the general member of $|C|$ cannot be irreducible, for otherwise the dimension of $|C|$ would be at most 1. But the general member of $|C|$ cannot be reducible either, for then $|C|$ would be composed with a pencil $|D|$, i.e. C would be homologous to $aD, a \geq 2$, and the only possibility left would be that a general member of $|C|$ would consist of two components, say C_1 and C_2, both homologous to D, so that K_XD would be 1, whereas on the other hand $K_X D \equiv D^2 \pmod 2$, and $D^2 = C^2/a^2 = 0$. So $|K_X|$ has no fixed part. Furthermore, the general member of $|K_X|$ is irreducible, since $|K_X|$ cannot be composed with a pencil, for $K_X^2 = 2$. For the same reason $|K_X|$ has at most one base point. But if $|K_X|$ had one, then f_1 would be a birational map onto $\mathbb{P}_2$ which is impossible. So f_1 has no base points, in other words it is a morphism onto $\mathbb{P}_2$. The fact that $K_X^2 = 2$ means that this morphism has degree 2.

(ii) Let $s : \mathbb{P}_2 \xrightarrow{\sim} Q \subset \mathbb{P}_9$ be a Segre embedding of $\mathbb{P}_2$ by $\mathcal{O}_{\mathbb{P}_2}(3)$. Then $(sf_1)^*(\Gamma(\mathcal{O}_Q(1)))$ is a linear subsystem of $\Gamma(\mathcal{K}_X^3)$. But both systems have dimension 10, hence sf_1 and f_3 only differ by an automorphism of $\mathbb{P}_9$. □

Remark. Exactly as in the preceding case we can find an alternative proof by showing that

$$R(X) \cong \mathbb{C}[x_0, x_1, x_2, y]/g(x_0, x_1, x_2, y),$$

where degree $x_i = 1$, degree $y = 2$ and g has total degree 8. Then f_1 comes from the inclusion $\mathbb{C}[x_0, x_1, x_2] \subset R(X)$. Since the respective quotient fields yield an extension of degree 2, we find that f_1 is indeed of degree 2. The argument for f_3 is similar.

The two classes of exceptions we just produced are the only ones. In fact we have

(7.3) **Proposition.** *Let X be a minimal surface of general type. If $n \geq 3$, then f_n is birational, except in the following cases:*
(i) $n = 4, K_X^2 = 1, p_g(X) = 2$;
(ii) $n = 3, K_X^2 = 2, p_g(X) = 3$ *and* $K_X^2 = 1, p_g(X) = 2$.

Because of Noether's inequality, one has only to consider the cases $p_g(X) \leq 3$ if $K_X^2 = 2$ and $p_g(X) \leq 2$ if $K_X^2 = 1$. Starting from these conditions all remaining possibilities with $p_g(X) = 1$ or 2 were treated by Bombieri in [Bom70], whereas Miyaoka proved the result for the case $K_X^2 = 1, p_g(X) = 0$. Finally, in [B-C] and [Cat81a] Bombieri and Catanese settled the cases $K_X^2 = 1, p_g(X) = 0$ for f_4 and $K_X^2 = 2, p_g(X) = 0$ for f_3.

During the last 15 years many results have been obtained about bicanonical maps (see [Cil97] for a survey). The main result concerning the base points of $|2K_X|$ is

(7.4) **Theorem.** *If $p_g \geq 1$, then f_2 is a morphism.*

Francia [Fr] proves that if $p_g \geq 1$, then $|2K_X|$ has no fixed components and it has no base points, with the possible exception of the case $p_g = q = 1$. Francia's technique is a refinement of Bombieri's method (which uses properties of 1-connectedness). Surfaces with $p_g = q = 1$ were studied by Catanese [Cat81b] and Catanese and Ciliberto [C-C91] and [C-C93]. It is a consequence of their analysis that $|2K_X|$ has no base points if $p_g = q = 1$

and $K_X^2 \leq 4$. The cases $K_X^2 \geq 5$ are taken care of by Theorem 6.1. The hypothesis $p_g \geq 1$ in the above theorem is essential.

Theorem 5.1 implies that f_2 is birational except possibly for the members of *finitely many* families of minimal surfaces of general type, for it follows immediately from Gieseker's theorem and Theorem 1.1 that there is only a finite number of families satisfying $K_X^2 <$ constant. As we have seen, f_2 is not birational if X has a pencil of genus 2 curves. We shall refer to this as the *standard case*. There are other cases where f_2 is not birational. A partial classification is given by the

(7.5) **Theorem.** *Let X be a minimal surface of general type whose bicanonical map is not birational and which does not represent the standard case. Then there are the following cases*

(i) $p_g = 6$ *and* $K_X^2 = 8, 9$ *or* $p_g = 5$ *and* $K_X^2 = 7, 8$ *or* $p_g = 4$ *and* $K_X^2 = 6, 7, 8$. *In all of these cases* $q = 0$,

(ii) $p_g = 3, q = 0$ *and* $5 \leq K_X^2 \leq 7$,

(iii) $p_g = 3$ *and* $q = 3$. *In this case X is the symmetric product of a curve of genus* 3.

All the examples of (i) and (ii) were already found by Du Val [DV]. The classification result for $p_g \geq 4$ was obtained in [Ci-F-M] and for $p_g = 3, q = 0$ in [Ci-M00]. The rest of the theorem is due to Catanese, Ciliberto and Mendes Lopes [C-C-M]. So far there is no complete classification available for $p_g \leq 2$. Partial results are in particular due to Xiao Gang [Xi86a], [Xi90] and Ciliberto and Mendes Lopes [Ci-M01]. In a different vein, we have a theorem due to Xiao Gang[Xi85c] (see also [ChenV]):

(7.6) **Theorem.** *Let X be a minimal surface of general type. If $p_2(X) > 2$ then the bicanonical map is generically finite.*

Finally a few words about the 1-canonical map. It has been studied systematically for the first time by Beauville in [Be79]. Before this paper appeared, only little was known (in particular Castelnuovo's inequality, see below). The main results of [Be79] are contained in the following

(7.7) **Theorem.** *Let X be a minimal surface of general type with $p_g(X) \geq 2$. Then there are the following possibilities for the (in general: rational) 1-canonical map f_1.*

A. *f_1 is composed with a pencil of curves of genus g. If $\chi(\mathcal{O}_X) \geq 21$, then $2 \leq g \leq 5$ and the pencil has no base points.*

B. $\dim f_1(X) = 2$.

Let Y be a minimal desingularization of $f_1(X)$

1) $p_g(Y) = 0$. *If* $\chi(\mathcal{O}_X) \geq 31$, *then* $\deg(f_1) \leq 9$.

2) *Y is of general type with $p_g(Y) \geq 4$, and the projection from Y onto $f_1(X)$ is a canonical map for Y. If $\chi(\mathcal{O}_X) \geq 14$, then* $\deg(f_1) \leq 3$.

The proof of this theorem rests heavily on the inequality $c_1^2 \leq 3c_2$. Gieseker's theorem and Theorem 1.1 imply that there is only a finite number of families of minimal surfaces of general type, satisfying $\chi(\mathcal{O}_X) <$ constant. So A) says in particular that the genus g of the fibre *always* satisfies $2 \leq g \leq 5$, except perhaps for the members of finitely many families. If A) holds, Xiao [Xi85b] proved:

(7.8) **Theorem.** *Assume that f_1 is composed with a pencil of curves and let b be the genus of the base curve of this pencil. Then either $b = q(X) = 1$ or $b = 0$ and $q(X) \leq 2$.*

Beauville has constructed minimal surfaces X, belonging to class A), with $g = 2$ and 3 and $\chi(\mathcal{O}_X)$ arbitrarily large. There exist surfaces belonging to class B1) with $\chi(\mathcal{O}_X)$ arbitrarily large, for which f_1 has degree 8 (and similarly for $\deg(f_1) = 2$ and 6). Every surface with $p_g = 0$ appears at least once as a surface Y in this class. For examples of surfaces of class B2) see [Be79], [Cat81b], [G-Z]. Beauville found an infinite series (see [Cat87a]). So far there is no complete classification of what can occur in case B2). For partial results see [M-P98], [M-P00]. Another important result, due to Horikawa [Hor76], Reid [Rei78], Beauville [Be79] and Debarre [Deb] is:

(7.9) **Theorem** (Castelnuovo's second inequality). *Let X be a minimal surface of general type, satisfying $K_X^2 < 3p_g(X) - 7$. Then f_1 is a (rational) map of degree 2 onto a ruled surface. If f_1 is birational then $K_X^2 \geq 3p_g + q - 7$ and if f_1 maps onto a curve then $K_X^2 \geq 3p_g - 3$. If moreover, the fibres have genus ≥ 3 and the image is a curve of genus b then the sharper inequality $K_X^2 \geq 4p_g + 4(b-1)$ holds.*

Castelnuovo proved $K_X^2 \geq 3p_g(X) - 7$, assuming that K_X is very ample.

Another approach to the study of the 1-canonical map is through the structure of the canonical ring (see [Cat87a] and [Cat97] and the references given there). More recent results on 1-canonical maps including also some interesting new examples with $p_g = q = 0, 1$ can be found in [Cat98]. I. Bauer [Bau] has made a detailed study of surfaces with $K^2 = 7$ and $p_g = 4$ via an analysis of the 1-canonical map; the corresponding moduli space is shown to have 3 components of dimension 36, 36 and 38.

Surfaces with Given Chern Numbers

8. The Geography of Chern Numbers

A. General existence results

In Sect. IV.9 we saw that, given any ordered pair (m, n) of integers with $n+m$ divisible by 12, there always exists a compact *almost complex* manifold of real dimension 4 with $c_1^2(X) = n$, $c_2(X) = m$. Combining the results of VI, Sect.1 with Theorem 1.1, we see that the same is not true if we require X to be a surface. In fact, since blowing up a point increases c_2 by 1 and increases c_1^2 by 1, we see that no pairs (m, n) with $n + m$ divisible by 12, satisfying $n > \max(2m, 3m)$, can be represented by a surface.

We now want to consider the Chern numbers for minimal surfaces of general type in more detail. In the light of Theorem 1.1 our first question is the following one:

If $D \subset \mathbb{Z}^2(m, n)$ is given by the inequalities $n > 0$, $n \leq 3m$ and $m \leq 5n + 36$, then for which "admissible pairs" $(m, n) \in D$, i.e., *pairs with $n + m \equiv 0 \bmod 12$, does there exist a minimal surface of general type X with $c_1^2(X) = n$ and $c_2(X) = m$?*

We divide D into two parts D_1 and D_2 given respectively by $n \leq 2m$ (signature ≤ 0) and $n > 2m$ (positive signature). The simplest examples, like complete intersections or double coverings of $\mathbb{P}_2$, practically always yield a point in D_1. Indeed, for a long time only very few examples with Chern pairs in D_2 were known.

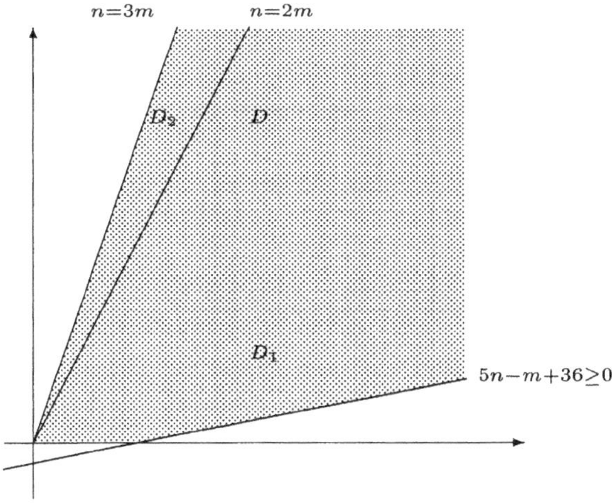

But, trivial as it is to find lots of pairs in D_1, it is a different matter to show that *all* (admissible) pairs in D_1 can be represented. It is even more difficult to fill up the region D_2. The combined efforts of Persson ([Per81], Theorem 2) for the region D_1 and Chen ([Chen87], [Chen91]) for the regions D_1 and D_2 gave rise to the following result:

(8.1) **Theorem.** *Given any admissible pair $(m, n) \in D$, there exists a minimal surface of general type X with $c_1^2(X) = n$, $c_2(X) = m$, except maybe for points on the finitely many lines $n - 3m + 4k = 0$ with $0 \leq k \leq 347$. In fact, for at most finitely many exceptions on these lines all admissible pairs in D_1 occur as Chern pairs.*

This theorem does not give surfaces on the borderline $n = 3m$. By Yau's theorem we know that they are quotients of the unit ball. As such we have implicit constructions by Borel (see V, Sect. 14). Explicit examples are Mumford's fake projective plane (V, Sect. Sect.1), and Hirzebruch's examples from V, Sect. 24. These provide many examples on the line $n = 3m$ (see for example V. Remark 24.0 (ii)). Using Hirzebruch's construction we prove the following theorem, originally due to Sommese, see [So84].

(8.2) **Theorem.** *Every rational point in* $[1/5, 3]$ *occurs as the slope of some (irregular) algebraic surface of general type.*

Proof. We only consider the interval $[2, 3]$, the remaining interval can be treated similarly. See e.g. [B-H-H], 156–157.
Let

$$s = c_1^2/c_2 \text{ (the Chern-slope) } .$$

The construction we use is based on the existence of a fibration $X \to S$ with connected fibres F over a curve S of genus ≥ 1. Then X is irregular (pull back 1-forms from the curve to the surface). We pull back this fibration by a ramified covering $S' \to S$ of degree d such that ramification only takes place at points where $X \to S$ has smooth fibres (of multiplicity 1). If the ramification divisor on S' has degree ρ the Chern-slope of the resulting irregular surface X' is easily calculated:

$$\begin{aligned} s(X') &= \frac{d \cdot c_1^2(X) - 2\rho \cdot (e(F)}{d \cdot e(X) - \rho \cdot e(F)} \\ &= s(X) + (2 - s(X)) \cdot \frac{-\rho \cdot e(F)}{d \cdot e(X) - \rho \cdot e(F)} \end{aligned} .$$

If now $g(S) \geq 1$ and $s(X) \geq 2$ every slope between 2 and $s(X)$ can be realized as follows: take

$$S' \xrightarrow{\alpha} S'' \xrightarrow{\beta} S$$

with α a double cover with $2y$ branch points and β an unramified cover of degree x. Note these do exist! Now $d = 2x$ and $\rho = 2y$ and we find

$$s(X') = s(X) + (2 - s(X)) \cdot \frac{-y \cdot e(F)}{x \cdot e(X) - y \cdot e(F)},$$

so if we are given a rational number p/q, $0 \leq p < q$ we can take $x = -(q - p)e(F)$ and $y = pe(X)$ and we see that the coefficient of $2 - s(X)$ is exactly p/q.
We apply this to Hirzebruch's surface, which has Chern slope 3 and which admits a fibration over a curve of genus 6 (see Remark V.24.0 (iv)). □

B. Simply connected surfaces

We may sharpen our question and impose restrictions on the surfaces, the most natural being simply-connectedness. Initial work for the domain D_1 was done in [Per81]:

(8.3) **Theorem.** *There exists a* $C > 0$ *defining a "forbidden sector"*

$$S := \{(m,n) \in D_1 \mid n \geq 8m - Cm^{2/3}\}$$

such that for any admissible pair $(m,n) \in D_1 \setminus S$ *there exists a simply connected minimal surface of general type with* $n = c_1^2(X)$ *and* $m = c_2(X)$. *In particular, all rational slopes between* $1/5$ *and* 2 *are realized as Chern slopes of simply connected surfaces.*

Inspired by a construction of G. Xiao, Chen found examples with slopes up to 2.700 (see Chen ([Chen87]) where one finds the following result.

(8.4) **Theorem.** *For any admissible pair of integers* $(m,n) \in D$ *with*

$$(352/716)m + C_1 m^{2/3} < n < (18644/6904)m - C_2 m^{2/3}$$
$$n > C_3,$$

where C_1, C_2 *and* C_3 *are positive constants there exists a simply connected minimal surface of general type with* $n = c_1^2(X)$ *and* $m = c_2(X)$.

Note that $352/716 = 0.4916$ and $18644/6904 = 2.700$ give two limit slopes between which this theorem asserts the existence of simply connected surfaces with this Chern slope. By Yau's theorem the upper bound is ≤ 3, but it is not known whether we can find slopes arbitrarily close to 3 or not.

We will see (Chap.. IX, Sect.1) that there are two topological distinct types of simply connected surfaces: those with even intersection form and those with odd form. Chen's and Persson's constructions very rarely produce surfaces with even intersection forms. In fact there is an obvious restriction for the Chern numbers

$$c_1^2 \equiv 2c_2 \bmod (48)$$

which is a consequence of Rochlin's theorem [Ro] stating that for even intersection forms the index is divisible by 16. In [P-P-X] it is shown that many admissible pairs in the region D_1 satisfying this restriction occur. In fact an analogue of Theorem 8.3 is valid and in particular all rational slopes between 1/5 and 2 are realized as Chern slopes of spin surfaces, i.e., simply connected surfaces with even intersection form.

For the region D_2 the results are less satisfactory although similar results hold as far as Chern slopes are regarded: a dense subset of $[2, b]$ with $b = 60068/22116 = 2,716042$ occurs as the set of slopes of spin surfaces.

C. Existence of special fibrations

One can pose other kinds of geographical questions, for example questions related to the existence of fibrations $f : X \to C$ of a surface X onto a curve C of genus b of a particular kind. Xiao Gang has undertaken a systematic study of the invariants of such fibrations. We have seen (Theorem III. 18.2) that for relatively minimal fibrations $\deg(f_*\omega_{X/C}) > 0$ unless f is locally trivial. Xiao in [Xi85a] introduces the "instability slope" $\lambda(f) = K^2_{X/C}/\deg(f_*\omega_{X/C})$ for fibrations f that are not locally trivial. Note that $K_{X/C} = K_X \otimes (f^*K_C)^{-1}$

and hence its self-intersection equals $c_1^2(X) - 8(b-1)(g-1)$, where g is the genus of the non-singular fibres. Riemann-Roch and Leray's spectral sequence give the equality

$$\deg(f_*\omega_{X/C}) = \chi(\mathcal{O}_X) - (b-1)(g-1)$$

and so $\lambda(f)$ can be expressed in b, g, c_1^2 and χ. By Theorem III. 11.4 $\lambda(f) \leq 12$. It is less trivial to find a lower bound. It was shown in [Xi87a] that for relatively minimal fibrations

$$4 - \frac{4}{g} \leq \lambda(f) \leq 12.$$

These inequalities can be refined. For this we refer to the survey article [As-Ko] and the references given there.

Returning to geographical questions, one can first ask for the existence of rational pencils ($C = \mathbb{P}_1$) of special curves. All surfaces in the region D_1 constructed in [Per81] have such a fibration of genus two curves. All surfaces in this region with even intersection form constructed in [P-P-X] admit a rational pencil of genus two, three or four curves. Conversely, Xiao [Xi87a] has shown that surfaces admitting genus two fibrations live in the region D_1. As to hyperelliptic fibrations, if $c_1^2 < \frac{1}{3}(c_2 - 40)$ results of Beauville ([Be79]) and Xiao Gang ([Xi87c]) imply that in this case the surface admits a unique pencil of hyperelliptic curves of genus 2 or 3.

As to more general fibrations, we have the following open conjecture ([Rei79]):

Reid's Conjecture. *A surface X with Chern slope $< \frac{1}{2}$ has either finite fundamental group or its fundamental group is commensurable with the fundamental group of a curve. More precisely, there exists a finite unramified cover X' and a fibration $f : X' \to C$ onto a curve such that $\ker(f_* : \pi_1(X') \to \pi_1(C))$ is finite.*

It is true for slope $< \frac{1}{3}$ (for $q = 0$ see [Rei79], for $q > 0$ see [Hor76]). See also [Xi87b], [Xi87c]. The conjecture is rather sharp, since J.H. Keum ([Ke]) has constructed a surface with $c_1^2 = 4$, $c_2 = 8$, $p_g = q = 0$ and π_1 having a normal abelian rank 4-subgroup of finite index. Related to Reid's conjecture there is

Severi's Problem. *For irregular surfaces with Chern slope $< \frac{1}{2}$, is the image of the Albanese map a curve?*

It is true when the Chern slope $< \frac{1}{3}$ (see [Hor76]). Conversely we have ([Siu87], [Be88]) the

Irrational Pencil Theorem. *If the fundamental group of a surface X admits a surjective map onto the fundamental group of a curve of genus $g \geq 2$ then there exists a fibration $f : X \to C$ where C has genus at least g.*

For other results concerning fibrations of surfaces see [Cat00, Cat01], [Ser92], [Ser96], [Y-M], [Zu].

D. Moduli spaces

We finish this section with a discussion of the few known results about the global structure of the Gieseker scheme $M_{m,n}$ parametrising surfaces with $c_1^2 = n$, $c_2 = m$.

There is a rough estimate on the dimension of a component M of $M_{m,n}$ due to Catanese [Cat84]:

$$10\chi - 2n \leq \dim M \leq 10\chi + 3n + 108, \quad \chi = \frac{1}{12}(n + m).$$

Another question: "how singular can the Gieseker scheme be?" has been addressed in [Cat89] where many examples are given of everywhere non-reduced Gieseker schemes.

Another obvious question: "how many connected or irreducible components can the Gieseker scheme have? " is in general difficult to answer. It is known however that for each given h and given parity (odd or even) there are admissible pairs (m, n) for which $M_{m,n}$ has at least h irreducible components, all of different dimensions, all parametrizing simply connected surfaces whose intersection form have the given parity ([Cat84], [Cat86], [Cat87b], [Man94]). In [Cat92] Catanese also provides upper bounds for the number of irreducible components of $M_{m,n}$ and the number of irreducible components corresponding to surfaces with $q = 0$. These bound are probably far from optimal (they are of the form $Kn^{Ln^{33}}$ respectively $K'n^{L'n^2}$ where K, L, K', L' are positive constants).

The first examples of differentiably distinct but oriented homeomorphic surfaces of general type are due to Moishezon and can be found in [F-M-M]. Ebeling [Ebe] found complete intersection surfaces which are homeomorphic but not diffeomorphic. These examples show that some of the connected components of the Gieseker scheme must correspond to differentiably distinct 4-manifolds. It can be shown that fixing a topological type there can be many connected components of $M_{m,n}$ corresponding to distinct differentiable structures all of the same given topological type. In fact, given any natural number d and a parity (even or odd), there are pairs (m, n) such that $M_{m,n}$ has at least d components realizing different smooth structures on the same simply connected 4-manifold whose intersection form has the given parity. See [Sal89] and [Sal91] where qualitative results can be found.

Surfaces parametrized by the same connected component are deformation equivalent (and hence diffeomorphic). Distinct connected components correspond to surfaces not deformation equivalent. Indeed, if we have a family of surfaces of general type with Chern numbers $c_1^2 = n$, $c_2 = m$ with connected base B, there is a holomorphic map $B \to M_{m,n}$ and so its image is in some path-component of $M_{m,n}$. Surfaces in different components may or may not be diffeomorphic to each other. Manetti [Man01] sharpened Salvetti's result as follows: for every $h > 0$ there are smooth 4-manifolds V with

$b_1(V) = 0$ such that $M_{m,n}$ has at least h components parametrizing surfaces diffeomorphic to V. In this article one finds the first examples of diffeomorphic but not deformation equivalent surfaces of general type. For simpler examples see [Cat01]. Catanese's examples are in fact certain Beauville surfaces (see Sect. 10 below).

9. Surfaces on the Noether Lines

On a Hirzebruch surface Σ_n (V, Sect. 4) every divisor D is homologous (even linearly equivalent) to $aC_n + bF$ (C_n a section with $C_n^2 = -n, F$ a fibre). In the sequel we shall call the ordered pair (a, b) the *type* of D.

Table 12.

c_1^2	p_g	Minimal resolution of the canonical model(= image of f_1)	Type of branch locus	Top. type of deformation class
2	3	$\mathbb{P}_2$	8	non-spin
4	4	Σ_0 Σ_2	$(6,6)$ $(6,12)$	non-spin
6	5	Σ_1 Σ_3	$(6,10)$ $(6,16)$	non-spin
8	6	$\mathbb{P}_2$ Σ_4	10 $(6,20)$	spin
8	6	Σ_0 Σ_2	$(6,8)$ $(6,14)$	non-spin
$8k+2$	$4k+3 (k \geq 1)$	$\Sigma_{2j+1} (0 \leq j \leq k)$	$(6, 4k+6j+8)$	non-spin
$8k-4$	$4k \quad (k \geq 2)$	$\Sigma_{2j} \quad (0 \leq j \leq k)$	$(6, 4k+6j+2)$	non-spin
$8k-2$	$4k+1 (k \geq 2)$	$\Sigma_{2j+1} (0 \leq j \leq k)$	$(6, 4k+6j+6)$	non-spin
$8k$	$4k+2 \quad (k \geq 2)$	$\Sigma_{2j} \quad (0 \leq j \leq k)$	$(6, 4k+6j+4)$	non-spin
$8k$	$4k+2 \quad (k \geq 2)$	Σ_{2k+2}	$(6, 10k+10)$	spin iff k odd

In [Hor76] Horikawa proved

(9.1) **Theorem.** *Let X be a minimal surface of general type with $c_1^2(X)$ even and $p_g(X) = \frac{1}{2}c_1^2(X)+2$. Then the 1-canonical map f_1 is a holomorphic map of degree 2 onto a 2-dimensional variety Y of degree $p_g - 2$ in $\mathbb{P}_{p_g-1}$. The minimal resolution Y' of Y is either $\mathbb{P}_2$ or a Hirzebruch surface, and there exists a 2-fold branched covering $g' : X' \to Y'$, branched along a curve of type (a, b), with simple singularities only, such that X is the minimal resolution of X' and the diagram*

$$\begin{array}{ccc} X & \xrightarrow[\text{res.}]{\text{min.}} & X' \\ {\scriptstyle f_1}\big\downarrow & & \big\downarrow{\scriptstyle g'} \\ Y & \xleftarrow[\text{res.}]{\text{min.}} & Y' \end{array}$$

is commutative. For each fixed value of $c_1^2(X) = k$ *there is a finite number of possibilities for* Y' *and a finite number of possibilities for* (a,b). *Conversely, starting from any of these* Y' *and any of these* (a,b), *the minimal resolution* X *of a 2-fold covering* $X' \to Y'$ *over a curve of type* (a,b) *with simple singularities only, is a minimal surface of general type with* $c_1^2(X) = k, p_g(X) = \frac{1}{2}c_1^2(X) + 2$.

All possibilities for Y' and (a,b) are listed in Table 12. Every row (separated by horizontal lines) contains exactly one deformation type whose topological type (spin or not) is indicated in the last column.

For surfaces X with $c_1^2(X)$ odd and $p_g(X) = \frac{1}{2}c_1^2(X) + \frac{3}{2}$, Horikawa's results are quite similar. Since, however, the situation is more complicated for a few low values of $c_1^2(X)$, we do not formulate a theorem, but restrict ourselves to a (slightly less precise) description of the different possibilities, referring to [Hor76] for details. (i) $c_1^2(X) = 1$. Then $p_g(X) = 2$, and $|K_X|$ is a pencil with one base point $x_0 \in X$. If $(\overline{X}, E) \to (X, x_0)$ is the blowing-up of X at x_0, then the rational map induced by f_2 on $\overline{X}\backslash E$ can be extended to a morphism $\overline{f}_2$ of degree 2 from $\overline{X}$ onto a quadratic cone $Y \subset \mathbb{P}_3$. Furthermore, there exists a 2-fold branched covering $g\colon X' \to \Sigma_2$, branched over a curve of type (6,10) with simple singularities, such that the diagram

$$\begin{array}{ccc} \overline{X} & \xrightarrow[\text{res.}]{\text{min.}} & X \\ {\scriptstyle \overline{f}_2}\big\downarrow & & \big\downarrow{\scriptstyle g} \\ Y & \xleftarrow[\text{res.}]{\text{min.}} & \Sigma_2 \end{array}$$

is commutative. Every curve of type (6,10) with at most simple singularities occurs in this way.

(ii) $c_1^2(X) = 3$. There are two types:

Type I. f_1 is a holomorphic map of degree 3 onto $\mathbb{P}_2$, and X is the minimal resolution of a 3-section in $\mathcal{O}_{\mathbb{P}_2}(2)$. The 3-sections thus occurring can be listed.

Type II. $|K_X|$ has one base point, and after blowing up this point f_1 extends to a map $\overline{f}_1 : \overline{X} \to \mathbb{P}_2$ of degree 2. There is again a commutative diagram

$$\begin{array}{ccc} \overline{X} & \xrightarrow{\text{min.res}} & \overline{X}' \\ \downarrow{\scriptstyle \bar{f}_1} & \swarrow{\scriptstyle g} & \\ \mathbb{P}_2 & & \end{array}$$

where the 2-fold covering $g : \overline{X}' \to \mathbb{P}_2$ is ramified over a curve of degree 10, decomposing into a line L and a curve of degree 9, having three simple triple points on L . Starting from such a curve the process can be reversed to obtain a minimal surface of general type X with $c_1^2(X) = 3$ and $p_g(X) = 3$.

The types I and II form a single deformation class. Horikawa's proof in [Hor76] can be simplified if we use a different description. Namely, it follows from his considerations that for a surface X of type I

$$R(X) = \mathbb{C}[x_0, x_1, x_2, y]/f(x_0, x_1, x_2, y)$$

with $\deg(x_i) = 1, \deg(y) = 2$ and $\deg(f) = 6$, whereas conversely a surface X with such a canonical ring is of type I, provided f is "generic".

On the other hand, as V. Iliev has shown, for a surface X of type II

$$R(X) = \mathbb{C}[x_0, x_1, x_2, y, z]/(f(x_0, x_1, x_2, y, z), g(x_0, x_1, x_2, y))$$

with $\deg(x_i) = 1, \deg(y) = 2, \deg(z) = 3, \deg(f) = 6, \deg(g) = 3$, while the converse is again true for "generic" f and g.

Since replacing g by $g + tz, t \neq 0$, in the above ring yields a ring of the form

$$\mathbb{C}[x_0, x_1, x_2, y]/f(x_0, x_1, x_2, y)$$

this shows that a type II surface deforms into a surface of type I.

(iii) $c_1^2(X) = 5$. Again there are two types, forming one deformation class (see [Hor75]).

Type I. k_1 is an embedding onto a quintic in $\mathbb{P}_3$ with at most rational double points. Conversely, the minimal desingularization of such a quintic is an X with $c_1^2(X) = 5, p_g(X) = 4$.

Type II. $|K_X|$ has one base point, and after blowing it up, f_1 extends to a holomorphic map of degree 2 onto either a smooth quadric or a quadratic cone in $\mathbb{P}_3$. As before, $\overline{X}$ is the minimal resolution of a 2-fold covering over either Σ_0 or Σ_2, the branch locus respectively being a curve of degree (6,7) and (6,13) with only simple singularities. Again all such curves occur.

(iv) $c_1^2(X) \geq 7$. Here $|K_X|$ has one base point, and after blowing up this point f_1 extends to a map of degree 2 onto either $\mathbb{P}_2$ or a Σ_d. From this point on the situation is analogous to the one in Theorem 9.1, but for the fact that in a well-determined way some non-simple singularities appear on the branch locus. To be more precise: at most two quadruple points appear. If one appears, after blowing up once, the proper transform has simple singularities only: if two of them appear, the same is to hold, except that the two quadruple points may become "infinitely near", i.e. they may coalesce to an octuple point which splits into two quadruple points after one blowing up. Again, all curves of this type appear. See Table 13 below. Every row (separated by horizontal rules) contains one deformation type. All surfaces are non-spin and so the oriented homeomorphism type completely determines the deformation type. The notation $+f$ means that one has to add a fibre to the curve of given type.

Remark 1. All the surfaces X which we just have described, and which are known as Horikawa-surfaces, are simply-connected ([Hor76], [Hor75]).

Remark 2. The simplest case with $c_1^2(X)$ even, namely $c_1^2(X) = 2$, occurred already before, in Proposition 7.2. It was well known long before Horikawa treated the general case ([Saf], Chap.VI,§3).

Table 13.

c_1^2	p_g	Minimal resolution of the canonical model	Type of branch locus	Quadruple points (on the fibre f)
7	5	Σ_1	$(8, 10)$	one
7	5	Σ_3 Σ_1	$(6, 17) + f$ $(6, 11) + f$	two
9	6	Σ_2	$(8, 14)$	none
9	6	Σ_0 Σ_2	$(6, 9) + f$ $(6, 15) + f$	two
$8k-1$	$4k+1$	Σ_{2j+1} $(0 \le j \le k-1)$	$(6, 4k+6j+7) + f$	two
$8k-1$	$4k+1$	Σ_{2k+1}	$(6, 10k+7) + f$	two
$8k-5$	$4k-1$	Σ_{2j+1} $(0 \le j \le k-1)$	$(6, 4k+6j+5) + f$	two
$8k+1$	$4k+2$	Σ_{2j} $(0 \le j \le k)$	$(6, 4k+6j+5) + f$	two
$8k-3$	$4k$	Σ_{2j} $(0 \le j \le k)$	$(6, 4k+6j+3) + f$	two

10. Surfaces with $q = p_g = 0$

The extremal points on the left hand side of D are the points on the line $c_1^2(X) + c_2(X) = 12$, i.e. the points $c_1^2(X) = a, c_2(X) = 12 - a$, with $1 \le a \le 9$. All these points can already be represented by surfaces X with $p_g(X) = q(X) = 0$ (if $c_1^2(X) = 1$, then automatically $p_g(X) = q(X) = 0$, since otherwise application of the unbranched covering trick would lead to a violation of Noether's inequality, but if $c_1^2(X) \ge 2$, this is no longer true). Such surfaces of general type have been studied for a long time, for they are very interesting from the point of view of Castelnuovo's criterion (Theorem VI.3.4): if $q(Y) = 0$ for the non-rational surface Y, then $P_2(Y) \geq 1$. Nowadays a large number of examples is known, they have been obtained by classical constructions, due to Godeaux, Campedelli, Burniat, as well as by some new methods. The fact that so many examples are known does by no means imply that the moduli scheme is known too. Still, some general results have been proved which we list below.

1. Surfaces with $p_g = q = 0$ and $c_1^2 = 1$ are nowadays called numerical Godeaux surfaces. One knows [Rei78], [Mi76] that $H_1(X, \mathbb{Z})$ is either $0, \mathbb{Z}/2\mathbb{Z}, \mathbb{Z}/3\mathbb{Z}, \mathbb{Z}/4\mathbb{Z}$ or $\mathbb{Z}/5\mathbb{Z}$ for these surfaces. Below we discuss a number of construction methods; the Godeaux method gives examples where the fundamental group can be $\mathbb{Z}/3\mathbb{Z}, \mathbb{Z}/4\mathbb{Z}$ or $\mathbb{Z}/5\mathbb{Z}$. Barlow gives a method yielding simply connected surfaces and surfaces with $\pi_1 = \mathbb{Z}/2\mathbb{Z}$. Recently Craighero and Gattazzo constructed numerical Godeaux surfaces by deforming a quintic surface (in projective three space) with certain types of singularities (see [C-G]). These surfaces are believed to be simply connected but so far we only know that $H_1 = 0$ (see [Dol-W]). Also Campedelli's method yields numerical Godeaux surfaces with $\pi_1 = \mathbb{Z}/2\mathbb{Z}$ and $\pi_1 = \mathbb{Z}/4\mathbb{Z}$ (see [O-P] and [We94]). There is a complete classification with description of moduli spaces for the cases $H_1(X, \mathbb{Z}) = \mathbb{Z}/3\mathbb{Z}, \mathbb{Z}/4\mathbb{Z}$ or $\mathbb{Z}/5\mathbb{Z}$ ([Rei78], [Mi77a]).
2. Nowadays surfaces with $p_g = q = 0$ and $c_1^2(X) = 2$ are called numerical Campedelli surfaces. We know that $|H_1(X, \mathbb{Z})| \leq 9$ ([Be79] and [Rei79]) and that neither the dihedral group nor the symmetric group of order 6 can occur as fundamental group ([Rei79], respectively. [Naie99]). There is a complete classification with description of moduli spaces for the cases $\pi_1(X) = \mathbb{Z}/2\mathbb{Z} \oplus \mathbb{Z}/2\mathbb{Z} \oplus \mathbb{Z}/2\mathbb{Z}, \mathbb{Z}/8\mathbb{Z}, \mathbb{Z}/2\mathbb{Z} \oplus \mathbb{Z}/4\mathbb{Z}$ or the quaternion group of order 8 (see[Rei79]).
   Besides the Campedelli method, there are other constructions giving such surfaces, notably certain double covers of Enriques surfaces ([Naie94], [Ke]).
3. For a recent survey of surfaces with $p_g = q = 0$, $c_1^2 \geq 2$ we refer to [M-P02], where the point of departure is the study of the bicanonical map; it discusses results by Xiao Gang [Xi85c], [Xi90], Langer [Langer], Mendes Lopes [ML], Mendes Lopes-Pardini [MP01a], [MP01b], [MP02a], [MP02b].

   We especially mention the results concerning moduli:

1) If $c_1^2 = 8$ and the bicanonical map (which is a morphism) has degree 2, the surface can be obtain by the Beauville method and belongs to exactly four different families. See [Pa].
2) If $c_1^2 = 6$ then the bicanonical (which is a morphism) has degree ≤ 4 and when equality holds the surface must be a Burniat surface. These surfaces yield a unirational irreducible 4-dimensional connected component of the moduli space. See [MP01].

We shall now list several of the aforementioned constructions. For more details we refer to the original papers or to Dolgachev's excellent survey [Dol81], which however covers only results up to 1977. For more recent surveys see [C-P] and [M-P02].

1. *The Godeaux Construction*

We have met this construction already in V, Sect. 15. The idea is to start with a minimal surface of general type Y with $\chi(Y) = a$, on which a group G of order a operates without fixed points. Then by Theorem I.7.4 the quotient will be a minimal surface of general type with $\chi(X) = 1$, and if moreover there are no holomorphic 2-forms on Y which are invariant under G, then $p_g(X) = q(X) = 0$. Apart from a smooth quintic in $\mathbb{P}_3$ (with $G = \mathbb{Z}/5\mathbb{Z}$ yielding surfaces with $c_1^2 = 1$) the construction can also be applied to an intersection of four quadrics in $\mathbb{P}_6$, with G one of the groups $\mathbb{Z}/2\mathbb{Z} \oplus \mathbb{Z}/2\mathbb{Z} \oplus \mathbb{Z}/2\mathbb{Z}, \mathbb{Z}/2\mathbb{Z} \oplus \mathbb{Z}/4\mathbb{Z}, \mathbb{Z}/8\mathbb{Z}$ or the quaternion group. The quotient X now has $c_1^2(X) = 2$, and $p_g(X) = q(X) = 0$. For recent classification results see [C-P].

2. *The Campedelli Construction*

The imposition of non-simple singularities on the branch curve of a 2-fold covering diminishes the values of c_1^2 and c_2 (compare V, Sect. 22). Starting from any reduced curve of degree 10 in $\mathbb{P}_2$, having six triple points with infinitely near ordinary triple points (i.e. after blowing up once the triple points become ordinary triple points) one can thus construct a surface X with $c_1^2(X) = 2, p_g(X) = q(X) = 0$. Campedelli ([Cam]) gave examples of such curves; his double planes are special Godeaux-type surfaces ([Pet76]). More surfaces with $c_1^2 = 2$ or 1 arise if certain other singularities occur on the branch locus, see for instance [O-P], [We94], [Sta].

3. *The Burniat Construction*

Starting point is a configuration formed by nine different lines $A_0, A_1, A_2, B_0, B_1, B_2, C_0, C_1, C_2$ in $\mathbb{P}_2$. The lines A_0, B_0, C_0 form a non-degenerate triangle Δ, whereas A_1, A_2 both pass through a common vertex of Δ; B_1, B_2 both pass through a second vertex and C_1, C_2 both pass through the third vertex of Δ. The surface X in question is obtained by first taking the double covering X_1 of $\mathbb{P}_2$, branched along $A_0+A_1+A_2+B_0+B_1+B_2$, then taking the double covering X_2 of X_1, branched along the inverse image of $C_0 + C_1 + C_2$ (of course, one has to verify that it exists), and finally desingularizing X_2 in a minimal way.

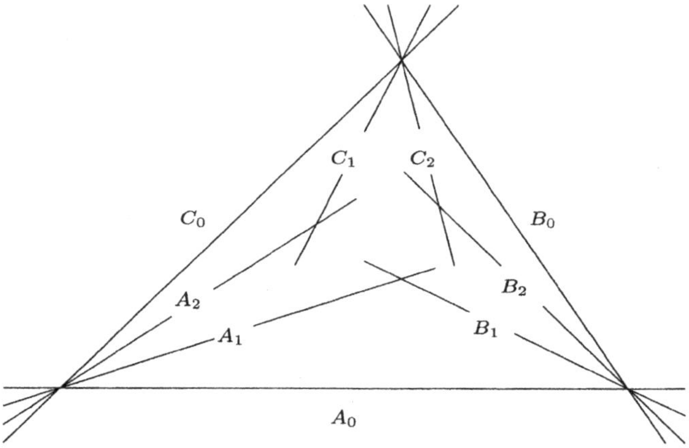

If m is the number of points through which there passes a line A_i, a line B_j and a line $C_k, 1 \leq i, j, k \leq 2$ (so $0 \leq m \leq 4$), then $c_1^2(X) = 6 - m$, whereas $p_g(X) = q(X) = 0$. For further details we refer to [Bu] and [Pet77].

4. *The Inoue Construction*

Let $Z = E_1 \times E_2 \times E_3$ be the product of three elliptic curves $E_i = \mathbb{C}/(\mathbb{Z}1 + \mathbb{Z}\tau_i), i = 1, 2, 3$, and Y an irreducible hypersurface on X with the following properties:

(i) Y is of type $(2, 2, 2)$, i.e. if $\rho_i : Z \to E_i$ is the projection, then $\mathcal{O}_Z(Y) \cong \prod_{i=1}^{3} \rho_i^*(\mathcal{L}_i)$, where $\mathcal{L}_i$ is a line bundle of degree 2;

(ii) $\mathcal{O}_Z(Y)$ is very ample;

(iii) as singularities Y has only n nodes, each of them in one of the 2-division points of Z;

(iv) Y is left invariant by the group G of automorphisms of Z, generated by

$$\begin{array}{lcl} (z_1, z_2, z_3) & \longrightarrow & (-z_1 + \frac{1}{2}, z_2 + \frac{1}{2}, z_3) \\ (z_1, z_2, z_3) & \longrightarrow & (z_1, -z_2 + \frac{1}{2}, z_3 + \frac{1}{2}) \\ (z_1, z_2, z_3) & \longrightarrow & (z_1 + \frac{1}{2}, z_2, -z_3 + \frac{1}{2}); \end{array}$$

(v) G acts without fixed points on Y', the minimal desingularization of the surface Y.

These requirements imply that $n \equiv 0 (\mathrm{mod}\, 4)$. The quotient $X = Y'/G$ has the invariants $p_g(X) = q(X) = 0, c_1^2(X) = 6 - \frac{n}{4}$. In [In94] Inoue constructs a family Y_c, depending on a parameter $c \in \mathbb{C}$, satisfying (i)-(v), such that $n = 0$ for general c, but $n = 4, 8, 12$ or 16 for special values of c. Using instead complete intersections in a product of 4 elliptic curves, he furthermore constructs surfaces on which $\bigoplus^5 \mathbb{Z}/2\mathbb{Z}$ acts freely; the quotient X can have $c_1^2(X) = 7, 6, 5$ and $p_g(X) = q(X) = 0$. While the first four examples can be shown to be Burniat surfaces, the surface of the last construction with $c_1^2(X) = 7$ certainly is not Burniat (and it is not clear for those with

$c_1^2(X) = 6, 5$). Finally, he produces a complete intersection inside a product of 4 elliptic curves on which $\bigoplus^4 \mathbb{Z}/2\mathbb{Z}$ acts freely yielding a surface of general type with $p_g = q = 0$, $c_1^2 = 8$.

5. *The Beauville Construction*

Let C_1, C_2 be two smooth connected compact curves of genus p_1, p_2 respectively, with $p_1, p_2 \geq 2$. Let $Y = C_1 \times C_2$ and G a finite group of order $(p_1 - 1)(p_2 - 1)$, operating on both C_1 and C_2, such that, but for the unit element, no element of G has a fixed point on both C_1 and C_2. This implies that the induced action of G on Y is fixed point-free. So if $X = Y/G$, then $c_1^2(X) = 8$, and if moreover $C_i/G \simeq \mathbb{P}_1$, then $p_g(X) = q(X) = 0$. We refer to [Be78], p. 159 and [Dol81] for explicit examples. Catanese ([Cat00], [Cat01]) has recently shown that either 1 or 2 connected components of the moduli space $M_{8,4}$ parametrize surfaces with the same fundamental group as X. This depends on how G acts. If we have 2 components complex conjugation exchanges the two components and so this yields examples of diffeomorphic surfaces that are not deformation equivalent. See the end of Sect. 8. The examples for which in addition $p_g = q = 0$ are all rigid in the sense that the component of the moduli space containing X is a point. See loc. cit. and [Bau-C] where many interesting new examples are constructed.

The fundamental group of these surfaces is always infinite: it has an explicit presentation as a central extension of G by the product $\pi_1(C_1) \times \pi_1(C_2)$.

6. *The Catanese Construction*

Let $Q \subset \mathbb{P}_3$ be a quintic with exactly twenty ordinary double points such that the following conditions are satisfied:

(i) the group $\mathbb{Z}/5\mathbb{Z}$ acts freely on Q, and hence freely on the minimal desingularization $\tilde{Q}$ of Q;

(ii) the line bundle corresponding to the sum of the twenty exceptional (-2)-curves on $\tilde{Q}$ is 2-divisible, so that there is a double covering $\tilde{Y}$ of $\tilde{Q}$, branched over the exceptional locus;

(iii) the $\mathbb{Z}/5\mathbb{Z}$-action on $\tilde{Q}$ can be lifted to $\tilde{Y}$.

If Y is obtained from $\tilde{Y}$ by blowing down the (-1)-curves lying over the twenty (-2)-curves, then $\mathbb{Z}/5\mathbb{Z}$ still acts freely on Y. The quotient $X = Y/(\mathbb{Z}/5\mathbb{Z})$ turns out to be minimal with $c_1^2(X) = 2, p_g(X) = q(X) = 0$. Catanese produced in [Cat81a] a 4-dimensional family of quintics satisfying the conditions above. His surfaces Y are simply-connected, as follows from the fact that one of them is the minimal model of a Hilbert modular surface (of a type more general then those described in Chap. V).

7. *The Barlow Construction*

Suppose we are given a simply-connected surface Y with $c_1^2(Y) = 10, p_g(Y) = 4$, admitting an action of the dihedral group $\mathbb{D}_5$, such that the normal subgroup $\mathbb{Z}/5\mathbb{Z}$ acts freely, whereas each of the five conjugate involutions in $\mathbb{D}_5$

has exactly four isolated fixed points. Then the desingularization X of the quotient $Y/\mathbb{D}_5$ is simply-connected and satisfies $c_1^2(X) = 1, p_g(X) = q(X) = 0$. In [Bw84], [Bw85] Barlow has constructed a 2-dimensional family of such Y's. In fact the base manifold of a versal deformation is smooth and of dimension 8; indeed the generic deformation does not contain (-2)-curves so that its canonical bundle is ample ([C-LeB] [Lee]).

In the following table we have collected many examples of surfaces with $p_g = q = 0$, giving for each type π_1, H_1 (in as far as they have been determined), at least one method of construction, and one or two references.

Table 14. Surfaces with $q = p_g = 0$

c_1^2	π_1	H_1	Method of construction	Moduli scheme	Reference
1	0	0	no. 7		[Bw84]
	?	0	sing. quintics		[C-G], [Dol-W]
	$\mathbb{Z}/2\mathbb{Z}$	$\mathbb{Z}/2\mathbb{Z}$	Barlow & Reid	8-dim	[Bw85]
	$\mathbb{Z}/2\mathbb{Z}$	$\mathbb{Z}/2\mathbb{Z}$	quot. of no. 4 with $c_1^2 = 5, 3$		[In94]
		$\mathbb{Z}/2$	no. 2		[We94], [We97]
	$\mathbb{Z}/3\mathbb{Z}$	$\mathbb{Z}/3\mathbb{Z}$	no. 1	irred.	[Rei78]
	$\mathbb{Z}/4\mathbb{Z}$	$\mathbb{Z}/4\mathbb{Z}$	no. 1	irred.	[Rei78]
		$\mathbb{Z}/4\mathbb{Z}$	no. 2		[O-P], [We94]
	$\mathbb{Z}/4\mathbb{Z}$	$\mathbb{Z}/4\mathbb{Z}$	double cover of Enr. surf.		[Naie94]
	$\mathbb{Z}/5\mathbb{Z}$	$\mathbb{Z}/5\mathbb{Z}$	no. 1	unirat. 8-dim	[Rei78] [Mi76]
2	$\mathbb{Z}/8\mathbb{Z}$	$\mathbb{Z}/8\mathbb{Z}$	no. 1	unirat. 6-dim	[Rei79]
	$\mathbb{Z}/4\mathbb{Z} \oplus \mathbb{Z}/2\mathbb{Z}$	$\mathbb{Z}/4\mathbb{Z} \oplus \mathbb{Z}/2\mathbb{Z}$	no. 1	unirat. 6-dim	[Rei79]
	$\mathbb{Z}/4\mathbb{Z} \oplus \mathbb{Z}/2\mathbb{Z}$	$\mathbb{Z}/4\mathbb{Z} \oplus \mathbb{Z}/2\mathbb{Z}$	double cover of Enr. surf.		[Naie94], [Ke]
	$(\mathbb{Z}/2\mathbb{Z})^3$	$(\mathbb{Z}/2\mathbb{Z})^3$	no. 1, no. 2	unirat. 6-dim	[Rei79], [Pet76]
	$\mathbb{H}^\dagger$	$(\mathbb{Z}/2\mathbb{Z})^2$	no. 1	unirat. dim = 6	[Rei79]
	$\mathbb{H}$	$(\mathbb{Z}/2\mathbb{Z})^2$	no. 3, no. 4		[In94], [Pet77]

†$\mathbb{H} = \{\pm 1, \pm \mathrm{i}, \pm \mathrm{j}, \pm \mathrm{k}\}$ with the usual relations (quaternion group of order 8)

Table 14. Continuation

c_1^2	π_1	H_1	Method of construction	Moduli scheme	Reference
2	$(\mathbb{Z}/2\mathbb{Z})^2$	$(\mathbb{Z}/2\mathbb{Z})^2$	quot. of no. 4		[In94]
	$\mathbb{Z}/5\mathbb{Z}$	$\mathbb{Z}/5\mathbb{Z}$	no. 6		[Cat81a], [Sup]
			no. 2		[Pet77]
	$(\mathbb{Z}/3\mathbb{Z})^2$	$(\mathbb{Z}/3\mathbb{Z})^2$	no. 1		[Xi85a]
3	$\mathbb{H} \oplus \mathbb{Z}/2\mathbb{Z}$	$(\mathbb{Z}/2\mathbb{Z})^3$	no. 3, no. 4		[In94], [Pet77]
	$(\mathbb{Z}/2\mathbb{Z})^3 \oplus \mathbb{Z}/4\mathbb{Z}$	$(\mathbb{Z}/2\mathbb{Z})^3 \oplus \mathbb{Z}/4\mathbb{Z}$	double cover of Enr. surf.		[Naie94], [Ke]
4	$\mathbb{H} \oplus (\mathbb{Z}/2\mathbb{Z})^2$	$(\mathbb{Z}/2\mathbb{Z})^4$	no. 3, no. 4		[In94], [Pet77]
	$\mathbb{Z}^4 \ltimes (\mathbb{Z}/2\mathbb{Z})^2$	$\mathbb{Z}/4\mathbb{Z} \oplus (\mathbb{Z}/2\mathbb{Z})^3$	double cover of Enr. surf.		[Naie94], [Ke]
5	$\mathbb{H} \oplus (\mathbb{Z}/2\mathbb{Z})^3$	$(\mathbb{Z}/2\mathbb{Z})^5$	no. 3, no. 4		[In94], [Pet77]
6	Extension of $(\mathbb{Z}/2\mathbb{Z})^3$ by $\mathbb{Z}^6$	$(\mathbb{Z}/2\mathbb{Z})^6$	no. 3, no. 4		[In94], [Pet77]
7			no. 4		[In94]
8	infinite	?	number theory (many examples) (Chap. V.20)		[Shav]
	infinite	$(\mathbb{Z}/5\mathbb{Z})^2$	no. 5 (many ex.)	1 point	[Be78], [Cat00], [Cat01], [Bau-C]
9	infinite	?	p-adic geometry (Chap.V, Sect 1)		[Mu79]

Chapter VIII. K3-Surfaces and Enriques Surfaces

What is said about the use of "surface" and "curve" (on a surface) at the beginning of Chap. VI also applies in this chapter.

In this chapter we consider in detail the class of K 3-surfaces and that of Enriques surfaces. We start with some notation and after that we state the main results in Sect. 2. In Chapt. IV, Sect. 3 we saw that K 3-surfaces are Kähler, a fact we use from the start. The main tool for studying moduli of K 3-surfaces is the period map and we describe these moduli spaces in terms of the corresponding period domains. This is done in Sect. 6–14 after we have proved some general facts concerning the geometry of divisors on K 3-surfaces and Kummer surfaces, collected in Sect. 3–5. The geometry of Enriques surfaces as discussed in Sect. 15–18 is then coupled with a study of the period map of their universal covers in order to arrive at a description of the moduli space in terms of certain classical bounded domains. See Sect. 19–21. We finish this chapter with more recent results for projective K 3-surfaces. First, we consider their moduli spaces. After this we discuss the construction of mirror families for K 3-surfaces. Next we present Mumford's proof that every K 3-surface contains a (possibly singular) rational curve and a 1-dimensional algebraic family of (in general singular) elliptic curves. Then we discuss enumerative results for rational curves and we finish with an application to hyperbolic geometry (related to the Green-Griffiths and Lang conjectures).

Introduction

1. Notation

Let X be a compact (connected) surface with $b_1(X)$ even, with cup product form $(\ ,\ )$ on $H^2(X, \mathbb{C})$. By Theorem IV. 2.14 the signature of $(\ ,\ )|H^{1,1}(X, \mathbb{R})$ is $(1, h^{1,1} - 1)$, so if $h^{1,1}(X) \geq 2$ the set $\{x \in H^{1,1}(X, \mathbb{R}) \mid (x, x) > 0\}$ consists of two disjoint connected cones, say $\mathcal{C}_X$ and $\mathcal{C}'_X$. Recall that the hypothesis $b_1(X)$ even implies that X is Kähler. Since the Kähler classes form a convex subcone of $\mathcal{C}_X \cup \mathcal{C}'_X$, they all belong to one of them, say $\mathcal{C}_X$, the positive cone. We next rephrase IV, Cor. 7.2 in our setting:

$$(1)\qquad \begin{cases} x, y \in \bar{\mathcal{C}}_X \text{ implies } (x,y) \geq 0 \text{ with strict inequality if either} \\ x \text{ or } y \text{ is contained in } \mathcal{C}_X. \end{cases}$$

Let $j : H^2(X, \mathbb{Z}) \to H^2(X, \mathbb{C})$ be induced by the inclusion, and let

$$\mathrm{NS}(X) = H^{1,1}(X, \mathbb{R}) \cap \operatorname{Im} j^*(H^2(X, \mathbb{Z})) \quad (\text{the Néron-Severi lattice}),$$
$$T_X = \mathrm{NS}(X)^\perp \cap \operatorname{Im} j^*(H^2(X, \mathbb{Z})) \quad (\text{the transcendental lattice}).$$

We shall give names to some special types of elements in $j^{-1}(\mathrm{NS}(X))$. So let $d \in j^{-1}(\mathrm{NS}(X))$. We call d divisorial if there exists at least one divisor D with $c_1(\mathcal{O}_X(D)) = d$. Then d is called effective if moreover D can be chosen effective and similarly for irreducible. The classes of the (-2)-curves are called the nodal classes in $\mathrm{NS}(X)$. An effective class which is *not* the sum of two other effective classes is called indecomposable. N.B. It may very well happen that an irreducible class is decomposable (for example the class of an irreducible conic in $\mathbb{P}_2$).

If X, X' are surfaces, an isomorphism of $\mathbb{Z}$-modules

$$H^2(X, \mathbb{Z}) \longrightarrow H^2(X', \mathbb{Z})$$

is called a Hodge-isometry if

i) it preserves the cup product (i.e., it is an isometry), and
ii) its $\mathbb{C}$-linear extension $H^2(X, \mathbb{C}) \longrightarrow H^2(X', \mathbb{C})$ preserves the Hodge-decomposition (IV, Sect. 2).

If X, X' are moreover Kähler surfaces, a Hodge-isometry is called effective, if it preserves the positive cones and induces a bijection between the respective sets of effective classes.

For any $d \in j^{-1}(\mathrm{NS}(X))$ with $(d, d) = -2$ the isomorphism

$$s_d : H^2(X, \mathbb{Z}) \longrightarrow H^2(X, \mathbb{Z})$$

defined by $s_d(x) = x + (x, d)d$ ($x \in H^2(X, \mathbb{Z})$) is a Hodge-isometry, because $d \in H^{1,1}$ implies that the $\mathbb{C}$-linear extension of s_d is the identity on $H^{2,0} \oplus H^{0,2}$ and preserves $H^{1,1}$. Its $\mathbb{R}$-linear extension to $H^2(X, \mathbb{R})$ is also denoted by s_d. Observe that $s_d(d) = -d$ and that s_d is just the orthogonal reflection in the hyperplane H_d orthogonal to d. We call s_d a Picard-Lefschetz reflection. The subgroup of $\operatorname{Aut}(H^{1,1}(X, \mathbb{R}))$ generated by the Picard-Lefschetz reflections is denoted by W_X. In case of a Kähler surface X we define the Kähler surface X we define the Kähler cone to be the convex subcone of the positive cone consisting of those elements which have positive inner product with any effective class in $\mathrm{NS}(X)$. The Kähler cone contains all Kähler classes.

We finally introduce some notation in connection with the relevant period domains. We put

$$L = -E_8 \oplus -E_8 \oplus H \oplus H \oplus H \quad (\text{cf. also I, Sect. 2}),$$
$$L_{\mathbb{C}} = L \otimes \mathbb{C} \quad \text{with } (\ ,\ ) \text{ extended } \mathbb{C}\text{-bilinearly},$$
$$L_{\mathbb{R}} = L \otimes \mathbb{R} \quad \text{with } (\ ,\ ) \text{ extended } \mathbb{R}\text{-bilinearly},$$

For $\omega \in L_{\mathbb{C}}$ we denote by $[\omega] \in \mathbb{P}(L_{\mathbb{C}})$ the corresponding line and set

$$\Omega = \{[\omega] \in \mathbb{P}(L_{\mathbb{C}}) \mid (\omega, \omega) = 0 \ , \quad (\omega, \bar{\omega}) > 0\}.$$

2. The Results

By definition a K 3-surface is a surface X with $\mathcal{K}_X$ trivial and $b_1(X) = 0$. In Sect. 8 we shall show that all K 3-surfaces are deformations of each other, so they are all diffeomorphic. In particular, since the K 3-surface considered by way of an example in V, Sect. 2 is simply-connected, all K 3-surfaces are simply-connected.

In Sect. 3 we shall show that $H^2(X, \mathbb{Z})$ is free of rank 22 and isometric to L. Since $p_g(X) = 1$, the choice of an isometry $\phi : H^2(X, \mathbb{Z}) \to L$ determines a line in $L_{\mathbb{C}}$ spanned by the $\phi_{\mathbb{C}}$-image of a nowhere vanishing holomorphic 2-form ω_X. The identity $(\omega_X, \omega_X) = 0$ and the inequality $(\omega_X, \bar{\omega}_X) > 0$ imply that this line, considered as a point of $\mathbb{P}(L_{\mathbb{C}})$, belongs to Ω (compare also IV, Sect. 4). This point is called the period point of the marked K 3-surface (X, ϕ).

The first main theorem states that two K 3-surfaces are isomorphic if and only if there are markings for them, such that the corresponding period points are the same. This is the Weak Torelli Theorem, which can also be stated as follows: *two* K 3-*surfaces* X, X' *are isomorphic if there exists a Hodge-isometry* $\phi : H^2(X', \mathbb{Z}) \to H^2(X, \mathbb{Z})$. In this form we shall prove it in Sect. 11. It is a consequence of the more refined Torelli theorem, which asserts that *if in addition* ϕ *is assumed to be effective, then it is induced by a unique isomorphism* $X \to X'$.

The second main theorem states that *all points of* Ω *occur as period points of marked* K 3-*surfaces* (Theorem 14.2).

An Enriques surface by definition is a surface Y with $\mathcal{K}_Y^{\otimes 2} = \mathcal{O}_Y, p_g(Y) = b_1(Y) = 0$ (compare VI, Sect. 1). We recall from V, Sect. 23 that every Enriques surface is projective. In Sect. 16 we shall show that they always admit at least one elliptic fibration over $\mathbb{P}_1$.

All Enriques surfaces turn out to be deformations of each other (Theorem 18.5), so they are all diffeomorphic. Since for an Enriques surface Y we have $p_g(Y) = 0$, we cannot form the period map. Instead we shall pass to the universal covering X of Y, which is easily shown to be a K 3-surface doubly covering Y (see Lemma 15.1). We then simply might take any marking for $H^2(X, \mathbb{Z})$ and consider the period point of X as a period point for Y. However, in doing this, we loose the extra information that X comes from an Enriques surface. So we consider instead only those markings $\phi : H^2(X, \mathbb{Z}) \xrightarrow{\sim} L$ for which $\phi \circ \sigma^* = \rho \circ \phi$, where $\sigma : X \to X$ is the involution interchanging the sheets of $X \to Y$ and where ρ is a certain involution, to be defined in Sect. 18.

The period points of the special marked K 3's then all belong to the intersection Ω^- of Ω with the linear subspace $\mathbb{P}(L^- \otimes \mathbb{C})$ of $\mathbb{P}(L_{\mathbb{C}})$, where $L^- = \{\ell \in L \mid \rho(\ell) = -\ell\}$. This intersection consists of two connected manifolds of dimension 10, each of which is isomorphic to a bounded domain of type IV. The discrete group Γ of automorphisms of L commuting with ρ acts properly and discontinuously on Ω and the quotient D is a quasi-projective variety, the **period domain.** For a fixed Enriques surface Y all of the previously considered special markings differ by elements of Γ, so they give the same point of D, the **period point** of Y.

The counterpart of the first main theorem for K 3-surfaces is valid in the following form: *two Enriques surfaces are isomorphic if and only if they have the same period point* (Theorem 21.2).

It is not true that all points of D occur as period points of Enriques surfaces. We have to delete certain hyperplanes from Ω^-, namely the hyperplanes $H_d = \{[\omega] \in \Omega \mid (\omega, d) = 0\}$ for $d \in L$, $(d,d) = -2$, $\rho(d) = -d$. In Sect. 20 we show that their images in D form finitely many algebraic hypersurfaces, and in Sect. 20.1 we finally prove that all points of D not lying on those, actually occur as period points for Enriques surfaces.

K 3-Surfaces

3. Topological and Analytical Invariants

To determine the isometry class of the cup product pairing, we need a tool from topology which we need later on as well:

(3.1) **Lemma.** *The cup product form of a complex surface X with $H_1(X,\mathbb{Z})$ free (this is the case for instance if X is simply connected) is even precisely when there exists a line bundle L with $K_X = L^{\otimes 2}$.*

Proof. Wu's formula ([MS p. 132]):

$$(w_2, c) \equiv (c,c) \bmod 2, \quad c \in H^2(X, \mathbb{Z}/2\mathbb{Z})$$

implies that the induced form with $\mathbb{Z}/2\mathbb{Z}$-coefficients is even (i.e. $(c,c) \equiv 0 \bmod 2$) precisely when $w_2 = 0$. But w_2 is the mod 2-reduction of $c_1 = c_1(K_X)$ and then the result follows since the universal coefficient theorem and the absence of torsion in $H_1(X,\mathbb{Z})$ imply that $H^2(X,\mathbb{Z}) \otimes \mathbb{Z}/2\mathbb{Z} = H^2(X, \mathbb{Z}/2\mathbb{Z})$. □

By Theorem IV. 2.7, (ii), we have that $b_1(X) = 0$ is equivalent to $q(X) = 0$. So we have $\chi(X) = 2$ and hence by Noether's formula I, (4) and Theorem I. 3.1 we obtain:

(3.2) **Proposition.** $c_1(X) = 0$, $c_2(X) = 24$, $\tau(X) = -16$ *for any* K 3-*surface X.*

(3.3) **Proposition.** *Let X be a* K 3-*surface. Then*
(i) $H^1(X,\mathbb{Z}) = H^3(X,\mathbb{Z}) = 0$, *and*

(ii) *$H^2(X,\mathbb{Z})$ is torsion-free of rank 22 and, when equipped with the cup product pairing, isometric to L.*

Proof. To prove (i) and the fact that $H^2(X,\mathbb{Z})$ is torsion-free it suffices to show that $H_1(X,\mathbb{Z})$ has no torsion. So suppose that $H_1(X,\mathbb{Z})$ has n-torsion. Then there exists a non-trivial unramified covering $Y \to X$ of degree n. Since $\mathcal{K}_1 \cong \mathcal{O}_Y$, it follows that $p_g(Y) = 1$. Noether's formula for Y now reads

$$2 - 2q(Y) = \chi(Y) = n\chi(X) = 2n ,$$

hence $n = 1$ and so $H_1(X,\mathbb{Z})$ is torsion-free. The preceding Lemma shows that the cup product form on $H^2(X,\mathbb{Z})$ is even. It is also unimodular (by Poincaré-duality) and indefinite (by Proposition 3.2), so (ii) follows from Theorem I. 2.8, for both L and $H^2(X,\mathbb{Z})$ have rank 22 and index -16.
□

(3.4) **Proposition.** *Let X be any* K 3*-surface. Then:*

$$h^{0,1}(X) = h^{1,0}(X) = h^{2,1}(X) = h^{1,2}(X) = 0 ,$$
$$h^{0,2}(X) = h^{2,0}(X) = 1 ,$$
$$h^{1,1}(X) = 20 .$$

Proof. Since $b_1(X) = 0$, we have $h^{0,1}(X) = h^{1,0}(X) = 0$ by Theorem IV. 2.7. This and Serre duality imply that $h^{2,1}(X) = h^{1,2}(X) = 0$. Secondly, $h^{2,0}(X) = h^{0,2}(X) = 1$, hence $h^{1,1}(X) = 20$ in view of the Hodge decomposition (Theorem IV. 2.10) and the fact that $b_2(X) = 22$. □

(3.5) **Corollary.** *For a* K 3*-surface X we have*

$$h^0(\mathcal{T}_X) = h^2(\mathcal{T}_X) = 0 \quad \text{and} \quad h^1(\mathcal{T}_X) = 20 .$$

In fact, the isomorphism $\mathcal{K}_X \simeq \mathcal{O}_X$ defines (via the exterior product) an isomorphism $\mathcal{T}_X \cong \Omega^1_X$.

Next, we study the Néron-Severi lattice of a K 3-surface.

(3.6) **Proposition.** *The map $c_1 : \mathrm{Pic}(X) \to H^2(X,\mathbb{Z})$ is injective, hence maps* Pic(X) *isomorphically onto the Néron-Severi lattice. In particular, an effective divisor is never homologous to zero.*

Proof. This follows from the cohomology exponential sequence (I, Sect. 6) since $H^1(\mathcal{O}_X) = 0$ by Proposition 3.4. □

(3.7) **Proposition.**
(i) *If $d \in \mathrm{NS}(X)$, $d \neq 0$, and $(d,d) \geq -2$, then d or $-d$ is effective.*(*)
(ii) *If d is irreducible, then $(d,d) \geq -2$ and equality holds precisely when d is a nodal class.*

(*) Since X is simply-connected, we can identify $H^2(X,\mathbb{Z})$ with its j-image.

(iii) *An indecomposable class is irreducible.*
(iv) *A nodal class is represented by only one effective divisor (the (−2)-curve) and in particular it is indecomposable.*

Proof.
(i) Let $d = c_1(\mathcal{L})$, $\mathcal{L} \in \mathrm{Pic}(X)$. Riemann-Roch yields the inequality

$$h^0(\mathcal{L}) + h^0(\mathcal{L}^\vee) \geq 2 + \frac{1}{2}(d, d) \tag{2}$$

and hence $\mathcal{L}$ or $\mathcal{L}^\vee$ has a non-trivial section.
(ii) Let D be an irreducible divisor. Since $\mathcal{K}_X \cong \mathcal{O}_X$ the adjunction formula implies that $D^2 \geq -2$, where equality holds if and only if D is a (-2)-curve.
(iii) By definition.
(iv) Let $d = c_1(\mathcal{O}_X(D)) = c_1(\mathcal{O}_X(D'))$ with D a (-2)-curve and D' effective. Since $DD' = -2$, we have that D is a component of D', hence $D' - D \geq 0$. Now by Proposition 3.6 an effective divisor cannot be homologous to 0 on X, so $D' = D$. □

(3.8) **Proposition.** *The set of effective classes on a* K 3*-surface is the semi group generated by the nodal classes and the integral points in the closure of the positive cone.*

Proof. Let d be an irreducible class. Since (,) is even by Proposition 3.3, (ii) we see that Proposition 3.7, (ii) implies that either d is nodal or $(d, d) \geq 0$. In the last case $d \in \overline{\mathcal{C}}_X \cup \overline{\mathcal{C}}'_X$, so $d \in \pm\overline{\mathcal{C}}_X$. Since $(d, \kappa) > 0$ for every Kähler class κ (Lemma I. 13.1) and since by (1), for every pair $x, y \in \overline{\mathcal{C}}_X$ we have that $(x, y) \geq 0$, it follows that $d \in \overline{\mathcal{C}}_X$ (*and not* $-d \in \overline{\mathcal{C}}_X$).

It remains to show that any class $d \in \overline{\mathcal{C}}_X \cap H^2(X, \mathbb{Z})$ is effective. Such a d is contained in the Néron-Severi lattice and by Proposition 3.7, (i) either d or $-d$ is effective. The second possibility is excluded as before. □

Next we set $\Delta = \{d \in \mathrm{NS}(X) \mid (d, d) = -2 \text{ and } d \text{ effective}\}$ and for every $d \in \Delta$ we let H_d be the hyperplane of fixed points of the Picard-Lefschetz reflection s_d corresponding to d. We call the connected components of $\mathcal{C}_X \setminus \bigcup_{d \in \Delta} H_d$ the chambers of $\mathcal{C}_X$. Every Kähler class is contained in the following chamber

$$\mathcal{C}_X^+ = \{y \in \mathcal{C}_X \mid (y, d) > 0 \quad \text{for all} \quad d \in \Delta\}.$$

(3.9) **Corollary.** *For a* K 3*-surface* $\mathcal{C}_X^+$ *is the Kähler cone.*

Proof. By definition the Kähler cone consists of all those $x \in \mathcal{C}_X$ such that $(x, d) > 0$ for all effective classes d. By Proposition 3.8 it is sufficient to test this for nodal classes and for $d \in \overline{\mathcal{C}}_X^+$, but for the latter we automatically have $(x, d) > 0$ by (1). □

(3.10) **Proposition.** *The Picard-Lefschetz reflections of a K 3-surface X leave the positive cone invariant, and the group W_X generated by them operates on this cone in a properly discontinuous fashion. The closure of the Kähler cone in the positive cone is a fundamental domain for W_X in the sense that any W_X-orbit in $\mathcal{C}_X$ meets it in exactly one point.*

Proof. Let $d \in \Delta$ and s_d, H_d as before. Since s_d preserves $(\, , \,)$, it preserves $\mathcal{C}_X \cup \mathcal{C}'_X$ so either preserves $\mathcal{C}_X$ or interchanges $\mathcal{C}_X$ and $\mathcal{C}'_X$. Since $s_d|H_d = \mathrm{id}$ and H_d necessarily meets $\mathcal{C}_X$ (because the cup product has signature (1,18) on H_d) the last possibility is excluded.

The subgroup G of $\mathrm{Aut}(H^{1,1}(X, \mathbb{R}), (\, , \,))$ preserving $\mathcal{C}_X$ acts transitively on the set of half-lines in $\mathcal{C}_X$ making this set into a homogeneous space, in fact, a Lobatchevski space. The isotropy group of a point being isomorphic to $O(19)$, hence compact, the group G operates in a proper discontinus way on $\mathcal{C}_X$. Since W_X is a subgroup of $\mathrm{Aut}(L)$, it is discrete in G and therefore acts properly and discontinuously on $\mathcal{C}_X$. It is a general fact that the fundamental domain of a discrete group generated by reflections in hyperplanes of a Lobatchevski space, is the closure of a chamber (compare [Bou68], Exercise to Chap. V, §4, and [Vin]). Since by Corollary 3.9 the Kähler cone is a chamber, the result follows. □

(3.11) **Proposition.** *Let X and X' be two K 3-surfaces and $\phi : H^2(X, \mathbb{Z}) \to H^2(X', \mathbb{Z})$ a Hodge-isometry. The following properties are equivalent:*
(i) *ϕ is effective,*
(ii) *ϕ maps the Kähler cone of X to the one of X',*
(iii) *ϕ maps one element of the Kähler cone of X into the Kähler cone of X'.*

Proof. The Kähler cone being defined in terms of effective classes and intersection properties only, the implication (i) $\Rightarrow$ (ii) is trivial. Also (ii) $\Rightarrow$ (iii) is obvious. As to implication (iii) $\Rightarrow$ (i) , we observe first of all that $\phi(\mathcal{C}_X) = \mathcal{C}_{X'}$, so by Proposition 3.8 we only have to show that if $d \in \mathrm{NS}(X)$ is nodal, then $d' = \phi(d)$ is effective. But if $x \in \mathcal{C}^+_X$ with $\phi(x) \in \mathcal{C}^+_{X'}$, then $(\phi(x), d') = (x, d) > 0$. So $-d'$ cannot be effective, and by Proposition 3.7, (i), the class d' will be effective. □

(3.12) **Corollary.** *If X is a projective K 3-surface, X' any K 3-surface, and $\phi : H^2(X, \mathbb{Z}) \to H^2(X', \mathbb{Z})$ an effective Hodge-isometry, then X' is projective and ϕ maps the class of an ample divisor on X to an ample class on X'.*

Conversely, if X and X' are projective and ϕ is a Hodge-isometry sending the class of an ample divisor on X to the class of an ample divisor on X', then ϕ is effective.

Proof. Integral points in the Kähler cone are just the classes of divisors $d \in \mathrm{NS}(X)$ for which $(d, d) > 0$ and $(d, c) > 0$ for all effective $c \in \mathrm{NS}(X)$. So by Corollary 6.4 these classes are classes of ample divisors. If ϕ is effective, it preserves such classes; in particular X' is projective. Conversely, if X, and X' are projective, then the set of integral points in the respective Kähler cones is non-empty and is preserved by ϕ, so Proposition 3.11 applies. □

4. Digression on Affine Geometry over $\mathbb{F}_2$

We start by recalling some elementary facts from affine geometry, for which we refer to [G-W]. By definition an affine space V over a field k is a set on which the additive group of a vector space $T(V)$ acts transitively and effectively. A choice of $v \in V$ determines the bijection ("choice of origin")

$$f_v : T(V) \longrightarrow V \quad (t \mapsto t(v)) .$$

If V and V' are affine spaces over k an affine-linear map is a map $\phi : V \to V'$ such that for some $v \in V$ the map $T(\phi) : T(V) \to T(V')$ defined by $T(\phi) = f_{\phi(v)}^{-1} \circ \phi \circ f_v$, is a k-linear vector space morphism. Then this is true for all $v \in V$ and $T(\phi)$ is independent of $v \in V$. It is called the linear map induced by ϕ. A map $A : W \to W'$ between k-vector spaces is called semi-linear if there exists an automorphism σ of k such that $A(\lambda w) = \sigma(\lambda)A(w)$ for all $\lambda \in k$, $w \in W$ and $A(w_1 + w_2) = A(w_1) + A(w_2)$ for all $w_1, w_2 \in W$. If in the definition for "affine-linear" we replace "k-linear" by "semi-linear" we obtain the definition of a semi-affine map. Its importance stems from the fundamental theorem of affine geometry which reads as follows:

Let V, V' be affine spaces over k of dimension ≥ 3, and let $\phi : V \to V'$ be a bijection having the property that W is an affine subspace of V if and only if $\phi(W)$ is an affine subspace of V'. Then ϕ is semi-affine.

In the sequel we need a special version of it for $k = \mathbb{F}_2$.

(4.1) **Lemma.** *Let V, V' be affine spaces over $\mathbb{F}_2$ and $\phi : V \to V'$ a bijection inducing a bijection between the respective sets of affine-linear functions on V and V'. Then ϕ is affine.*

Proof. If $\dim V = \dim V' \leq 2$, any bijection is easily checked to be affine. So it suffices to consider the case $\dim V = \dim V' \geq 3$, where we may apply the fundamental theorem of affine geometry. The assumptions of the lemma imply that ϕ maps hyperplanes to hyperplanes, so for any $W \subset V$ the image $\phi(W)$ is affine if and only if W is affine. It follows that ϕ is semi-affine, hence affine, since $\mathbb{F}_2$ does not have any automorphisms except the identity. □

(4.2) **Lemma.** *Let V be an affine space of finite dimension ≥ 2 over $\mathbf{F}_2$ and let U be a linear subspace of the vector space $\mathbb{F}_2^V$ of $\mathbb{F}_2$-valued functions on V. Suppose that U is invariant under the group* $\mathrm{Aut}(V)$ *of affine transformations of V. Then we have the following possibilities for U:*

(a) *U consists of the two constant functions (*i.e., *the polynomial functions of degree 0),*

(b) *U consists of the affine-linear functions (*i.e., *the polynomial functions of degree ≤ 1),*

(c) *U contains for any codimension-2 subspace $W \subset V$ its characteristic function χ_W (they generate the polynomial functions of degree ≤ 2).*

Proof. If U contains non-constant affine-linear functions, it contains all of them, since they form an $\mathrm{Aut}(V)$-orbit. But then U also contains the constants, for they are the sum of two non-constant affine-linear functions.

The non-constant affine-linear functions are exactly the characteristic functions for the affine hyperplanes: if f is affine-linear with zero-set the hyperplane W, then f is the characteristic function of the set $V \setminus W$, which is a hyperplane (we are working over $\mathbb{F}_2$) – and conversely $\chi_W = 1 + f$ where f vanishes exactly on W.

Now suppose that U contains a function g which is not affine-linear. We prove by induction on $n = \dim V \geq 2$ that U contains the characteristic functions of all codimension-2 subspaces.

For $n = 2$, we argue as follows. *Either* g is the characteristic function of one point and U contains all of those, *or* g is the characteristic function of three points and U contains all of those. In the first case we are done. In the second case we observe that U also contains the characteristic functions of *two* points (they are sums of characteristic functions of three points) and therefore also the characteristic functions of *one* point (they are sums of characteristic functions of two and three points).

For $n \geq 3$ we first observe that V contains a hyperplane V' such that $g|V'$ is not affine-linear. Indeed, for at least one hyperplane V'' we have $g|V'' \not\equiv 0$ or 1, since g is not linear. Hence either $g|V''$ is not affine-linear and we take $V' = V''$ *or* the zero set of $g|V''$ is a codimension-2 subspace $W \subset V''$. In the last case *either* $g|V \setminus W \equiv 1$ *or* $g|V'$ is not affine-linear for at least one hyperplane V' in the pencil through W. If $g|V \setminus W \equiv 1$ we may take for V' any hyperplane, not in the pencil through W, which has at least one point in common with W.

So we can apply the induction-hypothesis to $U' = \{f|V' \mid f \in U\}$. It follows that there exists a $g' \in U$, such that $g'|V' = \chi_{W'}$, where W' is an $(n-3)$-dimensional subspace of V'. Let $a \in W'$, $b \in V' \setminus W'$. The set $V \setminus V'$ is a hyperplane and therefore there exists a unique affine transformation t leaving $V \setminus V'$ pointwise fixed and sending a to b. Then t induces a translation by $b - a$ in V', carrying W' to an $(n-3)$-dimensional subspace $t(W')$ of V' which is disjoint from W'. So $g = (g' \circ t) + g' \in U$ is the characteristic function of $W = W' \cup t(W')$. Since W is the $(n-2)$-dimensional subspace of V generated by W' and b, the space U contains at least one characteristic function of a codimension-2 subspace. Since $\mathrm{Aut}(V)$ acts transitively on these subspaces, U contains all of these characteristic functions. □

(4.3) **Lemma.** *Let V be a finite-dimensional affine space over $\mathbb{F}_2$, $T(V)$ its vector space of affine translations and U_k the vector space of polynomial functions of degree $\leq k$. For fixed $v \in V$ and $g \in U_2$ the bilinear form $\mathcal{Q}(g)$ on $T(V)$ defined by*

$$\mathcal{Q}(g)(s,t) = g(f_v(s+t)) + g(v) + g(f_v(s)) + g(f_v(t))$$

is symplectic. It is independent of the choice of $v \in V$, and $\mathcal{Q}(g) = 0$ if and only if $g \in U_1$. The resulting $\mathbb{F}_2$-linear map

$$U_2/U_1 \longrightarrow \bigwedge^2 \mathrm{Hom}(T(V), \mathbb{F}_2) \quad (g \bmod U_1 \mapsto \mathfrak{Q}(g))$$

is an isomorphism.

Proof. Recall that a form $\mathfrak{Q}$ on a vector space W is symplectic if $\mathfrak{Q}(x,x) = 0$ for all $x \in W$. By [Bou59], III, §7, n° 4, prop. 7, p. 80, these forms form a vector space isomorphic to $\bigwedge^2 \mathrm{Hom}(W, \mathbb{F}_2)$. Since $\mathfrak{Q}(g)$ is obviously symplectic and independent of $v \in V$, the assignment $g \mapsto \mathfrak{Q}(g)$ defines a map $U_2 \to \bigwedge^2 \mathrm{Hom}(T(V), \mathbb{F}_2)$. The proof of the remaining assertions is straightforward. □

Remark. The choice of an origin $0 \in V$ identifies V with $T(V)$. If Z' is a codimension-2 affine subspace, we may write $Z' = z + Z$, with $z \in V$ and Z a linear subspace of $T(V)$. One verifies directly that in the above isomorphism $\chi_{Z'} \in U_2$ corresponds to the decomposable vector $f_1 \wedge f_2 \in \bigwedge^2 \mathrm{Hom}(T(V), \mathbb{F}_2)$ for which $(f_1 = 0) \cap (f_2 = 0) = Z$.

5. The Néron-Severi Lattice of Kummer Surfaces

In this section Y will be a complex analytic 2-torus and X its Kummer surface. Using the notations of V, Sect16 we have a commutative diagram

$$\begin{array}{ccc} \overline{Y} & \xrightarrow{\bar{p}} & X \\ {\scriptstyle\sigma}\downarrow & & \downarrow \\ Y & \xrightarrow[p]{} & Y/\langle 1, i \rangle. \end{array}$$

Let $\alpha = \bar{p}_! \circ \sigma^* : H^2(Y, \mathbb{Z}) \to H^2(X, \mathbb{Z})$.

(5.1) **Proposition.** *For all $x, y \in H^2(Y, \mathbb{Z})$ we have $(\alpha(x), \alpha(y)) = 2(x, y)$. In particular α is a monomorphism.*

Proof. Since every element $z \in H^2(\overline{Y}, \mathbb{Z})$ is invariant under the covering involution for $\overline{Y} \to X$ we can write $z = \bar{p}^*(w)$, $w \in H^2(X, \mathbb{Z})$, and the projection formula I, (1) shows that $\bar{p}^*(\bar{p}_! z) = 2z$. So $(\bar{p}_! z_1, \bar{p}_! z_2) = 2(z_1, z_2)$ for any pair $z_1, z_2 \in H^2(\overline{Y}, \mathbb{Z})$ and therefore $\bar{p}_!$ and α multiply the intersection numbers by 2. □

(5.2) **Proposition.** *For the complexification $\alpha_{\mathbb{C}}$ of α we have $\alpha_{\mathbb{C}}(H^{2,0}(Y)) = H^{2,0}(X)$.*

Proof. Since $h^{2,0}(X) = h^{2,0}(Y)$ it suffices to observe that any nowhere-zero holomorphic 2-form on Y is i-invariant. □

We proceed by introducing some more notation. We set

V : the set of point of order 2 on Y, equipped with its natural structure of 4-dimensional affine space over $\mathbb{F}_2$.

e_v : the class of $\bar{p}(\sigma^{-1}(v))$ in $H^2(X,\mathbb{Z})$, the distinguished nodal class of $v \in V$.

W : $\{e_v \mid v \in V\}$.

M : the smallest primitive sublattice of $H^2(X,\mathbb{Z})$ containing W.

$M^\vee$: the dual of M, i.e., $M^\vee = \{m' \in M\otimes_{\mathbb{Z}}\mathbb{Q} \mid (m',m) \in \mathbb{Z} \text{ for all } m \in M\}$ (see also I, Sect. 2).

Since $(w,w) = -2$ for $w \in W$ and $(w,w') = 0$ for $w' \neq w$ in W we have inclusions

$$\mathbb{Z}^W \subset M \subset M^\vee \subset (\frac{1}{2}\mathbb{Z})^W .$$

Let $(\frac{1}{2}\mathbb{Z})^W \to (\frac{1}{2}\mathbb{Z}/\mathbb{Z})^W = \mathbb{F}_2^W$ be the natural projection. So there results an epimorphism

$$r : (\frac{1}{2}\mathbb{Z})^W \to \mathbb{F}_2^W .$$

Explicitly, if $x = \sum_{v\in V} x_v e_v$ with $x_v \in \frac{1}{2}\mathbb{Z}$ then $r(x)(v) = 2x_v \bmod 2$. With these observations the next result is obvious.

(5.3) **Proposition.** *For any subset $V' \subset V$ the characteristic function $\chi_{V'} \in \mathbb{F}_2^V$ is precisely $r(\sum_{v\in V'} \frac{1}{2}e_v)$.*

Ultimately we want to show that $\operatorname{Im}\alpha = (\mathbb{Z}^W)^\perp$. In other words the lattice spanned by the distinguished nodal classes determines $\operatorname{Im}\alpha$ as its orthogonal complement. As a first step we observe:

(5.4) **Proposition.** *$\operatorname{Im}\alpha$ is of finite index in $(\mathbb{Z}^W)^\perp$.*

Proof. By construction $\operatorname{Im}\alpha \subset \operatorname{Im}(H^2(X,\bar{p}(\sigma^{-1}(V)),\mathbb{Z}) \to H^2(X,\mathbb{Z}))$, so $\operatorname{Im}\alpha \subset (\mathbb{Z}^W)^\perp$. Since $\operatorname{Im}\alpha$ and $(\mathbb{Z}^W)^\perp$ are both free $\mathbb{Z}$-modules of rank 6 (Proposition 5.1) the index $((\mathbb{Z}^W)^\perp : \operatorname{Im}\alpha)$ is finite. □

We compute this index, noting that by Lemma I. 2.1, (i) its square equals

$$d(\operatorname{Im}\alpha) \cdot (d(\mathbb{Z}^W)^\perp)^{-1} ,$$

where d denotes the determinant introduced in I, Sect. 2. Since

$$d(\operatorname{Im}\alpha) = 2^6 \quad\text{and}\quad d(M^\perp) = d(M)$$

by Lemma I 2.5, it suffices to compute $d(M)$. But $M \subset (\frac{1}{2}\mathbb{Z})^W$ as observed before and since $r(M) \cong M/\mathbb{Z}^W$ it follows that $d(M) = d(\mathbb{Z}^W) \cdot u^{-2}$ where u is the cardinality of $r(M)$. This shows the importance of the subspace $r(M)$. In fact, we shall presently show

(5.5) **Proposition.** *The subspace $U = r(M) \subset \mathbb{F}_2^V$ consists precisely of the affine-linear functions on V.*

As a corollary we infer that in the above computations $u = 2^5$ and so $[(\mathbb{Z}^W)^\perp : \operatorname{Im}\alpha]^2 = 2^6 \cdot 2^{-16} \cdot 2^{10} = 1$, i.e.,

(5.6) **Corollary.** $\operatorname{Im}\alpha = (\mathbb{Z}^W)^\perp$.

For the proof of Proposition 5.5 we need the following very geometric

(5.7) **Lemma.** *Let $t: V \to V$ be an affine transformation. Then there exists an isometry $\tau: H^2(X,\mathbb{Z}) \to H^2(X,\mathbb{Z})$ such that $\tau(e_v) = e_{t(v)}$ $(v \in V)$.*

Proof. Suppose $T: Y \to Y$ is a homeomorphism with the following properties

1) $T|V = t$,
2) T and i commute.
3) there exists a translation-invariant metric on Y such that on a suitably small ball centred at $v \in V$ the translation by $t(v) - v$ coincides with T.

Such a T lifts to a homeomorphism of $\overline{Y}$ onto itself commuting with the involution of $\overline{Y}$, and – mapping $\sigma^{-1}(v)$ onto $\sigma^{-1}(t(v))$ in an orientation-preserving manner – it descends to a homeomorphism of X onto itself, inducing an isometry of $H^2(X,\mathbb{Z})$ with the required properties.

So it suffices to construct such a T. This will be done by way of a homeomorphism of the universal covering $\mathbb{C}^2$ of Y onto itself, preserving the lattice $\Gamma = \bigoplus_{i=1}^{4} \mathbb{Z}e_i$ for which $\mathbb{C}^2/V \cong Y$. In case t is fixed point-free, it is a translation and for T we take the corresponding translation. If $v_0 \in V$ is a fixed point for t we may assume that $0 \in \mathbb{C}^2$ maps to v_0. Then $v_0 \in V$ serves as origin of the $\mathbb{F}_2$-space V and t is a linear map, given by

$$t\Big(\frac{1}{2}e_i\Big) = \sum_{j=1}^{4} t_{i,j}\Big(\frac{1}{2}e_j\Big)\,, \quad (t_{i,j} \in \mathbb{F}_2)\,.$$

Clearly t admits an $\mathbb{R}$-linear orientation-preserving extension T_0 to $\mathbb{C}^2$. Moreover, since t and hence also T_0 is of finite order, we may replace the euclidean metric on $\mathbb{C}^2$ by an equivalent T_0-invariant one, so that $T_0 \in \mathrm{SO}(4,\mathbb{R})$. Let $\varepsilon > 0$ be so small that for any two distinct $x, y \in \frac{1}{2}\Gamma$ the balls $B(x,\varepsilon)$ and $B(y,\varepsilon)$ are disjoint. The map T_0 takes $B(x,\varepsilon)$ orientation- and length-preserving onto $B(t(x),\varepsilon)$, so T_0 and the translation by $t(x)-x$ are homotopic, when restricted to $B(x,\varepsilon)$. This makes it possible to replace T_0 by $t(x) - x$ on every $B(x, \frac{1}{2}\varepsilon)$, while smoothly adjusting it to T_0 in the spherical shells $B(x,\varepsilon) \setminus B(x,\frac{1}{2}\varepsilon)$, $(x \in \frac{1}{2}\Gamma)$. Moreover, we can assume that the resulting homeomorphism of $\mathbb{C}^2$ onto itself commutes with $-\mathrm{id}$. Since it also respects Γ, it descends to a homeomorphism $T: Y \to Y$ with the required properties. $\square$

Now we give the proof of Proposition 5.5.

Proof. We first need an estimate on the cardinality of U. Since $\operatorname{Im}\alpha \subset M^\perp$ we have $2^6 = d(\operatorname{Im}\alpha) \geq d(M^\perp) = d(M)$ (the last equality follows from Lemma I. 2.5). Since $U \cong M/\mathbb{Z}^W$, the square of the cardinality of U equals $d(\mathbb{Z}^W)d(M)^{-1} \geq 2^{16} \cdot 2^{-6} = 2^{10}$, so U contains at least 2^5 elements.

By Lemma 5.7 the subspace U is invariant under the affine group of V. If U were to contain a function which is not affine-linear, according to

Lemma 4.2 the space U would contain the characteristic functions of *all* codimension-2 subspaces of V. Let V', V'' be two such subspaces having exactly one point in common. Since $\chi_{V'}, \chi_{V''} \in U$, it follows from Proposition 5.3 that $\sum_{v \in V'} \frac{1}{2} e_v$ and $\sum_{v \in V''} \frac{1}{2} e_v$ both belong to M. But their intersection product is $\frac{1}{2}$, so this is impossible. Again by Lemma 4.2, either U consists of the two constant functions or U contains all the affine-linear functions. The previous estimate rules out the first alternative, so the proposition is proved. □

Next, we give various interpretations of $H^2(Y, \mathbb{F}_2)$ in terms of the lattice M generated by the distinguished nodal classes of X. There is an isomorphism of $\mathbb{Z}$-modules

$$\beta : \mathrm{Hom}_{\mathbb{Z}}(\Gamma, \mathbb{Z}) \longrightarrow H^1(Y, \mathbb{Z}),$$

whose $\mathbb{F}_2$-reduction is an isomorphism of $\mathbb{F}_2$-vector spaces

$$\beta_2 : \mathrm{Hom}(T(V), \mathbb{F}_2) \longrightarrow H^1(Y, \mathbb{F}_2)$$

(we use that $T(V) = \Gamma/2\Gamma$ as an $\mathbb{F}_2$-vector space).

According to Proposition 5.1 the map α multiplies $(\ ,\ )$ by 2. Hence $(\mathrm{Im}\,\alpha)^\vee = \frac{1}{2}\,\mathrm{Im}\,\alpha$. Applying Corollary 5.6 we find:

$$\begin{aligned} H^2(Y, \mathbb{F}_2) &\cong H^2(Y, \mathbb{Z})/2H^2(Y, \mathbb{Z}) = \mathrm{Im}\,\alpha/2\ \mathrm{Im}\,\alpha \xrightarrow{\sim} \frac{1}{2}\,\mathrm{Im}\,\alpha/\,\mathrm{Im}\,\alpha \\ &= (\mathrm{Im}\,\alpha)^\vee/\,\mathrm{Im}\,\alpha = (M^\perp)^\vee/M^\perp. \end{aligned}$$

Since by I, formula (2) there is a natural isomorphism $(M^\perp)^\vee/M^\perp \to M^\vee/M$ and since $M^\vee/M \xrightarrow{\sim} r(M^\vee)/r(M)$ we obtain an isomorphism

$$\gamma : H^2(Y, \mathbb{F}_2) \longrightarrow r(M^\vee)/r(M).$$

For the proof of the next proposition we need an explicit representation for the image under $\gamma\circ(\beta_2 \wedge \beta_2)$ of a decomposable element. So let $z_i \in \mathrm{Hom}(\Gamma, \mathbb{Z})$ $(i = 1, 2)$ and $z \in \bigwedge^2 \mathrm{Hom}(T(V), \mathbb{F}_2)$ be the mod-2 reduction of $z_1 \wedge z_2$. Then Z, the mod-2 reduction of $\{z_1 = 0\} \cap \{z_2 = 0\}$, is a 2-codimensional linear subspace of $T(V)$ and we claim

$$\gamma(\beta_2 \wedge \beta_2)(z) = r\Big(\frac{1}{2} \sum_{v \in Z} e_v\Big) \mod r(M). \tag{3}$$

To establish (3), we first observe that the definition of α implies that the Poincaré-dual of $\alpha(\beta(z_1) \wedge \beta(z_2)) - \sum_{v \in Z} e_v$ is the class of the $\overline{p}$-image of the proper transform on $\overline{Y}$ of the cycle which on Y is given by $\{z_1 = 0\} \cap \{z_2 = 0\}$. This cycle is invariant under the covering involution, so the class of its image is 2-divisible in $H^2(X, \mathbb{Z})$, i.e.,

$$\frac{1}{2}\Big[\alpha(\beta(z_1) \wedge \beta(z_2)) - \sum_{v \in Z} e_v\Big] \in H^2(X, \mathbb{Z}).$$

This means that $\frac{1}{2}\alpha(\beta(z_1) \wedge \beta(z_2))$ and $\frac{1}{2}\sum_{v \in Z} e_v$ correspond under the isomorphism of I, (2) thereby proving (3).

(5.8) **Proposition.** *The subspace $r(M^\vee) \subset \mathbb{F}_2^V$ consists of all polynomial functions of degree ≤ 2. Hence the isomorphism of Lemma 4.3 gives a natural isomorphism of $\mathbb{F}_2$-vector spaces*

$$\delta : r(M^\vee)/r(M) \longrightarrow \mathrm{Hom}(T(V), \mathbb{F}_2).$$

This isomorphism is the inverse of $\gamma\circ(\beta \wedge \beta)$.

Proof. The subspace $r(M^\vee)$ contains $|r(M)| \cdot [r(M^\vee) : r(M)] = 2^{11}$ elements. On the other hand it is preserved by the affine group, hence by Lemma 4.2 contains the subspace U_2 of $\mathbb{F}_2^V$ of polynomial functions of degree ≤ 2 on V, which itself already consists of 2^{11} elements. So $U_2 = r(M^\vee)$ and we may apply Lemma 4.3.

The last assertion need only be checked for decomposable elements. By (3)and Proposition 5.3 we have that $\gamma\circ(\beta \wedge \beta)(z) = \chi_Z$, where z is the mod-2 reduction of $z_1 \wedge z_2 \in \bigwedge^2 \mathrm{Hom}(\Gamma, \mathbb{Z})$. On the other hand, the remark following Lemma 4.3 shows that $\delta(\chi_Z)$ is the mod-2 reduction of $z_1 \wedge z_2$. So indeed $\gamma\circ(\beta \wedge \beta)(z) = \delta^{-1}(z)$. □

Finally, we shall compare the Hodge structure on two Kummer surfaces $X = \mathrm{Km}(Y)$ and $X' = \mathrm{Km}(Y')$. For objects on Y', dashed symbols are used having similar meaning as the corresponding undashed ones for Y. Moreover, in addition to $U = r(M)$ we set

$$^\circ U = r(M^\vee).$$

(5.9) **Proposition.** *Let $X = \mathrm{Km}(Y)$, $X' = \mathrm{Km}(Y')$ and $\phi : H^2(X, \mathbb{Z}) \to H^2(X', \mathbb{Z})$ an isometry inducing a bijection $\nu : V \to V'$. Assume moreover that $H^2(Y, \mathbb{Z})$ (and hence $H^2(X, \mathbb{Z})$) contains effective classes and that ϕ is an effective Hodge-isometry. Then ϕ is induced by an isomorphism $X' \to X$.*

Proof. Since ϕ is an isometry sending M isometrically to M', the same holds for the orthogonal complements. So, by Proposition 5.1 and Corollary 5.6 we obtain an isometry

$$\psi : H^2(Y, \mathbb{Z}) \longrightarrow H^2(Y', \mathbb{Z}).$$

This is even a Hodge-isometry, by Proposition 5.2. We are going to show that the assumptions of Theorem V. 3.2 are satisfied. So we consider the mod-2 reduction ψ_2 of ψ. Now $U \mapsto U^\circ$ is contravariant, so by naturality of the isomorphisms γ and γ' we obtain a commutative diagram

$$\begin{array}{ccc} H^2(Y, \mathbb{F}_2) & \xrightarrow{\psi_2} & H^2(Y', \mathbb{F}_2) \\ \gamma\downarrow & & \downarrow\gamma' \\ {}^\circ U/U & \xleftarrow[{}^\circ\phi]{} & {}^\circ U'/U' \end{array} \tag{4}$$

where $^\circ\phi$ is induced by ϕ.

On the other hand, ν induces a map ν^*, making the following diagram commutative

$$\begin{array}{ccc} M & \xrightarrow{\phi|M} & M' \\ {\scriptstyle r}\big\downarrow & & \big\downarrow{\scriptstyle r'} \\ \mathbb{F}_2^V & \xleftarrow[\nu^*]{} & \mathbb{F}_2^{V'} . \end{array}$$

Since $r(M) = U$ and $r'(M) = U'$ consists of the affine-linear functions of V, V' respectively, an application of Lemma 4.1 shows that ν is affine-linear, inducing an isomorphism

$$\nu^\vee : \mathrm{Hom}(T(V'), \mathbb{F}_2) \longrightarrow \mathrm{Hom}(T(V), \mathbb{F}_2)$$

which, according to Proposition 5.8, fits into the following commutative diagram

$$(5) \qquad \begin{array}{ccc} \bigwedge^2 \mathrm{Hom}(T(V), \mathbb{F}_2) & \xleftarrow{\nu^\vee \wedge \nu^\vee} & \bigwedge^2 \mathrm{Hom}(T(V'), \mathbb{F}_2) \\ \wr\big\downarrow{\scriptstyle \delta} & & \wr\big\downarrow{\scriptstyle \delta'} \\ {}^\circ U/U & \xleftarrow{{}^\circ\phi} & {}^\circ U'/U' . \end{array}$$

By Proposition 5.8 we have $\delta^{-1} = \gamma\circ(\beta_2 \wedge \beta_2)$ and similarly for $(\delta')^{-1}$, so (4) and (5) imply that $\psi_2 = \chi_2^{-1} \wedge \chi_2^{-1}$ with $\chi_2 = (\beta')^{-1}\circ\nu^\vee\circ\beta$. The assumptions of Theorem V.3.2 are indeed satisfied, so we have that $\psi_2 = \pm f^*$ for some biholomorphic map

$$f : Y' \to Y .$$

In fact we have $\psi_2 = f^*$, since ϕ and hence ψ takes effective classes into effective classes.

Since composing f with a translation does not affect the induced homomorphisms in cohomology, we may assume that g is a group homomorphism. So f commutes with i. Next, we may compose f with a translation by a point of order 2, if necessary, such that the resulting map (still denoted by f) coincides with ν^{-1} for at least one $v' \in V'$. Now the new f still commutes with i and we obtain a biholomorphic map $g : X' \to X$ which by construction has the property that the induced map in 2-cohomology coincides with ϕ on $\mathrm{Im}\,\alpha$ and on $e_v(v = \nu^{-1}(v'))$. Clearly f induces an affine-linear isomorphism $V' \to V$, which differs from ν at most by a translation (they both induce the same isomorphism $\nu^\vee \wedge \nu^\vee$ on $\bigwedge^2 \mathrm{Hom}(T(V), \mathbb{F}_2)$). This means that the map induced by g^* on V and the map ϕ differ at most by a translation. Since $g^*(e_v) = \phi(e_v)$, this translation is the identity, so $g^*|M = \phi|M$ and therefore $g^* = \phi$. □

6. The Torelli Theorem for Kummer Surfaces

The weakness of Proposition 5.9 lies in the fact that one of its assumptions, namely that the set of distinguished nodal classes be preserved, is far too strong. In this section we show how to remove this assumption for projective Kummer surfaces.

(6.1) **Proposition.** *Let X be a K 3-surface containing sixteen disjoint (-2)-curves $C_1, \ldots, C_{16}$ such that $\mathcal{O}_X\left(\sum_{i=1}^{16} C_i\right)$ is 2-divisible in $\operatorname{Pic}(X)$. Then X has the structure of a Kummer surface with $\{C_1, \ldots, C_{16}\}$ as the set of distinguished (-2)-curves.*

Proof. Since $\mathcal{O}_X\left(\sum_{i=1}^{16} C_i\right)$ is 2-divisible in $\operatorname{Pic}(Y)$, by Proposition I. 18.1 there exists a double covering $\rho : Z \to X$ whose branch locus is $\sum_{i=1}^{16} C_i$. The curves $D_i = \phi^{-1}(C_i)$ are all (-1)-curves on Z and upon contracting them we get a smooth surface Y. The formulae of Lemma I. 7.1, (iii) and Theorem I. 8.1, (viii) for the behaviour of the canonical bundle under branched coverings and blowing up imply that $\mathcal{K}_Y \cong \mathcal{O}_Y$. For $e(Y)$ we find $e(Y) = e(Z) - 16 = 2(e(X) - 16) - 16 = 0$ upon applying V, formula (8) and Proposition 3.2. So Y is minimal, and in fact a torus by the Enriques classification. Moreover the involution on Z interchanging the sheets of ρ descends to an involution i on Y. Since the i-invariant part of $H^1(Y, \mathbb{Q})$ is canonically isomorphic to $H^1(X, \mathbb{Q}) = 0$, i acts as $-\operatorname{id}$ on $H^1(X, \mathbb{Q})$ and so $X = \operatorname{Km}(Y)$ by definition. □

Before stating the next theorem we have to make the following remark. If X and X' are K 3-surfaces and $\phi : H^2(X, \mathbb{Z}) \to H^2(X', \mathbb{Z})$ is an isometry preserving the Hodge-decomposition, projectivity of X implies projectivity of X'. Indeed, $\phi(\mathrm{NS}(X)) \subset \mathrm{NS}\,(X')$ and $\mathrm{NS}(X)$ contains an element d with $d^2 > 0$, hence $\phi(d) \in \mathrm{NS}\,(X')$ has positive norm and X' is projective by Theorem IV 6.2.

It follows that under the above assumptions it makes sense to speak of an effective Hodge-isometry, and we shall do this several times in the sequel.

(6.2) **Proposition** (Torelli theorem for projective Kummer surfaces). *Let X' be a K 3-surface and X a projective Kummer surface. Suppose we are given an effective Hodge-isometry $\phi : H^2(X, \mathbb{Z}) \to H^2(X', \mathbb{Z})$. Then $\phi = f^*$, where $f : X' \to X$ is a biholomorphic map.*

Proof. Let $\{c_1, \ldots, c_{16}\}$ be the set of distinguished nodal classes on X. Since ϕ and ϕ^{-1} preserve effective classes, they also preserve indecomposable classes. So, if we put $c_i' = \phi(c_i)$ $(i = 1, \ldots, 16)$ we know that $c_i' = c_1(\mathcal{O}_X(C_i'))$ for a unique (-2)-curve C_i'. Since $\sum_{i=1}^{16} c_i$ is 2-divisible in the Néron-Severi lattice of X, also $\sum_{i=1}^{16} c_i'$ and therefore, by Proposition 3.6, $\sum_{i=1}^{16} \mathcal{O}_X(C_i')$ is 2-divisible in $\mathrm{Pic}(X')$. From Proposition 6.1 we conclude that X' is a Kummer surface having $\{C_1', \ldots, C_{16}'\}$ as its distinguished set of (-2)-curves. Since X is projective, it contains an effective class orthogonal to $\{c_1, \ldots, c_{16}\}$. Such a class gives an effective class on the torus from which X is constructed. Hence all the hypotheses of Proposition 5.9 are satisfied and the result follows. □

(6.3) **Corollary** (Weak Torelli theorem for projective Kummer surfaces). *Let X' be a K3-surface and X be a projective Kummer surface. Suppose that there exists a Hodge-isometry $H^2(X,\mathbb{Z}) \to H^2(X',\mathbb{Z})$. Then X and X' are biholomorphically equivalent.*

Proof. By composing the Hodge-isometry with $-\mathrm{id}$, if necessary, we may assume that it preserves the positive cones. If we compose the given Hodge-isometry with an appropriate element of W_X, it will preserve the Kähler cones (Proposition 3.10), so the resulting Hodge-isometry is effective by Proposition 3.11 and we may apply Theorem 6.2 to obtain the corollary. □

7. The Local Torelli Theorem for K3-Surfaces

Let $\Delta = \{z \in \mathbb{C} \mid |z| < \varepsilon\}$ and $\gamma : \Delta \to \mathbb{P}(L_\mathbb{C})$ a holomorphic map with $\gamma(0) = \ell$ and tangent $\theta = \gamma'(0)$. Let $\tilde{\gamma} : \Delta \to L_\mathbb{C} \setminus \{0\}$ be a lifting of γ. Associating to ℓ the class of $\tilde{\gamma}'(0)$ in $L_\mathbb{C}/\ell$ gives a homomorphism $\ell \to L_\mathbb{C}/\ell$, which is independent of the lifting of γ. In fact it is uniquely determined by θ. We denote it by $h(\theta)$. Clearly $\theta \mapsto h(\theta)$ gives an injective map $h : \mathcal{T}_{\mathbb{P}(L_\mathbb{C})}(\ell) \to \mathrm{Hom}(\ell, L_\mathbb{C}/\ell)$, hence an isomorphism.

(7.1) **Proposition.** *For any $\ell = [\omega] \in \Omega$ the isomorphism h gives an identification of $\mathcal{T}_\Omega(\ell)$ with $\mathrm{Hom}(\ell, \ell^\perp/\ell)$.*

Proof. $\mathcal{T}_\Omega(\ell)$ is the subspace of those θ for which a lifting $\tilde{\gamma}$ satisfies $\tilde{\gamma}(\Delta) \subset \Omega$, i.e., for which $(\tilde{\gamma}(0), \tilde{\gamma}'(0)) = 0$, or equivalently $h(\theta) \in \mathrm{Hom}(\ell, \ell^\perp/\ell)$. □

Suppose that we have a family of K3-surfaces $p : Y \to S$ together with a trivialisation $\phi : p_{*2}\mathbb{Z}_Y \xrightarrow{\sim} L_S$. Consider its associated period mapping

$$\tau : S \to \Omega$$

defined by sending $s \in S$ to the complex line $\phi_\mathbb{C}(s)(H^{2,0}(Y_s))$. The map τ is holomorphic by IV, Sect. 4. We want to compute the induced map $(d\tau)_0$ on the tangent space $\mathcal{T}_S(0)$.

(7.2) **Proposition.** *Via the identification of Prop. 7.1, we have that*

$$(d\tau_0) : \mathcal{T}_S(0) \longrightarrow \mathrm{Hom}(H^{2,0}(Y_0), H^{1,1}(Y_0))$$

is the composition of the Kodaira-Spencer map and the homomorphism

$$\nabla : H^1(\mathcal{T}_{Y_0}) \longrightarrow \mathrm{Hom}(H^{2,0}(Y_0), H^{1,1}(Y_0))$$

obtained by the cup product

$$\bigcup : H^1(\mathcal{T}_{Y_0}) \otimes H^0(\Omega^2_{Y_0}) \longrightarrow H^1(\Omega^1_{Y_0}) .$$

Proof. By IV, Sect. 4 the bundle $\bigcup_s H^0(\Omega^2_{Y_s})$ is a holomorphic subbundle of $f_{*2}\mathbb{C}_X \otimes \mathcal{O}_S$. It admits a never vanishing holomorphic section $\omega(s)$ over S. We may assume that for all $s \in S$ the differentiable manifold underlying Y_s is the same, say Y. We identify Y with Y_0 and consider local holomorphic coordinates $\{z_1, z_2\}$ on Y_0 as differentiable coordinates on Y. Moreover, by [K-N-S] we may find holomorphic coordinates $\{w_1, w_2\}$ on Y_s such that $w_1 = w_1(z,s)$, $w_2 = w_2(z,s)$ (differentiably in z and holomorphically in s). Let $\Delta = \{t \in \mathbb{C} \mid |t| < \varepsilon\}$ and $s : \Delta \to S$ a holomorphic map, with $s(0) = 0$. We identify $(ds)_0(\partial/\partial t)$ with $\partial/\partial t$. If $\omega'(0) \equiv \partial/\partial t \omega(s(t))|_{t=0} \bmod (2,0)$-forms we have that $(d\tau)_0(\partial/\partial t) \in \mathrm{Hom}(\ell, \ell^{\perp}/\ell)$ $(\ell = \mathbb{C} \cdot \tau(0))$ associates to $\omega(0)$ the class $\phi_{\mathbb{C}}(0)\omega'(0)$.

Let $\omega(s)$ in local coordinates be given by $\omega(s) = f(z,s)dw_1 \wedge dw_2$. Then we have

$$\omega'(0) \equiv f(z,0)\left[\overline{\partial}(w_1'(z,0)) \wedge dz_2 + \overline{\partial}(w_2'(z,0)) \wedge dz_1\right] \bmod (2,0)\text{-forms.}$$

By [M-K] a Dolbeault representative for the Kodaira-Spencer class $\rho(\partial/\partial t)$ reads in local coordinates

$$\overline{\partial}(w_1'(z,0))\partial/\partial z_1 + \overline{\partial}(w_2'(z,0))\partial/\partial z_2 ,$$

and hence

$$\omega'(0) = \rho(\partial/\partial t) \cup \omega(0) . \qquad \square$$

(7.3) **Theorem.** (Local Torelli theorem) *The Kuranishi family for a* K3-*surface X is universal at all points in a small neighbourhood S of the point in the base corresponding to X. This base is smooth and of dimension* 20, *and the period map is a local isomorphism at each point of S.*

Proof. The first two assertions follow from Corollary 3.5 and Theorem I.10.5. It also follows that the Kodaira-Spencer map is an isomorphism at $s \in S$, so, the local injectivity follows from Proposition 7.2 and the fact that "$\cup$" is an isomorphism by Corollary 3.5. The last statement of the theorem follows, since $\dim S = 20 = \dim \Omega$. $\square$

8. A Density Theorem

We call a surface X exceptional, if its Picard number is maximal, i.e., rank $\mathrm{NS}(X) = h^{1,1}(X)$. This is equivalent to saying that $H^{2,0}(X) \oplus \overline{H^{2,0}(X)}$ is defined over $\mathbb{Q}$.

For a K 3-surface X with $H^{2,0}(X) = \mathbb{C} \cdot \omega_X$ we therefore have that X is exceptional if and only if the subspace $E(\omega_X)$ of $H^2(X, \mathbb{R})$ spanned by $\{\mathrm{Re}\,\omega_X, \mathrm{Im}\,\omega_X\}$ is defined over the rationals. Then we have (for the notation see Sect. 1):

$$E(\omega_X) = T_X \otimes \mathbb{R} \ ,$$

and T_X becomes an oriented euclidean lattice of rank 2 such that $(\ ,\ )|T_X > 0$.

If $X = \mathrm{Km}(Y)$, we have by Propositions 5.1 and 5.2 an injection $\alpha : H^2(Y, \mathbb{Z}) \to H^2(X, \mathbb{Z})$ preserving the Hodge decomposition, so X is exceptional if and only if Y is.

(8.1) **Proposition.** *Let T be a primitive oriented sublattice of rank 2 of L such that $(\ ,\ )|T > 0$ and $(x,x) \in 4\mathbb{Z}$ for all $x \in T$. Then there exists an exceptional Kummer surface X and an isometry $\phi : H^2(X, \mathbb{Z}) \to L$, mapping T_X isometrically onto T.*

Proof. Firstly, we construct a torus Y such that $T_Y \cong \frac{1}{2}T$. Let $\Gamma = \sum_{i=1}^{4} \mathbb{Z}e_i$ be an oriented $\mathbb{Z}$-module, i.e., we are given an isomorphism $\det : \bigwedge^4 \Gamma \to \mathbb{Z}$. Then the lattice

$$\bigwedge^2 \Gamma = \mathbb{Z}e_1 \wedge e_2 + \mathbb{Z}e_3 \wedge e_4 + \mathbb{Z}e_1 \wedge e_3 + \mathbb{Z}e_4 \wedge e_2 + \mathbb{Z}e_1 \wedge e_4 + \mathbb{Z}e_2 \wedge e_3 \ ,$$

when equipped with the bilinear form defined by $(x,y) = \det(x \wedge y)$ for $x, y \in \bigwedge^2 \Gamma$, is clearly isometric to $H \oplus H \oplus H$, so by Theorem I. 2.9, (i) we have an isometric primitive embedding of $\frac{1}{2}T$ into $\bigwedge^2 \Gamma$. Let T' be the image, $\{t_1, t_2\}$ an oriented orthogonal basis for $T' \otimes \mathbb{R}$ in $\bigwedge^2 \Gamma$ and let $\{t_3, t_4, t_5, t_6\}$ be a basis for the orthogonal complement of $T' \otimes \mathbb{R}$ in $\bigwedge^2 \Gamma \otimes \mathbb{R}$. Furthermore let $\{t_1^\vee, \ldots, t_6^\vee\}$ be the basis of $\mathrm{Hom}(\bigwedge^2 \Gamma, \mathbb{C})$ dual to $\{t_1, \ldots, t_6\}$. We provide the latter $\mathbb{C}$-vector space with a $\mathbb{C}$-bilinear form in the usual way, by demanding that the "evaluation"-map $\bigwedge^2 \Gamma \otimes \mathbb{C} \to \mathrm{Hom}(\bigwedge^2 \Gamma, \mathbb{C})$, i.e., the map given by $v \mapsto (v, -)$, is an isometry. The form thus obtained is nothing but the one defined by $\alpha \wedge \beta = (\alpha, \beta) \cdot \det$, for $\alpha, \beta \in \mathrm{Hom}(\bigwedge^2 \Gamma, \mathbb{C})$. If we put $\omega = t_1^\vee + i t_2^\vee$, it follows that $\omega \wedge \omega = 0$ and $\omega \wedge \bar{\omega} = 2\det$. The first equation shows that $\omega = \omega_1 \wedge \omega_2$ for $\omega_j \in \mathrm{Hom}(\Gamma, \mathbb{C})$ and the second one implies that if we map Γ into $\mathbb{C}^2$ via $\gamma \mapsto (\omega_1(\gamma), \omega_2(\gamma))$ we obtain a lattice of maximal rank. Identifying Γ with this lattice we obtain a 2-torus $Y = \mathbb{C}^2/\Gamma$ and a canonical isomorphism $\psi : \Gamma \xrightarrow{\sim} H_1(Y, \mathbb{Z})$ of oriented lattices, inducing an isomorphism $\psi^\vee \wedge \psi^\vee : H^2(Y, \mathbb{Z}) \to \mathrm{Hom}(\bigwedge^2 \Gamma, \mathbb{Z})$ of euclidean lattices. Since ω_1, ω_2 are exactly the coordinate functions on $\mathbb{C}^2$, they are complex-linear and so the complexification of $\psi^\vee$ maps $H^{1,0}(Y)$ to the subspace of $\mathrm{Hom}(\Gamma, \mathbb{C})$

spanned by the complex-linear functionals, i.e., $\mathbb{C}\omega_1 \oplus \mathbb{C}\omega_2$. Then $H^{2,0}(Y)$ corresponds to $\mathbb{C}\omega_1 \wedge \omega_2 = \mathbb{C}\omega$, $\mathrm{NS}(Y)$ to $\ker \omega$ and T_Y to T'.

If $X = \mathrm{Km}(Y)$, then, by Proposition 5.1, the map $\alpha : H^2(X, \mathbb{Z}) \to H^2(Y, \mathbb{Z})$ multiplies $(\ , \ )$ by 2 and its complexification sends $H^{2,0}(X)$ to $H^{2,0}(Y)$. It follows that T_X is isometric to T.

Let $\phi' : H^2(X, \mathbb{Z}) \xrightarrow{\sim} L$ be any isometry. Since T and $\phi'(T_X)$ are isometric by construction, we may apply Theorem I.2.9, (ii) to find an automorphism of L which sends $\phi'(T_X)$ to T in such a way that the ordering of two bases, coming from orientation, correspond. Composing ϕ' with this automorphism we get an isometry $\phi : H^2(X, \mathbb{Z}) \to L$ as desired. □

(8.2) **Proposition.** *The set of the rationally defined 2-planes $P \subset L_{\mathbb{R}}$ satisfying $(x, x) \in 4\mathbb{Z}$ for all $x \in P \cap L$ is dense in the Grassmannian $\mathrm{Gr}_2(L_{\mathbb{R}})$.*

For the proof we need

(8.3) **Lemma.** *Let $m, n \in \mathbb{N}$ and let M be a lattice containing a primitive vector e_0 with $(e_0, e_0) \equiv m \pmod n$. Then the set of lines $\ell \subset M_{\mathbb{R}}$ spanned by a primitive vector e with $(e, e) \equiv m \pmod n$ is dense in $\mathbb{P}(M_{\mathbb{R}})$.*

Proof. Let U be an open subset of $\mathbb{P}(M_{\mathbb{R}})$, $U \neq \emptyset$. It contains a rationally defined element $\ell = \mathbb{R} \cdot e$ where $e \in M$ may be chosen primitive. If $e = \pm e_0$ there is nothing to prove. So assume that this is not the case. Then e and e_0 are linearly independent. Let $f \in (\mathbb{R}e + \mathbb{R}e_0) \cap M = M'$ such that $\{e, f\}$ is a basis of M' and let $e_0 = ae + bf$, with $a, b \in \mathbb{Z}$. Since e_0 is primitive g.c.d.$(a, b) = 1$. Then, for any $N \in \mathbb{Z}$, we have that $e_N = e_0 + Nbe = (a + Nb)e + bf$ is also primitive. If N is a multiple of n we have $(e_N, e_N) \equiv (e_0, e_0) \equiv m \pmod n$ and if N is large enough $\mathbb{R}e_N = \mathbb{R}\big(e + \frac{1}{Nb} e_0\big)$ is a line in U. So U contains an element with the required properties. □

Proof of Proposition 8.2. Let U be an open subset of $\mathrm{Gr}_2(L_{\mathbb{R}})$, $U \neq \emptyset$. Since H contains a primitive vector of norm 4 so does $L \supset H$. By Lemma 8.3 there exists a $P' \in U$ containing a primitive vector e_1 with $(e_1, e_1) \equiv 4 \pmod 8$. Let M be the orthogonal complement of e_1 in L. Since by Theorem I. 2.9, (ii) there is an automorphism of L sending e_1 to a primitive vector in $H \subset L$ of length (e_1, e_1), we see that M contains a sublattice isometric to $H \oplus H \oplus (-E_8) \oplus (-E_8)$. Hence it contains a primitive vector of norm 64. Again Lemma 8.3 implies that M contains a primitive vector e_2 with $(e_2, e_2) \equiv 0 \pmod{64}$ such that the 2-plane spanned by e_1 and e_2 belongs to U.

We claim that for all $f \in P \cap L$ we have $(f, f) \in 4\mathbb{Z}$. Since

$$(e_1, e_1)f - (e_1, f)e_1 \in P \cap L$$

is orthogonal to e_1, it is an integral multiple ae_2 of e_2. So $(e_1, e_1)f = (e_1, f)e_1 + ae_2$ and taking norms we get

$$(e_1, e_1)^2 (f, f) = (e_1, f)^2 (e_1, e_1) + a^2 (e_2, e_2) \ .$$

Now (f, f) is even and if it were not 4-divisible the left hand side of the above equality would be divisible by 2^3, but by no larger power of 2. On the other hand the right hand side is either exactly divisible by 2^2 (in case (e_1, f) is odd) or by at least 2^6 (in case (e_1, f) is even). The contradiction shows that $(f, f) \in 4\mathbb{Z}$. □

(8.4) *Remark.* This proof shows that the following holds: given $x \in L$ with $(x, x) \in 4\mathbb{Z}$, the set of $y \in L_{\mathbb{R}}$ such that $(\mathbb{R}x + \mathbb{R}y) \cap L$ is a primitive rank-2 sublattice of L, all of whose vectors have norm in $4\mathbb{Z}$, is dense in $L_{\mathbb{R}}$.

(8.5) **Corollary.** *The period points of marked projective Kummer surfaces are dense in* Ω.

Proof. Let $G_2^+(L)$ be the set of oriented 2-planes of $L_{\mathbb{R}}$ on which $(\ ,\ )$ is positive definite, and consider the map $\pi : \Omega \to G_2^+(L)$ sending $\ell = [\omega]$ to the oriented 2-plane $E(\omega)$ with oriented basis $\{\mathrm{Re}\,\omega, \mathrm{Im}\,\omega\}$. This is a bijection, since if E is an oriented 2-plane on which $(\ ,\ )$ is positive, and $\{\omega_1, \omega_2\}$ an oriented orthogonal basis for it, the line $[\omega_1 + \mathrm{i}\omega_2] = [\omega]$ is in Ω and $E(\omega) = \mathbb{R}\omega_1 + \mathbb{R}\omega_2$.

So it suffices to show that the set of oriented 2-planes $E(\omega)$ belonging to marked projective Kummer surfaces is dense in the Grassmannian $\mathrm{Gr}_2(L)$, in which $G_2^+(L)$ is Zariski open. By Proposition 8.2 the set of $[\omega] \in \Omega$ such that (i) the 2-plane $E(\omega)$ is rationally defined, say $E(\omega) = P \otimes \mathbb{Q}$, and (ii) $(x, x) \in 4\mathbb{Z}$ for all $x \in P \cap L$, is dense in $\mathrm{Gr}_2(L)$. According to Lemma 8.1 any such $[\omega]$ is the image under the period map of a marked exceptional Kummer surface X. Since for such an X the Néron-Severi lattice contains elements with positive norm, X is projective. □

(8.6) **Corollary.** *Any two* K 3*-surfaces are diffeomorphic. In particular any* K 3*-surface is simply-connected.*

Proof. Since any two 2-tori are isomorphic as real Lie groups, the corresponding Kummer surfaces are diffeomorphic. If X is a K 3-surface, let $p : Y \to S$ be its Kuranishi deformation. Any trivialisation of $p_{*2}\mathbb{Z}_Y$ induces a period map which by Theorem 7.3 realises S as an open subset of Ω. By Remark 8.4 this open set contains period points of projective Kummer surfaces. So X is diffeomorphic to a Kummer surface. As was observed before, the last assertion is a consequence of the fact that any smooth surface of degree 4 in $\mathbb{P}_3$ is a simply-connected K 3-surface (Theorem V.2.1). □

Remark. In fact, the proof of Corollary 8.6 shows that any two K 3-surfaces are deformations of each other.

9. Behaviour of the Kähler Cone under Deformations

Let us recall some notation from Sect. 8 and at the same time introduce some more.

For $[\omega] \in \Omega$ we let $E(\omega)$ be the oriented 2-plane of $L_{\mathbb{R}}$ spanned by the oriented base $\{\mathrm{Re}\,\omega, \mathrm{Im}\,\omega\}$, viewed as an element of the manifold $G_2^+(L)$ of oriented 2-planes of $L_{\mathbb{R}}$ on which $(\ ,\ )$ is positive definite. Let

$$K\Omega = \{(\kappa, [\omega]) \in L_{\mathbb{R}} \times \Omega \mid (\kappa, E(\omega)) = 0\ ,\ (\kappa, \kappa) > 0\}$$

and for $(\kappa, [\omega]) \in K\Omega$ let $E(\kappa, \omega)$ be the oriented 3-space of $L_{\mathbb{R}}$ spanned by the oriented basis $\{\kappa, \mathrm{Re}\,\omega, \mathrm{Im}\,\omega\}$ and viewed as an element of the manifold $G_3^+(L)$ of oriented 3-planes of $L_{\mathbb{R}}$ on which $(\ ,\ )$ is positive definite. So we obtain a map

$$\begin{aligned} \Pi : K\Omega &\longrightarrow G_3^+(L) \\ (\kappa, [\omega]) &\mapsto E(\kappa, \omega)\ . \end{aligned}$$

(9.1) **Proposition.** *The map* Π *exhibits* $K\Omega$ *as an* $\mathrm{SO}(3)$*-fibre bundle over* $G_3^+(L)$ *with fibre* $\mathbb{R}^3 \setminus \{0\}$. *In fact* $\Pi^{-1}(E(\kappa, \omega)) \cong E(\kappa, \omega) \setminus \{0\}$. *The group* $\mathrm{Aut}(L_{\mathbb{R}}) \cong \mathrm{O}(3, 19)$ *acts* Π*-equivariantly on* $K\Omega$. *This action is proper.*

Proof. All assertions except the last one are immediately clear. To prove that $\mathrm{Aut}(L_{\mathbb{R}})$ acts properly, we observe that it acts properly on the base space (the stabiliser of a 3-plane is $\mathrm{O}(3) \times \mathrm{O}(19)$, hence compact) and so it also acts properly on $K\Omega$. □

(9.2) **Corollary.** *The set*

$$(K\Omega)^0 = \{(\kappa, [\omega]) \in K\Omega \mid (\kappa, d) \neq 0 \ \textit{ for every } d \in K \textit{ with } (d, d) = -2 \textit{ and } (\omega, d) = 0\}$$

is open in $K\Omega$.

Proof. $\mathrm{Aut}(L)$ is discrete in $\mathrm{Aut}(L_{\mathbb{R}})$, so Proposition 9.1 implies that it acts properly on $K\Omega$. A fortiori this holds for the group W generated by the reflections s_d, for $d \in L$, $(d, d) = -2$, $(d, \omega) = 0$, where $s_d \in \mathrm{Aut}(L)$ is defined as usual by $s_d(x) = x + (x, d)d$, $x \in L$. We obtain $(K\Omega)^0$ from $K\Omega$ by deleting all reflection hyperplanes H_d for s_d. Since W is discrete and acts properly on $K\Omega$, these form a locally finite, hence closed set: every point of $K\Omega$ has an open neighbourhood U such that $U \cap s_d(U) \neq \varnothing$ for only finitely many $s_d(U)$, so a fortiori only finitely many H_d meet U. □

(9.3) **Lemma.** *Let* $p : Y \to S$ *be a family of* K3*-surfaces. Then* $\bigcup_{s \in S} H^{1,1}(Y_s, \mathbb{R})$ *forms a real-analytic subbundle of* $p_{*2}\mathbb{R}_Y$ *in which the union of the Kähler cones is open.*

Proof. The assertion being local we may take for S any sufficiently small open polydisk, and replace p by a locally universal family, i.e., we may suppose that $S \subset \Omega$ and we may identify $H^2(Y_s, \mathbb{Z})$, $s \in S$ with L. Since $H^{1,1}(Y_s, \mathbb{R}) = E(s)^{\perp}$, this space varies in a real analytical way with s and the first assertion follows. Let $K_s \subset H^{1,1}(Y_s, \mathbb{R})$ be the Kähler cone of Y_s, and let $\kappa \in K_s$. If $\kappa_0 \in K_s$ is a Kähler class, the segment $[\kappa, \kappa_0]$ belongs to K_s. Let C_s be a compact connected neighbourhood of $[\kappa, \kappa_0]$. Since $K_s \times \{s\} \subset (K\Omega)^0$ by compactness and Proposition 9.1, we may find an open neighbourhood V of C_s inside the union $\bigcup_{s\in S} H^{1,1}(Y_s, \mathbb{R}) \subset K\Omega$ such that $V \subset (K\Omega)^0$. Moreover we may assume

(i) if $\kappa_1 \in V \cap H^2(Y_s, \mathbb{R})$, then $(\kappa_1, \kappa_1) > 0$, and
(ii) for every $s_1 \in S$, the set $V \cap H^{1,1}(Y_{s_1}, \mathbb{R})$ is connected.

According to [K-S62], Thm. 15 the set of Kähler classes is open in $\bigcup_{s\in S} H^{1,1}(Y_s, \mathbb{R})$. Since C_s contains a Kähler class, this implies that for some neighbourhood U of s in S the set $V \cap H^{1,1}(Y_{s_1}, \mathbb{R})$ contains a Kähler class. By (i), (ii) and Corollary 3.9 this set is contained in the Kähler cone of Y_{s_1}. The lemma follows. □

(9.4) **Proposition.** *Let $p : Y \to S$ and $p' : Y' \to S$ be two families of* K 3-*surfaces and let $\Phi : p'_{*2}\mathbb{Z}_{Y'} \xrightarrow{\sim} p_{*2}\mathbb{Z}_Y$ be an isomorphism of local systems such that $\Phi(s)$ is a Hodge-isometry for each $s \in S$. Then the set of $s \in S$ with the property that $\Phi(s)$ maps the Kähler cone of Y'_s to the Kähler cone of $Y_{s'}$ is open in S*

Proof. The conditions imposed on Φ imply that Φ induces bundle isomorphisms between the real-analytic bundles $\bigcup_{s\in S} H^{1,1}(Y'_s, \mathbb{R})$ and $\bigcup_{s\in S} H^{1,1}(Y_s, \mathbb{R})$. By Lemma 9.3 the set $s \in S$ for which an element in the Kähler cone of Y'_s is mapped into the Kähler cone of Y_s is open in S. But Proposition 3.11 teaches us that for such s the entire Kähler cone of Y'_s is mapped onto the Kähler cone of Y_s. □

10. Degenerations of Isomorphisms between K 3-Surfaces

(10.1) **Proposition.** *Let X be a* K 3-*surface and let $\{C_i, D_i\}_{i\in I}$ be a finite collection of curves on X, with $c_i = c_1(\mathcal{O}_X(C_i))$ and $d_i = c_1(\mathcal{O}_X(D_i))$.*

If the map $\phi : H^2(X, \mathbb{Z}) \to H^2(X, \mathbb{Z})$ defined by $\phi(x) = x + \sum_{i\in I}(c_i, x)d_i$, $x \in H^2(X, \mathbb{Z})$, is an effective Hodge-isometry, then $I = \varnothing$, i.e., $\phi = \text{id}$.

Proof. Let us take $x \in \mathcal{C}_X^+$, the Kähler cone of X. Then the numbers $r_i = (c_i, x)$ are positive and since ϕ is an isometry, we have

$$0 = (\phi(x), \phi(x)) - (x, x) = (\phi(x) - x, \phi(x) + x)$$
$$= \left(\sum_{i\in I} r_i d_i, \phi(x) + x\right) .$$

By Proposition 3.11, (iii) we have $\phi(x) \in \mathcal{C}_X^+$, hence $x + \phi(x) \in \mathcal{C}_X^+$ and so $(x + \phi(x), d_i) > 0$ for all $i \in I$, and the preceding equation shows that $I = \varnothing$. □

We now come to a crucial result, expressed by Theorem 10.6 below. It says that, under certain circumstances, isomorphisms between corresponding members of two sequences of K 3-surfaces, both converging to a limit K 3-surface, extend to these limiting surfaces.

Our point of departure is the following.

(10.2) **Data.**

i) Two families $p : Y \to S$, $p' : Y' \to S$ of K 3-surfaces over a polydisk S, centred at 0.
ii) A sequence $\{s_n\}$ in S converging to 0.
iii) An isomorphism $\Phi : p'_{*2}\mathbb{Z}_{Y'} \to p_{*2}\mathbb{Z}_Y$ of local systems.
iv) Isomorphisms $f_n : Y_{s_n} \to Y'_{s_n}$ such that the induced map in 2-cohomology coincides with $\Phi(s_n)$ for all $n \in \mathbb{N}$.

We consider the graphs Γ_n of f_n in $Y_{s_n} \times Y'_{s_n}$ as compactly supported cycles of $Y \times_S Y'$ and ask whether there exists a limit cycle Γ_0 in the topology introduced by Barlet in [Ba75]. In fact Barlet gives the set of compactly supported cycles on a fixed reduced analytic space Z the structure of a reduced analytic space $H(Z)$. Moreover, over $H(Z)$ there exists a universal family of compactly supported cycles of Z. Hence, if two points of $H(Z)$ are in the same connected component of $H(Z)$ they are cohomologous. A subset of $H(Z)$ is compact whenever the corresponding cycles all lie in a compact subset of Z and if moreover all volumes are uniformly bounded in a suitable but fixed metric on Z ([Ba78]). We apply this to the set consisting of the Γ_n.

(10.3) **Lemma.** *There exists a subsequence of $\{s_n\}$ such that the corresponding Γ_n converge to a limit cycle Γ_0 in Barlet's topology.*

Proof. We may assume that S is taken so small that

i) $Y \times_S Y'$ is differentiably trivial,
ii) each Y_s (Y'_s respectively,) carries a Kähler metric whose $(1,1)$-form θ_s (θ'_s respectively,) depends differentiably on s (see [K-S62]).

Let us take a metric on $Y \times_S Y'$ whose $(1,1)$-form θ on each fibre $Y_s \times Y'_s$ coincides with $p_1^*(\theta_s) + p_2^*(\theta'_s)$, where p_1 (p_2 respectively,) is the projection onto Y_s (Y'_s respectively).

If $[\Gamma_n] \in H^4(Y \times_S Y')$ is the dual cohomology class of the cycle Γ_n, then

$$\text{Vol}(\Gamma_n) = \frac{1}{2}\int_\Gamma (\theta \cup \theta) = \frac{1}{2}[\Gamma_n] \cup (\theta_{s_n} \cup \theta_{s_n}). \tag{6}$$

Since $Y \times_S Y'$ is differentiably trivial, the given isometry Φ in 10.2, (iii) is completely determined by its value at one point, i.e., if we choose a trivialisation $f : Y \times_S Y' \to Y_0 \times Y'_0 \times S$ and if $f(s)$ is the induced diffeomorphism $Y_s \times Y'_s \to Y_0 \times Y'_0$ we have that $\Phi(s) = (f(s)^*)^{-1}\Phi(0)f(s)^*$. Here we view $\Phi(s)$ as an element of $H^4(Y_s \times Y'_s, \mathbb{Z})$. In particular, $[\Gamma_n] = (f(s_n)^*)^{-1}\Phi(0)f(s_n)^*$, in view of the assumption 10.2, (iv). Therefore and because of (6) twice the volume of Γ_n is the value at s_n of the continuous function $(f(s)^{*-1}\Phi(0)f^*(s)) \cup \theta_s \cup \theta_s$, hence is uniformly bounded on any compact neighbourhood of $0 \in S$. □

(10.4) *Remark.* Two compactly supported cycles of $Y \times_S Y'$ in the same component of $H(Y \times_S Y')$ are cohomologous. In particular $[\Gamma_0] = \lim_{n\to\infty}[\Gamma_n]$ is the same as $\Phi(0) \in H^4(Y_0 \times Y_0', \mathbb{Z})$.

Before we proceed to the main Theorem 10.6 we make some general remarks. Any purely 2-dimensional cycle C on a product of two compact surfaces X and Y induces linear maps in homology

$$[C]_* : H_k(X) \to H_k(Y) \quad (k = 0, \ldots, 4)$$

as follows: if $p_2 : X \times Y \to Y$ denotes projection onto the second factor, $[C]_*\xi = p_{2*}([\xi \times Y] \cdot C)$ for all $\xi \in H_k(X)$.

By Poincaré-duality we obtain similar maps in cohomology, also denoted by $[C]_* : H^k(X) \to H^k(Y)$. We apply this to $\Gamma_n \subset Y_{s_n} \times Y'_{s_n}$ and $\Gamma_0 \subset Y_0 \times Y_0'$. Of course $\lim_{n\to\infty}[\Gamma_n]_* = [\Gamma_0]_*$ in all dimensions. In particular $[\Gamma_0]_*$ is an isomorphism. This motivates the hypotheses in the following

(10.5) **Proposition.** *Let X and Y be compact surfaces with $p_g(Y) \neq 0$. Let C be a purely 2-dimensional effective cycle on $X \times Y$, such that $[C]_*$ is an isomorphism in all dimensions. Then $C = \Delta + \sum_{i=1}^{N} C_i \times D_i$, where Δ gives a bimeromorphic correspondence and C_i (D_i respectively,) are curves on X (Y respectively).*

Proof. Since $[C]_*$ is an isomorphism in 4-homology, the degree of the projection $C \to Y$ must be one. So

$$C = \Delta + C',$$

where Δ projects bimeromorphically onto Y and C' maps to lower dimensional cycles. Similarly, *either* Δ projects onto a lower dimensional subvariety of X and $C' \to X$ has degree one *or* $\Delta \to X$ projects birationally. The first alternative is impossible, since in this case the image of $H^2(X,\mathbb{Z})$ in $H^2(Y,\mathbb{Z})$ via $[C]_*$ would entirely be contained in the Néron-Severi lattice. But $p_g(Y) \neq 0$, so this lattice is strictly smaller than $H^2(Y,\mathbb{Z})$ while $[C]_*$ is an isomorphism. It follows that Δ establishes a bimeromorphic correspondence $X \to Y$ and that $C' = \sum C_i \times D_i$ for some curves $C_i \subset X$ and $D_i \subset Y$. □

(10.6) **Theorem.** *Given two families $p : Y \to S$, $p' : Y' \to S$ of* K3-*surfaces over a polydisk S. Then the existence of a sequence and isomorphisms as in* 10.2, (ii)–(iv) *imply that Y_0 and Y_0' are isomorphic. If moreover $\Phi(0)$ is an effective Hodge-isometry, then a subsequence of $\{f_n\}$ converges uniformly in the Barlet topology to an isomorphism $f_0 : Y_0 \to Y_0'$ with the property that $(f_0)^* = \Phi(0)$.*

Proof. The first assertion is a consequence of Lemma 10.3 and Proposition 10.5. Indeed, passing to a subsequence if necessary, $\lim_{n\to\infty} \Gamma_n = \Gamma_0$ exists as a purely 2-dimensional cycle of $Y_0 \times Y_0'$ and by Proposition 10.5 the surfaces Y_0 and Y_0' are bimeromorphically equivalent, hence isomorphic, since both are minimal and non-ruled. (Apply Proposition III.4.6.) More precisely $\Gamma_0 = \Delta_0 + \sum_{i\in I} C_i \times D_i$, where Δ_0 is the graph of an isomorphism f_0 and the C_i's (the D_i respectively,) are curves on Y_0 (Y_0' respectively). By Remark 10.4 we have that $[\Gamma_0] = \Phi(0) \in H^4(Y_0 \times Y_0', \mathbb{Z})$. Let us for a moment identify Y_0 and Y_0' via f_0 and let us translate the identity $[\Gamma_0] = \Phi(0)$ into an identity between automorphisms of $H^2(Y_0, \mathbb{Z})$:

$$\Phi(0)x = x + \sum_{i\in I}([C_i], x)D_i, \quad x \in H^2(Y_0, \mathbb{Z}) .$$

So, if $\Phi(0)$ is effective, an application of Proposition 10.1 gives that $I = \emptyset$, i.e., $\Phi(0) = f_0^*$ as desired. □

(10.7) **Corollary.** *The group of automorphisms of a* K3*-surface X inducing the identity in $H^2(X, \mathbb{Z})$ is finite.*

Proof. Since $H^0(\mathcal{T}_X) = 0$ by Corollary 3.5, we have that the group of automorphisms G is discrete. It is also compact: let $g_n \in G$, $n \in \mathbb{N}$, and apply Theorem 10.6 to the trivial family $p = p'$, $\Phi^* = \mathrm{id}$, $\{s_n\}$ a sequence converging to $0 \in S$ and $f_n = g_n$; then a subsequence of g_n converges within G. □

11. The Torelli Theorems for K3-Surfaces

(11.1) **Theorem** (Torelli theorem). *Let X and X' be two* K3*-surfaces and suppose that there exists an effective Hodge-isometry $\phi : H^2(X', \mathbb{Z}) \to H^2(X, \mathbb{Z})$. Then $\phi = f^*$, with $f : X \to X'$ biholomorphic.*

Proof. Let $p : Y \to S$, $p' : Y' \to S'$ respectively, be the Kuranishi families for $X = Y_0$, $X' = Y_0'$ respectively, and let S and S' be small polydisks around 0. Then first of all we can choose a marking $\alpha : p_{*2}\mathbb{Z}_Y \xrightarrow{\sim} L_S$ and secondly we can extend ϕ to an isomorphism of local systems $\Phi : p'_{*2}\mathbb{Z}_{Y'} \to p_{*2}\mathbb{Z}_Y$, thereby obtaining a marking $\alpha\circ\Phi$ for the local system $p_{*2}\mathbb{Z}_Y$. The two markings determine period maps $\tau : S \to \Omega$, $\tau' : S' \to \Omega$ respectively, and, by construction, we have $\tau(0) = \tau'(0)$. By the local Torelli theorem 7.3 we may shrink S and S' if necessary in such a way that there is a biholomorphic map $q : S \to S'$ such that $\tau'\circ q = q$. The family which p' induces via q over S is of course still locally universal, so that we may as well assume that $S = S'$, $q = \mathrm{id}$ and (hence) $\tau = \tau'$.

Now by construction $\Phi(s)$ is a Hodge-isometry for every $s \in S$ and by Proposition 9.4 we may replace S by a smaller neighbourhood of $0 \in S$ such that not only $\Phi(0) = \phi$ but such that moreover $\Phi(s)$ maps the Kähler cone of Y'_s to the Kähler cone of Y_s for all $s \in S$. By Corollary 8.5 the period points of projective Kummer surfaces are dense in Ω, so we can find a sequence $\{s_n\}$ in S converging to 0 such that $\tau(s_n)$ is the period point of a projective Kummer surface. The Torelli theorem for projective Kummer surfaces 6.2 implies that $\Phi(s_n) = f_n^*$ for a biholomorphic map $f_n : Y_{s_n} \to Y'_{s_n}$. It now follows from Theorem 10.6 that $\Phi(0) = \phi$ is induced by an isomorphism $f : X \to X'$. □

(11.2) **Corollary** (Weak Torelli theorem.) *Let X and X' be two* K 3*-surfaces and suppose that there exists a Hodge-isometry $H^2(X', \mathbb{Z}) \to H^2(X, \mathbb{Z})$. Then X and X' are isomorphic.*

Proof. This is an immediate consequence of Theorem 11.1 and Proposition 3.10. □

(11.3) **Proposition.** *Let X be a* K 3*-surface. Any automorphism inducing the identity on $H^2(X, \mathbb{Z})$ is itself the identity.*

Proof. By Corollary 10.7 any such automorphism g has finite order, which we may assume to be prime, say ℓ. We shall show that the assumption $\ell \neq 1$ leads to a contradiction. Since g has finite order, locally around a fixed point $x \in X$ we may linearize the action ([Car]). Let (n_x, m_x) be the weights (III, Sect. 5). Since $g^*|H^2(X, \mathbb{C}) = \mathrm{id}$, for any nowhere vanishing holomorphic 2-form ω we have that $g^*(\omega) = \omega$, hence $\det(dg)_x = 1$, i.e., $n_x = -m_x$ and in particular $n_x \neq 0$, $m_x \neq 0$. This means that x is an isolated fixed point and we may apply Lefschetz' fixed point formula

$$\sum (-1)^k \mathrm{Tr}\, g^* | H^k(X, \mathbb{R}) = \mu ,$$

where μ is the number of fixed points. So $\mu = e(X) = 24$. Secondly, since $\det(1 - dg)_x \neq 0$ we can apply the holomorphic fixed point formula ([A-S], p. 567) to g :

$$2 = \sum_k (-1)^k \mathrm{Tr}(g^* | H^{0,k}(X)) = \sum_x \det(1 - (dg)_x)^{-1} ,$$

where the summation is over the μ fixed points x. Since

$$\det(1 - (dg)_x) = (1 - \lambda)(1 - \lambda^{-1}) = 4 \sin^2(\theta/2) \leq 4, \quad \lambda = \mathbf{e}(\theta)$$

it follows that $\mu \leq 8$, contradicting $\mu = 24$. □

(11.4) **Corollary.** *In Theorem* 11.1 *the isomorphism f is the unique isomorphism with $f^* = \phi$.*

12. Construction of Moduli Spaces

Let X_0 be any K 3-surface and let $p : X \to U$ be the Kuranishi family of X_0. We assume that U is taken small enough in order that
(i) p is the Kuranishi family for all $s \in U$ (see Theorem 7.3),
(ii) U is contractible.
By (ii) we can find a marking $\phi : p_{*2}\mathbb{Z}_X \xrightarrow{\sim} L_U$. Let $\tau : U \to \Omega$ be the period map determined by this marking. We assume that U is so small that
(iii) τ is an embedding.
This is again possible by Theorem 7.3. The last condition guarantees that no two marked K 3-surfaces $(X_s, \phi(s))$, $(X_{s'}, \phi(s'))$, $s, s' \in U$ are isomorphic. We consider all pairs $(p : X \to U, \phi)$ of marked Kuranishi families of K 3-surfaces, with base U satisfying the above conditions (i) up to (iii). Let us glue the U by identifying two point for which the corresponding marked K 3-surfaces are isomorphic. In this manner we obtain an analytic space M_1 such that every point has a neighbourhood isomorphic to some U. In particular M_1 is smooth (but possibly non-Hausdorff). The remark after I.10.6 implies that if, in glueing the U's, we glue the universal families over it, no two points in the same fibre come together so that over M_1 we obtain a marked family of K 3-surfaces. By construction it is a *universal* such family. Except for the fact that the base is non-Hausdorff, this proves the following theorem.

(12.1) **Theorem.** *There exists a universal marked family of* K 3*-surfaces. The base space is a non-Hausdorff "smooth analytic space" of dimension* 20.

As to the fact that the base is not Hausdorff we make the following remark.

(12.2) *Remark.* The following example, due to Atiyah ([At58]), shows that the base space M_1 is not a Hausdorff space. Consider the family of quartics $\{X_t\}$, $t \in \Delta = \{t \in \mathbb{C} \mid |t| < \varepsilon\}$, which in affine coordinates is given by

$$x^2(x^2 - 2) + y^2(y^2 - 2) + z^2(z^2 - 2) = 2t^2 .$$

For $t \neq 0$ the surface X_t is smooth and X_0 has a unique double point at $x_0 = (0, 0, 0)$ which is also the unique singular point of the total space X. As shown in V, Sect. 2 the surfaces X_t $(t \neq 0)$ are K 3-surfaces. The tangent cone at x_0 is $x^2 + y^2 + z^2 + t^2 = 0$, and one σ-process suffices to desingularize X, say $\sigma : \overline{X} \to X$. Let $E = \sigma^{-1}(x_0)$ be the exceptional manifold. The proper transform $\overline{X}_0$ of X_0 is a smooth K 3-surface meeting E $(\cong \mathbb{P}_1 \times \mathbb{P}_1)$ transversally in a (-2)-curve on $\overline{X}_0$ (a curve of bidegree $(1, 1)$ on E). Each of the rulings of E defines a contraction of E inside $\overline{X}$ onto this curve, and by an explicit calculation one shows that the resulting threefolds X_1 and X_2 are smooth. So we obtain two families $p_1 : X_1 \to \Delta$, $p_2 : X_2 \to \Delta$ of K 3-surfaces, identical over $\Delta \setminus \{0\}$. These families are not isomorphic, since otherwise there would be an automorphism of X extending the identity map on $X \setminus X_0$ but acting non-trivially on the tangent cone at x_0. Choose a marking for p_1. This induces a marking for p_2, since over $\Delta \setminus \{0\}$ the families p_1 and p_2 coincide. By construction the resulting period maps coincide over $\Delta \setminus \{0\}$, but the images of $\{0\}$ are different. This is only possible if M_1 is non-Hausdorff.

Associated to the universal family $p : Y \to M_1$ of Theorem 12.1, we have the period map (see IV, Sect. 4) $\tau_1 : M_1 \to \Omega$. The point $\tau_1(x)$ is called the period point of x.

We next define the notion of "marked pairs" consisting of a K 3-surface and a Kähler class on it. In the fibre bundle $K\Omega \to \Omega$ from Sect. 9 we take for every $[\omega] \in \Omega$ the cone $C_\omega = \{x \in L_{\mathbb{R}} \mid (x,\omega) = 0,\ (x,x) > 0\}$. Since by [He], Chap. IX, 4.3 $\mathrm{O}(a,b)$ has the homotopy type of $\mathrm{O}(a) \times \mathrm{O}(b)$, the exact homotopy sequence for the fibration $\mathrm{SO}(2) \times \mathrm{O}(1,19) \to \mathrm{O}(3,19) \to \Omega$ easily shows that Ω is connected and simply connected. So we may choose a connected component of C_ω, say C_ω^+ in such a way that it varies continuously with $[\omega] \in \Omega$. If now X is a K 3-surface and $\kappa \in H^{1,1}(X,\mathbb{R})$ any Kähler class, we say that (X,κ) is a marked pair if first of all X is marked, say $\phi : H^2(X,\mathbb{Z}) \xrightarrow{\sim} L$ and secondly $\phi_{\mathbb{C}}(\kappa) \in C_\omega^+$, where $[\omega] \in \Omega$ is the point $\tau_1(X,\phi)$.

Let $p : Y \to S$, $\phi : p_{*2}\mathbb{Z}_Y \to L_S$ be a marked family of K 3-surfaces and let $\kappa \in \Gamma(S, p_{*2}\mathbb{R}_Y \otimes_{\mathbb{R}} \mathcal{D}_S)$ have the property that for every $s \in S$ the value $\kappa(s)$ is a Kähler class such that $\phi(s)$ gives $(Y_s, \kappa(s))$ the structure of a marked pair. In this situation we speak of a family of marked pairs

Let M_2' be the total space of the real-analytic vector bundle $\bigcup_{t \in M_1} H^{1,1}(Y_t)$ (see Lemma 9.3) and let $M_2 \subset M_2'$ be the subset of Kähler classes in it. Then M_2 is an open subset of M_2' by [K-S62, Thm. 15], hence M_2 is a real-analytic manifold of dimension 60. It is not difficult to see that M_2 is a universal object for marked pairs (it is even a "fine moduli space" for such pairs) and recalling Corollary 9.2 we have a real-analytic map

$$\tau_2 : M_2 \to (K\Omega)^0 ,$$

defined as follows: if $\kappa \in H^{1,1}(Y_t)$, $(t \in M_1)$, we let $\tau_2(\kappa) = (\phi_{\mathbb{C}}(\kappa), \tau_1(t))$ – where ϕ is the marking for Y_t. We call this map the refined period map.

In this way the period map τ_1 and the refined period map τ_2, together with the forgetful maps

$$M_2 \to M_1, \quad (K\Omega)^0 \to \Omega$$

make the self-explanatory diagram

$$\begin{array}{ccc} M_2 & \xrightarrow[\tau_2]{} & (K\Omega)^0 \\ \downarrow & & \downarrow \\ M_1 & \xrightarrow[\tau_1]{} & \Omega \end{array}$$

commutative.

In view of this, the Torelli theorems 11.1 and 11.3 can be reformulated as follows.

(12.3) **Theorem.** *The map τ_2 is injective (in particular M_2 is Hausdorff).*

In the light of this result it is natural to ask for the image of τ_2, a question we deal with in Sect. 14.

13. Digression on Quaternionic Structures

Let X be a differentiable manifold equipped with a Riemannian metric g and an almost-complex structure I. The metric is hermitian if $g(IY, IZ) = g(Y, Z)$ for all $Y, Z \in \mathcal{T}_X(p)$, $p \in X$. Another way of saying that a hermitian metric g is a Kähler metric is that I is parallel with respect to the Levi-Civita connection. The existence of a Kähler metric implies that I is integrable ([K-N], Chap. IX, §4). We say that g is Kähler with respect to the complex structure I.

Let X be any differentiable manifold. An almost-quaternionic structure on X consists of two almost-complex structures I and J on X such that $I \circ J + J \circ I = 0$. Any almost-quaternionic structure $\{I, J\}$ defines an action of the algebra of quaternions on each tangent space $\mathcal{T}_X(p)$, $p \in S$, by setting $K = I \circ J$ and letting each quaternion $q = a + bi + cj + dk$ act as $Q = a + bI + cJ + dK$. Now Q is an almost-complex structure if and only if $a = 0$ and $b^2 + c^2 + d^2 = 1$. So for every almost-quaternionic structure $\{I, J\}$ there is associated a sphere $S(I, J)$ of almost complex structures.

If (X, g) is a Riemannian manifold, an almost-quaternionic structure $\{I, J\}$ on X is called quaternionic, if g is Kähler with respect to I and J. It is easily verified that g then is Kähler with respect to any of the almost-complex structures from $S(I, J)$. In particular all of those are integrable and we call $S(I, J)$ the sphere of complex structures associated to the quaternionic structure.

(13.1) **Lemma.** *Let X be a simply-connected complex manifold admitting a metric g, which is Kähler with respect to the underlying complex structure I.*

(i) *If the Ricci-tensor of g vanishes, the canonical bundle can be trivialised by a flat (hence holomorphic) section ω_I.*

(ii) *If moreover $\dim_{\mathbb{R}} X = 4$, then X can be given a quaternionic structure in the following way. We define $J_1 : \mathcal{T}_X(p) \to \mathcal{T}_X(p)$, $p \in X$ by $g(J_1 Y, Z) = \operatorname{Re} \omega_I(Y, Z)$, $Y, Z \in \mathcal{T}_X(p)$. Then for suitable $\lambda \in \mathbb{R}^*$ the operator $J = \lambda J_1$ is an almost-complex structure such that $\{g, I, J\}$ is a quaternionic structure on X.*

Proof. (communicated to us by Hitchin).

(i) The curvature of $\mathcal{K}_X$ is a multiple of the Ricci tensor ([K-N], Chap. IX, §4, §10). It follows that $\mathcal{K}_X$ is flat and since S is simply-connected it admits a flat section.

(ii) Let $\omega = \omega_I$. Since

$$\begin{aligned} g(I \circ J_1 Y, Z) &= -g(J_1 Y, IZ) = -\operatorname{Re} \omega(Y, IZ) \\ &= -\operatorname{Re} \omega(IY, Z) = -g(J_1 \circ IY, Z) \end{aligned}$$

(ω is of type $(2,0)$), it follows that $I \circ J_1 = -J_1 \circ I$. Now since J_1 is skew-adjoint, J_1^2 is self-adjoint with non-positive eigenvalues. Since $J_1 \neq 0$ we may find a non-zero eigenvalue, say $-\lambda^{-2}$ and we let V be its eigenspace. If we put $J = \lambda J_1$, then $J^2 = -\mathrm{id}$ on V. Since J^2 and I commute. I preserves V and $\{I, J, I \circ J\}$ gives V a quaternionic structure. So $\dim V \geq 4$ and hence $V = \mathcal{T}_X(p)$. Since $g(JY, JZ) = -g(Y, J^2 Z) = g(Y, Z)$ and since J is parallel, it follows that $\{g, I, J\}$ is a quaternionic structure on X. □

Since $\mathcal{K}_X$ is trivial for a K 3-surface X we can apply Theorem I.15.1 to any K 3-surface X. So for a given Kähler class κ there exists a unique Ricci-flat Kähler metric g (i.e., $\mathrm{Ric}(g) = 0$) and we can apply Lemma 13.1 to this g. We find

(13.2) **Theorem.** *Let X be a K 3-surface and κ a Kähler class. There exists a unique Ricci-flat Kähler metric g of class κ. If I denotes the complex structure on X, a second complex structure J on X exists such that $\{g, I, J\}$ is a quaternionic structure.*

The above metric g is called the canonical Kähler-Einstein metric of (X, κ). The quaternionic structure itself is not uniquely determined by (X, κ), but an easy calculation shows that if $\{g, I, J'\}$ is another quaternionic structure, then $J' \in S(I, J)$. So the sphere of complex structures is uniquely determined by (X, κ). We call it the canonical sphere of complex structures and denote it by $S(X)$. Now pick a non-zero tangent vector $Y \in \mathcal{T}_X(p)$. Since $g(Y, HY) = 0$ for any $H \in S(X)$, it follows immediately that $\{Y, IY, JY, KY\}$ is an orthogonal basis for the tangent space at p and in this order gives an orientation consistent with the orientation, induced by any of the complex structures $H \in S(X)$. Let $\{\varphi, \varphi_I, \varphi_J, \varphi_K\}$ be the dual basis for $\mathcal{T}_X^\vee(p)$. Recall ([K-N], §4) that the Kähler form $\kappa_H \in \Gamma(\mathcal{D}_X^{1,1})$ of g with respect to $H \in S(X)$ is defined by

$$g(Y, HZ) = \kappa_H(Y, Z) \ .$$

In terms of the above basis for $\mathcal{T}_X(p)$, it is easy to see that for suitable μ we have

$$\begin{aligned} \kappa_I(p) &= \varphi_I \wedge \varphi + \varphi_K \wedge \varphi_J \\ \kappa_J(p) &= \varphi_J \wedge \varphi + \varphi_I \wedge \varphi_K = \mu \operatorname{Re} \omega_I(p) \\ \kappa_K(p) &= \varphi_K \wedge \varphi + \varphi_J \wedge \varphi_I = \mu \operatorname{Im} \omega_I(p) \ . \end{aligned}$$

So we obtain:

(13.3) **Proposition.** *The subspace $E(\kappa_I, \omega_I)$ is spanned by κ_I, κ_J and κ_K* (cf. Sect. 9).

(13.4) *Remark.* Let X be a Kähler surface with given Kähler class κ and let H be one of the complex structures of the canonical sphere of complex structures. Let ω_H be a non-zero $(2,0)$-form, holomorphic with respect to H and let κ_H be the Kähler form with respect to H. Then the orthogonal complement of κ_H in $E(\kappa_I, \omega_I) = E(\kappa_H, \omega_H) = E$ is exactly $E(\omega_H)$ and since $\{\kappa_I, \kappa_J, \kappa_K\}$ is a basis of E the map $H \mapsto \omega_H$ is a surjection of the canonical sphere of complex structures onto the Grassmannian of oriented 2-planes contained in E. So we have

(13.5) **Proposition.** *The image of the refined period map (Sect. 12) consists of fibres of*

$$\Pi : K\Omega \to G_3^+(L) \ .$$

14. Surjectivity of the Period Map

In Sect. 12 we introduced the refined period map $\tau_2 : M_2 \to (K\Omega)^0$.

(14.1) **Theorem.** *The refined period map (for* K 3*-surfaces) is surjective.*

Proof. The proof consists of three steps.
Step 1: if $(\kappa, \omega) \in (K\Omega)^0$ such that $E(\kappa, \omega) \cap L$ contains a primitive rank-2 lattice M, all of whose vectors have a norm divisible by 4, then (κ, ω) belongs to the image of τ_2.

By Proposition 13.5 we may replace (κ, ω) by any other element in the fibre $\Pi^{-1}E(\kappa, \omega)$. We therefore may assume that $M \subset E(\omega)$. Then Proposition 8.1 implies the existence of a marked exceptional Kummer surface (X, ϕ) such that $\phi_{\mathbb{C}}(H^{2,0}(X)) = [\omega]$. By composing ϕ with an element of $\{\pm 1\} \times W_X$ we may suppose that $\phi^{-1}(\kappa)$ is in the Kähler cone of X (*cf.* Proposition 3.10). We have to show that it is in fact a Kähler class. Theorem IV.6.1 implies that the integral elements of the Kähler cone are classes of ample divisors, hence are Kähler classes. So all rational elements in the Kähler cone are Kähler classes. The fact that X is exceptional implies that $H^{1,1}(X, \mathbb{R})$ is defined over $\mathbb{Q}$ and hence the rational elements in the Kähler cone are dense in it. Since the set of Kähler classes also forms an open convex subcone of the Kähler cone, it follows that this subcone actually coincides with the Kähler cone.

Step 2: if $(\kappa, \omega) \in (K\Omega)^0$ is such that $E(\kappa, \omega) \cap L$ contains a primitive vector x with $(x, x) \in 4\mathbb{N}$, then $(\kappa, \omega) \in \text{Im}(\tau_2)$.

As in Step 1, we may suppose that $x \in E(\omega)$. Let

$$K \subset (K\Omega)^0 \cap \{L_{\mathbb{R}} \times [\omega]\} = C_\omega^0$$

be the component of C_ω^0 containing κ. Let $\eta \in K$ be chosen such that

$$M_\eta = (\mathbb{R} \cdot x + \mathbb{R} \cdot \eta) \cap L$$

is a rank-2 lattice, all of whose vectors have norm in $4\mathbb{Z}$. By Remark 8.4 those η are dense in K. By Step 1 we then have $(\eta, [\omega]) = \tau_2(X_\eta, \kappa_n, \phi_\eta)$ for some marked pair (X_η, κ_η). It follows from the injectivity of τ_2 (see Theorem 12.3) that for all different choices of η in K the isomorphism type of X_η is the same. So all of the η thus chosen via the marking ϕ_η, correspond to Kähler classes on the same K 3-surface, hence as in Step 1 they form a dense open convex subset in the Kähler cone of X_η. It follows that the entire Kähler cone of X_η consists of Kähler classes, in particular $\phi_\eta^{-1}(\kappa)$ is a Kähler class, i.e., $(\kappa, [\omega]) \in \operatorname{Im} \tau_2$.

Step 3: the final argument.

Let $(\kappa, [\omega]) \in (K\Omega)^0$ and let K be the connected component of C_ω^0 containing κ as in Step 2. If $\eta \in K$ is such that $E(\omega) + \mathbb{R} \cdot \eta$ contains an indivisible vector with norm in $4\mathbb{N}$, then $(\eta, [\omega]) \in \operatorname{Im} \tau_2$ by Step 2. Such η are dense in K by Remark 8.4. Since the cone $(K \times [\omega]) \cap (\operatorname{Im} \tau_2)$ is convex, it follows as in the proof of the preceding two steps, that $(\kappa, [\omega]) \in \operatorname{Im} \tau_2$. □

(14.2) **Corollary.** *Every point of Ω occurs as the period point of a marked* K 3-*surface.*

Enriques Surfaces

15. Topological and Analytic Invariants

In this section Y denotes an Enriques surface. By definition (V, Sect. 23) this means

$$\mathcal{K}_Y^{\otimes 2} = \mathcal{O}_Y \ , \quad \text{but } \mathcal{K}_Y \neq \mathcal{O}_Y,$$

and $q(Y) = 0$. We have seen (loc. cit.) that Y is always projective. The conditions on $\mathcal{K}_Y$ imply $c_1^2(Y) = 0$ and $p_g = 0$. So $\chi(\mathcal{O}_Y) = 1$ and Noether's formula becomes

$$e(Y) = 12.$$

Riemann-Roch for a line bundle $\mathcal{L}$ takes the form

$$\chi(\mathcal{L}) = 1 + \frac{1}{2} c_1^2(\mathcal{L}),$$

and for a rank-2 bundle $\mathcal{V}$ we have

$$\chi(\mathcal{V}) = 2 + \frac{1}{2} c_1^2(\mathcal{V}) - c_2(\mathcal{V}). \tag{7}$$

(15.1) **Lemma.** *Let Y be an Enriques surface. Then*

(i) $h^{1,0}(Y) = h^{0,1}(Y) = h^{2,0}(Y) = h^{0,2}(Y) = 0$, $h^{1,1}(Y) = 10$.

(ii) *The fundamental group of Y is $\mathbb{Z}/2\mathbb{Z}$ and the universal covering X of Y is a* K 3-*surface.*

(iii) *The intersection form on $H^2(Y, \mathbb{Z})_f = H^2(Y, \mathbb{Z})/$*torsion *is isometric to $(-E_8) \oplus H$.*

Proof.

(i) From the preceding remarks we conclude that

$$h^{1,0} = h^{0,1} = h^{2,0} = h^{0,2} = 0\ , \quad \text{so} \quad h^{1,1} = b_2 = e(Y) - 2 = 10\ .$$

(ii) By I, Sect.18 there exists an unramified double covering $\pi : X \to Y$ such that $\mathcal{K}_X = \pi^*(\mathcal{K}_Y) = \mathcal{O}_X$. Since $e(X) = 2e(Y) = 24$ and $p_g(X) = 1$, Noether's formula for X implies $q(X) = 0$. So by definition (VI, Sect. 1) X is a K 3-surface and therefore it is simply-connected (see Corollary 8.6).
(iii) Since $p_g = 0$, every $d \in H^2(Y, \mathbb{Z})$ is divisorial, say $d = \mathcal{O}_Y(D)$. Hence $d^2 = D^2 \equiv DK_Y \equiv 0 \bmod 2$. Since $\tau(Y) = \frac{1}{3}(c_1^2(Y) - 2c_2(Y)) = -8$ by the topological index theorem, this form is indefinite. The assertion then follows from Theorem I.2.8. □

(15.2) **Proposition.** *For every Enriques surface Y the map*

$$\mathrm{Pic}(Y) \xrightarrow[c_1]{} H^2(Y, \mathbb{Z}) = \mathbb{Z}^{10} \oplus \mathbb{Z}/2\mathbb{Z}$$

is an isomorphism.

Proof. The isomorphism is a consequence of the exponential sequence and of the vanishing of $H^i(\mathcal{O}_Y)$, $i = 1, 2$. □

When considering deformations of Y we need

(15.3) **Proposition.** *For every Enriques surface Y*

$$h^0(\mathcal{T}_Y) = h^2(\mathcal{T}_Y) = 0\ , \quad h^1(\mathcal{T}_Y) = 10.$$

Proof. The universal covering X of Y does not admit holomorphic vector fields by Corollary 3.5, so $h^0(\mathcal{T}_Y) = h^0(\mathcal{T}_X) = 0$. Since $\mathcal{K}_Y \otimes \mathcal{T}_Y^\vee = \mathcal{K}_Y^\vee \otimes \Omega_Y^1 = \mathcal{T}_Y$, by Serre duality we have $h^2(\mathcal{T}_Y) = h^0(\mathcal{T}_Y) = 0$, so $h^1(\mathcal{T}_Y) = 10$ by (7). □

16. Divisors on an Enriques Surface *Y*

In the sequel D will be a divisor on Y and $d \in H^2(Y, \mathbb{Z})$ its class.

(16.1) **Proposition.** *Let D be any divisor on the Enriques surface Y, $D \neq 0$, K_Y.*

(i) *If D is irreducible, then $D^2 \geq -2$ and $D^2 = -2$ if and only if D is a (-2)-curve.*
(ii) *If $D^2 \geq 0$, then either $|D| \neq \varnothing$ or $|-D| \neq \varnothing$ (but not both). If $D^2 \geq 0$ and $|D| \neq \varnothing$, then also $|K_Y + D| \neq \varnothing$.*
(iii) *If $D^2 \geq 0$, $|D| \neq \varnothing$, and if $|K_Y + D|$ contains a reduced connected divisor, then*

$$\dim |D| = \frac{1}{2} D^2\ .$$

Proof.

(i) By the adjunction formula $D^2 = \deg(\omega_D) \geq -2$, and $D^2 = -2$ implies that D is smooth rational.
(ii) The Riemann-Roch inequality implies

$$(8) \qquad h^0(\mathcal{O}_Y(D)) + h^0(\mathcal{O}_Y(K_Y - D)) \geq \frac{1}{2}D^2 + 1 .$$

So, if $D^2 \geq 0$, either $|D| \neq \varnothing$ or $|K_Y - D| \neq \varnothing$, but not both at the same time. Since $D^2 = (K_Y + D)^2$, the same argument shows that $|K_Y + D| \neq \varnothing$ or $|-D| \neq \varnothing$. So if $|D| \neq \varnothing$ the first alternative must hold.
(iii) The assumptions imply $h^0(\mathcal{O}_{D'}) = 1$ for some divisor $D' \in |K_Y + D|$. Since $h^1(\mathcal{O}_Y) = 0$ by definition, from the cohomology sequence of

$$0 \to \mathcal{O}_Y(-D') \to \mathcal{O}_Y \to \mathcal{O}_{D'} \to 0$$

we find $h^1(\mathcal{O}_Y(-D')) = 0$. Then $h^1(\mathcal{O}_Y(D)) = 0$ by Serre duality and we have equality in (8). □

As the intersection form on $H^2(Y, \mathbb{R})$ has signature $(1, 9)$, the set of classes $d \in H^2(Y, \mathbb{R})$ with $d^2 > 0$ is the union of two convex open cones, say $\mathcal{C}$ and $\mathcal{C}'$. Since

$$d_1 \cdot d_2 > 0 \text{ if both } d_1, d_2 \in \mathcal{C} ,$$
$$d_1 \cdot d_2 < 0 \text{ if } d_1 \in \mathcal{C},\ d_2 \in \mathcal{C}' ,$$

only one of them, say $\mathcal{C}$, contains ample classes. Let $\widetilde{\mathcal{C}} \in H^2(Y, \mathbb{Z})$ be the semi group of classes $d \neq 0$ mapping into the closure $\overline{\mathcal{C}}$.

(16.2) **Proposition.** *For a divisor D on Y of a non-torsion class $d \in H^2(Y, \mathbb{Z})$ the following statements are equivalent:*
a) $d \in \widetilde{\mathcal{C}}$,
b) $D^2 \geq 0$ *and* $|D| \neq \varnothing$.

Proof. If $d \in \widetilde{\mathcal{C}}$, then $D^2 \geq 0$ and either d or $-d$ is effective (Proposition 16.1, (ii)). But a class in $-\widetilde{\mathcal{C}}$ cannot be effective. □

Next, let $c \in H^2(Y, \mathbb{Z})$ be a class with $c^2 = -2$. Just as on a K 3-surface such a class defines a Picard-Lefschetz reflection

$$s_c : H^2(Y, \mathbb{Z}) \to H^2(Y, \mathbb{Z}) , \quad s_c(d) = d + (d \cdot c)c .$$

(16.3) **Proposition.** *The reflection s_c leaves $\widetilde{\mathcal{C}}$ invariant. In particular, if $d \in H^2(Y, \mathbb{Z})$ is effective with $d^2 \geq 0$, then so is $s_c(d)$.*

Proof. The proof is the same as that of the corresponding statement in Proposition 3.10. □

We consider $H^2(Y, \mathbb{Z})_f$ as a euclidean lattice. The decomposition $H^2(Y, \mathbb{Z})_f = -E_8 \oplus H$ is not unique of course. So $H^2(Y, \mathbb{Z})_f$ contains many hyperbolic planes H. Such a plane is generated by the images of classes $d_1, d_2 \in H^2(Y, \mathbb{Z})$ satisfying

(9) $$d_1^2 = d_2^2 = 0\ , \quad d_1 \cdot d_2 = 1\ .$$

(16.4) **Proposition.** *If Y is an Enriques surface and $d_1, d_2 \in H^2(Y, \mathbb{Z})$ satisfy (9), then either both d_1 and d_2, or both, $-d_1$ and $-d_2$, are effective.*

Proof. Both, d_1 and d_2, belong to $\widetilde{\mathcal{C}} \cup (-\widetilde{\mathcal{C}})$. Since $d_1 \cdot d_2 > 0$, the classes d_1 and d_2 cannot belong to different components of this double cone. The assertion now follows from Proposition 16.2. □

A pair of effective classes $d_1, d_2 \in H^2(Y, \mathbb{Z})$ satisfying (7) will be called an effective pair of generators of a hyperbolic plane.

17. Elliptic Pencils

In this section we show that every Enriques surface Y admits an elliptic fibration over $\mathbb{P}_1$. This means that there exists a holomorphic map $f : Y \to \mathbb{P}_1$, whose generic fibre is a smooth elliptic curve, or, equivalently, there exists a base point free rational pencil on Y, whose generic member is smooth elliptic. Such pencils will be called elliptic pencils in the sequel. Before proving the main result (Theorem 17.7) we make some simple observations concerning elliptic pencils on Enriques surfaces.

(17.1) **Lemma.** *Every elliptic fibration $f : Y \to \mathbb{P}_1$ of an Enriques surface Y has exactly two multiple fibres $2F$ and $2F'$, and*

$$\mathcal{K}_Y = \mathcal{O}_Y(F' - F) = \mathcal{O}_Y(F - F')\ .$$

Proof. The canonical bundle formula for an elliptic fibration (in fact Corollary V.12.3) yields

$$\mathcal{K}_Y = f^*\mathcal{O}_{\mathbb{P}_1}(-1) \otimes \mathcal{O}_Y\Big(\sum_{i=1}^{k}(r_i - 1)F_i\Big)\ ,$$

where r_iF_i, $i = 1, \ldots, k$, are the multiple fibres of f. So

$$\mathcal{O}_Y = \mathcal{K}_Y^{\otimes 2} = f^*\mathcal{O}_{\mathbb{P}_1}(-2) \otimes \mathcal{O}_Y\Big(\sum_{i=1}^{k}(2r_i - 2)F_i\Big)\ .$$

Restricting to F_i and using that $\mathcal{O}_{F_i}(F_i)$ has order exactly r_i in $\mathrm{Pic}(\mathcal{O}_{F_i})$ (Lemma III.8.3), we find that all r_i equal 2. From the preceding formula we infer $k = 2$. We put $F_1 = F$, $F_2 = F'$, whereupon the canonical bundle formula becomes:

$$\mathcal{K}_Y = f^*\mathcal{O}_{\mathbb{P}_1}(-1) \otimes \mathcal{O}_Y(F + F')\ .$$

The result follows, because $\mathcal{O}_Y(2F) = \mathcal{O}_Y(2F') = f^*\mathcal{O}_{\mathbb{P}_1}(1)$. □

The curves F, F' in Lemma 17.1 are called the half pencils. By an elliptic configuration on Y we mean a curve F appearing in Kodaira's table (V, Table 3) of singular fibres in elliptic fibrations, but not a multiple of such a fibre.

(17.2) **Lemma.** *Let $F \subset Y$ be a connected curve with $F^2 = 0$ and $F \cdot C \geq 0$ for all (-2)-curves C on Y. Then $F = mF_0$, $0 < m \in \mathbb{N}$, where F_0 is an elliptic configuration.*

Proof. By Proposition 16.2 we have $F \cdot C \geq 0$ for all effective curves C satisfying $C^2 \geq 0$. So Proposition 16.1, (i) and the assumption show $F \cdot C \geq 0$ for all effective curves C on Y. If $F = A + B$ is an effective decomposition, then $F \cdot A + F \cdot B = F^2 = 0$. This proves that $F \cdot C = 0$ for each irreducible component $C \subset F$. Now we apply Lemma I.2.10 (as in the proof of Zariski's lemma) to prove $D^2 \leq 0$ for all divisors D consisting of components of F, with $D^2 = 0$ only if $D = rF$, $r \in \mathbb{Q}$. In particular, if F contains more than one component, all its irreducible components are (-2)-curves. The assertion now follows just as in V. Sect 7. □

(17.3) **Lemma.** *Let F be an elliptic configuration on Y. Then either $\mathcal{O}_F(F) = \mathcal{O}_F$ and $|F|$ is an elliptic pencil, or $\mathcal{O}_F(F) \in \mathrm{Pic}(F)$ is a non-trivial 2-torsion element and $|2F|$ is an elliptic pencil of which F is a half pencil.*

Proof. By the adjunction formula $\deg(\omega_F|C) = 0$ for all irreducible components $C \subset F$. Since F has arithmetical genus 1, Riemann-Roch on F shows $h^0(\omega_F) = 1$. Then ω_F is trivial by Lemma II.12.2. Since $\omega_F = \mathcal{K}_Y \otimes \mathcal{O}_F(F)$, and $\mathcal{K}_Y^{\otimes 2} = \mathcal{O}_Y$, we have $\mathcal{O}_F(2F) = \mathcal{O}_F$.

If $\mathcal{O}_F(F)$ is already trivial, the sequence

$$0 \to \mathcal{O}_Y \to \mathcal{O}_Y(F) \to \mathcal{O}_F(F) \to 0$$

shows $h^0(\mathcal{O}_Y(F)) = 2$, and $|F|$ is an elliptic pencil.

If $\mathcal{O}_F(F)$ is non-trivial, then $h^0(\mathcal{O}_F(F)) = h^1(\mathcal{O}_F(F)) = 0$ and the above sequence shows $h^0(\mathcal{O}_Y(F)) = 1$ and $h^1(\mathcal{O}_Y(F)) = 0$. From the sequence

$$0 \to \mathcal{O}_Y(F) \to \mathcal{O}_Y(2F) \to \mathcal{O}_F(2F) \to 0$$

we then infer $h^0(\mathcal{O}_Y(2F)) = 2$ and $|2F|$ is an elliptic pencil. □

Remark. The two cases distinguished in Lemma 17.3 also differ with respect to the behaviour of F under the universal covering $\pi : X \to Y$. Since $\mathcal{O}_F(F) = \mathcal{K}_Y|F$, the inverse image $\pi^*(F) \subset X$ is connected if $\mathcal{O}_F(F) \neq \mathcal{O}_F$, and it decomposes into two connected components mapped isomorphically onto F, if $\mathcal{O}_F(F) = \mathcal{O}_F$. If $|2F_1|$ is an elliptic pencil on Y with half pencil F_1, then $|\pi^*(F_1)|$ is an elliptic pencil on the K3-surface X. The corresponding fibrations $g : X \to \mathbb{P}_1$, $f : Y \to \mathbb{P}_1$ form a commutative diagram

$$\begin{array}{ccc} X & \xrightarrow{\pi} & Y \\ g\downarrow & & \downarrow f \\ \mathbb{P}_1 & \longrightarrow & \mathbb{P}_1 \end{array},$$

where the induced map $\mathbb{P}_1 \to \mathbb{P}_1$ is a double covering, ramified over the image points $f(F_1), f(F_2)$ of the half pencils F_1, F_2 in $|2F_1|$. Let $\sigma : X \to X$ be the covering involution for π. This σ lifts to an involution $\sigma^* : \mathcal{O}_X(\pi^*(F_1)) \to \mathcal{O}_X(\pi^*(F_1))$ on the line bundle level. If $g_1, g_2 \in H^0(\mathcal{O}_X(\pi^*(F_1)))$ are "equations" for F_1, F_2, then it can be assumed that $\sigma^*(g_1) = g_1$. But this implies $\sigma^*(g_2) = -g_2$.

(17.4) **Lemma.** *Let E be an effective divisor on Y with $E^2 = 0$ and such that its class $e \in H^2(Y, \mathbb{Z})_f$ is primitive (i.e., $e = me'$, $m \in \mathbb{Z}$, $e' \in H^2(Y, \mathbb{Z})_f$ implies $m = \pm 1$). Then there are finitely many (-2)-curves $C_1, \ldots, C_k$ on Y with classes $c_1, \ldots, c_k \in H^2(Y, \mathbb{Z})$ such that $(s_{c_k} \circ \cdots \circ s_{c_1})(e)$ is the class of a half pencil in some elliptic fibration of Y.*

Proof. If there is a (-2)-curve C_1 on Y with $E \cdot C_1 = -n < 0$, we consider $e_1 = s_{c_1}(e) = e - nc_1$. By Proposition 16.3, there is an effective divisor E_1 with class e_1. If $E_2 \cdot C_2 < 0$ for some (-2)-curve C_2, we repeat this procedure and so on. With each reflection s_{c_i} the degree of E_{i-1} (with respect to an arbitrary projective embedding of Y) decreases by at least one. So after finitely many reflections this process will terminate. We arrive at a divisor F with class $(s_{c_k} \circ \cdots \circ s_{c_1})(e)$ satisfying $F \cdot C \geq 0$ for all (-2)-curves C on Y.

All the connected components F_i of F share the property $F_i^2 = 0$ and $F_i \cdot C \geq 0$. So Lemma 17.2 shows that $F = \sum m_i F_i$, where the F_i are elliptic configurations on Y. By Lemma 17.3 their classes $f_i \in H^2(Y, \mathbb{Z})_f$ are all proportional. Then primitivity shows that there is only one connected component F_1, which itself is a half pencil, and $m_1 = 1$. □

An obvious consequence of Lemma 17.4 is

(17.5) **Theorem.** *Every Enriques surface admits an elliptic fibration over the projective line.*

We shall need more:

(17.6) **Proposition.** *Given an elliptic fibration of the Enriques surface Y over $\mathbb{P}_1$, there is a 2-section G for this fibration with $G^2 \leq 0$. (Here, a 2-section means an irreducible curve G with $F \cdot G = 2$ for all the fibres F.)*

Proof. Let D_1 be one of the half pencils for the elliptic fibration and let $d_1 \in H^2(Y, \mathbb{Z})_f$ be its class. To show that d_1 is primitive, we assume $d_1 = md'$, $0 < m \in \mathbb{Z}$, $d' \in H^2(Y, \mathbb{Z})_f$. This d' is effective by Proposition 16.1, (ii). If D' is an effective divisor representing d', from Lemma 17.3 we infer that $h^0(\mathcal{O}_Y(D_1)) = 1$, so we have $D_1 = mD'$ with $m = 1$, because $2D_1$ is a fibre of multiplicity 2.

Since d_1 is primitive, the intersection product with d_1 defines a map of the unimodular lattice $-E_8 \oplus H$ onto $\mathbb{Z}$. So there is some d with $d \cdot d_1 = 1$. Since d^2 is an even integer, we may form $d_2 = d - \frac{d^2}{2} d_1$. Then $d_2^2 = 0$ and $d_1 \cdot d_2 = 1$. Now Proposition 16.4 implies the existence of some effective divisor D_2 with $D_2^2 = 0$ and $D_1 \cdot D_2 = 1$. Writing $D_2 = G + D$ with D consisting of all components contained in fibres, we see that there is some 2-section G. It remains to find such a 2-section with $G^2 \leq 0$.

So assume $G^2 \geq 2$. By Riemann-Roch $h^0(\mathcal{O}_Y(G)) \geq 2$. Since $\mathcal{O}_{D_1}(G)$ is an invertible sheaf of degree 1 on a reduced curve of genus 1, we have $h^0(\mathcal{O}_{D_1}(G)) = 1$. This implies that there is some divisor in $|G|$ containing D_1. Let this divisor be $D_1 + G' + R$ with G' a 2-section and R consisting of fibre components. Then

$$(G')^2 = (G - D_1 - R)^2 = G^2 - 2 - 2G \cdot R + (D_1 + R)^2 < G^2,$$

because $(D_1 + R)^2 \leq 0$ by Zariski's lemma. After repeating this procedure sufficiently many times, we obtain a 2-section with self-intersection ≤ 0. □

In view of Lemma 16.1, (i), there are two possibilities for the 2-section G which occurs in Proposition 17.6:
(i) G is a (-2)-curve, or
(ii) $G^2 = 0$, and G is an elliptic configuration by Lemma 17.2. Since $G \cdot F_1 = 1$ for a half pencil F_1 in the elliptic fibration given, the class of G in $H^2(Y, \mathbb{Z})_f$ is primitive. By Lemmas 17.3 and 17.1 it follows that $|2G|$ itself is an elliptic pencil, one of whose half pencils is G.
To formulate the final result of this section, we use the following

Definition. An Enriques surface is called special, if it carries an elliptic pencil together with a 2-section which is a (-2)-curve.

Then we have proved:

(17.7) **Theorem.** *Every Enriques surface admits an elliptic fibration. On a non-special Enriques surface there are two effective divisors D_1 and D_2 with $D_1^2 = D_2^2 = 0$, $D_1 \cdot D_2 = 1$, occurring as half pencils in two distinct elliptic pencils.*

18. Double Coverings of Quadrics

In this section we use elliptic pencils on the Enriques surface Y, as constructed in the preceding section, to represent Y in terms of a double covering of a quadric. Following Horikawa in [Hor78] we use this representation to prove e.g. the connectedness of the space of all Enriques surfaces.

So we assume first, that on Y we have two half pencils D_1, D_2 with $D_1 \cdot D_2 = 1$, as they exist on every *non-special* Y by Theorem 17.7. We denote by C_i, $i = 1, 2$, their inverse image $\pi^*(D_i) \subset X$. Both linear systems $|C_1|$ and $|C_2|$ are free from base points and fixed components. The same holds for the system $|C_1 + C_2|$ and the map $f_{C_1+C_2} = f_{\mathcal{O}_X(C_1+C_2)}$ is well-defined. The dimension $h^0(\mathcal{O}_X(C_1 + C_2))$ is computed easily: $|C_1 + C_2|$ contains reduced connected curves, so as in the proof of Proposition 16.1, (iii) we conclude

$$h^1(\mathcal{O}_X(C_1 + C_2)) = 0 \quad \text{and} \quad h^0(\mathcal{O}_X(C_1 + C_2)) = 2 + \frac{1}{2}(C_1 + C_2)^2 = 4,$$

because $C_1 \cdot C_2 = 2$. Hence we have $f_{C_1+C_2} : X \to \mathbb{P}_3$. The canonical map

$$H^0(\mathcal{O}_X(C_1)) \otimes H^0(\mathcal{O}_X(C_2)) \longrightarrow H^0(\mathcal{O}_X(C_1+C_2))$$

is injective, and even an isomorphism. This shows that the image of X in $\mathbb{P}_3$ is a non-singular quadric $Q = \mathbb{P}_1 \times \mathbb{P}_1$, where the two rulings of Q on X induce the elliptic fibrations defined by $|C_1|$ and $|C_2|$.

As $C_1 \cdot C_2 = 2$, the map $X \to Q$ is generically 2-to-1. The general elliptic curve in $|C_1|$ or $|C_2|$ is mapped 2-to-1 onto a projective line, so it is branched over four points. This shows that there is a branch curve $B \subset Q$ of bidegree $(4,4)$. Over $Q \setminus B$ the map $f_{C_1+C_2}$ is a non-ramified double covering, and the inverse images of the finitely many singularities of B are exceptional curves in X contracted to points under $f_{C_1+C_2}$. Also, all the singularities of B must be simple. Indeed, if not, then by Theorem III.7.2 a canonical divisor on X would be strictly negative, leading to the contradiction $p_g(X) = 0$ for the K 3-surface X.

Finally, the covering involution σ for $\pi : X \to Y$ acts on the whole situation. To be specific, we introduce bihomogeneous coordinates $(x_0 : x_1)$, $(y_0 : y_1)$ on Q as follows. Let D_1', $D_2' \subset Y$ be the half pencils adjoint to D_1, D_2 and let $C_1' = \pi^*(D_1')$, $C_2' = \pi^*(D_2')$. Let g_1, $g_1' \in H^0(\mathcal{O}_X(C_1))$ be equations for C_1, C_1' and g_2, $g_2' \in H^0(\mathcal{O}_X(C_2))$ equations for C_2, and C_2'. Then homogeneous coordinates $(x_0 : x_1)$, $(y_0 : y_1)$ respectively, on $\mathbb{P}_1$ can be chosen in such a way that the fibration defined by $|C_1|$ has equation $g_1 : g_1' = x_0 : x_1$, and that similarly $g_2 : g_2' = y_0 : y_1$ defines the fibration of the pencil $|C_2|$.

We introduce coordinates $(z_0 : z_1 : z_2 : z_3)$ on $\mathbb{P}_3$ such that Q is embedded by

$$z_0 = x_0 y_0, \quad z_1 = x_1 y_1, \quad z_2 = x_0 y_1, \quad z_3 = x_1 y_0.$$

Then $f_{C_1+C_2}$ is given by

$$z_0 = g_1 g_2, \quad z_1 = g_1' g_2', \quad z_2 = g_1 g_2', \quad z_3 = g_1' g_2 \ .$$

Using the remark after Lemma 17.3 we lift σ to an involution σ^* on $\mathcal{O}_X(C_1)$ with $\sigma^*(g_1) = g_1$, $\sigma^*(g_1') = -g_1$ and on $\mathcal{O}_X(C_2)$ such that $\sigma^*(g_2) = g_2$, $\sigma^*(g_2') = -g_2$.

So if we define the involution τ on $\mathbb{P}_3$ by

$$\tau(z_0 : z_1 : z_2 : z_3) = (z_0 : z_1 : -z_2 : -z_3) \ ,$$

then Q is τ-invariant with τ acting on Q by

$$\tau((x_0 : x_1), (y_0 : y_1)) = ((x_0 : -x_1), (y_0 : -y_1)).$$

If we let the generator of $\mathbb{Z}/2\mathbb{Z}$ act by σ on X and by τ on $\mathbb{P}_3$, then also $f_{C_1+C_2}$ is $\mathbb{Z}/2\mathbb{Z}$-equivariant. On Q the involution τ has the four fixed points

$$((1:0),(1:0)), \quad ((1:0),(0:1)), \quad ((0:1),(1:0)), \quad ((0:1),(0:1)) \ .$$

The branch curve B does not contain any of them. Indeed, if one of them were a smooth point of B, then the involution σ would leave the corresponding point on X invariant. This is impossible because σ has no fixed points. And if B had a simple curve singularity at a point x with $\tau(x) = x$, then σ would act as an involution on the A-D-E configuration in X lying over x. But any such involution has a fixed point, yielding again a contradiction. The polynomial of bidegree $(4,4)$ defining B is either invariant or anti-invariant under τ. The absence of fixed points on B shows that it must be invariant.

Altogether this proves:

(18.1) **Proposition** (Horikawa's representation of non-special Enriques surfaces). *Let Y be an Enriques surface and $D_1, D_2 \subset Y$ be two half pencils with $D_1 \cdot D_2 = 1$. If τ and σ are defined as above, then there is a τ-invariant bihomogeneous polynomial of bidegree $(4,4)$ in $(x_0 : x_1)$, $(y_0 : y_1)$, with zero-set B on the smooth quadric Q, such that the universal covering X of Y is the minimal resolution of the double covering of Q ramified over B. The curve B is reduced with at worst simple singularities and does not contain any fixed point of τ. The involution σ on X is induced by the involution τ on Q. The two rulings of Q define the two elliptic pencils $|\pi^*(D_1)|$, $|\pi^*(D_2)|$ on X.*

Theorem 18.1 has a converse: given a τ-invariant curve as in the theorem, the K 3-surface X and an Enriques surface $Y = X/\sigma$ can be constructed from it. This has already been shown in V, Sect. 23.

To obtain a representation of all Enriques surfaces Y, we still have to treat the *special* case. So we now assume that on Y we are given a half pencil D_1 and a (-2)-curve D_2 with $D_1 \cdot D_2 = 1$.

Again we put $C_1 = \pi^*(D_1)$. The elliptic pencil $|C_1|$ on X is free of base points and fixed components. The curve $\pi^*(D_2)$ decomposes into two disjoint (-2)-curves, say E_1 and E_2. On X we consider the linear system $|\pi^*(2D_1 + D_2)| = |2C_1 + E_1 + E_2|$. It contains reduced connected curves, so as in the non-special case we conclude

$$h^1(\mathcal{O}_X(2C_1 + E_1 + E_2)) = 0$$

and

$$h^0(\mathcal{O}_X(2C_1 + E_1 + E_2)) = 2 + \frac{1}{2}(2C_1 + E_1 + E_2)^2 = 4 .$$

Since $\mathcal{O}_{E_i}(2C_1 + E_1 + E_2) = \mathcal{O}_{E_i}$ we have an exact sequence

$$0 \longrightarrow H^0(\mathcal{O}_X(2C_1)) \longrightarrow H^0(\mathcal{O}_X(2C_1 + E_1 + E_2)) \longrightarrow H^0(\mathcal{O}_{E_1+E_2}) .$$

Together with $h^0(\mathcal{O}_X(2C_1)) = 3 < h^0(\mathcal{O}_X(2C_1 + E_1 + E_2))$ this implies that $|2C_1 + E_1 + E_2|$ has no base points. So $\varphi = f_{2C_1+E_1+E_2} : X \to \mathbb{P}_3$ is well-defined.

Next we choose sections $g_1, g_1' \in H^0(\mathcal{O}_X(C_1))$ as in the non-special case and a generator $e \in H^0(\mathcal{O}_X(E_1+E_2))$. Then $eg_1^2, eg_1g_1', eg_1'^2 \in H^0(\mathcal{O}_X(2C_1+E_1 + E_2))$ can be complemented to a basis of this space by some section f. On $\mathbb{P}_3$ coordinates $(z_0 : z_1 : z_2 : z_3)$ are introduced such that φ is given by

$$(z_0 : z_1 : z_2 : z_3) = (eg_1^2 : eg_1'^2 : eg_1g_1' : f) .$$

So $\varphi(X)$ is contained in the quadratic cone Q_0 with equation $z_0z_1 = z_2^2$. Both (-2)-curves E_1 and E_2 are contracted by φ, and their image is the vertex $(0:0:0:1)$ of the cone Q_0. On the general curve in $|C_1|$ there are two zeros of f, so φ is generically 2-to-1, mapping the members of $|C_1|$ onto the generators of the cone Q_0. Again there is a branch curve $B \subset Q_0$, not containing the vertex, which intersects the general line on Q_0 in four points. So B is the intersection of Q_0 with a surface of degree 4. As in the non-special case, B can have at most simple singularities.

Again we trace the action of the covering involution σ. As before it lifts to $\mathcal{O}_X(C_1)$ and hence to some σ^* on $\mathcal{O}_X(2C_1)$ such that $\sigma^*(g_1^2) = g_1^2$, $\sigma^*(g_1'^2) = g_1'^2$, $\sigma^*(g_1g_1') = -g_1g_1'$. We can lift σ to some σ^* on $\mathcal{O}_X(E_1+E_2)$ with $\sigma^*(e) = e$. So we have an induced involution, again denoted by σ^*, on the line bundle $\mathcal{O}_X(2C_1+E_1+E_2)$. Now f can be chosen as above, but additionally as an eigenvector for σ^*. Then

$$\sigma^*(eg_1^2) = eg_1^2, \quad \sigma^*(eg_1'^2) = eg_1'^2, \quad \sigma^*(eg_1g_1') = -eg_1g_1', \quad \sigma^*(f) = \pm f.$$

To determine the last sign, we first define τ on $\mathbb{P}_3$ by

$$\tau(z_0 : z_1 : z_2 : z_3) = (z_0 : z_1 : -z_2 : \pm z_3)\ ,$$

such that φ is $\mathbb{Z}/2\mathbb{Z}$-equivariant again. If τ left z_3 invariant, the plane $z_2 = 0$, consisting of fixed points under τ only, would intersect the branch curve B. As above we would find a fixed point for σ, that is, a contradiction. So we have the minus-sign, i.e., τ is the same involution as in the non-special case. Also B does not pass through any of the three fixed points of τ on Q_0: the vertex, and the two points $(1:0:0:0)$, $(0:1:0:0)$. And we find once more, that B is defined by some τ-invariant polynomial of degree 4.

All this proves the special-version of Theorem 18.1:

(18.2) **Proposition** (Horikawa's representation of special Enriques surfaces). *Let Y be an Enriques surface, $D_1 \subset Y$ a half pencil and $D_2 \subset Y$ a (-2)-curve with $D_1 \cdot D_2 = 1$. Then there exists a τ-invariant homogeneous polynomial of degree 4 in z_0, z_1, z_2, z_3, vanishing on the quadric cone $Q_0 = \{z_0z_1 = z_2^2\}$ exactly in the points of the curve B, such that the universal covering X of Y is the minimal resolution of the double covering of Q_0 ramified over B. The curve B is reduced with at most simple singularities and does not contain a fixed point of τ. The involution σ on X is induced by τ on Q_0. The system of generators on Q_0 defines the elliptic pencil $|\pi^*(2C_1)|$ on X, and $\pi^*(D_2)$ consists of the two (-2)-curves over the vertex of Q_0.*

Also Theorem 18.2 has a converse: given B, the surfaces X and Y can be constructed.

To apply the representations given above, we use the following notion:

(18.3) **Definition.** Any Enriques surface Y, which can be represented as in the construction of Theorem 18.1 is called general, if B is smooth.

This use of the word general is justified by the obvious fact, that almost every non-special surface is general, and by the following, less obvious observation.

(18.4) **Theorem.** *Every special Enriques surface Y is a deformation of general ones.*

Proof. Let Y be represented as in Theorem 18.2. Let $f(z_0, z_1, z_2, z_3)$ be some τ-invariant polynomial defining B. This f is not uniquely determined by B, the same curve B is defined by any

$$g = f + q_0 \cdot q \ ,$$

where $q_0 = z_0 z_1 - z_2^2$ is an equation for the cone Q_0 and q some τ-invariant quadratic polynomial. I.e., q is a linear combination of $z_0^2, z_0 z_1, z_1^2, z_2^2, z_2 z_3, z_3^2$. These six polynomials have no zero in common, so by Bertini's theorem we can choose q such that the quartic surface $g = 0$ is non-singular away from Q_0. We can deform Q_0 into $Q_t = \{z_0 z_1 - z_2(z_2 + t z_3) = 0\}$. Then Q_t is τ-invariant and non-singular for $t \neq 0$, and $B_t = Q_t \cap \{g = 0\}$ is smooth for general t. For small t, the curve B_t will not contain a fixed point of $\tau | Q$. Let $\mathcal{X}$ be the double covering of $\mathbb{P}_3$ branched over $\{g = 0\}$. The involution τ lifts to an involution σ on $\mathcal{X}$. Let X_t, $t \neq 0$, be the part of $\mathcal{X}$ over Q_t and X_0 the minimal resolution of the part of $\mathcal{X}$ over Q_0. By Brieskorn's theorem ([Bri]), X_0 is a deformation of the smooth surfaces X_t, $t \neq 0$ and small. The involution σ can be chosen without fixed points over X_0, hence over X_t for t small. Then $Y_t = X_t/\sigma$ is a deformation of $Y = Y_0$ and Y_t is general for small $t \neq 0$. □

The following is the main result of this section:

(18.5) **Theorem.** *Any two Enriques surfaces are deformations of each other.*

The proof follows from Theorem 18.4 and the next proposition.

(18.6) **Proposition.** *Any two general Enriques surfaces are deformations of each other.*

Proof. Let $(x_0 : x_1)(y_0 : y_1)$ be bihomogeneous coordinates on Q. The vector space of τ-invariant polynomials of bidegree $(4, 4)$ is spanned by the thirteen polynomials

$$\begin{aligned} &(x_0^k x_1^{2-k})^2 \cdot (y_0^\ell y_1^{2-\ell})^2 \ , \qquad && k, \ell = 0, 1, 2, \\ &x_0 x_1 y_0 y_1 \cdot (x_0^k x_1^{1-k})^2 (y_0^\ell y_1^{1-\ell})^2 \ , \qquad && k, \ell = 0, 1. \end{aligned}$$

Denote by P the corresponding projective space, considered as space of curves on Q, and by $P^0 \subset P$ the Zariski-open subset of smooth curves, not passing through any of the four fixed points of τ on Q. (From Bertini's theorem it follows immediately that $P^0 \neq \varnothing$). Let $\Gamma \subset P^0 \times Q$ be the hypersurface

$\{(F,x) \in P^0 \times Q \mid x \in F\}$, and $q : P^0 \times Q \to Q$ the projection. We can cover P^0 by open sets $\mathcal{U}$ such that $\Gamma \cap (\mathcal{U} \times Q)$ is determined by a section in $p^*(\mathcal{O}_Q(4,4)) = p^*(\mathcal{O}_Q(2,2)^{\otimes 2})$. So there is a double covering $\mathcal{X}_{\mathcal{U}}$ of $\mathcal{U} \times Q$ branched over the hypersurface $\Gamma \cap (\mathcal{U} \times Q)$. All the fibres X_t of $\mathcal{X}_{\mathcal{U}} \to \mathcal{U}$ are K 3-surfaces, double coverings of $\{t\} \times Q$, branched over the smooth curve $\Gamma \cap (\{t\} \times Q)$. The involution $\mathrm{id}_{\mathcal{U}} \times \tau$ lifts to an involution $\sigma_{\mathcal{U}}$ on $\mathcal{X}_{\mathcal{U}}$ without fixed points. Forming the quotient $\mathcal{X}_{\mathcal{U}}/\sigma_{\mathcal{U}}$ we obtain a family over $\mathcal{U}$ containing all the Enriques surfaces $X_t/(\sigma_{\mathcal{U}}|X_t)$. Covering P^0 by such open sets $\mathcal{U}$, we prove the assertion. □

19. The Period Map

First we introduce the following notation. As in Sect. 1-14 we put

$$L = -E_8 \oplus -E_8 \oplus H \oplus H \oplus H \ ,$$

the cohomology lattice of a K 3-surface. We define an involution $\rho : L \to L$ by

$$\rho(x \oplus y \oplus z_1 \oplus z_2 \oplus z_3) = (y \oplus x \oplus -z_1 \oplus z_3 \oplus z_2) \ .$$

Its $\mathbb{C}$-linear extension to $L_{\mathbb{C}}$ is denoted by $\rho_{\mathbb{C}}$. The ρ-(anti-)invariant sublattices of L are

$$L^+ = \{\ell \in L \mid \rho(\ell) = \ell\} \ , \quad L^- = \{\ell \in L \mid \rho(\ell) = -\ell\}.$$

Then L^+ is isometric to $-E_8(2) \oplus H(2)$ † and in particular

(10) $$\ell \in L^+ \ \text{implies} \ (\ell,\ell) \equiv 0 \bmod 4.$$

The unimodular lattice $L^+(\frac{1}{2}) = L^0$ is isometric to $-E_8 \oplus H$, the cohomology lattice of an Enriques surface. We moreover put

$$\begin{aligned} \Omega^- &= \mathbb{P}(L^- \otimes \mathbb{C}) \cap \Omega \\ \Gamma &= \mathrm{restr}_{L^-}\{g \in \mathrm{Aut}(L) \mid g\rho = \rho g\} \\ D &= \Omega^-/\Gamma \ . \end{aligned}$$

(19.1) **Lemma.** *Let $\pi : X \to Y$ be the universal covering of an Enriques surface Y, and let $\sigma : X \to X$ be the covering involution. Then there exists an isometry*

$$\phi : H^2(X,\mathbb{Z}) \to L$$

such that

$$\phi \circ \sigma^* = \rho \circ \phi \ .$$

Proof. Since by Proposition 18.6 all Enriques surfaces have the same deformation type, to prove the lemma it suffices to consider the particular

---

† $L(k)$ means that we keep the lattice, but that the bilinear form has been multiplied by k

Enriques surface Y of V, Sect. 23. The K 3-surface X doubly covering it contains two disjoint E_8-configurations, interchanged by the covering involution $\sigma : X \to X$. They generate a sublattice of $H^2(X, \mathbb{Z})$, and we take an identification of L with $H^2(X, \mathbb{Z})$, such that this sublattice becomes a direct summand $(-E_8) \oplus (-E_8)$ of L. Then σ^* and ρ coincide on it. Both σ^* and ρ operate on the orthogonal complement $H \oplus H \oplus H$. In general they are different there, but both $(+1)$-eigenspaces are isometric to $2H$. For σ^* this follows since $\pi^*(H^2(Y, \mathbb{Z})) = 2H \oplus -2E_8$ is the $(+1)$-eigenspace, and for ρ it can be checked easily. Now by Theorem I.2.9 we may compose σ^* with an automorphism of $H \oplus H \oplus H$ such that these two eigenspaces coincide. But then the (-1)-eigenspaces coincide, since for any involution of a euclidean lattice the (-1)-eigenspace is the orthogonal complement of the $(+1)$-eigenspace. It follows that σ^* and ρ coincide on a lattice of finite index in $H^2(X, \mathbb{Z})$ and hence on all of $H^2(X, \mathbb{Z})$. □

(19.2) **Definition.** A marked Enriques surface is a pair (Y, ϕ) with Y an Enriques surface and $\phi : H^2(X, \mathbb{Z}) \to L$ an isometry satisfying $\phi \circ \sigma^* = \rho \circ \phi$, as in Lemma 19.1.

Similarly we may speak of a marked family of Enriques surfaces. If we have a family $p : Y \to S$ of Enriques surfaces over a contractible base S, we may form the universal covering $q : X \to S$, and an isometry $\phi(s) : H^2(X_s, \mathbb{Z}) \to L$ extends to a unique marking $\phi : q_{*2}\mathbb{Z}_X \xrightarrow{\sim} L_S$. If, moreover $\phi(s) \circ \sigma_s^* = \rho \circ \phi(s)$ (where σ_s is the covering involution of $X_s \to Y_s$) then this relation holds for all points of S (two markings coinciding at $s \in S$ coincide over all of S).

A marked family defines a period map. In fact the resulting marked family of K 3-surfaces determines a period map

$$\tilde{\tau} : S \to \Omega\ .$$

The extra relation $\phi(s) \circ \sigma_s^* = \rho \circ \phi(s)$ implies that the image of $\tilde{\tau}$ belongs to $\Omega^- = \{[\omega] \in \Omega \mid \rho_{\mathbb{C}}(\omega) = -\omega\}$. Indeed, if ω_s is a holomorphic 2-form on X_s we have $\omega_s^*(\omega) = -\omega_s$ (since on Y_s there is no holomorphic 2-form), which translates into $\rho_{\mathbb{C}}(\omega_s') = -\omega_s'$, where $\omega_s' = \phi_{\mathbb{C}}(\omega_s)$. We let

$$\tau : S \to \Omega^-$$

be the resulting period map.

(19.3) **Theorem** (Local Torelli theorem). *The Kuranishi family for an Enriques surface Y_0 is universal at all points in a small neighbourhood U around the point corresponding to Y_0. This base is smooth and has dimension 10. The period map is a local isomorphism at each point of U.*

Proof. Since $h^0(\mathcal{T}_Y) = h^2(\mathcal{T}_Y) = 0$ and $h^1(\mathcal{T}_Y) = 10$, the first assertions follow from Theorem I.10.5. To prove that the period map is a local isomorphism, we may assume the base U to be contractible. Let us denote by $p : Y \to U$

the universal family and by $q : X \to U$ its universal covering. We have a commutative diagram

$$\begin{array}{ccccccccc} 0 & \longrightarrow & q^*\mathcal{T}_u|X_0 & \longrightarrow & f^*\mathcal{T}_Y|X_0 & \longrightarrow & f^*\mathcal{T}_{Y_0} & \longrightarrow & 0 \\ & & \| & & \downarrow & & \downarrow & & \\ 0 & \longrightarrow & q^*\mathcal{T}_u|X_0 & \longrightarrow & \mathcal{T}_X|X_0 & \longrightarrow & \mathcal{T}_{X_0} & \longrightarrow & 0 \end{array}$$

where $f : X \to Y$ is the covering map. By definition of the Kodaira-Spencer map (I, Sect. 10) this diagram yields a commutative diagram

$$\begin{array}{ccc} & & H^1(\mathcal{T}_{Y_0}) \\ & \nearrow^{\rho_p} & \\ \mathcal{T}_u(0) & & \downarrow \\ & \searrow_{\rho_q} & \\ & & H^1(\mathcal{T}_{X_0}) \, . \end{array}$$

The vertical arrow embeds $H^1(\mathcal{T}_{Y_0})$ into $H^1(\mathcal{T}_{X_0})$ as the subspace of its σ_0-invariants, so ρ_q is injective and we may view q as a subfamily of the Kuranishi deformation of X_0. Marking the family of Enriques surfaces we obtain a commutative diagram of period maps

$$\begin{array}{ccc} & & \Omega^- \\ & \nearrow^{\tau} & \\ U & & \downarrow \\ & \searrow_{\tilde{\tau}} & \\ & & \Omega \end{array}$$

and since $\tilde{\tau}$ is locally injective by Theorem 7.3, it follows that τ is locally injective, hence locally bijective since $\dim \Omega^- = 10$. □

20. The Period Domain for Enriques Surfaces

A bounded domain of type IV in $\mathbb{C}^n$ is given by $\{z \in \mathbb{C}^n \mid |(z,z)|^2 + 1 - 2(z,\bar{z}) > 0,\ |(z,z)| < 1\}$, where $(z,z') = \sum_{i=1}^{n} z_i z_i'$.

(20.1) **Lemma.** *Let V be a real vector space of dimension $n+2$ equipped with a symmetric bilinear form $(\ ,\ )$ of signature $(2,n)$. The set*

$$\Omega(V) = \{[\omega] \in \mathbb{P}(V \otimes \mathbb{C}) \mid (\omega,\omega) = 0,\ (\omega,\overline{\omega}) > 0\}$$

is biholomorphically equivalent to a disjoint union of two copies of a bounded domain of type IV in $\mathbb{C}^n$. In fact, if a basis of V is taken such that $(\ ,\ )$ has diagonal form $\mathbb{1}_2 \oplus -\mathbb{1}_n$, the two connected components are distinguished by *the sign of* $\mathrm{Im}(\omega_0/\omega_1)$, *where* $\{\omega_0, \ldots, \omega_n\}$ *are the coordinate functions on* $V \otimes \mathbb{C}$ *with respect to this basis.*

For a proof we refer to [Pi].

Putting $n = 10$ we find that the period domain Ω^- consists of two connected components, each of which is a bounded domain of type IV in $\mathbb{C}^{10}$.

(20.2) **Proposition.** *The involution* $\lambda = \mathbb{1} \oplus \mathbb{1} \oplus -\mathbb{1} \oplus \mathbb{1} \oplus \mathbb{1}$ *of* L *commutes with* ρ *and the induced involution of* Ω^- *interchanges the two connected components.*

Proof. Clearly $\lambda \circ \rho = \rho \circ \lambda$. To prove the second statement we use the following basis for L^- (which incidentally shows that $L^- = -E_8(2) \oplus H \oplus H(2)$). If $\{e_1, \ldots, e_8\}$ respectively $\{e'_1, \ldots, e'_8\}$, are bases for the first, respectively the second copy of $-E_8$ and $\{f_i, g_i\}$ $(i = 1, 2, 3)$ for the i-th copy of H we take

$$(11) \qquad \{e_1 - e'_1, e_2 - e'_2, \ldots, e_8 - e'_8, f_1, g_1, f_2 - f_3, g_2 - g_3\}$$

and observe that in this basis $\lambda(f_1) = -f_1$, $\lambda(g_1) = -g_1$ and $\lambda = \mathbb{1}$ on the orthogonal complement of $\mathbb{Z}f_1 + \mathbb{Z}g_1$. Of course this basis cannot be used to apply Lemma 20.1. However, with respect to the basis (11) the sign in question is nothing but the sign of $\mathrm{Im}(\omega_8 + \omega_9/\omega_{10} + \omega_{11})$, if $[\omega] = (\omega_0 : \cdots : \omega_{11})$. Then λ obviously changes this sign. □

(20.3) **Lemma.** *The group* $\Gamma = \{g|L^- \mid g \in \mathrm{Aut}(L), g \circ \rho = \rho \circ g\}$ *is of finite index in* $\mathrm{Aut}(L^-)$, *hence* Γ *is an arithmetic subgroup of* $\mathrm{Aut}(L^- \otimes \mathbb{R})$.

Proof. Let $g \in \mathrm{Aut}(L^-)$ and suppose $g \equiv \mathrm{id} \bmod 2L$. Using the basis (11), it is easy to check that an extension $h \in \mathrm{Aut}(L)$ exists with $h|L^+ = 1$. By construction $h \circ \rho = \rho \circ h$, hence $g \in \Gamma$. So Γ contains a congruence subgroup of $\mathrm{Aut}(L^-)$, and hence Γ is of finite index in $\mathrm{Aut}(L^-)$. □

(20.4) **Corollary.** *The group* Γ *acts properly discontinuously on* Ω^- *and the complex space* $D = \Omega^-/\Gamma$ *admits the structure of a quasi-projective variety.*

Proof. The first assertion follows from [Bou71], Chap. III.4.2 and the second from [B-B]. □

(20.5) **Proposition.** *If* $d \in L^-$, $(d, d) = -2$, *then no point of*

$$H_d = \{[\omega] \in \Omega \mid (\omega, d) = 0\}$$

can be the period point of a marked Enriques surface.

Proof. Suppose (Y, ϕ) is a marked Enriques surface such that its period point belongs to H_d. Then the class $\delta = \phi^{-1}(d)$ is in the Néron-Severi lattice of X, the universal covering of Y. So, by Proposition 3.8, (i) the class $\pm\delta$ is effective. But no effective class can be anti-invariant as is $\pm\delta$. □

(20.6) **Proposition.** *There are only finitely many* Γ*-equivalence classes of elements* $d \in L^-$ *with* $(d,d) = -2$.

Proof. The group Γ is of finite index in $\mathrm{Aut}(L^-)$, so we may replace Γ by $\mathrm{Aut}(L^-)$. Let Γ_1 be the subgroup of $\mathrm{Aut}(L')$ with $L' = -E_8(2) \oplus H \oplus H \supset L^-$ which preserves L^-. Since restriction defines an embedding $\Gamma_1 \hookrightarrow \mathrm{Aut}(L^-)$, we may replace $\mathrm{Aut}(L^-)$ by Γ_1. This latter group contains the congruence subgroup $\{g \in \mathrm{Aut}(L') \mid g \equiv \mathbb{1} \bmod 2L'\}$ of $\mathrm{Aut}(L')$, and we may replace Γ_1 by $\mathrm{Aut}(L')$. We claim that all elements $d' \in L'$ satisfying $(d',d') = -2$ are conjugate to each other under the action of $\mathrm{Aut}(L')$. If L' were unimodular, this would follow from Theorem I.2.9. The proof of this theorem, as presented in [L-P], p. 156 can easily be modified to cover our case (for details, see [Hor78], II, p. 223). □

(20.7) **Corollary.** *The union* $\bigcup_d H_d/\Gamma$ *for* $d \in L^-$, $(d,d) = -2$ *consists of finitely many irreducible algebraic hypersurfaces in* $D = \Omega^-/\Gamma$, *so*

$$D^0 = D \setminus \Big(\bigcup_d H_d\Big)/\Gamma$$

is quasi-projective.

Proof. Each H_d consists of two connected components isomorphic to a bounded domain of type IV. Since Γ operates properly and discontinuously, the transforms of H_d form a locally finite collection of connected parts of hyperplanes because these hyperplanes are fixed points of reflections belonging to Γ (*cf.* the proof of Corollary 9.2). In virtue of Proposition 20.6 the same holds for the union of all H_d. So its image in D consists of finitely many irreducible (analytic) hypersurfaces. The Baily-Borel compactification D^* has the property that $\dim D^* \setminus D = 1$ ([Pi], §4, Lemme 1). A theorem of Remmert and Stein ([R-S], Satz 13) then implies that all of the above hypersurfaces extend to hypersurfaces in D^*. These are algebraic, by Chow's theorem I.19.2, since D^* is projective ([B-B]). So D^0, like D, is quasi-projective. □

21. Global Properties of the Period Map

Firstly, we shall prove a Torelli theorem for Enriques surfaces. As a preliminary we prove:

(21.1) **Proposition.** *If* X *is the universal covering of an Enriques surface,* σ *the corresponding involution and* $\ell \in \mathrm{NS}(X)$ *with* $\sigma^*(\ell) = \ell$, $(\ell,\ell) > 0$ *and* $(\ell, d) \neq 0$ *for all* $d \in \mathrm{NS}(X)$ *with* $(d,d) = -2$, *then there exists* $w \in W_X$ *commuting with* σ^*, *such that* $\pm w(\ell)$ *is the class of an ample divisor.*

Proof. By replacing ℓ by $-\ell$, if necessary, we may assume that ℓ belongs to the positive cone. In particular ℓ and all of its W_X-images are effective. If ℓ itself is the class of an ample divisor we take $w = \mathrm{id}$. If not, we proceed as follows. Let $\pi : X \to Y/\{1, \sigma\}$ be the quotient map and let $m = \pi_*(\ell)$. If ℓ is not the class of an ample divisor, this also holds for m, since ℓ is invariant. So there exists an irreducible class $e_1 = c_1(\mathcal{O}_Y(E_1))$ with $(m, e_1) \leq 0$. The curve $\pi^{-1}(E_1)$ cannot be irreducible, since Proposition 3.7, (i) implies that it would be nodal, whereas σ-invariant curves by (10) have self-intersection divisible by 4. So $\pi^{-1}(E_1)$ consists of two irreducible curves D_1 and $\sigma(D_1)$. These are (-2)-curves, again by Proposition 3.7, (i). If these curves were to meet, then $(D_1 + \sigma(D_1))^2 \geq 0$. So the class of $D_1 + \sigma(D_1)$ belongs to $\overline{\mathcal{C}}_X$, and has strictly positive intersection with any element of $\mathcal{C}_X$, such as ℓ. But by construction this is not the case. It follows that D_1 and $\sigma(D_1)$ are disjoint. So for their classes d_1 and $d_2 = \sigma^*(d_1)$ we have $(d_1, d_2) = 0$, and the product of the corresponding Picard-Lefschetz reflections commutes with σ^*. Moreover $(\ell, d_2) = (\ell, \sigma^*(d_1)) = (\sigma^*\ell, d_1) = (\ell, d_1) \leq 0$, hence < 0, by assumption. By induction we obtain a finite ordered set of nodal classes $d_1, \ldots, d_{2k-1}, d_{2k} = \sigma^*(d_{2k-1})$ such that the product $s_{d_{2k}} \circ s_{d_{2k-1}} \circ \cdots \circ s_{d_1}$ of the corresponding reflections commutes with σ^*. If $w_j = s_{d_j} \circ s_{d_{j-1}} \circ \cdots \circ s_{d_1}$, we moreover have

$$(w_{r-1}(\ell), d_r) = -\alpha_r < 0 \; .$$

We shall show that this process terminates after finitely many steps, namely as soon as

$$\ell_r = w_r(\ell)$$

is the class of an ample divisor. To this end observe that

$$\ell_r = s_r(\ell_{r-1}) = \ell_{r-1} - \alpha_r d_r \quad (\alpha_r > 0) \; .$$

So, if L_k is a divisor representing ℓ_k, we find

$$\dim |L_k| \leq \dim |L_{k-1}| \; .$$

If the process did not terminate, $\dim |L_k|$ would stabilise from $k = N$ on and then $\sum_{k>N} \alpha_k D_k$ would be contained in $|L_N|$ as fixed part. This clearly is impossible. So the process stops at some stage r, when ℓ_r is the class of an ample divisor. □

(21.2) **Theorem** (Global Torelli theorem for Enriques surfaces). *The isomorphism class of an Enriques surface is uniquely determined by its period point.*

Proof. Let Y, Y' respectively, be two Enriques surfaces, X, X' respectively, their universal coverings and σ, σ' respectively, the corresponding involutions. Choose markings ϕ, ϕ' respectively, for X, X' respectively, such that $\rho \circ \phi = \phi \circ \sigma^*$ and similarly for ϕ'. If the period points of (Y, ϕ) and (Y', ϕ') are the same, the isometry

$$\psi = \phi^{-1} \circ \phi' : H^2(X', \mathbb{Z}) \to H^2(X, \mathbb{Z})$$

is a Hodge-isometry with $\psi \circ (\sigma')^* = \sigma^* \circ \psi$. If $\ell' \in H^2(X', \mathbb{Z})$ is the class of an ample divisor, invariant under σ', we may apply Proposition 21.1 to

$\ell = \psi(\ell')$ and replace ψ by $\psi_1 = \pm w \circ \psi$ such that still $\psi_1 \circ (\sigma')^* = \sigma^* \circ \psi_1$, but now with $\psi_1(\ell)$ ample. By Proposition 3.11 ψ_1 is an effective Hodge-isometry and by the Torelli theorem 11.1, it is induced by an isomorphism $g : X' \to X$. Since $g \circ \sigma' \circ g^{-1} \circ \sigma$ induces the identity on $H^2(X', \mathbb{Z})$, by Proposition 11.3 we have that it is itself the identity, i.e., $g \circ \sigma' = \sigma \circ g$ and hence g induces an isomorphism between Y and Y'. □

Next, we shall show that all points of D^0 correspond to marked Enriques surfaces. The next lemma plays a central role in this proof.

(21.3) **Lemma.** *Let X be a projective* K 3*-surface and let $\phi : H^2(X, \mathbb{Z}) \to L$ be a marking such that the period point of (X, ϕ) belongs to Ω^-.*
(i) *The involution $j = \phi^{-1} \circ \rho \circ \phi$ is a Hodge-isometry.*
(ii) *Either for some $d \in \mathrm{NS}(X)$, $(d, d) = -2$, $w \in W_X$ we have*

$$w^{-1} \circ j \circ w(d) = -d$$

or, there exists $w \in W_X$ such that

$$w^{-1} \circ j \circ w = w(j)$$

is effective.

Proof.
(i) Obvious.
(ii) Upon replacing j by $-j$, if necessary, we may assume that j preserves the positive cone. Choose a class ℓ of an ample divisor on X. Then $j(\ell)$ as well as $w(j)(\ell)$ is effective for all $w \in W_X$. If $j(\ell)$ is already the class of an ample divisor we take $w = \mathrm{id}$ and we are done. If not, there exists an irreducible class d with $(j(\ell), d) \le 0$, which is then necessarily nodal by Proposition 3.8. We also observe that $d' = -j(d)$ is effective. Indeed, if not, then Proposition 3.7, (i) implies that $j(d)$ would also be effective and the inequality $(\ell, j(d)) > 0$ would contradict $(\ell, j(d)) = (j(\ell), d) \le 0$.

We let s, s' respectively, be the Picard-Lefschetz reflections in d, d' respectively, and observe that

$$s \circ j \circ s(\ell) = j(\ell) - (\ell, d)d' - (\ell, s(d'))d$$
$$s' \circ j \circ s'(\ell) = j(\ell) - (\ell, d')d - (\ell, s'(d))d' \ .$$

So, if $s(d')$ is effective the coefficients of both d and d' in $s(j)(\ell)$ are strictly negative and we can take $d_1 = d$. If $s'(d)$ is effective, the same holds for the coefficients of d and d' in $s'(j)(\ell)$ and we set $d_1 = d'$. If neither $s(d')$ nor $s'(d)$ is effective, $s(d') + s'(d) = (1 + (d, d'))(d + d')$ not being effective then implies $(d, d') < 0$. Since Proposition 3.7, (i) implies that $-s(d') = -d' - (d, d')d$ is effective, it follows that $d = d'$, i.e., $j(d) = -d$ and the first alternative holds.

Next we put $\ell_1 = s_1(j)(\ell)$. If ℓ_1 is the class of an ample divisor we are done. If not, we proceed as before with $j_1 = s_1(j)$ instead of ℓ_1. Inductively we find effective (-2)-classes $d_1, \ldots, d_r$, such that, if we let

$$w_r = s_r \circ s_{r-1} \circ \cdots \circ s_1$$

be the product of the corresponding Picard-Lefschetz reflections and

$$\begin{aligned} w_r(j) &= j_r \\ j_r(\ell) &= \ell_r \\ d'_r &= -j_r(d_r) \quad \text{(an effective class!)} \end{aligned}$$

then we *either* have $d'_r = d_r$ for some r, *or* for all $r \geq 1$ we have $\ell_{r+1} = \ell_r - \beta_r d_r - \gamma_r d'_r$, with strictly positive β_r and γ_r. As in the proof of Proposition 21.1 the inequalities $\beta_r > 0$ and $\gamma_r > 0$ force the process to stop at some class ℓ_r of an ample divisor. Then j_r is an effective Hodge-isometry by Corollary 3.12. □

(21.4) **Theorem.** *All the points of the variety D^0, introduced in Corollary 20.7, occur as period points of Enriques surfaces.*

Proof. The surjectivity of the period map for K 3-surfaces (Theorem 14.1) implies that for a given $[\omega]$ with image in D^0, there is a K 3-surface X and a marking

$$\phi : H^2(X, \mathbb{Z}) \to L \ , \quad \text{with} \ \ \phi_{\mathbb{C}}(H^{2,0}(X)) = [\omega] \ .$$

The surface is projective by Theorem IV.6.2, since for any $\ell' \in L^+$ with $(\ell', \ell') > 0$ the class $\ell = \phi^{-1}(\ell')$ is the class of a divisor with positive self-intersection. We apply Lemma 21.3 to change the marking. So let $\psi = w \circ \phi$ be a new marking with w as in Lemma 21.3.

Since $[\omega] \in D^0$ the first alternative in Lemma 21.3 cannot occur (Proposition 20.5). So

$$j = \pm \psi^{-1} \circ \rho \circ \psi$$

is an effective Hodge-isometry. Then Theorem 11.1 implies that there exists an automorphism

$$\sigma : X \to X \ , \quad \sigma^* = j \ ;$$

Since $j^2 = \mathrm{id}$, Proposition 11.3 implies that σ is an involution. Let us compute its Lefschetz number, i.e.,

$$L(\sigma) = \sum (-1)^j \mathrm{Tr}(\sigma^* | H^j(X, \mathbb{R})) \ .$$

Since on $H^2(X, \mathbb{R})$, Tr σ^* = Tr ρ up to sign, we find:

$$L(\sigma) = 2 \mp 2 \ . \tag{12}$$

Similarly, for the holomorphic Lefschetz number

$$L_{\mathrm{hol}}(\sigma) = 1 + \mathrm{Tr}(\sigma^* | H^{2,0}(X))$$

we find

$$L_{\mathrm{hol}}(\sigma) = 1 \mp 1 \ .$$

Applying the holomorphic Lefschetz formula for an involution ([A-S,] Prop. 4.8)

$$L_{\text{hol}}(\sigma) = \frac{1}{4} \cdot \#(\text{isolated fixed points}),$$

we find that the number of isolated fixed points of μ equals 0 or 8.

On the other hand, the usual Lefschetz fixed point formula reads ([Ue76], Lemma 1.6):

$$L(\sigma) = \mu + \sum_{j=1}^{t} e(F_j) \ ,$$

where $F_1, \dots, F_t$ are the fixed curves of σ, so combining this with (12) we find

$$\sum_{j=1}^{t} e(F_j) = -4 \quad (\text{if } \mu = 8) \text{ or } 0 \ (\text{if } \mu = 0) \ .$$

If $\mu = 8$ we blow up X at the eight fixed points of σ, and obtain an involution $\tilde{\sigma}$ on the blown-up $\widetilde{X}$. Let $Z = \widetilde{X}/\{1, \tilde{\sigma}\}$. The fixed locus of $\tilde{\sigma}$ consists of the exceptional locus E and a curve F, the proper transform of $\bigcup F_j$. The canonical bundle formula for double coverings (V, Sect. 22) and blow-ups (Theorem I.9.1, (viii)) show that $\mathcal{K}_Z$ lifts on X to $\mathcal{O}_X(-F)$, and so $p_g(Z) = 0$. Since $\sigma^*|H^{2,0}(X) = \text{id}$, there is at least one non-zero holomorphic 2-form on Z. This contradiction show that $\mu = 0$. So the (+)-sign holds in the definition of j and hence

$$(13) \qquad \sigma^* \circ \psi = \psi \circ \rho \ .$$

Since by (10) all σ^*-invariant curves have self-intersection divisible by 4, the quotient surface $Z = X/\{1, \sigma\}$ must be minimal (any (−1)-curve would lift to a σ^*-invariant (−2)-curve). Suppose that the fixed point set of σ is non-empty. Then $P_2(Z) = 0$, since $\mathcal{K}_Z$ lifts on X to $\mathcal{O}_X(-F)$. Moreover $q(Z) \leq q(X) = 0$. Applying Castelnuovo's criterion VI.3.4, we conclude that Z is rational. Since $e(Z) = \frac{1}{2}e(X) = 12$ and for a minimal rational surface $e = 3$ or 4 (VI, Theorem 1.1), we obtain a contradiction. So $F = 0$, σ is a fixed point-free involution and Z is an Enriques surface. Because of (13) the map ψ is a marking for Z in the sense of Definition 19.2, and so the image of $[\omega]$ in D^0 is the period point of Z. □

Special Topics

22. Projective K3-surfaces and Mirror Symmetry

As before, we let L be the K3-lattice. Fix a *primitive* vector $\ell \in L$ with $\ell^2 > 0$. The roots orthogonal to ℓ are denoted by

$$\Delta_\ell = \{d \in L \mid d^2 = -2,\ (d,\ell) = 0\}.$$

Recall that the period domain Ω for K 3-surfaces is a certain 20-dimensional open subset of the quadric inside $\mathbb{P}(L_\mathbb{C})$ consisting of isotropic lines. Inside this projective space non-zero elements $\lambda \in L$ define hyperplanes

$$H_\lambda = \{[\omega] \in \mathbb{P}(L_\mathbb{C}) \mid ([\omega],\lambda) = 0\}$$

and we introduce

$$\Omega_\ell = \Omega \cap H_\ell$$

and

$$\Omega_\ell^{\mathrm{pol}} = \Omega_\ell \setminus \bigcup_{d\in\Delta_\ell} H_d \cap \Omega_\ell.$$

In Sect. 12 we introduced the period point $\tau_1(X,\phi) \in \Omega$ of a marked K 3-surface $(X, \phi : H^2(X,\mathbb{Z}) \xrightarrow{\sim} L)$. Surjectivity of the period map for marked K 3-surfaces tells us that each point in Ω_ℓ corresponds to a marked K 3-surface and ℓ corresponds to a divisor. Composing the marking with a suitable isometry of L we can assume that ℓ corresponds to a class in the closure of the ample cone. For period points in $\Omega_\ell^{\mathrm{pol}}$ this class is in the interior of this cone, i.e., corresponds to an ample divisor. This explains the notation and motivates the following definition.

(22.1) **Definition.** A K 3-surface X has a polarization of type ℓ if for some marking ϕ the class $\phi^{-1}(\ell) \in H^2(X,\mathbb{Z})$ is the class of an ample divisor.

So, such K 3-surfaces are exactly the K 3-surfaces admitting a marking such that the period point belongs to $\Omega_\ell^{\mathrm{pol}}$.

(22.2) **Remarks.**
1) If $\ell^2 = (\ell')^2 = 2k$, there is an isometry of L sending ℓ to ℓ' (Theorem I. 2.9 (ii)) and so the polarization type of a K 3-surface depends only on the positive integer k: a K 3-surface X has a polarization of type k if there exists an ample primitive class on X with self-intersection $2k$.

Choosing $\ell = ke + f$, where $\{e, f\}$ is a basis of the first hyperbolic summand of L, we see that its orthogonal complement is isometric to $(-2k) \oplus H \oplus H \oplus -E_8 \oplus -E_8$. Then, as in Sect. 20, one can see that the domain Ω_ℓ consists of two connected components, each of which is a domain of type IV in $\mathbb{C}^{19}$. Moreover, the involution of $L = H \oplus H \oplus H \oplus -E_8 \oplus -E_8$ which is minus the identity on the second hyperbolic summand and the identity on the other summands preserves ℓ and interchanges the two connected components.
2) On every projective K 3-surface one can find an ample divisor L whose class is primitive. So all projective K 3-surfaces occur in that part of the universal marked family that lies over the union $\bigcup_{k\in\mathbb{N}}(\Omega^{\mathrm{pol}})^0_{\ell_k}$, where one chooses for any natural number k a primitive vector ℓ_k of norm squared $2k$ and one connected component $(\Omega^{\mathrm{pol}})^0_{\ell_k}$ of the domain $\Omega^{\mathrm{pol}}_{\ell_k}$.

The group $\mathrm{O}(L,\ell)$ of isometries of L fixing the class ℓ acts on the domains Ω_ℓ and $\Omega_\ell^{\mathrm{pol}}$. Exactly as in Sect. 20, we can prove:

(22.3) **Theorem.** *There is an isometry in the group* $O(L,\ell)$ *interchanging the two connected components of* Ω_ℓ. *The group* $O(L,\ell)$ *acts properly and discontinuously on* Ω_ℓ. *The union* $\left(\bigcup_{d\in\Delta_\ell} H_d\right)/O(L,\ell)$ *consists of finitely many irreducible hypersurfaces and*

$$D_\ell = \Omega_\ell^{\text{pol}}/O(L,\ell)$$

is a connected quasi-projective variety.

Suppose that (X,ϕ) and (X',ϕ') are two K 3-surfaces of polarization type ℓ. We assume that the markings are chosen such that $\tau_1(X,\phi)$, $\tau_1(X',\phi)\in\Omega_\ell^{\text{pol}}$ and that they define the same point in D_ℓ. By the Torelli theorem X and X' must be isomorphic. It follows that D_ℓ is a moduli space for the isomorphism classes of K 3-surfaces of polarization type ℓ. In fact, in view of Remark 22.2 we have

(22.4) **Corollary.** *Fix a positive integer* k. *The moduli space of (isomorphism classes of)* K 3-*surfaces with a polarization of type* k *is a connected* 19-*dimensional quasi-projective variety isomorphic to* D_ℓ *where* $\ell\in L$ *is any vector with* $\ell^2=2k$.

Next, we want to fix a sublattice $M\subset L$ of signature $(1,t)$ instead of a vector. We say that X is a **K 3-surface of type** M if there exists a marking $\phi: H^2(X,\mathbb{Z})\to L$ such that all elements of $\phi^{-1}(M)$ are divisors on X. Observe that the cone of positive elements inside $M_{\mathbb{R}}$ consists of two connected components. Fix one, by definition the positive cone C_M. Deleting from this cone the hyperplanes H_d, where d belongs to the set of roots $\Delta_M=\{d\in M\mid d^2=-2\}$ we fix one of the connected components, the chamber C_M^{pol}. By definition ϕ is an M-**polarization** if all divisors in $\phi^{-1}(C_M^{\text{pol}})$ are ample. In this case X is called M-**polarized**. Since M contains elements of positive self intersection, X is automatically projective. The case of a K 3-surface with a polarization of type ℓ can be viewed as the special case $t=0$. Generalizing that case, the K 3-surfaces of type M have period points in

$$\Omega_M=\{[\omega]\in\Omega\mid([\omega],m)=0,\ \forall m\in M\}$$

which this time is a union of two copies of a domain of type IV in $\mathbb{C}^{19-t}$. As in the case $t=0$, to obtain M-polarized K 3-surfaces, we have to remove certain hyperplanes from the domain Ω_M corresponding to the roots in L orthogonal to M. This set of roots is denoted Δ_M. As before, we put

$$\Omega_M^{\text{pol}}=\Omega_M\setminus\bigcup_{d\in\Delta_M}H_d\cap\Omega_M$$

and the group $O(L,M)$ of isometries of L preserving M acts properly and discontinuously on Ω_M. The quotient

$$D_M=\Omega_M^{\text{pol}}/O(L,M)$$

is a moduli space of isomorphism classes of K 3-surfaces with a polarization of type M. It has at most two connected components, it is quasi-projective and of dimension $19 - t$.

We now discuss mirror symmetry for K 3-surfaces. Originally mirror symmetry has been introduced for certain families of Calabi-Yau threefolds, i.e., threefolds with trivial canonical bundle. We refer to the book [C-K] and the references given there. Since K 3-surfaces are two-dimensional Calabi-Yau's, it is natural to try to find analogues for mirror phenomena for K 3-surfaces. Here we follow Dolgachev's exposition [Do98].

The *first ingredient of mirror symmetry* involves the definition of a mirror family. We first want to define the mirror lattice of certain sublattices $M \subset L$ of signature $(1, t)$:

(22.5) **Definition.** A sublattice M as above is called m-admissible if the orthogonal complement inside L splits off a summand N isometric to $H(m)$. In other words, we have an orthogonal splitting

$$M^{\perp} = N \oplus \tilde{M}, \quad N \cong (H(m))$$

with $\tilde{M}$, the mirror of M, a lattice of signature $(1, 18 - t)$.

It should be noted that Nikulin has found a purely arithmetic criterion for a sublattice of L to be m-admissible. The K 3-surfaces corresponding to points of $D_{\tilde{M}}$ are the $\tilde{M}$-polarized K 3-surfaces and these are the mirrors of the M-polarized K 3-surfaces. In fact (see [Do98], Theorem 5.1), for $m = 1$, this is really a reflexive correspondence.

Now, in the theory of mirror symmetry, families of K 3-surfaces are needed. For these we take the restrictions $\mathcal{F}_M = \{X_M \to \Omega_M^{\mathrm{pol}}\}$ to M of the universal marked family of K 3-surfaces constructed in Sect. 12. We call $\mathcal{F}_{\tilde{M}}$ the mirror family of $\mathcal{F}_M$.

Note that for a generic member of the $(19-t)$-dimensional family $\mathcal{F}_M$ the Néron-Severi group is isomorphic to M and so has rank $1 + t$. The generic member of the $(1 + t)$-dimensional mirror family has Picard number $19 - t$. This is analogous to what happens in mirror symmetry for Calabi-Yau threefolds.

Next, we want to introduce in our setting a *second ingredient in mirror symmetry*, the Yukawa-coupling. Let X be any K 3-surface. We have seen (Sect. 7) that the differential of the period map on the Kuranishi space for X can be identified with the cup product $\nabla : H^1(X, \mathcal{T}_X) \to \mathrm{Hom}(H^{2,0}(X), H^{1,1}(X))$. Using also the cup product $\nabla' : H^1(X, \mathcal{T}_X) \to \mathrm{Hom}(H^{1,1}(X), H^{0,2}(X))$ we obtain a symmetric pairing

$$\begin{aligned} Y : S^2 H^1(X, \mathcal{T}_X) &\to \mathrm{Hom}(H^{0,2}(X), H^{0,2}(X)) \cong \mathbb{C} \\ (\theta, \theta') &\mapsto \nabla'(\theta) \circ \nabla(\theta'), \end{aligned}$$

the Yukawa coupling. The fact that it is symmetric follows from the fact that for $\theta, \theta' \in H^1(X, \mathcal{T}_X)$, $\xi \in H^{0,2}$ we have $(\theta' \cup \xi) \cup \theta = \theta \cup (\xi \cup \theta')$.

We want to find an explicit form for this coupling. We choose a marking ϕ for X so that we can identify the Kuranishi space of X with the germ of Ω at the period point $[\omega] = \tau_1(X, \phi)$ and so that its tangent space $H^1(X, \mathcal{T}_X)$ becomes identified with $\mathrm{Hom}(\mathbb{C}\omega, L_{\mathbb{C}}/\mathbb{C}\omega)$. If ϕ is an ℓ-polarization we can make this more explicit. First of all, we restrict the Kuranishi space to the ℓ-polarized K 3's. The tangent space T to the latter is $\mathrm{Hom}(\mathbb{C}\omega, (\mathbb{C}\ell)^{\perp}/\mathbb{C}\omega)$. We choose an isotropic vector $f \in \ell_{\mathbb{R}}^{\perp}$ and we introduce the 19-dimensional real vector space $T_{\mathbb{R}} = \{x \in \ell_{\mathbb{R}}^{\perp} \mid (x, f) = 0\}/\mathbb{R}f$. Then T becomes canonically isomorphic to the complexification of $T_{\mathbb{R}}$ through the isomorphism

$$\begin{aligned} \mathrm{Hom}(\mathbb{C}\omega, (\mathbb{C}\ell)^{\perp}/\mathbb{C}\omega) &\xrightarrow{\sim} \{x \in \ell_{\mathbb{C}}^{\perp} \mid (x, f) = 0\}/\mathbb{C}f \\ h &\mapsto [h(\omega)' - (h(\omega)', \omega)] \bmod \mathbb{C}f, \end{aligned}$$

where we have chosen $\omega \in L_{\mathbb{C}}$ such that $(\omega, f) = 1$; moreover, $h(\omega)' \in \omega_{\mathbb{C}}^{\perp}$ represents $h(\omega)$. The Yukawa coupling thus becomes the canonical pairing induced from $L_{\mathbb{C}}$.

We want to refine this for M-polarized K 3's. So we assume that ϕ is an M-polarization so that in particular $\phi_{\mathbb{C}}^{-1}(M) \subset H^{1,1}(X)$. We introduce

$$\begin{aligned} H^1(X, \mathcal{T}_X)_M &= \{\theta \in H^1(X, \mathcal{T}_X) \mid (\nabla\theta, m) = 0 \ \forall m \in M\} \\ H^{1,1}(X)_M &= H^{1,1}(X)/\phi_{\mathbb{C}}^{-1}(M). \end{aligned}$$

The differential to the period map restricted to M-polarized K 3's yields an isomorphism

$$H^1(X, \mathcal{T})_M \xrightarrow{\sim} \mathrm{Hom}(H^{2,0}(X), H^{1,1}(X)_M)$$

and the Yukawa coupling restricts to

$$Y : S^2 H^1(X, \mathcal{T}_X)_M \to \mathrm{Hom}(H^{2,0}(X), H^{0,2}(X)) \cong \mathbb{C}.$$

As before we can make it more canonical at the expense of having to choose an isotropic vector f in the orthogonal complement of $M_{\mathbb{R}}$. Indeed, setting

$$N = M^{\perp},$$

and using

$$V_f = (\mathbb{R}f)_{N_{\mathbb{R}}}^{\perp}/\mathbb{R}f,$$

a real vector space of signature $(1, 18 - t)$, the same argument as before now yields:

(22.6) **Proposition.** *The Yukawa coupling on the tangent space of the Kuranishi space at* $[\omega] \in \Omega_M$ *can be identified with the quadratic form on the complexification of* V_f *induced by the bilinear form on* L. *This identification depends on the choice of an isotropic vector* $f \in N_{\mathbb{R}}$ *and a vector* $\omega \in L_{\mathbb{C}}$ *with* $(\omega, f) = 1$ *representing* $[\omega]$.

A *third ingredient in mirror-symmetry* is a tube domain realization for the Kuranishi space of X in terms of the Néron-Severi group of the mirror. We first explain the construction of the tube domains. We start with the real vector space V_f of signature $(1, 19-t)$ and we assume that $t < 19$ so that the form is indefinite. Let

$$C_f \subset V_f$$

be one of the connected components of the cone of elements with positive square. Fix x_0 with $(x_0, f) = 1$ and introduce the affine space

$$W_f = x_0 + V_f.$$

The tube domain realization of Ω_M is then given by

$$\Omega(W_f, C_f) = W_f + \mathrm{i}C_f \subset (W_f)_{\mathbb{C}}$$

and we have the well known (see [Do98], Theorem 4.2)

(22.7) **Lemma.** *The (rational) projection* $\mathbb{P}(N_{\mathbb{C}}) \to \mathbb{P}(N_{\mathbb{C}}/\mathbb{C}f)$ *induces an isomorphism*

$$\Omega_M^0 \xrightarrow{\sim} \Omega(W_f, C_f),$$

where Ω_M^0 *is any connected component of* Ω_M.

We now pass to tube domains related to the Néron-Severi groups associated to any isotropic vector $g \in (T_X)_{\mathbb{R}} = \mathrm{NS}(X)_{\mathbb{R}}^{\perp}$, the real vector space associated to the transcendental lattice. In the above construction replace V_f and W_f by

$$V(X)_g = \{x \in \mathbb{R}g + \mathrm{NS}(X)_{\mathbb{R}} \mid (x, g) = 0\} \cong \mathrm{NS}(X)_{\mathbb{R}}$$
$$W(X)_g = \{x \in \mathbb{R}g + \mathrm{NS}(X)_{\mathbb{R}} \mid (x, g) = 1\}.$$

Recalling (Sect. IV.9) that $\mathrm{NS}_+(X)$ is the real cone inside $\mathrm{NS}(X)_{\mathbb{R}}$ spanned by the effective divisors, we form the tube domain

$$T(X)_g = W(X)_g + \mathrm{i}\,\mathrm{NS}_+(X).$$

Let (Y, ϕ) now be any $\tilde{M}$-polarized K 3-surface and choose an isotropic vector in the orthogonal complement of $M \oplus \tilde{M}$ and put $\phi^{-1}(f) = g$. Observe now that the marking defines embeddings

$$\phi^{-1}(V_f) \hookrightarrow \mathrm{NS}(Y)_{\mathbb{R}} = V(Y), \quad \phi^{-1}(W_f) \hookrightarrow W(Y)_g$$

which for generic Y from the mirror family are isomorphisms. Combining this with the preceding lemma, we thus have constructed embeddings

$$\tau_M : \Omega_M^0 \hookrightarrow T(Y)_g = W(Y)_g + \mathrm{i}\,\mathrm{NS}_+(Y)_{\mathbb{R}}$$

which for generic Y of the mirror are isomorphisms.

This tube domain realization tells us how to calculate the Yukawa coupling: observe that $V_f = \tilde{M}_{\mathbb{R}}$ so that Prop. 22.6 implies:

(22.8) **Proposition.** *The Yukawa coupling on $H^1(X, \mathcal{T}_X)$ gets (in general non-canonically) identified with the quadratic form on the complexification of the mirror-lattice $\tilde{M}$.*

Remark. There is one case where the above identification is canonical, namely when M is 1-admissible (Def. 22.5). In this case $M^\perp = H \oplus \tilde{M}$ and two generators (f, g) of the hyperbolic summand make it possible to define a canonical representative

$$\omega = g - \frac{1}{2}(z, z)f + z$$

for $[\omega]$ corresponding to z in the tube domain realization. This implies that all of the above identifications preserve the integral structures so that the Yukawa coupling is the complexification of this integral quadratic form. This still depends on the choice of a basis for the hyperbolic summand of $M^\perp$, but a different choice leads to an isometric quadratic form.

This is analogous to what happens for Calabi-Yau threefolds, where the Yukawa coupling (defined in a similar fashion) contains, however, much more information related to the geometry of rational curves on the mirror.

Examples.

1. When $M = (2m)$, we have m-polarized K3-surfaces forming a 19-dimensional family with mirror a 1-dimensional family. When $m = 1$, this mirror family can be described as the minimal resolution of singularities of the family of non-singular quartics in $\mathbb{P}_4$ which is given by the equations

$$u_0 u_1 u_2 u_3 - u_4^4 = 0, \quad u_0 + u_1 + u_2 + u_3 + 4\lambda u_4 = 0, \quad \lambda^4 \neq 1.$$

   See loc. cit Example 8.1.
2. With the notation of Sect. 19 we put $M = L^+ = H(2) \oplus -E_8(2)$, then $M^\perp = L^- = H \oplus H(2) \oplus -E_8(2)$ and so we can peel off H getting $\tilde{M} = H(2) \oplus -E_8(2)$ so that M is its own mirror. The moduli space Ω_M^{pol} in this case parametrizes the double covers of Enriques surfaces and so in this sense Enriques surfaces are self-mirrored.
3. The mirror family of the 3-dimensional family of Kummer surfaces of principally polarized abelian surfaces turns out to be a suitable 17-dimensional family of hypersurfaces of degree $(2, 2, 2)$ in $\mathbb{P}_1 \times \mathbb{P}_1 \times \mathbb{P}_1$. See loc. cit. Example 9.2.

23. Special Curves on K3-Surfaces

Existence results

In this subsection we are going to present Mumford's proof [M-M] of the following result:

(23.1) **Theorem.** *Every projective* K 3*-surface contains a rational curve (in general singular) and a* 1*-dimensional family of (in general singular) elliptic curves.*

The idea of the proof is quite simple.

Every projective K 3-surface X has an ample divisor C with $C^2 = 2g-2$ and whose cohomology class is primitive. The linear system $|C|$ will be fixed point free in general, of dimension g, and its general member is an irreducible smooth curve of genus g. Since it is one condition on curves in $|C|$ to acquire a double point, one expects that there are (in general finitely many) curves in the system with exactly g nodes and at least a one-dimensional system of curves in the system with exactly $g-1$ nodes. The former are rational curves and the latter have elliptic desingularizations. The problem of course is that this presupposes that the condition to acquire a node is uniform on points, which is not at all obvious.

To circumvent the difficulty, one first constructs a K 3-surface with a linear system $|C_0|$, $C_0^2 = 2g-2$, whose generic member is a smooth genus g curve and whose special member C_0 is reducible and has exactly $g+1$ double points. We then consider the deformations preserving the line bundle $\mathcal{O}(C_0)$ and show that C_0 deforms to a nodal rational curve in all directions. Since for the generic such deformation the Néron-Severi group is generated by one class, this must be the class of the deformed C_0 and moreover this class ℓ is an ample class. So this proves the theorem for the generic point in D_ℓ^{pol}. A specialization argument, using that D_ℓ^{pol} is connected (Corollary 22.4), then proves the result for all points in D_ℓ^{pol}.

Before we enter into the details of the construction, we gather a few facts concerning linear systems on K 3-surfaces that we use in the sequel.

(23.2) **Lemma.** *Let D be a reduced connected divisor on a* K 3*-surface X. Then $h^1(\mathcal{O}_X(D)) = h^2(\mathcal{O}_X(D)) = 0$ and $h^0(\mathcal{O}_X(D)) = \frac{1}{2}D^2 + 2$. Moreover, if the linear system $|D|$ has neither a fixed part nor fixed points, then the generic member D_λ is a smooth connected curve whose genus g is given by the formula*

$$D^2 = 2g-2. \tag{14}$$

The morphism $f_D : X \to \mathbb{P}_g$ restricts to the canonical map $D_\lambda \to \mathbb{P}_{g-1}$.

Proof. The fact that D is reduced and connected implies $h^0(\mathcal{O}_D) = 1$ and so the cohomology sequence for

$$0 \to \mathcal{O}_X(-D) \to \mathcal{O}_X \to \mathcal{O}_D \to 0$$

shows that $h^1(\mathcal{O}_X(D)) = h^1(\mathcal{O}_X(-D)) = 0$. Since we also have $h^2(\mathcal{O}_X(D)) = h^0(\mathcal{O}_X(-D)) = 0$, Riemann-Roch gives the stated expression for $h^0(\mathcal{O}_X(D))$.

Recall (Sect. II.1) that the canonical bundle is given by $\omega_D = \mathcal{O}_D(D)$ and the arithmetic genus of D by formula (14). Bertini implies the remaining assertions. □

We are going to apply the preceding Lemma in the following special case:

Example. Assume that D consists of two smooth rational curves D_1 and D_2 meeting transversally in $g+1$ points. Since $D_1^2 = D_2^2 = -2$, we have $D^2 = 2g-2$. We claim that $|D|$ has neither a fixed part nor fixed points so that the preceding Lemma can be applied. Indeed, the normalisation consists of two copies of $\mathbb{P}_1$, say E_k, $k = 1, 2$ and we have $P_j^k \in E_k$, $j = 0, \ldots, g$, $k = 1, 2$ corresponding to the intersection points. The canonical system $|\omega_D|$ comes from the Rosenlicht-differentials (cf. Chap. II 6a), i.e., those meromorphic differentials on $E_1 \cup E_2$ having poles at these points with opposite residues at each of the pair of points $\{P_j^1, P_j^2\}$, $j = 0, \ldots, g$. Since one can always find a Rosenlicht differential which does not vanish at any given point of $E_1 \cup E_2$ the system $|\omega_D|$ does not have fixed points on D and hence $|D|$ itself cannot have fixed points on X.

1. Construction of a special K 3*-surface with nef nodal curves.*

Consider an elliptic curve E and another elliptic curve F related to E by an isogeny $\lambda : E \to F$ of degree $d = 2(g+1)+1$. Let $p \in E$ and $q \in F$ be 2-torsion points. We may assume that $\lambda(p) = q$. Now consider the Kummer surface $X_0 = \mathrm{Km}(E \times F)$. In addition to the 16 rational curves coming from the 2-torsion of $E \times F$ there are rational curves which are the images of the 4 "vertical" curves {2-torsion point} $\times F$ and the 4 "horizontal" curves $E \times$ {2-torsion point}. We shall only use the image S coming from $E \times q$. The isogeny λ gives another rational curve $G \subset X_0$ coming from its graph Γ_λ on $E \times F$. We have $\Gamma_\lambda.(E \times q) = d = 2(g+1)+1$. The curves Γ_λ and $E \times q$ meet transversally and have exactly one 2-torsion point in common, and so on X_0 the curves G and S meet transversally in $GS = g+1$ points. Taking $C_0 = G + S$ we are exactly in the situation of the preceding example. We have seen there that $|C_0|$ is free of base points, that dim $|C_0| = g$ and that the generic member is a smooth curve of genus g.

For later reference we denote the points of intersection of G and S by $P_0, \ldots, P_g$.

2. Construction of the deformation.

Let $p : X \to U$ be the Kuranishi deformation of X_0. We know (Theorem 7.3) that U is smooth and 20-dimensional. We may assume that U is contractible. Then we can choose a trivialization $\phi : R^2 p_* \mathbb{Z} \xrightarrow{\sim} L$ inducing markings $\phi_t : H^2(X_t, \mathbb{Z}) \xrightarrow{\sim} L$ for all fibres X_t of p over $t \in U$. The marking induces a period map

$$\tau : U \longrightarrow \Omega,$$

which is a biholomorphism onto its image. Let $\ell = \phi_0\left(c_1(\mathcal{O}(C_0))\right)$ and, using the notation of the beginning of Sect. 22, define

$$U_\ell = \tau^{-1}(\Omega_\ell).$$

This is a 19-dimensional submanifold $U_\ell \subset U$ passing through 0. The Kuranishi family restricts to $p : X|U_\ell \to U_\ell$. The global section $\phi^{-1}(\ell)$ of $R^2 p_* \mathbb{Z}|U_\ell$ defines a $(1,1)$-class in $H^2(X) \simeq H^2(X_0)$ which is the class of a line bundle $\mathcal{L}$

on $X|U_\ell$. On X_0 we have $h^1(\mathcal{L}_0) = h^1(\mathcal{O}(C_0)) = 0$ and so by semi-continuity, possibly after passing to a smaller contractible neighbourhood of 0, we may assume that this holds in U_ℓ. Then $R^0p_*\mathcal{L}$ becomes a trivial bundle and we take the associated projective bundle $V \cong U_\ell \times \mathbb{P}_g \to U_\ell$. Pulling back the Kuranishi family gives $X_V \to V$ with $X_V \cong X|U_\ell \times \mathbb{P}_g$. The divisor $C_0 \in |C_0| = \mathbb{P}H^0(X_0, \mathcal{L}_0)$ is the fibre at $(0, [C_0]) \in V$ of a relative divisor $C \subset X_V$ over V whose fibre over $(u, [C_u])$ is the divisor $C_u \subset X_u$. The double points $P_0, \ldots, P_g \in X_0$ can be considered as lying on C and we are going to show:

Claim. There is a germ of a subvariety $V_\delta \subset V$ of codimension δ passing through $(0, [C_0])$ such that for $t \in V_\delta$, the divisor C_t has at least δ double points. Such divisors form a subvariety in $|C_t|$ of dimension δ. Moreover, the projection $V_\delta \to U_\ell$ maps surjectively onto a neighbourhood of $0 \in U_\ell$.

Replace V by a sufficiently small coordinate neighbourhood centred at $v_0 := (0, [C_0])$, say with coordinates $v = (v_1, \ldots, v_{19+g})$. We construct the subvariety $V_\delta \subset V$ as follows. The family $X_V \to V$, locally around P_j is a product $X^{(j)} \times V$ in which C is a deformation of the double point P_j of the curve $C_0 \cap X^{(j)}$. The universal deformation of an ordinary double point is given by the family $xy = t$ over the unit t-disk Δ. By universality, there is a holomorphic map $g_j : X^{(j)} \times V \to \Delta$ which induces $C^{(j)}$ from the universal family. Pulling back x, y and t respectively, we find functions x_j, y_j on $X^{(j)} \times V$ and g_j on V respectively such that the local equation of $C^{(j)}$ in $X^{(j)} \times U$ is given by

$$x_j y_j - g_j(v) = 0, \quad g_j(v_0) = 0$$

The functions $(x_j, y_j, v_1, \ldots, v_{19+g})$ give local coordinates on X_V centred at $P_j \times \{v_0\}$. We define $V_\delta \subset V$ by the equations $g_1 = \cdots = g_\delta = 0$. This variety has codimension $\leq \delta$ and it intersects $\{0\} \times \mathbb{P}_g \subset V$ in a subvariety $C_{\delta,0}$ of curves in $|C_0|$ having at least δ nodes. Near $[C_0]$ the curves corresponding to this intersection are small deformations of C_0 and hence are reduced. Recall that C_0 has two rational components which intersect in $(g+1)$ points. If we consider small deformations of C_0 which keep g of the double points and smooth the remaining double point, then C_0 deforms to an irreducible curve C_t with g double points whose normalization is then a $\mathbb{P}_1$. Since K 3-surfaces have Kodaira dimension 0 they do not contain positive dimensional families of rational curves and hence $C_{g,0}$ has dimension 0. This means that V_g intersects $\{0\} \times \mathbb{P}_g$ in a zero-dimensional subvariety which implies that V_δ intersects $\{0\} \times \mathbb{P}_g$ in a variety of codimension exactly δ (and not less). So V_δ has dimension $19 + g - \delta$. Now consider the projection $V_\delta \to U_\ell$. It has to be surjective, since if not, all fibres would have dimension $> g - \delta$, while the fibre over 0 has dimension $g - \delta$.

3. Completion of the proof using the Claim.

Over each point t in a suitably small open neighbourhood of $0 \in U_\ell$, the above claim implies that $(V_g)_t$, respectively $(V_{g-1})_t$, gives a rational curve on X_t, respectively a one-dimensional system of elliptic curves. Taking for t a generic point, the Néron-Severi group has rank one and is generated by a (rational) multiple of the class of C_t. In particular must be *ample*. This shows that such a t belongs in fact to the period domain Ω_ℓ^{pol} introduced at the beginning of Sect. 22. This completes the proof for a generic K 3-surface in this period domain.

Now any projective K 3-surface X_0, being the limit of a 1-parameter family of K 3-surfaces X_t each containing a rational curve C_t itself contains a rational curve. To see this, remark that the irreducible rational curve C_t can be chosen to depend smoothly on t, $t \neq 0$. But then the limit curve is also a (possibly reducible) rational curve. Using the fact that X_0 has no positive dimensional family of rational curves the same argument shows that X_0 has a 1-dimensional family of elliptic curves.

Enumerative results

In the previous subsection, we saw that every projective K 3-surface has at least one rational curve and we explained that we expect that there are at most finitely many rational curves in a given linear system.

Yau and Zaslow [Y-Z] predicted a closed formula giving the number n_g of rational curves on a K 3-surface in a linear system of genus g curves; this number is claimed to be the coefficient of the q-expansion involving the modular form $\Delta(q) = q\prod_{n\geq 1}(1-q^n)^{24}$:

$$\sum_{g\geq 0} n(g)q^g = \frac{q}{\Delta(q)}, \tag{15}$$

i.e., the prediction is $n(g) = n_g$. The exact definition of the numbers n_g was not made clear however in loc. cit. When all rational curves that occur in the system are irreducible with ordinary nodes the number should be the actual number of rational curves (without multiplicities). If there are worse singularities, one expects that multiplicities have to be defined. When a curve in the system is reducible and one or more components are rational it is not even clear whether such curves should be counted with *positive* multiplicity. If not, cancellation might occur and the interpretation of n_g as the number of rational curves would be problematic.

If we assume that all curves in the system are irreducible, the preceding difficulties have been solved by Beauville [Be99] and Fantechi et. al. [F-G-vS]. This is the generic case (since a generic projective K 3-surface has Picard number 1). For reducible curves the situation is much more subtle and one has to interpret the numbers n_k as "family Gromov-Witten invariants". See [LeJ], [Liu] and [B-L]. For an enumerative application see [Gat].

We are not going to give the proof of this result, which is outside the scope of this book, but we are going to explain some of the main ideas and ingredients of Beauville's proof. Let us first explain why $n_1 = 24$. The linear system is an elliptic pencil in this case ($g = 1$). The Euler number of such a pencil is the sum of the Euler numbers of the singular fibres. It is 1 for a rational curve with 1 node, it is 2 for a cusp, etc. Defining these as the multiplicity with which to count a rational curve, we thus get indeed $n_1 = e(X) = 24$. Moreover, the generic such surface only has elliptic pencils whose singular fibres are precisely rational curves with one node, justifying the count.

The idea now is to generalize this for a K 3-surface with a g-dimensional linear system $|C| \simeq \mathbb{P}_g$ of genus g curves by looking at the so-called compactified Jacobian family $\bar{J} \to \mathbb{P}_g$, whose fibre over $[C_t]$ is Rego's compactified Jacobian $\bar{J}_t$ of C_t [Re]. In fact, one takes here the degree g Jacobian parametrising degree g divisors. One then proceeds by showing:

1. The Euler number of $\bar{J}_t$ is zero unless C_t is rational. In the latter case, it is equal to 1 if C_t has only nodal singularities;
2. the Euler number of the total space $\bar{J}$ equals the coefficient of q^g in the q-expansion of (15);
3. each rational curve contributes positively and rational curves with only nodes have multiplicity one in the formula.

We are not going to say more about 1) and 3). But we comment briefly on the computation of $e(\bar{J})$. It is done in three steps. First we remark that $\bar{J}$ is a projective manifold of dimension $2g$ birational to the manifold $X^{[g]}$ parametrising 0-dimensional subschemes of X having length g. This we see as follows. Consider the set $U \subset \bar{J}$ consisting of pairs (C_t, L) of genus g curves C_t and line bundles of degree g on it with $h^0(C_t, L) = 1$. The unique section of L defines a divisor of degree g on $C_t \subset X$ which we consider as a length g subscheme in X. Since dim $|C_t| = g$ and since it is g conditions to pass through g points, for a generic pair (C_t, L) this subscheme will be contained in a unique curve of $|C_t|$, namely C_t. In this way get a birational isomorphism between U and its image in $X^{[g]}$. Secondly, we use a result due to Batyrev [Bat99] stating that birationally isomorphic Calabi-Yau manifolds (manifolds with trivial canonical bundle such as $\bar{J}$ and $X^{[g]}$) must have the same Betti-numbers. Finally, we observe that Göttsche's formula [Gts] precisely says that the Euler number of $X^{[g]}$ is the coefficient in the q-expansion of (15).

24. An Application to Hyperbolic Geometry

The basic facts from hyperbolic geometry can be found in Kobayashi's book [Kob]; we shall only briefly discuss the necessary background. We start by giving the definition of the Kobayashi pseudo-metric on the tangent space at $x \in X$ of any complex manifold X. Fix a direction $\xi \in T_x X$ and consider holomorphic unit discs $f : \Delta \to X$ centred at x such that $f_*(d/dz)_0$ is proportional to ξ, say

$$\lambda(f) f_*(d/dz)_0 = \xi.$$

Then we define $\delta_K(\xi) = \inf(\lambda(f))$, where f ranges over such maps. This defines a pseudo-metric in the sense that $\delta_K(\xi) \geq 0$ and that $\delta_K(\lambda\xi) = |\lambda|\delta_K(\xi)$ and so integrating yields a pseudo-metric d_K on X. This is the **Kobayashi pseudo-metric**. If it is a metric we say that X is **hyperbolic**. It follows from this definition that a manifold cannot be hyperbolic if it admits a non-constant holomorphic map $\mathbb{C} \to X$. Indeed, if it would, there would exist disks of arbitrary large radius mapping to X and by scaling down this gives a map $\Delta \to X$ from the unit disk such that the image of standard tangent vector at the origin has arbitrary small image in X. The converse is also true; this is due to Brody [Brod].

We can do a similar thing for maps of the unit ball of any dimension to X and in particular for n-balls , where n is the dimension of X. In this case, we consider the proportionality of the standard volume element at the origin of the unit n-ball and a generator of $\Lambda^n T_x X$. We then arrive at the notion of the **Kobayashi pseudo-measure**. If this is a true measure outside a proper analytic subvariety, then X is called **measure hyperbolic**. We saw that non-constant maps $\mathbb{C} \to X$ prevent X from being hyperbolic. The same argument shows that an n-dimensional manifold X cannot be measure hyperbolic if there exists a holomorphic map $f : \mathbb{C} \times \Delta^{n-1} \to X$ which is non-degenerate in the sense that at some point of $\mathbb{C} \times \Delta^{n-1}$ the Jacobian of f is non-zero. From this observation it follows that any hyperbolic manifold is measure hyperbolic (see also [Kob], Ch. IX).

Let us turn to projective manifolds X. If such a manifold admits a rational or an elliptic curve it cannot be hyperbolic. If an algebraic version of Brody's theorem were true, the converse would also hold. Kobayashi [Kob] and Lang [Lan] have suggested that a slightly stronger hypothesis, would indeed imply that X is hyperbolic: every morphism $f : Z \to X$ from an abelian variety Z to X should be constant (observe that a non-constant map from $\mathbb{P}_1$ to X automatically gives a non-constant map from an elliptic curve to X, so we need not consider rational curves separately). This is completely open, even for surfaces.

Another problem is to recognize hyperbolic manifolds algebraically in an intrinsic fashion. For curves the situation is clear: a compact Riemann surface is hyperbolic if and only if its genus is ≥ 2, i.e., if and only if it is of general type (see [Kob], Ch. IV.4). It is too much to ask for this to generalize to higher dimensions since surfaces of general type can very well have rational and elliptic curves, preventing them from being hyperbolic. Lang conjectured therefore in [Lan] that a projective manifold of general type should be hyperbolic if all of their subvarieties are of general type. And if not, one expects that the source of non-hyperbolicity is algebraic in the following sense: any entire holomorphic curve $\mathbb{C} \to X$ in a projective manifold X of general type must be **algebraically degenerate**, i.e., has its image in a proper algebraic subvariety. This is the content of another celebrated conjecture due to Green-Griffiths [Gr-Gf] and Lang (loc. cit.).

Returning to surfaces of general type, the truth of the conjecture would imply that there are only finitely many rational and elliptic curves. This has been proven by Bogomolov [Bog77] when $c_1^2 > c_2$. McQuillan [McQ] proves the full conjecture: in this case entire holomorphic curves are algebraically degenerate. For the remaining surfaces of general type there are partial results ([Lu-Mi]).

One can ask oneself the supposedly easier question as to when projective manifolds are measure hyperbolic. One knows that general type implies measure hyperbolic [Kob-O]. We want to show that the converse is true for surfaces.

(24.1) **Theorem.** *An algebraic surface is measure hyperbolic if and only if it is of general type. In particular any hyperbolic surface must be of general type.*

Proof. As we remarked above, for a given surface X not of general type, it suffices to find a non-degenerate map

$$f : \mathbb{C} \times \Delta \to X.$$

This can be proved using classification (Theorem VI.1.1), the preceding result about elliptic curves on K 3-surfaces (Theorem 23.1) and the result that all Enriques surfaces have elliptic fibrations (Theorem VIII. 17.5). Indeed, an algebraic surface from the classes 1)-3) in Theorem VI.1.1 admits a pencil of rational curves and hence there exists a non-degenerate map as above. Since tori are covered by $\mathbb{C}^2$, this is also true for them. All the other algebraic classes then admit an algebraic family of elliptic curves. In particular, over some disk Δ in the parameter space there will be a smooth elliptic fibration. By the description of stable fibrations (Ch. V, §8) such a fibration is the quotient of $\mathbb{C} \times \Delta$ by a certain group of automorphisms. In particular, there is a non-degenerate holomorphic map $\mathbb{C} \times \Delta \to X$, thereby completing the proof of the theorem. □

The somewhat mysterious name "K 3" has been explained by A. Weil in the comment on his final report on contract AF 18 (603)-57: "ainsi nommées en l'honneur de Kummer, Kähler, Kodaira et de la belle montagne K2 au Cachemire" (*cf.* [Wei80], p. 546).

The classical results on Kummer surfaces, which go back to the beginning of this century, are presented in the final chapter of [G-H78a].

The idea of studying deformations of K 3's via the periods of their holomorphic 2-forms stems from A. Andreotti ([Ad]) and A. Weil. The local Torelli-property is due to them (unpublished). Proofs have been given by Kodaira ([Ko66], part. I, Theorem 17) for arbitrary K 3's and by G.N. Tjurina ([Saf], Chap. IX, Theorem 2) who needed the Kähler assumption.

The famous conjectures:

(i) all K 3-surfaces constitute one connected family,

(ii) all K 3-surfaces are kählerian,
(iii) the period map is surjective,
(iv) (a form of) the global Torelli theorem holds,
were made by Andreotti and Weil (see [Wei80]).

It should be said that Weil's definition of a K 3-surface differs from ours in that he calls any surface "K 3" if it carries the differentiable structure of a smooth quartic surface in $\mathbb{P}_3$. Adopting our definition, (i) was proved independently by Kodaira ([Ko66], part I, Theorem 19) and – under the Kähler assumption – by G.N. Tjurina ([Saf], Chap. IX, Theorem 7).

Conjecture (ii) was proved by Siu [Siu83] (Sect. 14). As stated before, we avoid this by making use of an *a priori* proof for the fact that b_1 even implies Kähler.

Conjecture (iv) was solved in the affirmative for projective K 3's by Piatečkii-Shapiro and Šafarevič in [Pi-S], but their proof, although in principle correct, contained several gaps and errors, some of which are rather serious. These were later corrected by M. Rapoport and independently by T. Shioda. The results have not been published in full (see however [Shi]). A detailed and corrected account of the original proof can be found in [L-P], in which paper also a simplified version is presented of the work of D. Burns and M. Rapoport on the period map for Kähler K 3's ([B-R]). Our exposition is a partly rewritten version of this last work. Conjecture (iii) was first solved for special classes of algebraic K 3-surfaces by J. Shah ([Sha76], [Sha80], [Sha81]) and independently by E. Horikawa ([Hor77]). Then V. Kulikov ([Kul]) gave a proof for projective K 3's (without restriction), but his proof needed clarification at several points, subsequently provided by U. Persson and H. Pinkham in [P-P]. Relying on these results and making essential use of the Atiyah-Hitchin-Yau results – as presented in Sect.13 – A. Todorov gave a proof of the surjectivity for the period map for Kähler K 3's ([To]). The proof we give does not use the surjectivity for projective K 3's and is due to E. Looijenga ([Lo]).

Enriques surfaces bear the name of their inventor. He constructed these surfaces to give examples of non-rational surfaces with $q = p_g = 0$ ([Enr49]). Many of his assertions were proved rigorously (in all characteristics except 2) by M. Artin in his Harvard thesis (not published). Several of his ideas have been used freely in Sect. 17. The idea to use double coverings of quadrics to show that any two Enriques surfaces are deformations of each other, seems to be new (Artin, following Enriques, uses instead the fact that all Enriques surfaces can be represented as sixth-degree surfaces in $\mathbb{P}_3$ passing doubly through the edges of a tetrahedron). But it remains nevertheless true that our treatment of Enriques surfaces stays close to Horikawa's.

Results on the period map were first obtained by E. Horikawa in [Hor78]. Our proof of the injectivity is basically the same as his, except for some simplifications. The surjectivity we prove however in a considerably different and shorter way, by making use of the corresponding statement for K 3-surfaces. (Our proof does not make use of degenerations, so stays entirely within the realm of non-singular surfaces.)

We briefly comment on some recent developments. The compactified moduli space has been studied by Sterk [Ste]. An algebraic approach to the moduli of

Enriques surfaces has been undertaken by Cossec and Dolgachev. See [Co-D] and the references in it. The moduli space of Enriques surfaces is rational as shown by Kondō [Kon94]. Using automorphic forms Borcherds shows in [Borc] that it is Zariski-open in an affine variety. For other recent results see [Al]. Automorphisms of K 3-surfaces have been investigated by Nikulin [Ni79], Mukai [Muk] and Kondō [Kon98], while automorphism groups of Enriques surfaces have been studied by Barth-Peters [B-P], Dolgachev [Dol84], Kondō [Kon86], Mukai-Namikawa [M-N] and Nikulin [Ni84]

The material in Sect. 22 is based on [Do98]. The proof that every K 3-surface has a rational curve and a one-dimensional system of elliptic curves is due to Mumford; we give the proof from [M-M]. The application to hyperbolic geometry appears in [Gr-Gf].

Chapter IX . Topological and Differentiable Structure of Surfaces

Topology of Simply Connected Compact Complex Surfaces

In the first subchapter we review Freedman's topological classification of 4-manifolds as well as the 11/8-conjecture which predicts which unimodular forms can be represented by smooth 4-folds and we discuss the implications for compact complex surfaces.

The second and third parts deal with Donaldson invariants and Seiberg-Witten invariants respectively. These are treated from an axiomatic point of view and we give two striking examples of how the theory of the previous chapters together with these new invariants can be applied to obtain spectacular results about the differential topology of 4-manifolds.

We are aware of the fact that for most important applications the Seiberg-Witten invariants suffice. But we feel that in a book on surfaces the Donaldson invariants should also be treated.

1. Freedman's Results

Let V be a simply-connected, oriented compact topological 4-manifold. The cup product provides the lattice $H^2(V,\mathbb{Z})$ with a unimodular quadratic form S_V. The fact that the pairing is unimodular follows directly from Poincaré duality.

(1.1) **Theorem.** (M. Freedman, [Frm]). *The map*

$$\left\{\begin{array}{l}\text{homeomorphism types of compact, oriented}\\ \text{and simply connected topological 4-manifolds}\end{array}\right\} \longrightarrow \left\{\begin{array}{l}\text{unimodular}\\ \text{quadratic forms}\end{array}\right\},$$

given by $V \mapsto S_V$ is surjective, one-to-one for S_V even and two-to-one for odd forms. In the last case the two manifolds V and V' with $S_V = S_{V'}$ can be distinguished by the fact that for one of them, say V, the 5-manifold $V \times S^1$ admits a differentiable structure, whereas $V' \times S^1$ does not.

(1.2) **Corollary.** *The oriented homeomorphism type of a simply connected differentiable manifold V is completely determined by S_V.*

The theorem says in particular that, given a unimodular form S, there exists a topological manifold V with $S_V \cong S$. However, in general there is no *differentiable* such manifold, and if if it exists, it will in general not be unique.

Recall that the signature $(b^+(S), b^-(S))$ is the first basic invariant of any non-degenerate form such as S_V. Equivalently, we can give its rank and its index

$$\tau(S) = b^+(S) - b^-(S).$$

For unimodular forms we have another invariant, its parity. The main result of this section states that for complex surfaces these three determine the homeomorphism type:

(1.3) **Theorem.** *Two simply connected complex surfaces whose cup product forms have the same parity and signature are homeomorphic as oriented manifolds. In other words, the oriented homeomorphism type of any simply connected complex surface X is completely determined by the rank, index and parity of S_X.*

Proof. We know from Theorem I, 2.8 that the isomorphism type of an *indefinite* unimodular quadratic form is completely determined by its rank, index and parity. So, in view of the preceding corollary, we shall be done as soon as we have proved

The only simply-connected complex surface X for which S_X is definite, is $\mathbb{P}_2$.

By Table 10 in Chap. VI, we know that there are only two types of simply-connected non-algebraic surfaces, namely (blown up) K 3-surfaces and (blown up) elliptic surfaces.. By the results of Chapt. VIII we know that S_X is indefinite when X is a K 3-surface, and in the case of an elliptic surface X, we can for example say that X is Kähler by Theorem IV, 3.1, hence a fibre is not homologous to zero, whereas its square is, hence S_X is indefinite.

If X is algebraic, and S_X is definite, it is automatically positive definite, in other words $b^-(X) = 0$. From

$$\begin{aligned} b^+(X) + b^-(X) &= b_2(X) = c_2(X) - 2, \\ b^+(X) - b^-(X) &= \frac{1}{3}(c_1^2(X) - 2c_2(X)), \end{aligned}$$

we find $c_1^2(X) - 5c_2(X) + 6 = -6b^-(X) = 0$. If $c_2(X) \geq 4$ we would find $c_1^2(X) > 3c_2(X)$. But for all simply connected algebraic surfaces we have $c_1^2(X) \leq 3c_2(X)$ (this is the inequality from Theorem VII.4.1). There still remains the case $c_2(X) = 3$, i.e., $b_2(X) = 1$, but in this case X is even algebraically isomorphic to $\mathbb{P}_2$ by Theorem V.1.1. □

If we drop the condition that X is simply connected, then there exist non-algebraic surfaces with a negative-definite intersection form (some surfaces of class VII, all blown-ups of surfaces of class VII, all blown ups of secondary Kodaira surfaces). On the other side we have:

(1.4) **Proposition.** *If X is a complex surface with S_X positive definite, then X is an algebraic surface with $c_1^2(X) = 9$, $c_2(X) = 3$. In particular $b_1(X) = 0$ and if the fundamental group is finite, X is algebraically isomorphic to $\mathbb{P}_2$.*

Proof. Using once again $c_2(X) - 2 + 2b_1 = b_2 = b^+ = \frac{1}{3}(c_1^2 - 2c_2)$ we find $c_1^2 = 5c_2 - 6 + 6b_1$ and hence

$$(1) \qquad c_1^2 - 3c_2 = -6 + 6b_1 + 2c_2 = 2b^+ + 2b_1 - 2.$$

For $c_2 < 0$ there are only blown-ups of ruled surfaces and these do not have a definite intersection form.

Now note that

$$b_2 = c_2 + 2(b_1 - 1) =$$
$$b^+ = \frac{1}{3}(c_1^2 - 2c_2) > 0.$$

From the last equation we see that if $c_2 \geq 0$ we have $c_1^2 > 0$ and so X is algebraic. In fact, by Table 10, X is rational or of general type and we must have $c_2(X) > 0$. By Theorem VII.4.1 the inequality $c_1^2 \leq 3c_2$ holds. Hence (1) implies that $b_1 = 0$ and $b_2 = b^+ = 1$ and so $c_2 = 3$, $c_1^2 = 9$ as claimed. □

2. Representability of Unimodular Forms

As we have already observed, all forms can be represented by compact, oriented and simply connected topological 4-manifolds. The question arises: which forms can be represented by differentiable 4-manifolds? Which ones by, say, algebraic surfaces? We consider these questions for indefinite forms. Such a form is isometric to

$$F^{p,q} := p\langle \mathbb{1} \rangle \oplus q\langle -\mathbb{1} \rangle, \quad \tau = p - q \quad \text{(odd forms)}$$
$$E^{(p,\pm q)} := pH \oplus q(\pm E_8), \quad \tau = \pm 8q \quad \text{(even forms)}.$$

The odd form $F^{p,q}$ is represented by the differentiable manifold $V^{p,q}$ which is the connected sum of p projective planes and q projective planes with reversed orientation:

$$V^{(p,q)} = \underbrace{\mathbb{P}_2 \# \cdots \mathbb{P}_2}_{p} \# \underbrace{\overline{\mathbb{P}}_2 \ldots \# \overline{\mathbb{P}}_2}_{q}.$$

But the situation is quite different in the algebraic case. For p even, $V^{(p,q)}$ does not even admit an almost complex structure. As to the case p odd, take for example $p = 3$. Let X be an algebraic K 3-surface, blown up in one point. Then the rank of the (odd) form S_X is 23 and its index -17. So X is homeomorphic to $V^{(3,20)}$. We claim that, on the other hand $F^{(3,1)}$ cannot be realized by a simply connected algebraic (or complex) surface. Suppose that it were, say $S_Y = F^{(3,1)}$. Then, from the standard formulae for c_2 and τ we find $c_2(Y) = 6$ and $c_1^2(Y) = 18$. So $c_1^2(Y) = 3c_2(Y)$ and by I, 15.5 it must be a quotient of the unit 4-ball. But this is impossible since Y is simply-connected.

Once $F^{(p,q)}$ is represented by an algebraic surface, then so is $F^{(p,q+1)}$ (just perform a blow-up). So, if $F^{p,q}$ is represented, there is a minimal value $q(p)$ such that $F^{(p,q(p))}$ is representable by an algebraic surface. What this function $q(p)$ is, remains an open problem.

Now we turn to even forms. A quadric has intersection form H, whereas a K 3-surface has the form $E^{(3,-2)}$. Taking direct sums, we can now differentiably realize $E^{(p,\pm q)}$ with $p = 3a + b$ and $\pm q = 2a$ copies of $\pm E_8$. But this leaves the representability of the form $E^{(p,q)}$ with $(p, q) \neq (3a + b, \pm 2b)$ by a differentiable manifold completely open. A first restriction is Rochlin's theorem ([Ro]):

(2.1) **Theorem.** *For differentiable 4-manifolds with even intersection form the index is divisible by 16.*

The following conjecture says that no other indefinite even forms can be realized by a differentiable 4-fold but those above, i.e., the forms $E^{(3a+b,\pm 2b)}$:

(2.2) **The 11/8 Conjecture.** *For every simply connected oriented differentiable 4-manifold V with even intersection form we have the inequality*

$$b_2 \geq \frac{11}{8}|\tau|.$$

Indeed, if the conjecture holds for some V, then, first of all, by Rochlin's theorem the index would be divisible by 16 and we can take $a := |\tau|/16$ copies of a K 3-surface (if the index is negative) or of a K 3 with orientation reversed (if $\tau \geq 0$) and of

$$b := \frac{1}{16}(8b_2 - 11|\tau|)$$

copies of a quadric. The fact that b is an integer follows from the fact that the unimodular intersection form is isometric to a direct sum of copies of H and $\pm E_8$ so that b_2 is even. The 11/8 Conjecture just says that this number should be non-negative.

We can show that the conjecture is true for complex surfaces:

(2.3) **Proposition.** *The 11/8 Conjecture is true for simply connected complex surfaces with even cupproduct form.*

Proof. For almost complex surfaces we have $b_2 = c_2 - 2$ and $\tau = 1/3(c_1^2 - 2c_2)$ and we find that the 11/8-conjecture boils down to

$$\begin{cases} 48b = 11c_1^2 + 2c_2 - 48 \geq 0 & \text{if } \tau \leq 0 \\ 48b = -11c_1^2 + 46c_2 - 48 \geq 0 & \text{if } \tau \geq 0. \end{cases}$$

In the non-algebraic case, by Theorem VI.1.1 we only have to look at K 3-surfaces, for which the conjecture is true, and to minimal elliptic surfaces (even cup form implies minimality). In the last case we have $c_1^2(X) = 0$ and so by Rochlin's theorem c_2 must be divisible by 24, whereas simple-connectivity implies $c_2(X) \geq 2$. Hence the conjecture is true in this case too.

For minimal algebraic surfaces, by the classification results (see Chap. VI, Table 10) we have $c_1^2 \geq 0$ and $c_2 \geq 3$ and so if the index is negative, the conjecture follows (the left hand side being divisible by 48 it suffices to show that it is greater than -48). And if the index is positive, we use the inequality $c_1^2 \leq 3c_2$ (Theorem VII.4.1) to see that in this case the inequality holds. □

Finally, we note the following consequence of the previous results.

(2.4) **Theorem.** *Every simply connected algebraic surface is homeomorphic to either a connected sum of copies of* $\mathbb{P}_2$ *and* $\overline{\mathbb{P}}_2$ *or a connected sum of 2-quadrics and* K 3*-surfaces (respectively* K 3*-surfaces with orientation reversed).*

Donaldson Invariants

3. Introduction

As far as differentiable structures on a compact, oriented and (for simplicity) simply connected topological manifold M with $\dim(M) \neq 4$ are concerned, the situation was well understood before Donaldson's work. Already in the early fifties Moise proved that for $\dim(M) \leq 3$ there is (up to equivalence) one differentiable structure on M. And for $\dim(M) \geq 5$ if was shown by Smale and others in the course of the sixties and seventies that, given the Pontryagin classes of M, there is at most a finite number of differentiable manifolds with the same homotopy type as M. But in the case of 4-dimensional manifolds the question was, apart from the non-existence results which follow from Freedman's work as discussed in section 1, completely open until Donaldson came up with his pioneering work. It has lead to a wealth of new results, and in particular it follows that if $\dim(M) = 4$, then there can be infinitely many inequivalent differentiable structures on a given topological manifold M. There is no need to fix the Pontryagin class p_1 here, since it is equal to 3 times the index. Though Donaldson theory deals with 4-manifolds in general, effective calculations can only be carried out in the case of algebraic surfaces. This is due to the fact that certain differential geometric objects (anti self-dual connections) can be interpreted in this case as purely algebro-geometric objects, namely stable vector bundles, which have been studied very much during the past decades.

This is certainly not the place to explain Donaldson's methods in any detail; but we feel that a modern book about surfaces must at least demonstrate how, starting from the general framework for the differentiable case, complex surfaces are used to obtain completely new results about differentiable structures. We shall do this by working out one of the most striking applications, and prove that on a specific topological 4-fold there exists an infinity of algebraic structures which are inequivalent as differentiable manifolds.

4. The Donaldson Invariant, a Bird's Eye View

Restricting ourselves to the original Donaldson invariant, we only give an outline of the general theory, without dealing with proofs or background.

Let M be a compact, oriented and simply connected differentiable 4-fold with $b_2^+ = 1$, and g some Riemannian metric on M. The metric g induces metrics on the bundle of k-forms and if v_g is the volume form, we recall that the Hodge $*$-operator, sending a k-form to one of complementary degree $4-k$, is defined by the formula

$$\alpha \wedge *\beta = g(\alpha, \beta) v_g.$$

So $*$ acts on 2-forms as an involution and the $+1$ and (-1)-eigenspaces respectively, define a splitting into self-dual and anti self-dual 2-forms. On the level of de Rham cohomology one obtains a 1-dimensional space of classes of self-dual 2-forms, say

$$H^+(M, \mathbb{R}) = \mathbb{R}[\omega_g]. \tag{2}$$

The point of departure is the complex 2-bundles on M with structure group SU(2) and certain connections on them. For such a bundle $c_1 = 0$ (since its determinant is trivial) and for this exposition we shall only consider the unique SU(2)-bundle $\mathcal{V}$ with $c_2(\mathcal{V}) = 1$. Recall that a connection on $\mathcal{V}$ is an $\mathbb{R}$-linear operator $d_A : \mathcal{V} \to \mathcal{V} \otimes \mathcal{D}_M^1$ satisfying Leibniz' rule $d_A(fs) = df \otimes s + f \otimes d_A s$ for any section s of $\mathcal{V}$ and any C^∞ function f on M. Here A is merely a label. Now, from d_A one constructs its curvature $F_A = d_A \circ d_A$, a 2-form with values in the endomorphism bundle of $\mathcal{V}$. The Chern classes being polynomials in the curvature imposes certain topologically restraints on the connections we are considering. Since $c_1(\mathcal{V})$ is a multiple of the class represented by the trace of F_A, it is automatically zero since we are dealing with an SU(2)-bundle whose curvature assumes values in the adjoint bundle of $\mathfrak{su}(2)$-endomorphisms of $\mathcal{V}$. And so the only topological constraint comes from the second Chern class. Let us next explain what it means for a connection to be anti self-dual. The Hodge $*$-operator induces an operation on the adjoint bundle, denoted by the same symbol, and we say that the connection A is anti-self dual if $*F_A = -F_A$. These may or may not exist and indeed, the main object of study is the moduli space of all anti self-dual connections on $\mathcal{V}$ taken modulo the group of all bundle automorphisms of $\mathcal{V}$ (this is the gauge group):

$$\mathcal{M}_g = \{\text{anti self-dual connections on } \mathcal{V}\}/\mathrm{Aut}(\mathcal{V}).$$

First of all it is then proved that for a general metric g this moduli space is embedded in the space $\mathcal{B}$ of all irreducible connections (those which are not sums of connections on two rank-1 subbundles of $\mathcal{V}$) modulo the gauge group, as a 2-dimensional differentiable submanifold. This is because we have chosen $\mathcal{V}$ with $c_1 = 0$ and $c_2 = 1$; in general one would have found a manifold of real dimension $2(4c_2 - c_1^2 - 3)$.

On $M \times \mathcal{B}$ there is a "universal bundle", i.e., a U(2)-bundle $\mathfrak{U}$ with a partial connection in the M-direction, such that $\mathcal{U}|M \times [A]$ is isomorphic to $\mathcal{V}$, and such that the partial connection yields a connection A in the equivalence class of $[A]$. Now if $\mathcal{M}_g$ would be compact and if we could provide it with a natural orientation, then we would have on the one hand the element $m_g \in H_2(\mathcal{B}, \mathbb{Z})$ represented by $\mathcal{M}_g$ and on the other hand $c_2(\mathcal{U}) \in H^4(M \times \mathcal{B}, \mathbb{Z})$. Now there is the slant product

$$H_2(\mathcal{B}, \mathbb{Z}) \times H^4(M \times \mathcal{B}, \mathbb{Z}) \longrightarrow H^2(M, \mathbb{Z}),$$
$$(\beta, \gamma) \longmapsto \beta/\gamma$$

defined as in [Sp], p. 286.

It would be natural to try $m_g/c_2(\mathcal{U}) \in H^2(M, \mathbb{Z})$ as an invariant of the differential manifold M. But this does not work, already because $\mathcal{M}_g$ is not compact in most cases. Still, the basic idea turns out to be right, and the main step of the whole construction consists of finding the proper compactification for $\mathcal{M}_g$ (or at least some substitute for it in terms of homology) as well as an orientation. The element of $H^2(M, \mathbb{Z})$ that finally emerges is ρ_g, a **Donaldson element** associated to the general metric g.

Next, we consider in $H^2(M, \mathbb{R})$ the positive cone (compare Sect. IV.7)

$$\mathcal{C}_M = \{x \in H^2(M, \mathbb{R}) \mid x^2 > 0\}.$$

For every $a \in H^2(M, \mathbb{Z})$ with $a^2 = -1$ we consider its wall $W_a = a^\perp \cap \mathcal{C}_M$. The connected components of $\mathcal{C}_M \setminus \bigcup W_a$ are called **chambers.** If g is sufficiently general, then the self-dual form ω_g is not contained in any wall, and if we attach to ω_g the Donaldson element $\rho_g \in H^2(M, \mathbb{Z})$, then it can be proved that if ω_g and $\omega_{\tilde{g}}$ are in the same chamber, then $\rho_g = \rho_{\tilde{g}}$. So, varying the metric, this defines a well-defined function on (certain) chambers with values in $H^2(M, \mathbb{Z})$. It turns out that the values on different chambers are related by a universal formula independent of the C^∞-structure (this is the formula (3) below). Using this formula, the map extends to all chambers yielding **the Donaldson invariant** whose properties are stated below (Theorem 4.2). Before this stage is reached, however, many more, serious difficulties have to be overcome.

The link with algebraic geometry comes from the relation between anti self-dual connections and stable bundles. The notion of stability depends on the choice of an ample line bundle H. Let us recall that a rank 2 vector bundle $\mathcal{V}$ on a projective surface X is H-**stable** (in the sense of Maruyama) if for any line bundle L with a non-trivial homorphism $\mathcal{O}_X(L) \to \mathcal{O}_X(\mathcal{V})$ one has $2LH < c_1(\mathcal{V})H$. In our case $c_1(\mathcal{V}) = 0$ and since we are on a surface, one can easily show that stability then simply means that any locally free rank one subsheaf of $\mathcal{O}_X(\mathcal{V})$ has strictly negative degree. We set apart an immediate consequence which we use later on:

(4.1) **Lemma.** *Let $\mathcal{V}$ be a stable bundle on an algebraic surface X, then $\mathcal{V}$ has no holomorphic sections.*

By a result of Maruyama [Maru75] the set of H-stable rank 2 vector bundles with given Chern classes (c_1, c_2) on X admits the structure of a quasi-projective variety. In fact it is a coarse moduli space for such bundles. In our case $c_1(\mathcal{V}) = 0$ and $c_2(\mathcal{V}) = 1$ and the Riemann-Roch theorem (see formula I, (8)) yields

$$\chi(X, \mathcal{V}) = \chi(X, \mathcal{O}_X) = 1.$$

This guarantees that our moduli space space, which will be denoted by $\mathcal{M}_X^H$, is even fine, i.e., admits a universal bundle $\mathfrak{U}$ over $X \times \mathcal{M}_X^H$. Indeed, by [Maru78], p. 598 a sufficient condition for this to be true is that the numbers $2H^2$, $\chi(X, \mathcal{V})$ and $H^2 + c_1 H - HK_X$ are relatively prime.

The Kobayashi-Hitchin correspondence describes how the two moduli spaces are related. Although this correspondence is valid in much more generality (see [L-T]) we shall describe it only for holomorphic vector bundles $\mathcal{V}$ on projective surfaces X with $c_1(\mathcal{V}) = 0$. Moreover, for simplicity, H will be a very ample line bundle on X and for g we take the Hodge metric (Ch. I.19) associated to the corresponding embedding of X in projective space. Recall that any holomorphic bundle $\mathcal{V}$ equipped with a hermitian metric h admits a unique metric connection whose $(0,1)$-part gives the holomorphic structure on E, the Chern connection. If its curvature form is a multiple of the identity we say that h is Hermite-Einstein. The condition $c_1(\mathcal{V}) = 0$ implies that in this case the curvature form is automatically anti-self dual. It is a deep result (due to Donaldson) that $\mathcal{V}$ is H-stable if and only if $\mathcal{V}$ admits a Hermite-Einstein metric. In our situation we thus obtain a map

$$I_g : \mathcal{M}_X^H \longrightarrow \mathcal{M}_g.$$

This gives the Kobayashi-Hitchin correspondence: the map I_g can be refined to give an isomorphism of real-analytic spaces. Moreover, if universal bundles exist, such as in our situation where $\mathcal{V}$ is a 2-bundle with $\chi(\mathcal{V}) = 1$, these correspond under I_g. For details we refer to [L-T].

For a serious treatment of the Donaldson invariant we refer to [Don86], [O-V86], [Kot].

After this summary introduction we state the precise result we shall need.

(4.2) **Theorem.** *Let M be a compact, oriented and simply-connected differentiable 4-manifold with $b^+(M) = 1$.*

Then there exists a map

$$\rho = \rho_M : \mathcal{C}_M \setminus \bigcup_{a^2=-1} W_a \to H^2(M, \mathbb{Z})$$

with the following properties :

(i) ρ *is constant on chambers;*

(ii) $\rho(-x) = -\rho(x)$;

(iii) *if* x_1, *and* x_2 *are in the same component of* $\mathcal{C}_M$ *(which has two components), then*

$$\rho(x_2) = \rho(x_1) + 2\sum a, \tag{3}$$

where the sum is taken over all $a \in H^2(M, \mathbb{Z})$ *with* $a^2 = -1$ *and* $ax_1 < 0$, $ax_2 > 0$;

(iv) *if* $f : M \to N$ *is an orientation preserving diffeomorphism then*

$$f^*(\rho_N(y)) = \rho_M(f^*(y))$$

for all $y \in \mathcal{C}_N$.

(v) *Now assume that* $M = X$ *is an algebraic surface and* H *is ample. Assume that the moduli space of all holomorphic* H*-stable vector bundles on* X *with* $c_1 = 0$, $c_2 = 1$ *is a smooth, reduced algebraic curve. Let* $h = c_1(H) \in H^2(X, \mathbb{Z})$. *Then*

$$\rho_X(h) = -c_1(X) + 2\sigma_!(c_2(\mathcal{U})),$$

where $\mathcal{U}$ *is the universal bundle on* $X \times \mathcal{M}$, $\sigma : X \times \mathcal{M} \to X$ *the projection and where* $\sigma_! : H^4(X \times \mathcal{M}, \mathbb{Z}) \to H^2(X, \mathbb{Z})$, *as usual, denotes the Gysin homomorphism,* i.e., *the map Poincaré-dual to the homology homomorphism.*

5. Infinitely many Homeomorphic Surfaces which are not Diffeomorphic

In the projective plane $\mathbb{P}_2(x)$ we take two smooth cubics in general position (in particular there are no reducible cubics in their pencil). Let $f_i(x) = 0$, $i = 1, 2$ be their respective equations, and $p_1, \ldots, p_9$ their intersection points. On $\mathbb{P}_2(x) \times \mathbb{P}_1(y_1, y_2)$ we consider the smooth surface X_0 given by $f_1(x)y_2 - f_2(x)y_1 = 0$. The projection of X_0 onto $\mathbb{P}_2$ is nothing but the inverse of the blowing up of $\mathbb{P}_2$ in $p_0, \ldots, p_9$, whereas the projection $\pi : X_0 \to \mathbb{P}_1(y_1, y_2)$ exhibits X_0 as an elliptic fibre space with irreducible fibres only.

Next we introduce the Dolgachev-surfaces X_q, $q \geq 3$ and q prime. See [Dol81], where a detailed discussion can be found. Their construction is very simple: X_q is obtained from X_0 by applying logarithmic transformations in two smooth fibres of π of order 2 and q respectively. The resulting surface is simply connected by [Ue86], p. 639. The canonical bundle formula (V, 12.1) holds for every relatively minimal elliptic surface. So we can apply it to X_q, finding in particular that $K^2_{X_q} = 0$. Since $c_2(X_q) = c_2(X_0) = 12$, the Todd-Hirzebruch formula implies

$$\chi(X_q, \mathcal{O}_{X_q}) = 1. \tag{4}$$

By Cor. V, 12.3, we then have

$$K_{X_q} = -F + F_2 + (q-1)F_q, \tag{5}$$

where F denotes a general fibre, and F_2 and F_q are the fibres of multiplicity 2 and q respectively. In particularly we have $CK_{X_q} \geq 0$ for any irreducible curve C on X_q and so X_q is minimal. From $b_1(X_q) = 0$ it follows that X_q is Kähler and hence $h^{1,0}(X_q) = h^{0,1}(X_q) = 0$. By (4) we find $p_g(X_q) = 0$, and hence $b^+(X_q) = 1$. Furthermore $p_g = h^{0,2}(X_q) = 0$ implies that every element of $H^2(X_q, \mathbb{Z})$ can be represented by a holomorphic line bundle. Since $b^+(X_q) = 1$, there is a holomorphic line bundle $\mathcal{L}$ on X_q with $c_1^2(\mathcal{L}) > 0$, and so X_q is algebraic (Theorem IV. 6.2).

To determine the topological type, we first remark that all surfaces X_q have the same signature, because $b_2(X_q) = e(X_q) - 2 = 10$ and

$$\tau(X_q) = \frac{1}{3}\left(c_1^2(X_q) - 2c_2(X_q)\right) = -8.$$

By Rochlin (Theorem 2.1) this shows that the cup product form is odd. In particular, by Theorem 1.3, all surfaces X_q are homeomorphic. The fact that the cup product is odd can also be deduced from Lemma VIII. 3.1 if one knows that the canonical class, as given by (5), is not 2-divisible. This will follow from the explicit expression for K_{X_q} which we shall now derive. We first introduce some terminology. We shall call an effective divisor D vertical if $DF = DK_X = 0$. Clearly, up to linear equivalence, every vertical divisor is of type $aF + bF_2 + cF_q$, $a \geq 0$, $b = 0$ or 1, $0 \leq c \leq q-1$. Since $(2,q) = 1$, we can find an integral combination $G = \alpha F_2 + \beta F_q$ representing $(1/2q)F$. Its fundamental class $[G]$ is a primitive class in $H^2(X_q, \mathbb{Z})$. Indeed, if for some $\lambda \in \mathbb{Q}$, $0 < \lambda < 1$ the class λG again is represented by a divisor, by the Riemann-Roch formula either λG or $K - \lambda G$ must be effective. The first alternative clearly cannot hold and so $K - \lambda G$ must be effective and hence vertical, say $K - \lambda G = aF + bF_2 + cF_q$, and we would have $a + (b-1)/2 + (c+1-\lambda/2)/q = 0$ and in particular $\lambda \in \mathbb{Z}$, a contradiction. This shows that

$$K_{X_q} = (-2q + q + 2(q-1))G = (q-2)G, \quad [G] \text{ primitive}, \tag{6}$$

and hence is not 2-divisible.

The main result is as follows:

(5.1) Theorem. *The surfaces X_q and X_r (index $q = 0$, $r = 0$ included) are diffeomorphic if and only if $q = r$.*

This theorem will follow from Theorem 4.2 above. So we have to study the stable vector bundles $\mathcal{V}$ on X_q with $c_1(\mathcal{V}) = 0$, $c_2(\mathcal{V}) = 1$ (stability with respect to a suitable ample line bundle). Before we do this we prove the following fact, to be used in the course of the proof.

(5.2) **Proposition.** *$h^0(bF_2 + cF_q) = 1$ for $b = 0$ or 1, $0 \leq c \leq q - 1$.*

Proof. For any fibre F_m of multiplicity m by Lemma III, 8.3 the restriction $\mathcal{O}_X(F_m)|F_m$ is a torsion bundle of order precisely m. This gives the result when $b = 0$. For $b = 1$, you look at the cohomology sequence for the exact sequence

$$0 \to \mathcal{O}_X(cF_q) \to \mathcal{O}(cF_q + F_2) \to \mathcal{O}_{F_2}(F_2) \to 0. \qquad \square$$

If there is no danger of confusion we shall often write X instead of X_q, etc.

After these preparations we can start with the **proof of Theorem 5.1.** We divide it into three parts:

I. Construction of the moduli space.

First we consider the case of X_0. We fix some ample divisor L. Riemann-Roch yields

$$h^0(\mathcal{V}) + h^0(\mathcal{V}^* \otimes \mathcal{O}_X(K_X)) \geq 1 .$$

But $\mathcal{V} \cong \mathcal{V}^* \otimes \Lambda^2\mathcal{V} \cong \mathcal{V}^*$ (on X homological and linear equivalence are the same, so $c_1(\mathcal{V}) = 0$ means $\Lambda^2\mathcal{V} = \mathcal{O}_X$)). So $h^0(\mathcal{V}) + h^0(\mathcal{V} \otimes \mathcal{O}_X(K_X)) \geq 1$. The proper transform of any cubic through $p_1, \ldots, p_9$ yields a divisor in $-K_X$, so we must have $h^0(\mathcal{V}) \geq 1$. But this violates stability of $\mathcal{V}$ by Lemma 4.1. So in this case the moduli space of L-stable 2-bundles with $c_1 = 0$, $c_2 = 1$ is empty for any ample L.

We come to the case of X_q, $q \geq 3$. Let $\mathcal{V}$ be any L-stable 2-bundle on X_q with $c_1(\mathcal{V}) = 0$, $c_1(\mathcal{V}) = 1$, where $L = L_0 + nK_X$, L_0 a fixed suitably ample line bundle and n large (this will only play a role at the very end of the proof). At this point we recall that the sum of an ample and a nef divisor is ample; this is a direct consequence of Prop. IV.7.5.

As before, we conclude from the stability of $\mathcal{V}$ that $h^0(\mathcal{V}) = 0$, so Riemann-Roch gives $h^0(\mathcal{V} \otimes \mathcal{O}_X(K_X)) \geq 1$. If $s \in \Gamma(X, \mathcal{V} \otimes \mathcal{O}_X(K_X))$, then there is a maximal divisor D on which this section vanishes, i.e., s induces in $\mathcal{V} \otimes \mathcal{O}_X(K_X - D))$ a section $\tilde{s}$ with isolated zeroes only. This means that there is a non-trivial homomorphism from $\mathcal{O}_X(D - K_X)$ into $\mathcal{V}$. Using the stability of $\mathcal{V}$ we easily find that $D = cF_q$, $0 \leq c \leq (q-3)/2$.

Given $\mathcal{V}$, the value of c we thus find may depend on s. From here on we assume that c is *as large as possible* for this $\mathcal{V}$.

Since $c_2(\mathcal{V} \otimes \mathcal{O}_X(K_X - D)) = c_2(\mathcal{V}) + c_1(\mathcal{V})(K_X - D) = 1$, the section $\tilde{s}$ vanishes transversally in one point x_0. We want to prove that $x_0 \in F_2$. By construction we have

$$(7) \quad 0 \longrightarrow \mathcal{O}_X \longrightarrow \mathcal{V} \otimes \mathcal{O}_X(K_X - D) \longrightarrow \mathcal{J}_{x_0} \otimes \mathcal{O}_X(2K_X - 2D) \longrightarrow 0 .$$

Let $(\widetilde{X}, E)$ be obtained from (X, x_0) by blowing up x_0, and let $g : \widetilde{X} \to X$ be the projection. On $\widetilde{X}$ the sequence (7) becomes an exact sequence of *bundles*

$$(8) \; 0 \longrightarrow \mathcal{E} \longrightarrow g^*(\mathcal{V} \otimes \mathcal{O}_X(K_X - D)) \longrightarrow \mathcal{E}^* \otimes g^*(\mathcal{O}_X(2K_X - 2D)) \longrightarrow 0 ,$$

with $\mathcal{E} = \mathcal{O}_{\tilde{X}}(E)$. Restriction to E gives an extension of $\mathcal{E}^*$ by $\mathcal{E}$, which is the trivial bundle. This shows that the extension does not split and so the extension-class in $H^1(\mathcal{E}^2 \otimes g^*(\mathcal{O}_X(2D - 2K_X)))$ is non-trivial. Now $\deg(\mathcal{E}|E) = -1$, hence $h^1(E, \mathcal{E}) = h^0(E, \mathcal{E}) = 0$. This we use in the long exact sequence for

$$0 \to g^*(\mathcal{O}_X(2D - 2K_X)) \to \mathcal{E} \otimes g^*(\mathcal{O}_X(2D - 2K_X)) \to \mathcal{O}_E(\mathcal{E}) \to 0$$

to see that $h^1(g^*(\mathcal{O}_X(2D - 2K_X))) = h^1(\mathcal{E} \otimes g^*(\mathcal{O}_X(2D - 2K_X)))$. By Leray $h^1(g^*(\mathcal{O}_X(2D - 2K_X))) = h^1(\mathcal{O}_X(2D - 2K_X))$. We show that the last group vanishes. By Riemann-Roch this will follow if we can show that $h^0(\mathcal{O}_X(2D - 2K_X)) = 0$ and $h^2(\mathcal{O}_X(2D - 2K_X)) = 1$. The first follows since $2D - 2K_X$ is a negative multiple of F_q. For the second assertion we use Serre duality $h^2(2D - 2K_X) = h^0(3K_X - 2D)$ together with

$$3K_X - 2D = F_2 + (q - 3 - 2c)F_q.$$

Then Lemma 5.2 completes the assertion.

We now tensor the preceding sequence with $\mathcal{E}$. By the above argument $h^1(\mathcal{E}^2 \otimes g^*(\mathcal{O}_X(2D - 2K_X))) \neq 0$ and we also have $h^1(E, \mathcal{E}^2) = 1$. We conclude that

$$H^1(\mathcal{E}^2 \otimes g^*(\mathcal{O}_X(2D - 2K_X))) \longrightarrow H^1(\mathcal{E}^2|E) \cong \mathbb{C}$$

is an isomorphism and so

$$h^1(\mathcal{E}^2 \otimes g^*(\mathcal{O}_X(2D - 2K_X))) = 1. \tag{9}$$

Consequently, the next map

$$H^2(\mathcal{E} \otimes g^*(\mathcal{O}_X(2D - 2K_X))) \longrightarrow H^2(\mathcal{E}^2 \otimes g^*(\mathcal{O}_X(2D - 2K_X)))$$

is injective. Since $\mathcal{K}_{\tilde{X}} = g^*(\mathcal{O}_X(K_X)) \otimes \mathcal{E}$, we have that

$$H^0(\mathcal{E}^* \otimes g^*(\mathcal{O}_X(3K_X - 2D))) \longrightarrow H^0(g^*(\mathcal{O}_X(3K_X - 2D)))$$

is surjective by Serre duality. This means that every section of $\mathcal{O}_X(3K_X - 2D)$ vanishes at x_0. But we have seen that $3K_X - 2D = F_2 + (q - 3 - 2c)F_q$ and it follows that in any case $x_0 \in F_2 \cup F_q$.

If $2c = q - 3$, then $x_0 \in F_2$. To exclude the possibility that $x_0 \in F_q$ for $0 \le c < (q - 3)/2$, we proceed as follows.

The sequence (7) yields

$$0 \to \mathcal{O}_X(-F_q) \longrightarrow \mathcal{V} \otimes \mathcal{O}_X(K_X - D - F_q) \longrightarrow \mathcal{J}_{x_0} \otimes \mathcal{O}_X(2K_X - 2D - F_q) \to 0\,.$$

Now $2K_X - 2D - F_q = (q - 3 - 2c)F_q$, so $h^0(\mathcal{J}_{x_0}(2K_X - 2D - F_q)) = 1$ if $x_0 \in F_q$ and $c \neq (q - 3)/2$.

Since $h^0(-F_q) = 0$ and $h^2(-F_q) = h^0(K_X + F_q) = h^0(F_2) = 1$, we conclude from Riemann-Roch that $h^1(-F_q) = 0$. Hence the preceding sequence then shows that $h^0(\mathcal{V} \otimes \mathcal{O}_X(K_X - D - F_q)) = 1$, i.e., there is a non-trivial homomorphism from $\mathcal{O}_X(D+F_q)$ to $\mathcal{V}\otimes\mathcal{O}_X(K_X)$, which would violate the maximality of D. So, indeed, $x_0\in F_2$.

We next show that x_0 is uniquely determined. Since $2K_X-2D=(q-c-2)F_q$, by Lemma 5.2, it has a unique section vanishing along F_q and since $x_0\in F_2$ it follows that $h^0(\mathcal{J}_{x_0}(2K_X-2D))=0$. Hence, by (7), $H^0(\mathcal{V}\otimes\mathcal{O}_X(K_X-D))$ is 1-dimensional and indeed x_0 is uniquely determined.

So far we have constructed a map

$$\alpha:\left\{\begin{array}{l}L\text{-stable 2-bundles } \mathcal{V} \text{ on } X_q \text{ with}\\ c_1(\mathcal{V})=0, c_2(\mathcal{V})=1\end{array}\right\}\longrightarrow\coprod_{0\le c\le\frac{q-3}{2}}F_2$$

by attaching to any $\mathcal{V}$ and every integer c with $0\le c\le\frac{q-3}{2}$ a point $x_0\in F_2$.

The map α is injective. In fact, let $\alpha(\mathcal{V})=\alpha(\mathcal{W})$. Then on $\widetilde{X}$ we obtain two extensions of the same two line bundles, namely $\mathcal{E}$ and $\mathcal{E}^*\otimes g^*(\mathcal{O}_X(2K_X-2D))$. The restriction of both to E is trivial, so to show that $\mathcal{V}$ and $\mathcal{W}$ are isomorphic, it is sufficient to show that $H^1(\widetilde{X},\mathcal{E}^2\otimes g^*(\mathcal{O}_X(2D-2K_X)))=1$. But we have seen this already ((9)). Therefore $g^*(\mathcal{V})$ and $g^*(\mathcal{W})$ are isomorphic and hence $\mathcal{V}\cong\mathcal{W}$.

The surjectivity of α can be proved by reversing the preceding procedure, with very much the same kind of arguments.

So, at least set theoretically, the moduli space of L-stable 2-bundles $\mathcal{V}$ on X_q with $c_1(\mathcal{V})=0$, $c_2(\mathcal{V})=1$ consists of $(q-1)/2$ copies of F_2.

As we observed before, the moduli space of such bundles $\mathcal{V}$ exists, even as a fine moduli space. The local structure of the moduli space is governed by the associated bundle $\mathcal{E}nd(\mathcal{V})$. Indeed, the Zariski tangent space at the point given by $\mathcal{V}$ is $H^1(\mathcal{E}nd(\mathcal{V}))$ and the moduli space is reduced and smooth at this point as soon as $h^2(\mathcal{E}nd(\mathcal{V}))=0$. In our case this can be proved in the following way. Since $\mathcal{E}nd(\mathcal{V})=\mathcal{H}om(\mathcal{V},\mathcal{V})=\mathcal{V}\otimes\mathcal{V}^*=\mathcal{V}\otimes\mathcal{V}$, we have to show that $h^2(\mathcal{V}\otimes\mathcal{V})=h^0(\mathcal{V}\otimes\mathcal{V}\otimes\mathcal{O}_{X_q}(K_{X_q}))=0$. To that purpose we consider the sequence (7) twisted by $V\otimes\mathcal{O}_{X_q}(D)$:

$$0\longrightarrow\mathcal{V}\otimes\mathcal{O}_{X_q}(D)\longrightarrow\mathcal{V}\otimes\mathcal{V}\otimes\mathcal{O}_{X_q}(K)\longrightarrow\mathcal{J}_{x_0}\otimes\mathcal{V}\otimes\mathcal{O}_{X_q}(2K_{X_q}-D))\longrightarrow 0\,.$$

This sequence shows that it is sufficient to prove that $h^0(\mathcal{V}\otimes\mathcal{O}_X(D))=h^0(\mathcal{J}_{x_0}\otimes\mathcal{V}\otimes\mathcal{O}_{X_q}(2K_{X_q}-D))=0$. The vanishing of $h^0(\mathcal{V}\otimes\mathcal{O}_X(D))$ follows from the sequence

$$0\longrightarrow\mathcal{O}_{X_q}(2D-K_{X_q})\longrightarrow\mathcal{V}\otimes\mathcal{O}_{X_q}(D)\longrightarrow\mathcal{J}_{x_0}(K)\longrightarrow 0$$

since we proved above that $h^0(\mathcal{O}_{X_q}(2D-K_{X_q}))=0$. As to $h^0(\mathcal{J}_{x_0}\otimes\mathcal{V}\otimes\mathcal{O}_{X_q}(2K_{X_q}-D))$, the restriction of $\mathcal{V}\otimes\mathcal{O}_{X_q}(K_X-D)$ to F_2 has one simple zero at x_0, so this restriction is isomorphic to $\mathcal{O}_{F_2}(x_0)\oplus\mathcal{O}_{F_2}(-x_0)$. Now $K_{X_q}|F_2=F_2|F_2$, and we have used several times now that this a non-trivial line bundle of order 2. Hence $\mathcal{V}\otimes\mathcal{O}_{F_2}(2K_{X_q}-D)\cong\mathcal{O}_{F_2}(x_1)\oplus\mathcal{O}_{F_2}(-x_1)$, with $x_1\neq x_0$. If the restriction $H^0(\mathcal{V}\otimes\mathcal{O}(2K_{X_q}-D))\longrightarrow H^0(\mathcal{V}\otimes\mathcal{O}_{F_2}(2K_{X_q}-D))$ is injective, then all non-zero sections of $\mathcal{V}\otimes\mathcal{O}_{F_2}(2K_{X_q}-D)$ have zeros at

x_1 and none at x_0. Hence $H^0(\mathcal{J}_{x_0} \otimes \mathcal{V} \otimes \mathcal{O}_{X_q}(2K_{X_q} - D)) = 0$. So we shall be done as soon as we have proved that $H^0(\mathcal{V} \otimes \mathcal{O}_{X_q}(2K_{X_q} - D - F_2)) = 0$. But this follows from (7) in yet another disguise:

$$0 \longrightarrow \mathcal{O}_{X_q}(K_{X_q} - F_2) \longrightarrow \mathcal{V} \otimes \mathcal{O}_{X_q}(2K_{X_q} - D - F_2) \longrightarrow \mathcal{J}_{x_0}(2K_{X_q} - 2D - F_2) \longrightarrow 0$$

since $h^0(\mathcal{O}_{X_q}(K_{X_q} - F_2)$ obviously vanishes, and $3K_{X_q} - 2D - F_2 = (q - 3 - 2c)F_2$, so $h^0(\mathcal{J}_{x_0}(3K_{X_q} - 2D - F_2))$ vanishes too.

At this stage it is very plausible that $\mathcal{M}$ is nothing but the disjoint union of $(q-1)/2$ copies of F_2. This is actually the case. For details see [OV86], p. 368.

As we have seen before, $\mathcal{M} = \mathcal{M}^L_{X_q}$ is even a fine moduli space, i.e., on $\mathcal{M} \times X$ there is a universal bundle. If the restriction of this universal bundle to the c-th copy of F_2, $0 \le c \le (q-3)/2$, is denoted by $\mathcal{U}^c$, then on $F_2 \times X$ the universal extension is part of a sequence

$$(10)\quad 0 \to \pi_2^*(\mathcal{O}_{X_q}(D - K_{X_q})) \to \mathcal{U}^c \to \mathcal{J}_\Gamma \otimes \pi_2^*(\mathcal{O}_{X_q}(K_{X_q} - D) \otimes \pi_1^*(\mathcal{L}^{(c)})) \to 0,$$

where $\pi_1 : F_2 \times X \to F_2$, $\pi_2 : F_2 \times X \longrightarrow X$ are the projections, Γ the graph of the embedding of F_2 in X, and $\mathcal{L}^{(c)}$ a line bundle on F_2. The restriction of this sequence to $\{p\} \times X$ yields the bundle $\mathcal{V} = \mathcal{U}^{(c)}|\{p\} \times X$ which is represented by p.

II. Calculation of the Donaldson element.

In the case of X_0, where as we proved, $\mathcal{M}$ is empty, we simply find $\rho(L) = -c_1(X_0) = -3g + \sum_{i=1}^9 e_i$, where g is the inverse image of the class of a line, and $e_1, \ldots, e_9$ the blow-ups of $p_1, \ldots, p_9$. Also, $L + n_0 K_{X_0}$ is not contained in any wall for n large enough. Intersection with any e_i shows that $\rho(L)$ is a primitive class.

We turn to the case $q \ge 3$. Our first task is the calculation of $c_2(\mathcal{U})$. From (10) we infer that $c_1(\mathcal{U}^c) = \pi_1^*(c_1(\mathcal{L}^{(c)})$. Writing (10) as

$$0 \to \mathcal{O}_{F_2 \times X_q} \to \mathcal{U}^c \otimes \pi_2^*(K_{X_q} - D) \to \mathcal{J}_\Gamma \otimes \mathcal{O}_{X_q}(2K_X - 2cF_q) \otimes \pi_1^*(\mathcal{L}^{(c)})) \to 0$$

we see that $\mathcal{U}^c \otimes \pi_2^*(\mathcal{O}_{X_q}(K_{X_q} - D))$ has a section vanishing transversally on Γ (it is transversal, since its restriction to any fibre $p \times X$ is transversal). By Prop. I, 5.2 we have

$$\begin{aligned}[\Gamma] &= c_2\Big(\mathcal{U}^c \otimes \pi_2^*(\mathcal{O}_{X_q}(K_X - cF_q))\Big) \\ &= c_2(\mathcal{U}^c) + c_1(\pi_1^*(\mathcal{L}^{(c)}))c_1(\pi_2^*(\mathcal{O}_{X_q}(K_{X_q} - D)))\end{aligned}$$

and

$$\begin{aligned}\pi_{2*}\left(c_2(\mathcal{U}^c)\right) &= \pi_{2*}([\Gamma]) - \deg\left(\mathcal{L}^{(c)}\right)(K_{X_q} - D) \\ &= F_2 - \deg\left(\mathcal{L}^{(c)}\right)(K_{X_q} - D)\ .\end{aligned}$$

We claim that $\deg(\mathcal{L}^{(c)}) = 0$. In fact,

$$\begin{aligned}\deg(\pi_1^*(\mathcal{L}^{(c)})|\Gamma) &= \deg(c_1(\mathcal{U}^c)|\Gamma)\\ &= \deg c_1\big((\mathcal{N}_{\Gamma|F_2\times X})\otimes(\text{line bundle of degree } 0)\big)\end{aligned}$$

(because $\mathcal{U}^c \otimes \pi_2^*(\mathcal{O}_{X_q}(K_{X_q}-D))|\Gamma = \mathcal{N}_{\Gamma|F_2\times X}$)

$$= \deg c_1(\mathcal{T}_{F_2\times X}|\Gamma = \deg \mathcal{T}_X|F_2 = 0\ .$$

(Note that $\mathcal{L}$ need not be trivial !).

Summing up for $c = 0, 1, \ldots, (q-1)/3$ we find $\pi_{2*}(c_2(\mathcal{U})) = (q-1)/2\cdot F_2$. So by property (v) of the Donaldson invariant, using the formula (6), we finally get

(11) $$\rho(L) = K_{X_q} + (q-1)F_2 = (q^2-2)G.$$

III. The final step.

Firstly, we show that X_0 is not diffeomorphic to any X_q, $q \geq 3$. Suppose we have an (orientation-preserving) diffeomorphism $f : X_0 \to X_q$. Let L_0 and L_q be ample classes on X_0 and X_q respectively. In general, $f^*(L_q)$ and L_0 will be in different chambers of $\mathcal{C}_{X_0}$, and not much can be concluded. However, since the cupproduct form has rank 10 and is indefinite, we can apply a theorem of Wall ([Wall], Cor. to Theorem 2) saying that given two chambers in $\mathcal{C}_{X_0}$, there is an orientation-preserving diffeomorphism from X_0 onto itself, carrying the first chamber into the other or its opposite. Indeed, a chamber together with its opposite gives a fundamental domain for the group generated by the reflections in the walls (compare Prop. VIII. 3.7) and Wall's result states that any reflection can be realized by a diffeomorphism. So we may assume that L_0 and $f^*(L_q)$ are in the same or in opposite chambers. This means that $\rho_{X_0}(L) = \pm f^*(\rho_{X_q}(L_q))$. But, as we have seen in part II, $\rho_{X_0}(L)$ is primitive, whereas $f^*(\rho_{X_q}(L_q))$ is divisible by (q^2-2). So f cannot exist.

Now we compare X_q and X_r, $q, r > 0$. On X_q, we take $L = L_0 + nK_{X_q}$ where L_0 is a fixed ample class, and $n > 0$ (since K_{X_q} is nef, it follows from IV, 7.5 that $L_0 + nK_{X_q}$ is ample).

(5.3) **Proposition.** *The chamber containing $L_0 + nK_{X_q}$ is independent of n, provided n is large enough.*

Proof. It is easily verified that $L_0 + nK_{X_q}$ is not contained in any wall for n large enough. The divisors $L_0 + nK_{X_q}$ and $L_0 + mK_{X_q}$ are clearly in the same component of $\mathcal{C}_{X_q}$, so we can apply Theorem 4.2,(iii):

$$\rho(L_0 + nK_{X_q}) = \rho(L_0 + mK_{X_q}) + 2\sum A,$$

with $A^2 = -1$, $A(L_0 + nK_{X_q}) < 0$, $A(L_1 + mK_{X_q}) > 0$. From $\rho(L_0 + nK_{X_q}) = \rho(L_0 + mK_{X_q})$ we find $\sum A = 0$, but this is incompatible with $A(L_0 + nK_{X_q}) < 0$ unless there are no A at all, that is, and $L_0 + mK_{X_q}$ are in the same chamber (two points are in the same chamber if and only if they are on the same side of any wall). $\square$

(5.4) **Proposition.** *If $f : X_q \to X_r$ is a diffeomorphism, then for n and m large $L_0 + nK_{X_q}$ and $f^*(L_1 + mK_{X_r})$ are in the same chamber.*

Proof. First suppose that $L_0 + nK_{X_q}$ and $f^*(L_1 + mK_{X_r})$ are in the same component of $\mathcal{C}_{X_q}$. Then Theorem 4.2, (iii) gives

$$\rho(f^*(L_1 + mK_{X_r})) = \rho(L_0 + nK_{X_q}) + 2\sum A\ ,$$

with $A^2 = -1$, $A(L_0 + nK_{X_q}) < 0$, $A(f^*(L_1 + mK_{X_r})) > 0$. Substituting the values of ρ we have found at the end of the last section, this implies

$$\frac{2r}{r-2} f^*(K_{X_r}) = \frac{2q}{q-2} K_{X_q} + 2\sum A\ .$$

Intersecting with K_{X_q} we find

$$\frac{2r}{r-2}(f^*K_{X_r}, K_{X_q}) = 2\sum AK_{X_q}\ .$$

From $A(L_0 + nK_{X_q}) < 0$ it follows that, for n large, $AK_{X_q} < 0$ for all A under consideration. So $f^*(K_{X_r})K_{X_q} \leq 0$.

On the other hand, by assumption $L_0 + nK_{X_q}$ and $f^*(L_1 + mK_{X_r})$ are in the same component of $\mathcal{C}_{X_q}$. So $K_{X_q} + (1/n)L_0$ and $f^*(K_{X_r}) + (1/m)f^*(L_1)$ are in the same component of $\mathcal{C}_{X_q}$, and therefore, if we take n and m large, $f^*(K_{X_r})$ and K_{X_q} are in the same component of $\mathcal{C}_{X_q}$. This implies $(K_{X_q}, f^*K_{X_r}) \geq 0$ (compare Cor. IV, 7.2). Hence $(K_{X_q}, f^*K_{X_r}) = 0$ and there cannot be any A, for $AK_{X_q} \equiv A^2 (\mathrm{mod}\, 2)$.

If $mK_{X_q} + L_0$ and $f^*(mK_{X_r} + L_1)$ are in different components of $\mathcal{C}_{X_q}$, then we can apply property (ii) and repeat the argument with $-(nK_{X_q} + L_0)$ instead of $nK_{X_q} + L_0$. □

(5.5) **Theorem.** *The surfaces X_q and X_r are diffeomorphic if and only if $q = r$.*

Proof. By (11) we know that the divisibility of $\rho(L)$ is exactly $(q^2 - 2)$. If X_q and X_r are diffeomorphic, then these should be equal and thus $q = r$. □

Remark. Donaldson proved ([Don86]) that X_0 and X_3 are not diffeomorphic. Later Friedman and Morgan (with a different, though related method, see [F-M88a]) and independently Okonek and Van de Ven ([O-V86]) proved the general result. Our proof here is similar to that of the last two authors.

6. Further Results obtained by the Donaldson Method

Using variations and refinements of the Donaldson invariant, many other results about the differentiable structure of algebraic surfaces have been obtained. We list some of them, but for all details as well as a systematic treatment of the subject, we must refer to the original papers, to Friedman and Morgan's book [F-M94], and to surveys like [O-V90].

1. Friedman and Morgan have extended the result above to the case of blown-up surfaces X_q: if X_q, blown up in k points, is diffeomorphic to X_r blown up in k points, then $q = r$. The proof requires more than the calculation of Donaldson invariants.
2. Kotschik ([Kot]) and Okonek and Van de Ven ([O-V89]) independently proved: the Barlow surface (see VII, Sect. 10) blown up in one point is (homeomorphic but) not diffeomorphic to X_0. It trivially follows that the Barlow surface itself is not diffeomorphic to $\mathbb{P}_2$ blown up in eight points. The Barlow surface blown up in one point is also not diffeomorphic to any other surface X_q ([O-V89]).
3. Both Moishezon ([F-M-M]) and Salvetti ([Sal89]) constructed examples of surfaces of general type which are homeomorphic but not diffeomorphic.
4. Very important is the result of Friedman and Morgan ([FM90]) that on a compact, oriented differentiable 4-manifold there is only a finite number of algebraic structures up to deformation.
5. Donaldson ([Don90]) proved that if an algebraic surface is diffeomorphic to a connected sum $M_1\#M_2$ with $b_2(M_1) > 0$ and $b_2(M_2) > 0$ then either the cup form of M_1 or that of M_2 is $\mathbb{Z}$-equivalent to $\langle -\mathbb{1}\rangle \oplus \cdots \oplus \langle -\mathbb{1}\rangle$. Combining this result with the fact that every complex surface with even first Betti number is the deformation of an algebraic surface, we obtain as a corollary: the oriented differentiable 4-fold $\mathbb{P}_2^{(1)}\#\mathbb{P}_2^{(2)}\#\cdots\#\mathbb{P}_2^{(k)}\#\overline{\mathbb{P}}_2^{(1)}\#\cdots\#\overline{\mathbb{P}}_2^{(\ell)}$ carries a complex structure if and only if $k = 1$ ($\overline{\mathbb{P}}_2$ denotes $\mathbb{P}_2$ with orientation reversed).

Results like 1, 2 and 4 strongly point into the direction of the Van de Ven conjecture: two surfaces of different Kodaira dimension are never diffeomorphic. In other words in complex dimension 2 the Kodaira dimension is a differentiable invariant. This conjecture has now been proved and we shall deal with it in the next subchapter.

Seiberg-Witten Invariants

7. Introduction

In 1995, E. Witten came up with a new type of differentiable invariants for 4-manifolds. Like the Donaldson invariants, to which they are closely related, these new invariants are particularly important for algebraic surfaces. In this case too we only want to show, by way of an example, how these new invariants, together with the theory as it has been developed in the preceding chapters of this book, can be used to obtain results about the differentiable structure of surfaces.

We start with a very short description of the Seiberg-Witten invariants; for details we have to refer to [Wit], [Mor], [F-M97], [O-T].

Since Spin^c-structures play a major role in the theory, let us begin by recalling the definition of the relevant groups. We identify $\mathbb{R}^4$ with the quaternions $\mathbb{H} = \{x_0 + \mathrm{i}x_1 + \mathrm{j}x_2 + \mathrm{k}x_3 \mid x_j \in \mathbb{R}\}$. The unit sphere in $\mathbb{H}$ (with quaternion-multiplication) is isomorphic to SU(2) by way of the map

$$(x_0, x_1, x_2, x_3) \mapsto \begin{pmatrix} x_0 + \mathrm{i}x_1 & x_2 + \mathrm{i}x_3 \\ -x_2 + \mathrm{i}x_3 & x_0 - \mathrm{i}x_1 \end{pmatrix}.$$

Multiplication of quaternions defines the homomorphism

$$\begin{aligned} \psi : \mathrm{SU}(2) \times \mathrm{SU}(2) &\to \mathrm{SO}(4) \subset \mathrm{Aut}(\mathbb{H}) \\ (A, B) &\mapsto \{X \mapsto AX\bar{B}\}, \end{aligned}$$

a $2:1$-cover with kernel $\mathbb{Z}/2 = \pm\mathbb{1}$. This shows that $\mathrm{SU}(2)\times\mathrm{SU}(2) = \mathrm{Spin}(4)$, the **spinor group**, by definition the universal cover of SO(4). The **complex spinor group**

$$\mathrm{Spin}^c(4) = \mathrm{Spin}(4) \times_{\mathbb{Z}/2} S^1$$

thus fits into an exact sequence

$$(12) \qquad 1 \to \mathrm{U}(1) \to \mathrm{Spin}^c(4) \xrightarrow{r} \mathrm{SO}(4) \to 1.$$

Let M be a compact, connected, oriented differentiable manifold of dimension 4, which is provided with a Riemannian metric g. We shall assume that $\pi_1(M) = 1$. This is by no means necessary for the general theory, but it will simplify certain things and the assumption holds in the case we shall treat in detail. If there exist elements in $H^2(M, \mathbb{Z})$ whose reduction mod 2 is the Stiefel Whitney class $w_2(M)$, then the structure group SO(4) of the tangent bundle can be lifted to $\mathrm{Spin}^c(4)$ by way of the homomorphism r of the exact sequence (12). Such a homomorphism is called a **Spin^c-structure**. In general there is more than one lifting, and we fix one. From the previous description of the complex spinor group, we see that $\mathrm{Spin}^c(4)$ is isomorphic to the subgroup of $\mathrm{U}(2) \times \mathrm{U}(2)$ consisting of all pairs (u_1, u_2) with $\det(u_1) \cong \det(u_2)$. The projections of $\mathrm{U}(2) \times \mathrm{U}(2)$ onto the first and second factor yield two U(2)-bundles on M, the bundles $\Sigma_c^{\pm}$ of $+$-spinors and $-$-spinors respectively. The **spinor bundle** Σ_c is their direct sum. These $\pm$-spinor bundles have the same determinant bundle $\mathcal{L}_c$ with $c_1(\mathcal{L}_c) = c \in H^2(M, \mathbb{Z})$, $c \equiv w_2(M) \bmod 2$. In this way a $1{-}1$ correspondence

$$\{\mathrm{Spin}^c(4)\text{-structures}\} \longleftrightarrow \left\{ \begin{array}{l} \text{elements } \ c \in H^2(M, \mathbb{Z}) \\ \text{with } \ c \equiv w_2(M) \bmod 2 \end{array} \right\}$$

is obtained. (If $\pi_1(M) \neq 1$, this correspondence is in general not one-to-one.)

The given Spin^c-structure is known to induce an isomorphism

$$\gamma_+ : \mathcal{T}_M^\vee \otimes \mathbb{C} \xrightarrow{\sim} \mathrm{Hom}_{\mathbb{C}}(\Sigma_c^+, \Sigma_c^-)$$

which is used to define the following two homomorphisms

$$\begin{aligned} \gamma : \mathcal{T}_M^\vee &\to \mathrm{End}(\Sigma_c) \\ \xi &\mapsto \begin{pmatrix} 0 & -\gamma_+(\xi)^* \\ \gamma_+(\gamma) & 0 \end{pmatrix} \end{aligned}$$

and

$$\begin{aligned} \Gamma : \Lambda^2\mathcal{T}_M^\vee &\to \mathrm{End}(\Sigma_c) \\ \xi \wedge \eta &\mapsto \frac{1}{2}[\gamma(\xi), \gamma(\eta)]. \end{aligned}$$

The choice of a unitary connection a in $\mathcal{L}_c$ (considered as a unitary bundle) induces a connection $\mathcal{D}^0(\Sigma_c) \xrightarrow{d_a} \mathcal{D}^1(\Sigma_c)$ on the spinor bundle. Clearly, γ can also be seen as a homomorphism $\mathcal{T}_M^\vee \otimes \Sigma_c \to \Sigma_c$ and therefore induces $\mathcal{D}^1(\Sigma_c) \xrightarrow{\gamma^*} \mathcal{D}^0(\Sigma_c)$. The associated Dirac-operator is then defined as the composition

$$\not{D}_a : \mathcal{D}^0(\Sigma_c) \xrightarrow{d_a} \mathcal{D}^1(\Sigma_c) \xrightarrow{\gamma^*} \mathcal{D}^0(\Sigma_c).$$

Consider now a pair (a, ψ), where as before a is a unitary connection in $\mathcal{L}_c$, and ψ a section in Σ_c^+. Recall that given any hermitian vector bundle (E, h), we can identify the dual of E with its complex conjugate and so for any smooth section s of E we may interpret $s \otimes \bar{s}$ as an endomorphism of E. Applying this remark to the section ψ of Σ_c^+, we obtain an endomorphism of the latter bundle. With this convention we have the monopole equations for (a, ψ)

$$\begin{cases} \not{D}_a \psi = 0 \\ \Gamma(F_a^+) = 2(\psi \otimes \bar{\psi})_0. \end{cases}$$

Here the subscript 0 means that we take the traceless part of the endomorphism in question and F_a^+ is the self-dual part on the curvature F_a, considered as a section of $\Lambda^2\mathcal{T}_M^\vee$. We want to emphasize that to understand the meaning of these equations (physical) gauge theory is needed.

On the space of all pairs (a, ψ) the gauge group of unitary transformations of $\mathcal{L}_c$ operates in a natural way, leaving the set of solutions of the monopole equations invariant. The quotient is the Seiberg-Witten space $SW^g(c)$ for the metric g and the class c.

8. Properties of the Invariants

We list some properties of Seiberg-Witten spaces which we shall need.

(SW 1) For a general perturbation

$$\begin{cases} \not{D}_a\psi = 0 \\ \Gamma(F_a^+ + i\mu) = 2(\psi \otimes \bar{\psi})_0 \end{cases}$$

of the monopole equations the space of solutions modulo the gauge group, $SW^{g,\mu}(c)$, is compact.

This is already an enormous advantage compared to the Donaldson invariants, where compactification of the moduli spaces is often one of the major problems. Another complication in the Donaldson case is the existence of reducible connections. This difficulty appears here too, if pairs (a, ψ) with a non-trivial stabiliser (in the gauge group) occur. It turns out that this is equivalent to a being self-dual and $\psi = 0$. Following Okonek and Teleman we shall say that a metric is **good** for c if there are no such pairs in the space of solutions.

(SW 2) If g is good for c, then for general enough μ, $S^{g,\mu}(c)$ is smooth of dimension

$$\frac{1}{4}(c^2 - 2e(M) - 3\tau(M)) \quad \text{(the expected dimension).}$$

If μ is small enough, then any two of these moduli spaces are cobordant as non-oriented manifolds. They are cobordant to $S^g(c)$ itself if this space is compact, smooth and of the expected dimension. If the expected dimension is strictly negative, then the statement above has to be interpreted as to say that $S^{g,\mu}(c) = \varnothing$.

By far the most important case for the applications is the case $c^2 - 2e(M) - 3\tau(M)) = 0$, so $SW^{g,\mu}(c)$ consists of a finite number of points, counted with multiplicity. In the sequel we shall mainly restrict ourselves to this case, and we define the **Seiberg-Witten number** $W(g, c)$ as the number of these points mod 2 (remember that we are dealing with non-oriented cobordisms).

At this point, we are faced with two problems:
(1) How to calculate the numbers $W(g, c)$?
(2) How do these numbers depend on g (for given c)?

As to (1), effective calculations are only available if our manifold M is an algebraic surface X. Let g be a Hodge metric on X. The self dual class $[\omega_g]$ (see (2)) can be chosen as the corresponding Kähler class or its negative. Let c be an algebraic element of $H^2(X, \mathbb{Z})$ with $c \equiv w_2(X) \bmod 2$ and $c[\omega_g] < 0$. Then $S^g(c)$ can be smoothly identified with a complex projective space, namely $\mathbb{P}(\Gamma(X, \mathcal{M}))$, where $\mathcal{M}$ is a holomorphic line bundle on X which is described in the following way. Since X is simply-connected, there is one divisor class C with $c_1(\mathcal{O}_X(C)) = c$. Since C and K_X have the same reduction mod 2, $K_X + C$ is divisible by 2, and since $\pi_1(X) = 1$, there is exactly one divisor class M with $2M = K_X + C$. Then $\mathcal{M} = \mathcal{O}_X(M)$. Furthermore, if $\mathbb{P}(P(X, \mathcal{M}))$ consists of effective divisors $D \neq 0$, then $S^g(c)$ will be smooth and of the expected dimension if for all those divisors D we have $h^1(D, \mathcal{O}_X(D)) = 0$.

We formulate all this separately for the case of Seiberg-Witten numbers.

(SW 3) Let X be a simply-connected algebraic surface, g a Hodge metric on X and $c \in H^2(X, \mathbb{Z})$ an algebraic class with $c = w_2(X) \bmod 2$ and $c[\omega_g] < 0$. Assume that the expected dimension of $S^g(c)$ is zero. Then $W(g, c) \equiv 1 \bmod 2$ if there is an effective divisor $M \neq 0$, with $2M = K_X + C$ (where $\mathcal{O}_X(C) = c$), such that $h^1(M, \mathcal{O}_X(M)) = 0$, whereas $W(g, c) = 0$ if $\Gamma(X, \mathcal{M}) = 0$.

We come to point (2), the dependence of the Seiberg-Witten numbers on the metric. If $b^+(M) \geq 2$, then there is no problem. In this case every two good metrics can be joined by a smooth family of such metrics, and so $W(c) = W(g, c)$ is well defined independent of the metric. But if $b^+(M) = 1$, we again have to take into account that the value of $W(g, c)$ may depend on a chamber structure, very much like in the Donaldson case.

Let $\mathcal{C}_M = \{x \in H^2(M, \mathbb{R}) \mid x^2 > 0\}$ be the positive cone. Here M is a differentiable fourfold (as always compact, connected and simply-connected) with $b^+(M) = 1$. The cone $\mathcal{C}_M$ consists of two connected components. By choosing $p \in \overline{\mathcal{C}}_M$, $p \neq 0$ we can label them $\mathcal{C}^+$ and $\mathcal{C}^-$ by requiring $xp > 0$ and $xp < 0$ respectively. Let g be a Riemannian metric on M. This time we take for $[\omega_g]$ (see (2)) a generator of the 1-dimensional line of self-dual harmonic forms, such that $[\omega_g] \in \mathcal{C}^+$. For $c \in H^2(M, \mathbb{R})$, $c \equiv w_2(M) \bmod 2$ and $c^2 = 2e(M) + 3\tau(M)$ we consider the wall $xc = 0$. It divides $\mathcal{C}^+$ into at most two chambers, $\mathcal{C}^+ \cap \{xc > 0\}$ and $\mathcal{C}^+ \cap \{xc < 0\}$, and g is called c-**good** if $[\omega_g]$ is contained in either of them (and not in the wall).

(SW 4) Let the situation be as above. If g_1 and g_2 are two c-good metrics, such that $[\omega_{g_1}]$ and $[\omega_{g_2}]$ are in the same chamber with respect to c, then $W(g_1, c) = W(g_2, c)$.

9. Surfaces Diffeomorphic to a Rational Surface

We shall use the following auxiliary result.

(9.1) **Theorem.** *On every rational surface X there exists a Kähler metric g with the property that for all $c \in H^2(X, \mathbb{Z})$ with*
 i) $c \equiv w_2(X) \,(\mathrm{mod} 2)$, $c^2 = 3\tau(X) + 2e(X)$
 ii) *g good for c,*
we have $W(g, c) = 0$.

Proof. A theorem of N. Hitchin [Hit] says that on every rational surface there exists a Kähler metric of positive total curvature, i.e., a Kähler metric g with $k_X[\omega_g] < 0$, where $[\omega_g]$ is the Kähler class and $k_X = c_1(\mathcal{O}_X(K_X))$. We claim that this g meets the requirements of the theorem. In fact, this can easily be derived from (SW 3).

Since g is good for c, we have $c[\omega_g] \neq 0$. We distinguish between the cases $c[\omega_g] < 0$ and $c[\omega_g] > 0$. Our claim will follow from the last statement of (SW 3) as soon as we prove that $\Gamma(X, \mathcal{M}) = 0$. If not so, then $m[\omega_g] \geq 0$ ($m = c_1(\mathcal{M})$), and we obtain the contradiction

$$0 < (2m - k_X)[\omega_g] = c[\omega_g] < 0.$$

It follows that $\Gamma(X, \mathcal{M})$ vanishes indeed in this case. If $c[\omega_g] > 0$ one only has to replace c by $-c$. □

We are now all set for the application.

(9.2) **Theorem.** *If an algebraic surface X is diffeomorphic (as an oriented manifold) to a rational surface, then X is rational itself.*

Proof. We start by making some simple observations. Let X be diffeomorphic to a rational surface Y. Then X is simply-connected since all rational surfaces are simply-connected. Also $b^+(X) = 1$, since $b^+ = 1$ for every rational surface.

Since non-rational surfaces of Kodaira dimension $-\infty$ are not simply-connected (Theorem VI.1.1), it will be sufficient to derive a contradiction from the fact that $\text{kod}(X) \geq 0$. We may even assume that $\text{kod}(X) \geq 1$, for otherwise, by the same theorem, X would be either a blown-up torus, a blown-up bi-elliptic surface, a blown-up Enriques surface or a blown-up K 3-surface. But the first three are not simply-connected (see the explanation in Sect. VI.1) whereas $b^+ = 3$ for a blown-up K 3-surface (see Table 10 in Sect. VI.1). From $\text{kod}(X) \geq 1$ it follows that if X_{min} is the minimal model of X, then $K_{X_{\text{min}}}$ represents a non-zero element of $H^2(X, \mathbb{R})$ and $K^2_{X_{\text{min}}} \geq 0$.

We may assume that X is obtained from X_{min} by blowing up in k different points, and we may also assume (to obtain our contradiction) that $k \geq 2$. Let $f : X \to X_{\text{min}}$ be the projection and let $E_1, \ldots, E_k$ be the exceptional curves. We set $E_I = \bigcup_{i=1}^{\ell} E_i$, $E_{II} = \bigcup_{j=\ell+1}^{k} E_j$, with $1 \leq \ell \leq k-1$. The roles to be played by E_I and E_{II} are completely symmetric.

We fix a very ample divisor class H on X_{min}. By Nakai's criterion Cor. IV.6.4 we know that $nf^*(H) - \sum_{i=1}^{k} E_i$ is very ample for $n \geq n_0$. Let g_1 be the corresponding Hodge metric.

Next we set

$$C_I = -K_X + 2E_I$$
$$C_{II} = -K_X + 2E_{II}.$$

Then C_I and C_{II} have the following properties.

(i) $C_I \equiv C_{II} \equiv w_2(X) \pmod 2$.

(ii) $[\omega_{g_1}]C_I < 0$ and $[\omega_{g_1}]C_{II} < 0$ for $n \geq n_1$.

Indeed, denoting the cohomology classes of the different divisors by the corresponding small letter, we have

$$\begin{aligned}[\omega_{g_1}]C_I &= (nf^*(h) - \sum_{i=1}^{k} e_i)(-k_X + 2e_I) \\ &= (nf^*(h) - \sum_{i=1}^{k} e_i) \cdot (-f^*(k_{\min}) - \sum_{i=1}^{k} e_i + 2\sum_{j=1}^{\ell} e_j) \\ &= -nf^*(hk_{\min}) + \text{ constant for } n \geq n_1\end{aligned}$$

and since $\mathrm{kod}(X) \geq 1$ a multiple of $k_{\min}$ can be represented by an effective divisor so that $hk_{\min} > 0$).

From here on we assume $n \geq \max(n_0, n_1)$.

(iii) g_1 is good for C_I and C_{II}. This follows from (ii).

(iv) $C_I^2 = C_{II}^2 = k_X^2$, hence $C_I^2 = C_{II}^2 = k_X^2 = 2e(X)+3\tau(X)$ by the signature formula (Theorem I.3.1).

It follows that the expected dimensions of $S^{g_1,\mu_1}(C_I)$ and $S^{g_1,\mu}(C_{II})$ are both zero.

In order to apply (SW 3) we have to find the effective divisors M with $2M = K_X + C_I$. Visibly $M = E_I$. Since $H^1(E_i, \mathcal{O}_X(E_i)) = 0$ we conclude by (SW 3) that $W(g_1, C_I) \equiv W(g_1, C_{II}) \equiv 1 \,(\mathrm{mod} 2)$.

We come to the final argument. By way of the given diffeomorphism between X and the rational surface Y we transpose the Hitchin metric on Y, considered as a Riemannian metric, to X and call this transposed metric g_2. We label the components of the positive cone $\mathcal{C}_X$ by setting $\mathcal{C}^+ = \{x \in \mathcal{C}_X \mid xf^*(k_{\min}) > 0\}$. This is possible since $k_{\min} \neq 0$ and $k_{\min}^2 \geq 0$ as we observed earlier. We choose a generator $[\omega_{g_2}]$ of the line of self-dual harmonic forms with respect to g_2 such that $[\omega_{g_2}] \in \mathcal{C}^+$ and claim: $[\omega_{g_1}]$ and $[\omega_{g_2}]$ belong either to the same chamber with respect to C_I or to the same chamber with respect to C_{II} (or both). If not, then $[\omega_{g_2}]C_I \geq 0$ and $[\omega_{g_2}]C_{II} \geq 0$ by property (ii) above. Since there is an orthogonal decomposition $H^2(X, \mathbb{R}) = f^*H^2(X_{\min}, \mathbb{R}) \oplus \mathbb{R}e_1 \oplus \cdots \oplus \mathbb{R}e_k$, we have for suitable $\lambda_i \in \mathbb{R}$ and $\omega \in H^2(X_{\min}, \mathbb{R})$:

$$[\omega_{g_2}] = f^*(\omega) + \Sigma\lambda_i e_i \; .$$

Then $[\omega_{g_2}]C_I \geq 0$ and $[\omega_{g_2}]C_{II} \geq 0$ yield

$$-f^*(\omega)f^*(k_{\min}) - \sum_{i=1}^{\ell} \lambda_i + \sum_{j=\ell+1}^{k} \lambda_j \geq 0 \; ,$$

$$-f^*(\omega)f^*(k_{\min}) + \sum_{i=1}^{\ell} \lambda_i - \sum_{j=\ell+1}^{k} \lambda_j \geq 0 \; ,$$

hence $f^*(\omega)f^*(k_{\min}) \leq 0$, and $[\omega_{g_2}]f^*(k_{\min}) \leq 0$, which contradicts $[\omega_{g_2}] \in \mathcal{C}^+$.

This contradiction shows in particular that g_2 is good for at least one of the two, C_I or C_{II}. Therefore, by Theorem 9.1 either $W(g_2, C_I) = 0$

or $W(g_2, C_{II}) = 0$. On the other hand, by (SW 4) we have $W(g_1, C_I) = W(g_2, C_I)$ or $W(g_1, C_{II}) = W(g_2, C_{II})$. But we have seen that $W(g_I, C_I) \equiv W(g_1, C_{II}) \equiv 1 \pmod 2$. So our assumption $\mathrm{kod}(X) \geq 0$ leads to a contradiction and the theorem is proved. □

(9.3) **Corollary.** *If an algebraic surface X (considered as an oriented manifold) is diffeomorphic to a surface Y with* $\mathrm{kod}(Y) = -\infty$, *then* $\mathrm{kod}(X) = -\infty$ *too.*

Proof. In view of Theorem 9.2 and the classification theorem (Theorem VI.1.1) it only remains to be shown that if Y is a blown-up ruled surface of base genus at least one then the same holds for X. We shall show that this is true even if we only assume that X and Y are homeomorphic (as oriented manifolds).

Now $\chi(X) = \chi(Y) \leq 0$, since χ is a topological invariant. By VI, Table 10 this excludes already the case $\mathrm{kod}(X) = 2$. The case $\mathrm{kod}(X) = 0$ is easily excluded too, because (blown up) tori and K 3-surfaces have $p_g = 1$, whereas (blown-up) bi-elliptic surfaces and Enriques surfaces have torsion in H_1. In the case $\mathrm{kod}(X) = 1$, we know (VI, Table 10) that $\chi(X) \geq 0$, so we only have to prove that a blown-up $\mathbb{P}^1$-bundle over an elliptic curve is not homeomorphic to a properly elliptic surface.

$\mathrm{Alb}(X)$ is an elliptic curve. In view of the fact that X does not admit maps onto curves of genus ≥ 2 and the universal property of the Albanese torus, we know that the Albanese map f is connected. It factorises through the minimal model $X_{\min}$:

$$X \longrightarrow X_{\min} \xrightarrow{\tilde{f}} \mathrm{Alb}(X)$$

(with the curved arrow $f: X \to \mathrm{Alb}(X)$ below)

If the general fibre of $\tilde{f}$ is a rational curve, then so is every fibre (apply Prop. V.4.3) and X is a blown-up elliptic ruled surface, hence $\mathrm{kod}(X) = -\infty$.

Now let the general fibre of $\tilde{f}$ have genus at least 1. The surface $X_{\min}$ is a minimal elliptic surface, so $c_1^2(X_{\min}) = 0$. Since $\chi(X_{\min}) = \chi(X) = \chi(Y) = 0$, we find $c_2(X_{\min}) = 0$. According to Prop. III.11.4 and Remark III.11.5 this leaves only two possibilities: either $X_{\min}$ is a bundle with fibre a curve of genus ≥ 2, or it is an elliptic fibre space with no singular fibres but multiple elliptic curves. In both cases the universal covering is a 4-cell. For the second case we need Prop. III.9.1. Hence $\pi_2(X_{\min}) = 0$, and $\pi_2(X)$ is the direct sum of k copies of $\mathbb{Z}$, where k is the number of blown-ups, needed to come from $X_{\min}$ to X. On the other hand, the second homotopy group of an elliptic ruled surface blown-up ℓ times is the direct sum of $\ell + 1$ copies of $\mathbb{Z}$. This follows by inspecting the exact homotopy sequence of the elliptic fibration. However, since $e(X_{\min}) = e(Y_{\min}) = 0$ we have $k = \ell$. The contradiction $k = k + 1$ shows that $\mathrm{kod}(X) = -\infty$ and X is indeed a blown-up elliptic ruled surface. □

The Van de Ven conjecture says that the Kodaira dimension is a differentiable invariant. This generalizes the previous theorem and it has meanwhile been proved. In fact one knows the following:

(9.4) **Theorem.** *Let X and X' be two kählerian complex surfaces which are diffeomorphic by way of an orientation preserving diffeomorphism. Then $P_n(X) = P_n(X')$ for all $n \geq 1$. In particular* $\mathrm{kod}(X) = \mathrm{kod}(X')$.

A (partially incomplete) proof of this theorem using Donaldson invariants was first published by Brussee ([Brus]). Using Seiberg-Witten invariants M. Dürr recently found an elegant proof [Dürr].

For the non-kählerian case one has:

(9.5) **Theorem.** *The plurigenera of a non-kählerian complex surface (and hence also the Kodaira dimension) are determined by the isomorphism class of its fundamental group.*

For a proof see [F-M-M], Theorem S3, [Lö] and [Dürr], Theorem 0.0.3 where the assumption of the existence of a global spherical shell can be dropped.

Bibliography

[Ad] Andreotti, A.: On the complex structure of a class of simply connected manifolds, in *Algebraic geometry and topology*, Princeton Univ, Press (1957), 58–77.

[Ae] Aeppli, A.: Modifikationen von reellen und komplexen Mannigfaltigkeiten, Comment. Math. Helv. **32** (1957), 217–301.

[Al] Allcock, D.: The period lattice for Enriques surfaces, Math. Ann. **317** (2000), 483–488.

[A-K] Altmann, A., S. Kleiman: *Introduction to Grothendieck duality theory*, Lect. Notes Math., Springer Verlag, Berlin etc. **146** (1970).

[An60] Artin, M.: On Enriques surfaces, thesis, Harvard (1960).

[An62] Artin, M.: Some numerical criteria for contractability of curves on algebraic surfaces, Am. J. Math. **84** (1962), 485–496.

[An66] Artin, M.: On isolated rational singularities of surfaces, Am. J. Math. **88** (1966), 129–136.

[Ar] Arakělov, S. Ju: Families of algebraic curves with fixed degeneracies, Math. USSR Izv. **5** (1971), 1277–1302.

[A-S] Atiyah, M. F. and I. Singer: The index of elliptic operators III, Ann. Math. **87** (1968), 546–604.

[As-Ko] Ashikaga, T. and K. Konno: Global and local properties of pencils of algebraic curves, *Algebraic geometry, Azumino 2000*, Preprint (2001).

[At57] Atiyah, M. F.: Vector bundles over an elliptic curve, Proc. Lond. Math. Soc. **7** (1957), 414–452.

[At58] Atiyah, M. F.: On analytic surfaces with double points, Proc. Royal Soc., Ser. A. **247** (1958), 237–244.

[At69] Atiyah, M. F.: The signature of fibre bundles, in *Global analysis*, Univ. Tokyo Press, Tokyo (1969), 73–84.

[Ba75] Barlet D.: Espace analytique réduit des cycles analytiques complexes compacts, in *Sém. Norguet, Vol. I*, Springer Lect. Notes in Math. **482** (1975), 1–158.

[Ba78] Barlet D.: Convexité de l'espace des cycles, Bull. Soc. Math. France **106** (1978), 373–397.

[Bau] Bauer, I.: *Surfaces with* $K^2 = 7$ *and* $p_g = 4$, Mem. Amer. Math. Soc. **152** (2001).

[Bau-C] Bauer, I., F. Catanese : Some new surfaces with $p_g = q = 0$, To appear in the Fano UMI-memorial volume (2003).

[**Bad**] Bădescu, L.: *Algebraic surfaces*, Springer Verlag, Berlin etc. (1999), translation of: *Suprafeţe algebriche*, Edit. Acad, rep. Soc. Romania, Bucarest (1981).

[**Bar**] Barth, W.: Two projective surfaces with many nodes, admitting the symmetries of the icosahedron, J. Algebr. Geom. **5** (1996), 173–186.

[**Bat99**] Batyrev, V.: Birational Calabi-Yau manifolds have equal Betti numbers, in *New trends in algebraic geometry, Warwick 1996*, ed. K. Hulek et. al., Cambr. Univ. Press (1999), 1–11.

[**B-B**] Baily, W. L. and A. Borel: Compactification of arithmetic quotients of bounded symmetric domains, Ann. Math. **84** (1966), 442–528.

[**Bc**] Borcea, C.: Some remarks on deformations of Hopf manifolds, Rev. Roum. Math. Pures Appl. **26** (1981), 1287–1294.

[**B-C**] Bombieri, E. and F. Catanese: The tricanonical map of a surfaces with $K^2 = 2$, $p_g = 0$, in *C.P.Ramanujam, a tribute*, Springer Verlag, Berlin etc. (1978), 279–290.

[**Be78**] Beauville, A.: *Complex algebraic surfaces,* Cambridge Univ. Press (1983) (translation of *Surfaces algébriques complexes*, Astérisque **54** Soc. Math. France, Paris (1978)).

[**Be79**] Beauville, A.: L'application canonique pour les surfaces de type général, Invent. Math. **55** (1979), 121–140.

[**Be83**] Beauville, A.: Surfaces K3, Sém. Bourbaki **609** (1982/83).

[**Be85**] Beauville, A.: Toutes les surfaces K3 sont kähleriennes, Astérisque **126** (1985), 137–140.

[**Be88**] Beauville, A.: Annulation du H^1 et systèmes paracanoniques sur les surfaces, J. f. reine u. angew. Math. **388** (1988), 149–157.

[**Be99**] Beauville, A.: Counting rational curves on K3 surfaces, Duke Math. J. **97** (1999), 99–108.

[**B-F**] Bagnera, G. and M. de Franchis: Le superficie algebrique, de quali ammettono una rappresentazione parametrica mediante funzione iperellip-tiche di due argomenti, Mem. Soc. Ital. delle Scienze II Ser. **15** (1908), 251–343.

[**B-Fl**] Braun, R. and G. Fløystad: A bound for the degree of smooth surfaces in $\mathbb{P}^4$ not of general type, Compos. Math. **93** (1994), 211–229.

[**B-Ha**] Borel, A. and A. Haefliger: La classe d'homologie fondamentale d'un espace analytique, Bull. Soc. Math. Fr. **89** (1961), 461–513.

[**B-H-H**] Barthel, G., F. Hirzebruch and T. Höfer: *Geradenkonfigurationen und Algebraische Flächen,* Aspects of Mathematics **D4**, Vieweg (1987).

[**B-Hu**] Bombieri, E. and D. Husemoller: Classification and embeddings of surfaces, in *Algebraic geometry Arcata 1974*, A.M.S. Proc. Symp. Pure Math. **29** (1975), 329–420.

[**B-J**] Bröcker, T. and K. Jänich: *Einführung in die Differentialtopologie,* Heidelberger Taschenbücher, Springer Verlag, Berlin etc. **143** (1973).

[**Bl**] Blanchard, A.: Sur les variétés analytiques comlexes, Ann. Sci. Éc. N. S. **73** (1958), 157–202.

[B-L] Bryan, J. and N. C. Leung: The enumerative geometry of K 3 surfaces and modular forms, J. Amer. Math. Soc. **13** (2000), 371–410.

[B-M77] Bombieri, E. and D. Mumford: Enriques classification II, in *Complex analysis and algebraic geometry*, Iwanami Shoten & Cambr. Univ. Press (1977), 23–42.

[B-M76] Bombieri, E. and D. Mumford: ibid. III, Invent. Math. **35** (1976), 197–232.

[Bog76] Bogomolov, F. A.: Classification of surfaces of class VII_0 with $b_2 = 0$, Math. USSR Izv. **40** (1976), 255-269 (1977).

[Bog77] Bogomolov, F.: Families of curves on a surface of general type, Sov. Math. Dokl. **18** (1977), 294–1297.

[Bog79] Bogomolov, F.: Holomorphic tensors and vector bundles on projective varieties, Math. USSR Izv. **13** (1979), 499–555.

[Bog83] Bogomolov, F. A.: Surfaces of class VII_0 and affine geometry, Math. USSR Izv. **21** (1983), 31–74.

[Bom70] Bombieri, E.: The pluricanonical map of a complex surface, in *Several Complex Variables I, Maryland 1970*, Springer Lect. Notes in Math. **155** (1970), 35–87 .

[Bom73] Bombieri, E.: Canonical models of surfaces of general type, Publ. Math. IHES **42** (1973), 171–219.

[Bor] Borel, A.: Compact Clifford-Klein forms of symmetric spaces, Topology **2** (1963), 111–222.

[Borc] Borcherds, R.: The moduli space of Enriques surfaces and the fake Monster Lie superalgebra, Topology **35** (1996), 699–710.

[Bou71] Bourbaki, N.: *Topologie générale, Chap. 1–4,* Hermann, Paris (1971).

[Bou59] Bourbaki, N.: *Algèbre, Chap. 1–3,* Hermann, Paris (1959).

[Bou68] Bourbaki, N.: *Groupes et algèbres de Lie, Chap 4, 5, 6, Hermann, Paris,* (1968).

[B-P] Barth, W. and C. Peters: Automorphisms of Enriques surfaces, Invent. Math. **73** (1983), 383–411.

[Br] Berger, M: *Géométrie 1. Action des groupes, espaces affines et projectifs,* Cedic/Nathan, Paris (1977).

[B-R] Burns, D. and M. Rapoport: On the Torelli problem for Kählerian K3-surfaces, Ann. Sc. ENS. **8** (1975), 235–274.

[Bra] Brand, R.: *Parallelizability of compact complex surfaces,* thesis, Leiden (1980).

[Bre] Bredon, G.: *Sheaf theory,* McGraw-Hill, New York (1967).

[Bri] Brieskorn, E.: Singular elements of semi simple algebraic groups, Actes Congr. Int. Math. Nice **2** (1970), 279–284.

[Bro] Brotherton, N.: Some paralellizable manifolds not admitting a complex structure, Bull. Lond. Math. Soc. **10** (1978), 303–304.

[Brod] Brody, R.: Compact manifolds in hyperbolicity, Trans. Am. Math. Soc. **235** (1978), 213–219.

[Brus] Brussee, R.: The canonical class and the C^∞ properties of Kähler surfaces, New York J. Math. **2** (1996)103–146 (electronic).

[**B-So**] Beltrametti, M. and A. J. Sommese: Zero cycles and k-th order embeddings of smooth projective varieties, Symp. Math. **XXXII** (1991), 33–48.

[**Bu**] Burniat, P.: Sur les surfaces de genre $P_{12} > 0$, Ann. Math. Pura Appl. **71** (1966), 1–24.

[**Buch**] Buchdahl, N.: On compact Kähler surfaces, Ann. Inst. Fourier **49** (1999), 287–302.

[**B-S**] Bănică and C., Stănăşilă: *Algebraic methods in the global theory of complex spaces,* John Wiley & Sons, New York (1976).

[**Bus**] Del Busto, G. F.: Bogomolov's instability and Kawamata-Viehweg vanishing, J. Alg. Geom. **4** (1995), 693–700.

[**Bw84**] Barlow, R.: Some new surfaces with $p_g = 0$, Duke Math. J. **51** (1984), 889–904.

[**Bw85**] Barlow, R.: A simply connected surface of general type with $p_g = 0$, Invent. Math. **79** (1985), 293–302.

[**Cam**] Campedelli, L.: Sopra alcuni piani doppi notevoli con curva di diramazioni del decimo ordine, Atti Acad. Naz. Lincei **15** (1932), 536–542.

[**Car**] Cartan, H.: Quotients d'un espace analytique par un groupe d'automorphismes, in *Algebraic geometry and topology*, Princeton Univ.Pr. (1957), 90–102.

[**Cas98**] Castelnuovo, G.: Sulle superficie di genere zero, Mem. Soc. Ital. Sci. II Ser **10** (1898), 102–126=*Mem. Scelte*, Zanichelli, Bologna (1937), 307–334.

[**Cas05**] Castelnuovo, G.: Sulle superficie avente il genere aritmetico negativo, Rend. Circ. Math. Palermo **20** (1905)= *Mem. Scelte*, Zanichelli, Bologna (1937), 501–507.

[**Cat81a**] Catanese, F.: Pluricanonical mappings of surfacs with $K^2 = 1, 2$, $q = p_g = 0$, in *CIME 1977 Algebraic Surfaces*, Liguori, Napoli (1981), 249–266.

[**Cat81b**] Catanese, F.: On a class of surfaces of general type. In: *CIME 1977 Algebraic surfaces*, Liguori, Napoli (1981), 269–284.

[**Cat81c**] Catanese, F.: Babbage's conjecture, contact of surfaces, symmetrical determinantal varieties and applications, Invent. Math. **63** (1981), 433–466.

[**Cat84**] Catanese, F.: On the moduli space of surfaces of general type, J. Diff. Geom. **19** (1984), 483–516.

[**Cat86**] Catanese, F.: Connected components of moduli spaces, J. Diff. Geom **24** (1986), 395–399.

[**Cat87a**] Catanese, F.: Canonical rings and "special" surfaces of general type, Proc. Sympos. Pure Math. **46** (I), Amer. Math. Soc. Providence, RI (1987), 175–194.

[**Cat87b**] Catanese, F.: Automorphisms of rational double points and moduli spaces of surfaces of general type, Comp. Math. **61** (1987), 81–102.

[**Cat89**] Catanese, F.: Everywhere non reduced moduli spaces, Invent. Math. **98** (1989), 293–310.

[Cat90] Catanese, F.: Footnotes to the theorem of Reider, in *Algebraic geometry*, Springer Verlag, Berlin etc. , Lect. Notes Math. **1417** (1990), 67–74.

[Cat92] Catanese, F.: Chow varieties, Hilbert schemes, and moduli spaces of surfaces of general type, J. Alg. Geom. **1** (1992), 561–590.

[Cat97] Catanese, F.: Homological algebra and algebraic surfaces, Proc. Sympos. Pure Math. **62**, Part 1, Amer. Math. Soc. Providence, RI (1997), 3–56.

[Cat98] Catanese, F.: Singular bidouble covers and the construction of interesting algebraic surfaces. In *Algebraic geometry: Hirzebruch 70 (Warsaw, 1998)*, Contemp. Math **241** (1999), Amer. Math. Soc., Providence, RI, 97–120.

[Cat00] Catanese, F.: Fibred surfaces, varieties isogenous to a product and related moduli spaces, Am. J. Math. **122** (2000), 1–14.

[Cat01] Catanese, F.: Moduli spaces of surfaces and real structures, Preprint (2001).

[C-C91] Catanese, F. and C. Ciliberto: Surfaces with $p_g = q = 1$, Symposia Math. **32** (1991), 49–79.

[C-C93] Catanese, F. and C. Ciliberto: Symmetric products of elliptic curves and surfaces of general type with $p_g = q = 1$, J. Alg. Geom. **2** (1993), 389–411.

[C-C-M] Catanese, F., C. Ciliberto and M. Mendes Lopes: On the classification of irregular surfaces of general type with nonbirational bicanonical map, Trans. Am. Math. Soc. **350** (1998), 275–308.

[C-E] Castelnuovo, G. and F. Enriques: Die algebraichen Fächen vom Gesichtspunkt der birationalen Transformationen aus, Enzycl. Math. Wiss. $\mathbf{III}_2$ (1914), 677–768.

[C-F-H-R] Catanese, F., M. Franciosi, K. Hulek and M. Reid: Embeddings of curves and surfaces, Nagoya Math. J. **154** (1999), 185–220.

[C-G] Craighero P.C. and R. Gattazzo: Quintic surfaces of $\mathbb{P}^3$ having a nonsingular model with $q = p_g = 0$, $P_2 \neq 0$, Rend. Sem. Mat. Univ. Padova **91** (1994), 187–198.

[Chen87] Chen, Z.: On the geography of surfaces (simply connected minimal surfaces with positive index), Math. Ann. **277** (1987), 141–164.

[Chen91] Chen, Z.: The existence of algebraic surfaces with pre-assigned Chern numbers, Math. Z. **206** (1991), 241–254.

[Chm] Chmutov, S. V.: Examples of projective surfaces with many singularities., J. Algebraic Geom. **1** (1992), 191–196.

[Ci-F-M] Ciliberto, C., P. Francia, and M. Mendes Lopes: Remarks on the bicanonical map for surfaces of general type, Math. Z. **224** (1997), 137–166.

[Cil97] Ciliberto, C.: The bicanonical map for surfaces of general type, Proc. Sympos. Pure Math. **62**, Part 1, Amer. Math. Soc. Providence, RI (1997), 57–84.

[**Ci-M00**] Ciliberto, C. and M. Mendes Lopes: On regular surfaces of general type with $p_g = 3$ and non-birational bicanonical map, J. Math. Kyoto Univ. **40** (2000), 79–117.

[**Ci-M01**] Ciliberto, C. and M. Mendes Lopes: On surfaces with $p_g = q = 2$, Preprint math AG **0012211** (2001).

[**Ck**] Cook, M.: An improved bound for the degree of smooth surfaces in $\mathbb{P}^4$ not of general type, Compos. Math. **102** (1996), 141–145.

[**C-K**] Chow, W. L. and K. Kodaira: On analytic surfaces with two independent meromorphic functions, Proc. Nat. Ac. Sci. USA **38** (1952), 319–325.

[**Cl**] Clemens, H. et. al.: *Seminar on degeneration of algebraic varieties,* Inst. Adv. Study, Princeton (1969).

[**C-LeB**] Catanese, F. and C. LeBrun: On the scalar curvature of Einstein manifolds, Math. Res. Lett. **4** (1997), 843–854.

[**Cl-K-M**] Clemens, H., J. Kollár and S. Mori: *Higher dimensional complex geometry,* Astérisque **166** (1988).

[**Co-D**] Cossec, F. and I. Dolgachev: *Enriques surfaces I,* Birkh. Verlag (1989).

[**C-P**] Catanese, F. and R. Pignatelli: On simply connected Godeaux surfaces, in *Complex analysis and algebraic geometry* (T. Peternell and F.-O. Schreyer, eds.), Walter de Gruijter, Berlin (2000), 117–153.

[**C-V**] Calabi, E. and E. Vesentini: On compact, locally symmetric Kähler manifolds, Ann. Math. **71** (1960), 472–507.

[**Da**] Dabrowski, K.: Moduli spaces for Hopf surfaces, Math. Ann. **259** (1982), 201–226.

[**Dd**] Dold, A.: *Lectures on algebraic topology*, Grundl. Math. Wiss. **200**, Springer Verlag, Berlin etc. (1972).

[**Deb**] Debarre, O.: Inégalités numériques pour les surfaces de type général, Bull. Soc. Math. Fr. **110** (1982), 319–346; Addendum, Bull. Soc. Math. Fr. **111**, (1983), 301.

[**Del68**] Deligne, P.: Théorème de Lefschetz et critère de dégénérescence de suites spectrales, Publ. Math. IHES **35** (1968), 107–126.

[**Del71**] Deligne, P.: Théorie de Hodge II, Publ. Math. IHES **40** (1971), 5–58.

[**Dem**] Demazure, M.: A, B, C, D, E, F etc. in *Séminaire sur les singularités de surfaces*, Springer Lect. Notes in Math. **777** (1982), 222–227.

[**Dema**] Demailly, J.-P.: Regularization of closed positive currents and intersection theory, J. Alg. Geom. **1** (1992), 361–409.

[**Dema97**] Demailly, J.-P.: *Complex analytic and algebraic geometry,* `http://www-fourier.ujf-grenoble.fr/~demailly/manuscripts/agbook.ps.gz` (1997).

[**Dema-P**] Demailly J.-P. and M. Paun: Numerical characterization of the Kähler cone of a compact Kähler manifold, math.AG/ **0105176** (2001).

[**D-E-S**] Decker, W., L. Ein and F.-O. Schreyer: Construction of surfaces in $\mathbb{P}_4$, J. Alg. Geom. **2** (1993)185–237.

[**Dl84**] Dloussky, G.: Structure des surfaces de Kato, Mem. Soc. Math. Fr., suppl. Bull. Soc. Math. France **112** (1984).

[Dl88] Dloussky, G.: Une construction élémentaire des surfaces d'Inoue-Hirzebruch, Math. Ann. **280** (1988), 663–682.

[D-M] Deligne, P. and D. Mumford: The ireducibility of the space of curves of given genus, Publ. Math. IHES **36** (1969), 75–109.

[Dol81] Dolgachev, I.: Algebraic surfaces with $q = p_g = 0$, in *CIME "Algebraic surfaces"*, Liguori Edit., Naples (1981), 97–216.

[Dol84] Dolgachev, I.: On automorphisms of Enriques surfaces, Invent. Math. **76** (1984), 163–177.

[Dol96] Dolgachev, I.: Mirror symmetry for lattice polarized $K3$ surfaces, in *Algebraic geometry, 4*, J. Math. Sci. **81** (1996), 2599–2630.

[Dol-W] Dolgachev, I. and C. Werner: A simply connected numerical Godeaux surface with ample canonical class, J. Alg. Geom. **8** (1999), 737–764. Erratum ibid. **10** (2001), 397.

[Don86] Donaldson, S.K.: Irrationality and the h-cobordism conjecture, J. Diff. Geom. **24** (1986), 275–341.

[Don90] Donaldson, S. K.: Polynomial invariants for smooth 4-manifolds, Topology **29** (1990), 257–315.

[D-O-T] Dloussky, G., K. Oeljeklaus and M. Toma: Class VII_0 surfaces with b_2 curves, math. CV **021010** (2002).

[D-S] Decker, W. and F.-O. Schreyer: Non-general type surfaces in $\mathbb{P}_4$: Some remarks on bounds and constructions, J. Symbolic Comput. **29** (2000), 545–582.

[Dürr] Dürr, M.: *Seiberg-Witten theory and the C^∞-classification of complex surfaces,* Dissertation, Zürich (2002).

[DV] Du Val, P.: On surfaces whose canonical system is hyperelliptic, Canadian J. Math. **4** (1952), 204–221.

[Dy] Douady, A.: Le problème des modules pour les variétés analytiques complexes, Sém. Bourbaki 1964/65 Exp. **277** (1965).

[Ebe] Ebeling, W.: An example of two homeomorphic nondiffeomorphic complete intersection surfaces, Invent. Math. **99** (1990), 651–654.

[Ek] Ekedahl, T.: Canonical models of surfaces of general type in positive characteristic, Publ. Math. IHES **67** (1988), 97–144.

[En81] Enoki, I.: Surfaces of class VII_0 with curves, Tôhoku Math. J. **33** (1981), 453–492.

[En82] Enoki, I.: On compactifiable complex analytic surfaces, Invent. Math. **67** (1982), 189–212.

[Enr14] Enriques, F.: Sulla classificazione delle superficie algebriche e particolarmente sulle superficie di genere $p^1 = 1$ (2 notes), Atti Acc. Lincei V Ser. **23**^{1} (1914).

[Enr49] Enriques, F.: *Le superfichie algebriche,* Zanicelli, Bologna (1949).

[E-P] Ellingrud, G. and C. Peskine: Sur les surfaces lisses de $\mathbb{P}_4$, Invent. Math. **95** (1989), 1–11.

[Es] Esnault, H.: Classification des variétés de dimension 3 et plus (d'après T. Fujita, S. Iitaka, Y. Kawamata, K. Ueno, E. Viehweg), Sém. Bourbaki exp. **568** (1980/81), Springer Lect. Notes in Math. **901** (1981), 111–131.

[F] Fischer, G.: *Complex analytic geometry,* Springer Lect. Notes in Math. **901** (1976).

[F-F] Fischer, G., O. Forster: Ein Endlichkeitssatz für Hyperflächen auf kompakten komplexen Raümen, J. f. reine u. angew. Math. **306** (1979), 88–93.

[F-G] Fischer W. and H. Grauert: Lokal triviale Familien kompakter komplexen Mannigfaltigkeiten, Nachr. Akad. Wiss. Göttingen, II. Math. Phys. Kl. (1965), 89–94.

[F-G-vS] Fantechi, B., L. Göttsche and D. van Straten: Euler number of the compactified Jacobian and multiplicity of rational curves., J. Alg. Geom. **8** (1999), 115–133.

[F-K] Forster, O. and K. Knorr: Ein Beweis des Grauertschen Bildgarbensatzes nach Ideen von B. Malgrange, Manuscr. Math. **5** (1971), 19–44.

[F-L] Fulton B. and R. Lazarsfeld.: Connectivity and its applications to algebraic geometry, in *Algebraic geometry,* Lect. Notes Math. **862**, Springer Verlag, Berlin etc. (1981), 26–92.

[F-M88a] Friedman, R. and J. Morgan: On the diffeomorphism type of certain algebraic surfaces I, II, J. Diff. Geom. **27** (1988), 297–398.

[F-M88b] Friedman, R. and J. Morgan: Algebraic surfaces and 4-manifolds: some conjectures and speculations, Bull. Am. J. Math. **18** (1988), 1–19.

[F-M90] Friedman, R. and J. Morgan: Complex versus differentiable classification of algebraic surfaces, Topology Appl. **32** (1989), 135–139.

[F-M94] Friedman, R. and J. Morgan: *Smooth four-manifolds and complex surfaces,* Springer Verlag, Berlin, etc. (1994).

[F-M97] Friedman, R. and J. Morgan: Donaldson and Seiberg-Witten invariants of algebraic surfaces, in *Algebraic geometry—Santa Cruz 1995*, Proc. Sympos. Pure Math., Amer. Math. Soc., Providence, RI **62-1** (1997), 85–100.

[F-M-M] Friedman, R., Moishezon, B. and J. Morgan: On the C^∞ invariance of the canonical classes of certain algebraic surfaces, Bull. Amer. Math. Soc. (N.S.) **17** (1987), 283–286.

[Fr] Friedman, R.: A new proof of the global Torelli theorem for K3 surfaces, Ann. Math. **120** (1984), 237–270.

[Fra] Francia, P.: On the base points of the bicanonical system, Symposia Math. **32** (1991), 141–150.

[Frm] Freedman, M.: The topology of 4-manifolds, J. Diff. Geo. **17** (1982), 357–454.

[Fu] Fujiki, A.: Kählerian normal complex spaces, Tôhoku Math. J. **35** (1983), 101–118.

[Fult] Fulton, W.: *Intersection theory,* Springer Verlag, Berlin etc. (1984).

[Ga77] Gauduchon, P.: Le théorème d'excentricité nulle, C. R. Ac. Sci. Paris, Sér. A. **285** (1977), 387–390.

[**Ga85**] Gauduchon, P.: Les métriques standard sur une surface complexe compacte a premier nombre de Betti pair, Astérisque **126** (1985), 129–135.

[**Gat**] Gathmann, A.: The number of plane conics 5-fold tangent to a given curve, math. AG **0202002** (2002).

[**Ge**] Geppert, H.: Die Klassifikation algebraischer Flächen, Jahresbericht deutscher Mathematiker Verein **41** (1932), 18–39.

[**Gf**] Griffiths, P.: Periods of integrals on algebraic manifolds I, Am. J. Math. **90** (1968), 568-626, ibid. II, Am. J. Math. **90** (1968), 805–865.

[**G-H78a**] Griffiths, P. and J. Harris: *Principles of algebraic geometry,* John Wiley & Sons, New York etc. (1978).

[**G-H78b**] Griffiths, P. and J. Harris: Residues and zero-cycles on algebraic varieties, Ann. Math. **108** (1978), 461–505.

[**Gi**] Gieseker, D.: Global moduli for surfaces of general type, Invent. Math. **43** (1977), 233–282.

[**Gk57a**] Grothendieck, A.: Sur la classification des fibrés holomorphes sur la sphère de Riemann, Am. J. M. **79** (1957), 121–138.

[**Gk57b**] Grothendieck, A.: Sur quelques points d'algèbre homologique, Tôhoku Math. J. **9** (1957), 119–221.

[**Go**] Godement, R.: *Théorie des faisceaux,* Hermann, Paris (1964).

[**Gr60**] Grauert, H.: Ein Theorem der analytischen Garbentheorie und die Modulraüme komplexer Strukturen, Publ. Math. IHES **5** (1960), 460–472.

[**Gr62**] Grauert, H.: Über Modifikationen und exzeptionelle analytischen Mengen, Math. Ann. **146** (1962), 331–368.

[**G-R55**] Grauert, H. and R. Remmert: Zur Theorie der Modifikationen I, stetige und eigentliche Modifikationen komplexen Räume, Math. Ann. **129** (1955), 274–296.

[**G-R58**] Grauert, H. and R. Remmert: Komplexer Raümen, Math. Ann. **136** (1958), 245–318.

[**G-R71**] Grauert, H. and R. Remmert: *Analytischen Stellenalgebren,* Grundl. Math. Wiss **176**, Springer Verlag, Berlin etc. (1971).

[**G-R77**] Grauert, H. and R. Remmert: *Theorie der Steinschen Räumen,* Grundl. Math. Wiss **227**, Springer Verlag, Berlin etc. (1977).

[**Gr-Gf**] Green, M. and Ph. Griffiths: Two applications of algebraic geometry to entire holomorphic mappings, in *The Chern Symposium 1979,* Proc. Int. Symp. Berkely, CA, 1979, Springer Verlag, Berlin, etc. (1980), 41–74.

[**Gts**] Göttsche, L.: The Betti numbers of the Hilbert scheme of points on a smooth projective surface, Math. Ann. **86** (1990), 193–207.

[**Gx31**] Godeaux, L.: Sur une surface algébrique de genre zéro et de bigenre deux, Atti Acad. Naz. Lincei **14** (1931), 479–481.

[**Gx49**] Godeaux, L.: Sur la construction de surfaces non rationelles de genres zéro, Bull. Acad. Roy. Belg. **45** (1949), 688–693.

[**G-V**] Geer, G. van der and A. Van de Ven: On the minimality of certain Hilbert modular surfaces, in *Complex analysis and algebraic geometry,* Iwanami Shoten & Cambr. Univ. Press (1977), 137-150.

[**G-W**] Gruenberg, K. and A. J. Weir: *Linear geometry,* Graduate texts **49**, Springer Verlag, Berlin etc. (1977).

[**G-Z**] Geer, G. van der and D. Zagier: The Hilbert modular group for the field $\mathbb{Q}(\sqrt{13})$, Invent. Math. **42** (1978), 93–134.

[**Ha66**] Hartshorne, R.: Ample vector bundles, Publ. Math. IHES **29** (1966), 319–350.

[**Ha77**] Hartshorne, R.: *Algebraic geometry,* Springer Verlag, Berlin, etc. (1987).

[**He**] Helgason, S.: *Differential geometry and symmetric spaces,* Acad. Press, New York, London (1962).

[**Hik64**] Hironaka, H.: Resolution of singularities of an algebraic variety over a field of characteristic zero I, II, Ann. Math. **79** (1964), 109–326.

[**Hik71**] Hironaka, H.: Bimeromorphic smoothing of complex analytic spaces, Preprint, Math. Inst. Warwick (1971).

[**Hir53**] Hirzebruch, F.: Über vierdimensionale Riemannsche Flächen mehrdeutiger analytischen Funktionen von zwei komplexen Veranderlichen, Math. Ann. **126** (1953), 1–22.

[**Hir58a**] Hirzebruch, F.: Automorphe Formen und der Satz von Riemann-Roch, in *Symp. Intern. Topol. Algéb. 1956 , México*, Univ. de México (1958), 129–144.

[**Hir58b**] Hirzebruch, F.: Komplexe Mannigfaltigkeiten, in *Proc. Int. Congr. Math. 1958*, Cambridge Univ. Press (1960).

[**Hir66**] Hirzebruch, F.: *Topological methods in algebraic geometry,* Grundlehren der math. Wiss. **131** 3rd edition, Springer Verlag, Berlin etc. (1966).

[**Hir73**] Hirzebruch, F.: Hilbert modular surfaces, Enseign. Math. **19** (1973), 183–281.

[**Hir78**] Hirzebruch, F.: Modulflächen und Modulkurven zur symmetrischen Hilbertschen Modulgruppe, Ann. Sc. ENS. **11** (1978), 101–166.

[**Hir83**] Hirzebruch, F.: Arrangements of lines and algebraic surfaces, in *Arithmetic and geometry*, Progr. Math., Birkhäuser Verlag **36** (1983), 113–140.

[**Hir84**] Hirzebruch, F.: Chern numbers of algebraic surfaces, an example, Math. Ann. **266** (1984), 351–356.

[**Hit**] Hitchin, N.: On the curvature of rational surfaces, Proc. Symp. Pure Math., Stanford **27** (1975).

[**H-K**] Hirzebruch, F. and K. Kodaira: On the complex projective spaces, Journ. de Math. Pures Appl. **36** (1957), 201–216.

[**H-L**] Harvey, R. and H. B. Lawson Jr.: An intrinsic characterization of Kähler manifolds, Invent. Math. **74** (1983), 261–295.

[**Hol80**] Holzapfel, R. P.: A class of minimal surfaces in the unknown region of surfaces geography, Math. Nachr. **98** (1980), 221–232.

[**Hol81**] Holzapfel, R. P.: Invariants of arithmetic ball quotients, Math. Nachr. **103** (1981), 117–153.

[**Hol86**] Holzapfel, R. P.: Chern numbers of algebraic surfaces-Hirzebruch's examples are Picard-modular surfaces, Math. Nachr. **126** (1986), 255–273.

[**Hop48**] Hopf, H.: Zur Topologie der komplexen Mannigfaltigkeiten, in *Studies and Essays presented to R. Courant*, Interscience, New York (1948), 167–185.

[**Hop54**] Hopf, H.: Schlichte Abbildingen und lokale Modifikationen 4-dimensionaler komplexer Mannigfaltigkeiten, Comm. Math. Helv. **29** (1954), 132–156.

[**Hor75**] Horikawa, E.: On deformations of quintic surfaces, Invent. Math. **31** (1975), 43–85.

[**Hor76**] Horikawa, E.: Algebraic surfaces of general type with small c_1^2, I, Ann. Math. **104** (1976), 357–387. II Invent. Math. **37** (1976), 121–155. III Invent. Math. **47** (1978), 209–248. IV Invent. Math. **50** (1979), 103–128, V J. Fac. Sci. Univ. Tokyo, Sec. A. Math. **283** (1981), 745–755.

[**Hor77**] Horikawa, E.: Surjectivity of the period map of K 3-surfaces of degree 2, Math. Ann. **228** (1977), 113–146.

[**Hor78**] Horikawa, E.: On the periods of Enriquess surfaces, I, Math. Ann. **234** (1978), 73–108. II, Math. Ann. **235** (1978), 217–246.

[**Hu**] Hulek, K: The Horrocks-Mumford bundle. In *Vector bundles in algebraic geometry (Durham 1993)*, Cambr. Univ. Press, Cambridge, Lond. Math. Soc. Lect. Notes **208** (1995)139–177.

[**H-V74**] Hirzebruch, F. and A. Van de Ven: Hilbert modular surfaces and the classification of algebraic surfaces, Invent. Math. **23** (1974), 1–29.

[**H-V79**] Hirzebruch, F. and A. Van de Ven: Minimal Hilbert modular surfaces with $p_g = 3$, $K^2 = 2$, Am. J. Math. **101** (1979), 132–148.

[**H-W**] Hardy, G. and E. Wright: *Introduction to the theory of numbers*, 4th edition, Clarendon Press, Oxford (1965).

[**H-Z**] Hirzebruch, F. and D. Zagier: Classification of Hilbert modular surfaces, in *Complex analysis and algebraic geometry*, Iwanami-Shoten, Tokyo (1977), 43–77.

[**Ii69**] Iitaka, S.: Deformations of compact complex surfaces I, in *Global analysis*, Princeton Un. Press. (1969), 267–272.

[**Ii70**] Iitaka, S.: Deformations of compact complex surfaces II, J. Math. Soc. Japan **22** (1970), 247–261.

[**Ii71**] Iitaka, S.: Deformations of compact complex surfaces III, J. Math. Soc. Japan **23** (1971), 692–705.

[**Il81**] Illiev, V.: Surfaces with $p_g = 3$ and $K^2 = 3$, I, Serdika **6** (1980), 352–362, ibid. II Serdika **7**(1981), 390–395.

[**Il84**] Illiev, V.: A note on certain surfaces, Bull. Lond. Math. Soc. **16** (1984), 135–138.

[**Il88**] Illiev, V.: On the canonical ring of Horikawa surfaces, I., Trans. Am. Math. Soc. **309** (1988), 309–324.

[**In74**] Inoue, M.: On surfaces of class VII_0, Invent. Math. **24** (1974), 269–310.

[**In77**] Inoue, M.: New surfaces with no meromorphic functions, Proc. Int. Congress, Vancouver (1974), 423–426. II, in *Complex analysis and algebraic geometry*. Iwanami-Shoten, Tokyo (1977), 91–106.

[**In94**] Inoue, M.: Some new surfaces of general type, Tokyo J. Math. **17** (1994), 295-319.

[**Is83**] Ishida, M.: The irregularity of Hirzebruch's examples of surfaces of general type with $c_1^2 = 3c_2$, Math. Ann. **262** (1983), 407–420.

[**Is88**] Ishida, M.: An elliptic surface coverd by Mumford's fake projective plane, Tôhoku Math. J. **40** (1988), 367-398.

[**J**] James, I. D.: On Witt's theorem for unimodular quadratic forms, Pac. J. Math. **26** (1968), 303–316.

[**J-Y**] Jost, J. and S.-T. Yau: Harmonic mappings and Kähler manifolds, Math. Ann. (1983), 145–166.

[**Ka78a**] Kato, Ma.: Topology of Hopf surfaces, J. Math. Soc. Japan **27** (1978), 222–238.

[**Ka78b**] Kato, Ma.: Compact complex manifolds containing "global spherical shells" I, in *Int. symp. alg. geometry Kyoto 1977*, Kinokuniya, Tokyo (1978), 45–84.

[**Kaw**] Kawamata, Y.: A generalization of Kodaira-Ramanujam's vanishing theorem, Math. Ann. **261** (1982), 43–46.

[**Kaw99**] Kawamata, Y.: Deformations of canonical singularities. , J. Amer. Math. Soc. **12** (1999), 85–92.

[**Ke**] Keum, J.: Some new surfaces of general type with $p_g = 0$, Thesis, University of Ann Arbor (1988).

[**Ke-L**] Keum, J and Y. Lee: Fixed locus of an involution acting on a Godeaux surface, Math. Proc. Cambridge Philos. Soc. **129** (2000), 205–216.

[**Ki-V**] Kiehl, R. and J.-L Verdier: Ein einfacher Beweis des Kohärenzsatzes von Grauert, Math. Ann. **195** (1971), 24–50.

[**K-M**] Kollár, J. and S. Mori: Classification of three-dimensional flips, J. Amer. Math. Soc. **5** (1992), 533–703.

[**K-N**] Kobayashi, S. and K. Nomizu: *Foundations of differential geometry I,* John Wiley & Sons, New York (1963). Ibid. II (1969).

[**K-N-S**] Kodaira, K., L. Nirenberg and D. Spencer: On the existence of deformations of complex structures, Ann. Math. **68** (1958), 450–459.

[**Ko54**] Kodaira, K.: On Kähler varieties of restricted type (an intrinsic characterisation of algebraic varieties), Ann. Math. **60** (1954), 28–48.

[**Ko60**] Kodaira, K.: On compact complex analytic surfaces I, Ann. Math. **71** (1960), 111–152. II, Ann. Math. **77** (1963), 563–626. III, Ann. Math. **78** (1963), 1–40.

[**Ko63**] Kodaira, K.: On stability of compact submanifolds of complex manifolds, Ann. Math. **85** (1963), 79–94.

[**Ko66**] Kodaira, K.: On the structure of complex analytic surfaces I, Am. J. Math. **86** (1966), 751–798. II, Am. J. Math. **88** (1966), 682–721. III, Am. J. Math. **90** (1969), 55–83. IV ibid., 1048–1066.

[**Ko67**] Kodaira, K.: A certain type of irregular algebraic surfaces, J. Anal. Math. **19** (1967), 207–215.

[**Ko68**] Kodaira, K.: Pluricanonical systems on algebraic surfaces of general type, J. Math. Soc. Japan **20** (1968), 170–192.

[Ko70] Kodaira, K.: On homotopy K 3-surfaces, in *Essays on Topology and Related Topics*, Springer Verlag , Berlin etc. (1970), 58–69.

[Kob] Kobayashi, S.: *Hyperbolic manifolds and holomorphic mappings,* Marcel Dekker, New York (1970).

[Kob-O] Kobayashi, S., T. Ochiai: Mappings into compact complex manifolds with negative first Chern class, J. Math. Soc. Japan **23** (1971), 137–148.

[Kon86] Kondō, S.: Enriques surfaces with finite automorphism groups, Japan. J. Math. (New Ser.) **12** (1986), 191–282.

[Kon94] Kondō, S.: The rationality of the moduli space of Enriques surfaces, Compositio Math. **9** (1994), 159–173.

[Kon98] Kondō, S.: Niemeier lattices, Mathieu groups, and finite groups of symplectic automorphisms of $K3$ surfaces. With an appendix by Shigeru Mukai., Duke Math. J. **92** (1998), 593–603.

[Kot] Kotschik, D.: On manifolds homeomorphic to $\mathbb{C}P^2\#8\overline{\mathbb{C}P^2}$, Invent. Math. **95** (1989), 591–600.

[Kr] Krasnov, V.: Compact complex surfaces without meromorphic functions, Math. Zametki **17** (1975), 119–122.

[Kul] Kulikov, V.: Degenerations of K 3-surfaces and Enriques surfaces, Ivz. Akad. Nauk. SSSR, Ser. Math. **41** (1977), 1008–1042.

[Kur] Kurke, H.: *Vorlesungen über algebraische Flächen,* Teubner, Leipzig, Teubner Texte zur Mathematik **43** (1982).

[Ks] Kas, A.: On deformations of a certain type of irregular algebraic surface, Am. J. Math. **90** (1968), 789–804.

[K-S58] Kodaira, K. and D. Spencer: A theorem of completeness for complex analytic fibre spaces, Acta Math. **100** (1958), 281–294.

[K-S62] Kodaira, K. and D. Spencer: On deformations of complex analytic structures III. Stability theorems for complex analytic structures, Ann. Math. **75** (1962), 536–577.

[La] Lamotke, K.: Die homologie isolierter Singularitäten, Math. Z. **143** (1975), 27–44.

[Lama] Lamari, A.: Courants kählériens et surfaces compactes, Ann. Inst. Fourier **49** (1999), 263–285.

[Lan] Lang, S.: Hyperbolic and diophantine analysis, Bull. Amer. Math. Soc. **14** (1986), 159–205.

[Langer] Langer, A.: Pluricanonical systems on surfaces with small K^2, Internat. J. Math. **11** (2000), 379–392.

[Le] Lefschetz, S.: *L'analysis situs et la géométrie algébrique,* Gauthier-Villars, Paris (1924).

[LeJ] Lee, J.: Family Gromov-Witten invariants for Kähler surfaces, math. SG **0209402** (2002).

[Lee] Lee, Y.: Bicanonical pencil of a determinantal Barlow surface, Trans. Am. Math. Soc. **353** (2001), 893–905.

[Lel] Lelong, P.: Intégration sur un ensemble analytique complexe, Bull. Soc. Math. France **85** (1957), 239–262.

[**Lel68**] Lelong, P.: Plurisubharmonic functions and positive differential forms, Gordon & Breach (1968).

[**Lev**] Levine, M.: Pluri-canonical divisors on Kähler manifolds, Invent. Math. **74** (1983), 293–303.

[**Lg79**] Lang, W. E.: Quasi-elliptic surfaces in characteristic 3, Ann. Sc. Norm. Sup. **12** (1979), 473-520.

[**Lg80**] Lang, W. E.: On the Euler number of algebraic surfaces in characteristic p, Am. J. Math. **102** (1980), 511–516.

[**Lg83**] Lang, W. E.: Examples of surfaces of general type with vectorfields, in *Arithmetic and geometry II*, Progress in Math. **36**, Birkhäuser Verlag (1983), 167–174.

[**Li**] Lipman, J.: Introduction to resolution of singularities, in *Algebraic geometry, Arcata 1974*, Proc. Symp. Pure Math. **24** (1975), 187–230.

[**Liu**] Liu, Ai-Ko: Family blow up formula, admissable graphs and the enumeration of singular curves, I, J. Diff. Geom. **56** (2000),381–579.

[**Liv**] Livné, R. A.: On certain covers of the universal elliptic curve, Thesis, Harvard (1981).

[**Lj**] Łojasiewicz, S.: Triangulation of semianalytic sets, Ann. Sc. Norm. Pisa, III Ser. **18** (1964), 449–474.

[**L-T**] Lübke, M. and A. Teleman: *The Kobayashi-Hitchin correspondence,* World Scientific, Singapore etc. (1995).

[**Lo**] Looijenga, E.: A Torelli theorem for Kähler-Einstein K 3-surfaces, in *Geometry sympos. Utrecht 1980*, Springer Lect. Notes in Math. **894** (1981), 107–112.

[**Lö**] Lönne, M.: The homotopy invariance of the plurigenera of non-kählerian compact complex surfaces, Schriftenreihe des Forschungsschwerpunkts Komplexe Mannigfaltigkeiten **213** (1994), Preprint.

[**L-P**] Looijenga, E. and C. Peters: Torelli theorems fo Kähler K 3-surfaces, Compos. Math. **42** (1981), 145–186.

[**Lu-Mi**] Lu, S.S.Y. and Y. Miyaoka: On hyperbolicity and the Green-Griffiths conjecture for surfaces, in *Geometric complex analysis*, ed. J. Noguchi et al., World Scientific Publ. Co. (1996), 401–408.

[**Maa**] Maaß, H.: *Siegel modular forms and Dirichlet series,* Springer Lect. Notes in Math. **216** (1971).

[**Man94**] Manetti, M.: On some components of moduli space of surfaces of general type, Compos. Math. **92** (1994), 285–297.

[**Man01**] Manetti, M.: On the moduli space of diffeomorphic algebraic surfaces, Invent. Math. **143** (2001), 29–76.

[**Mar**] Martens, H.: A new proof of Torelli's theorem, Ann. Math. **78** (1963), 107–111.

[**Maru75**] Maruyama, M.: Stable vector bundles on an algebraic surface, Nagoya Math. J. **58** (1975), 25–68.

[**Maru78**] Maruyama, M.: Moduli of stable sheaves, II, J. Math. Kyoto Univ. **18** (1978), 557–614.

[**May**] Mayer, A.: Families of K 3-surfaces, Nagoya Math. J. **48** (1972), 1–17.

[McQ] McQuillan, M.: Diophantine approximations and foliations, Publ. Math. IHES **87** (1998).

[M-H] Milnor, J. and D. Husemoller: *Symmetric bilinear forms,* Erg. der Math.**73**, Springer Verlag, Berlin etc. (1973).

[Mi74a] Miyaoka, Y.: Extension theorems for Kähler metrics, Proc. Japan Acad. **50** (1974), 407–410.

[Mi74b] Miyaoka, Y.: Kähler metrics on elliptic surfaces, Proc. Japan Acad. **50** (1974), 533–536.

[Mi76] Miyaoka, Y.: Tricanonical maps of numerical Godeaux surfaces, Invent. Math. **34** (1976), 99–111.

[Mi77a] Miyaoka, Y.: On numerical Campedelli surfaces, in *Complex analysis and algebraic geometry*, Iwanami-Shoten & Cambr.Univ.Press (1977), 112–118.

[Mi77b] Miyaoka, Y.: On the Chern numbers of surfaces of general type, Invent. Math. **42** (1977), 225–293.

[Mi80] Miyaoka, Y.: On the Mumford-Ramanujam vanishing theorem on a surface, in *Journées de géométrie algébrique d'Angers (juillet 1979) Algebraic geometry Angers 1979*, Sijthoff-Noordhoff (1980), 239–248.

[Mi84] Miyaoka, Y.: The maximum number of quotient singularities on surfaces with given numerical invariants, Math.Ann. **268** (1984), 159–172.

[Mil58] Milnor, J.: On simply connected 4-manifolds, in *Symp. Int. Top. Alg. México*, Univ. Mexico (1958), 122–128.

[Mil63] Milnor, J.: *Morse theory,* Princeton Univ. Press Ann. Math. Studies **51** (1963).

[Mir] Miranda, R.: On canonical surfaces of general type with $K^2 = 3\chi - 10$, Math. Z. **198** (1988), 83–94.

[Miy] Miyanishi, M.: *Open algebraic surfaces,* American Mathematical Society , Providence, RI (2001).

[M-K] Morrow, J. and K. Kodaira: *Complex manifolds,* Holt, Rinehart and Winston, New York etc. (1971).

[ML] Mendes Lopes, R.: The degree of the generators of the canonical ring of surfaces of general type with $p_g = 0$, Arch. Math. **69** (1997), 435–440.

[M-M] Mori, S. and S. Mukai: The uniruledness of the moduli space of curves of genus 11, Springer Lect. Notes in Math. **1016** (1983), 334–353.

[M-N] Mukai, S. and Y. Namikawa: Automorphisms of Enriques surfaces which act trivially on the cohomology groups, Invent. Math. **77** (1984), 383–397.

[Mo78] Mostow, G. D.: Existence of a non-arithmetic lattice in $SU(2,1)$, Proc. Nat. Acad. Sci. USA **86** (1978), 3029–3033.

[Mo80] Mostow, G. D.: On a remarkable class of polyhedra in complex hyperbolic space, Pac. J. Math. **86** (1980), 171–276.

[Mor] Morgan, J.: The Seiberg-Witten equations and applications to the topology of smooth four-manifolds, Princeton, NJ: Princeton Univ. Press, Mathematical Notes **44** (1996).

[**M-P98**] Mendes Lopes, M. and R. Pardini: Irregular canonical double covers, Nagoya Math. J. **152** (1998), 203–230.

[**M-P00**] Mendes Lopes, M. and R. Pardini: Triple canonical surfaces of minimal degree, Internat. J. Math. **11** (2000), 553–587.

[**M-P01a**] Mendes Lopes, M. and R. Pardini: A connected component of the moduli space of surfaces of general type with $p_g = 0$, Topology **40** (2001), 977–991.

[**M-P01b**] Mendes Lopes, M. and R. Pardini: The bicanonical map of surfaces with $K^2 \geq 7$, Bull. London Math. Soc. **33** (2001), 977–991, part II, to appear.

[**M-P02a**] Mendes Lopes, M. and R. Pardini: Enriques surfaces with eight nodes, Math. Z. (2002), to appear.

[**M-P02b**] Mendes Lopes, M. and R. Pardini: A survey on the bicanonical map of surfaces with $p_g = 0$ and $K^2 \geq 2$, (2002), preprint.

[**M-S**] Milnor, J. and J. Stasheff: *Characteristic classes,* Princeton Univ. Press, Ann. Math. Studies **75** (1974).

[**M-Si**] Mostow, G.D. and Y. T. Siu: A compact Kaehler surface not covered by the ball, Ann. Math. **112** (1980), 321–360.

[**M-T**] Moishezon, B. and M. Teicher: Simply connected algebraic surfaces of positive index, Invent. Math **89** (1987), 601–644.

[**Mu61**] Mumford, D.: Topology of normal singularities and a criterion for simplicity, Publ. Math. IHES **36** (1961), 229–246.

[**Mu62**] Mumford, D.: The canonical ring of an algebraic surface, Ann. Math. **76** (1962), 612–615.

[**Mu66**] Mumford, D.: *Lectures on curves on an algebraic surface,* Princeton University Press, Annals of Math. Studies **59** (1966).

[**Mu67**] Mumford, D.: Pathologies III, Am. J. Math. **89** (1967), 94–104.

[**Mu69**] Mumford, D.: Enriques classification of surfaces in charactistic p I, in *Global analysis*, Princeton Princeton U.P. (1969), 325–339.

[**Mu77**] Mumford, D.: *Stability of projective varieties,* Monographie de l'Enseignement Mathématique **24** (1977) 74pp.

[**Mu79**] Mumford, D.: An algebraic surface with K ample $K^2 = 9$, $p_g = q = 0$, Am. J. Math. **101** (1979), 233–244.

[**Muk**] Mukai, S.: Finite groups of automorphisms of $K3$ surfaces and the Mathieu group, Invent. Math. **94** (1988), 183–221.

[**Na84**] Nakamura, I.: On surfaces of class VII_0 with curves, Invent. Math. **78** (1984), 393–444.

[**Na90**] Nakamura, I.: On surfaces of class VII_0 with curves. II, Tôhoku Math. J. **42** (1990), 475-516.

[**Naie94**] Naie, D.: Surfaces d'Enriques et une construction de surfaces de type général avec $p_g = 0$, Math. Z. **215** (1994), 269–280.

[**Naie99**] Naie, D.: Numerical Campedelli surfaces cannot have the symmetric group as the algebraic fundamental group, J. London Math. Soc. (2) **59** (1999), 813–827.

[Ni75] Nikulin, V.: On Kummer surfaces, Izv. Akad. Nauk SSR, ser.Math. **39** (1975), 278–293.

[Ni79] Nikulin, V.: Finite groups of automorphisms of Kählerian $K3$ surfaces. (Russian), Trudy Moskov. Mat. Obshch. **38** (1979), 75–137.

[Ni84] Nikulin, V.: Description of automorphism groups of Enriques surfaces. (Russian), Dokl. Akad. Nauk SSSR **277** (1984), 1324–1327.

[Ny] Nygaard, N.: Closedness of regular 1-forms on algebraic surfaces, Ann. Sc. ENS. **12** (1979), 33–45.

[O-P] Oort, F. and C. Peters: A Campedelli surface with torsion group $\mathbb{Z}/2$, Indag. Math. **43** (1981), 399–407.

[O-S-S] Okonek, C., M. Schneider and H. Spindler: *Vector bundles on complex projective spaces,* Birkhäuser Verlag (1980).

[O-T] Okonek, C. and A. Teleman: Recent developments in Seiberg-Witten theory and complex geometry, in *Several complex variables (Berkeley, CA, 1995–1996)*, Cambridge Univ. Press, Cambridge, Math. Sci. Res. Inst. Publ. **37** (1999), 391–428.

[O-T-Z] Oeljeklaus, K., M. Toma and D. Zaffran: Une caractérisation des surfaces d'Inoue-Hirzebruch, Ann. Inst. Fourier **51** (2001), 1243–1257.

[O-V86] Okonek, C. and A. Van de Ven: Stable vector bundles and differentiable structures on certain elliptic surfaces, Invent. Math. **86** (1986), 357–370.

[O-V89] Okonek, C. and A. Van de Ven: Γ-type invariants associated to $PU(2)$-bundles and the differentiable structure of Barlow's surface, Invent. Math. **95** (1989), 601–614.

[O-V90] Okonek, C. and A. Van de Ven: Stable bundles, instantons and C^∞-structures on algebraic surfaces in *Several complex variables VI,* W. Barth, R. Narasimhan (Eds), Springer Verlag, Berlin etc. (1990), 197–249.

[Pa] Paršin, A. N.: Algebraic curves over function fields I, Math. USSR, Izv. **2** (1968), 1145–1176.

[Per81] Persson, U.: On Chern invariants of surfaces of general type, Comp. Math. **43** (1981), 3–58.

[Per87] Persson, U.: An introduction to the geography of surfaces of general type. In: Proc. Symp.Pure Math. **(46-1)** *Algebraic geometry, Bowdoin 1985*, A.M.S., Providence RI (1987), 195–220.

[Pet76] Peters, C.: On two types of surfaces of general type with vanishing geomeric genus, Invent. Math. **32** (1976), 33–47.

[Pet77] Peters, C.: On certain examples of surfaces with $p_g = 0$ due to Burniat, Nagoya Math. J. **66** (1977), 109–119.

[Pi] Piatečkii-Shapiro, I. I.: *Geometry of classical domains and automorphic functions,* Fizmatgiz, Moscow (1961) French trans., Dunod, Paris (1966).

[Pi-S] Piatečkii-Shapiro, I. I. and I. R. Šafarevič: A Torelli theorem for algebraic surfaces of type K 3, Math. USSR, Izv. **5** (1971), 547–588.

[**P-P**] Persson, U. and H. Pinkham: Degeneration of surfaces with trivial canonical bundle, Ann. Math. **113** (1981), 45–66.

[**P-P-X**] Persson, U., C. Peters and G. Xiao, Gang: Geography of spin surfaces, Topology **35** (1996), 845–862.

[**P-S**] Picard, E. and G. Simart: *Théorie des fonctions algébriques de deux variables indépendents, I, II,* Gauthier-Villars, Paris (1987, 1906).

[**Ra**] Raynaud, M.: Familles de fibrés vectoriels sur une surface de Riemann, Sem. Bourbaki **316** (1966).

[**Ram**] Ramanujam, C. P.: Remarks on the Kodaira vanishing theorem, J. Ind. Math. Soc. **36** (1972), 41–51.

[**Re**] Rego, C. J.: The compactified Jacobian, Ann. sc. ENS **13** (1980), 211–223.

[**Rei77**] Reid, M.: Bogomolov's theorem $c_1^2 \leq 4c_2$ in *Proc. Int. Symp. Alg. geometry Kyoto, 1977*, Kinokuniya, Tokyo (1977), 623–642.

[**Rei78**] Reid, M.: Surfaces with $p_g = 0$, $K^2 = 1$, J. Fac. Sc., Univ. Tokyo **25** (1978), 75–92.

[**Rei79**] Reid, M.: π_1 for surfaces with small K^2, in *Algebraic geometry, Copenhagen, 1978*, Springer Lect. Notes in Math. **732** (1979), 534–544.

[**Reider**] Reider, I.: Vector bundles of rank 2 and linear systems on algebraic surfaces, Ann. Math. **127** (1988), 309–316.

[**Rem56**] Remmert, R.: Meromorphe Funktionen in kompakten komplexen Räumen, Math. Ann. **132** (1956), 277–288.

[**Rem57**] Remmert, R.: Holomorphe und meromorphe Abbildungen komplexer Räume, Math. Ann. **133** (1957), 328–370.

[**Rh**] Rham, G. de: *Variétés différentiables,* Hermann, Paris (1960).

[**Ro**] Rohlin, V. A.: A new result in the theory of 4-dimensional manifolds (Russian), Dokl. Akad. Nauk. SSSR **84** (1952), 221–224.

[**R-R**] Ramis, J. P. and G. Ruget: Complexe dualisant et theoremes de dualité en géométrie analytique complexe, Publ. IHES. **38** (1971), 77–91.

[**R-R-V**] Ramis, J .P., G. Ruget and J.-L. Verdier: Dualité relative en géométrie analytique complexe, Invent. Math. **13** (1971), 261–283.

[**R-S**] Remmert, R. and K. Stein: Über die wesentlichen Singularitäten analytischer Mengen, Math. Ann. **126** (1953), 263–306.

[**R-V60**] Remmert, R. and A. Van de Ven: Zwei Sätze über die komplex projektive Ebene, Nieuw Arch. Wisk. **3** (1960), 147–157.

[**R-V63**] Remmert, R. and A. Van de Ven: Zur Funktionentheorie homogener komplexer Mannigfaltigkeiten, Topology **2** (1963), 137–157.

[**Saf**] Şafareviç, I. R. et al.: *Algebraic surfaces,* Proc. of the Steklov Institute **75** (1965). Translation A. M. S., Providence RI (1967).

[**Sai**] Saint-Donat, B.: Projective models of K3-surfaces, Am. J. Math. **96** (1972), 602–639.

[**Sak80**] Sakai, F.: Semi-stable curves on algebraic surfaces and logarithmic pluricanonical maps, Math. Ann. **254** (1980), 89–120.

[**Sak82**] Sakai, F.: Enriques classification of normal Gorenstein surfaces, Am. J. Math. **104** (1982), 1233–1241.

[Sak83] Sakai, F.: D-dimensions of algebraic surfaces and numerically effective divisors, Comp. Math. **48** (1983), 101–118.

[Sak84] Sakai, F.: Weil divisors on normal surfaces, Duke Math. J. **51** (1984), 877–888.

[Sal89] Salvetti, M.: On the number of non-equivalent differentiable structures on 4-manifolds, Manuscr. Math. **63** (1989), 157–171.

[Sal91] Salvetti, M.: A lower bound for the number of differentiable structures on 4-manifolds, Boll. U.M.I. (7) **5-A** (1991), 33–40.

[Schae] Schaefer, H. H.: *Topological vector spaces*, Second edition, with M. P. Wolff, Springer Verlag, Berlin etc. (1999).

[Se55a] Serre, J.-P.: Un théorème de dualité, Comm. Math. Helv. **29** (1955), 9–26.

[Se55b] Serre, J.-P.: Faisceaux algébriques cohérents, Ann. Math. **61** (1955), 197–278.

[Se56a] Serre, J.-P.: Sur la dimension homologique des anneaux et des modules noethériens, in *Proc. Int. Symp. Alg. Number Theory, Tokyo and Nikko 1955,* Kasai, Tokyo, (1956), 175–189.

[Se56b] Serre, J.-P.: Géometrie algébrique et géométrie analytique, Ann. Inst. Fourier **6** (1956), 1–42.

[Se59] Serre, J.-P.: *Groupes algébriques et corps de classes,* Hermann, Paris (1959).

[Se73] Serre, J.-P.: *A course in arithmetic,* Springer Verlag, Berlin etc. (1973).

[Se79] Serre, J.-P.: *Local fields,* Springer Verlag, Berlin etc. (1979).

[Sei87a] Seiler, W. K.: Global moduli for polarised elliptic surfaces, Comp. Math. **62** (1987), 187–213.

[Sei87b] Seiler, W. K.: Global moduli for polarised elliptic surfaces with a section, Comp. Math. **62** (1987), 169–186.

[SemP] Séminaire Palaiseau 1978: *Première classe de Chern et courbure de Ricci: preuve de la conjecture de Calabi,* Astérisque **58** (1978).

[Ser92] Serrano, F.: Fibred surfaces and moduli, Duke Math. J. **67** (1992), 407–421.

[Ser96] Serrano, F.: Isotrivial fibred surfaces, Ann. Mat. Pura Appl. (4) **171** (1996), 63–81.

[Sev] Severi, F.: Some remarks on the topological classification of surfaces, in *Studies presented to R. von Mises*, Acad. Press, New York (1954).

[Sha76] Shah J.: Surjectivity of the period map in the case of quartic surfaces and sextic double planes, Bull. AMS. **82** (1976), 716–718.

[Sha80] Shah J.: A complete moduli space for K3-surfaces of degree 2, Ann. Math. **112** (1980), 485–510.

[Sha81] Shah J.: Degenerations of K3-surfaces of degree 4, Trans. Am. Math. Soc. **263** (1981), 271-308.

[Shav] Shavel, I. M.: A class of algebraic surfaces of general type constructed from quaternion algebras, Pac. J. Math. **76** (1978), 221–245.

[ShB91a] Shepherd-Barron, N. I.: Unstable vector bundles and linear systems on surfaces in characteristic p, Invent. Math. **106** (1991), 243–262.

[**ShB91b**] Shepherd-Barron, N. I.: Geography for surfaces of general type in positive characteristic, Invent. Math. **106** (1991), 263–274.

[**Shi**] Shioda, T.: The period map of abelian surfaces, J. Fac. S. Univ.Tokyo **25** (1978), 47–59.

[**Shif**] Shiffmann, B: Extension of positive line bundles and meromorphic maps, Invent. Math. **15** (1972), 332–347.

[**Siu83**] Siu, Y. T.: Every K3-surface is Kähler, Invent. Math. **73** (1983), 139–150.

[**Siu87**] Siu, Y. T.: Strong rigidity for Kähler manifolds and the construction of bounded holomorphic functions in *Discrete groups and analysis*, R. Howe ed., Birkh. Verlag (1987), 124–151.

[**Siu98**] Siu, Y. T.: Invariance of plurigenera, Invent. Math. **134** (1998), 661–673.

[**S-I**] Shioda, T. and H. Inose: On singular K3-surfaces in *Complex analysis and algebraic geometry*, Cambr. Univ. Press (1977), 117–136.

[**S-M**] Shioda, T. and N. Mitani: Singular abelian surfaces and binary quadratic forms, in *Classification of algebraic varieties and complact complex manifolds, Mannheim, 1974*, Springer Lect. Notes in Math. **412** (1974), 259–287.

[**So79**] Sommese, A.: Hyperplane sections of projective surfaces I -The adjunction mapping, Duke Math. J. **46** (1979), 377–401.

[**So84**] Sommese, A.: On the density of ratios of Chern numbers of algebraic surfaces, Math. Ann. **268** (1984), 207–222.

[**Sp**] Spanier, E.: *Algebraic topology,* Springer Verlag , Berlin etc. (1966).

[**St**] Stein, K.: Analytische Zerlegungen komplexer Räume, Math. Ann. **132** (1956), 63–93.

[**Sta**] Stagnaro, E.: On Campedelli branch loci, Ann. Univ. Ferrara Sez. VII (N.S.) **43** (1997), 1–26.

[**Ste**] Sterk, H.: Compactifications of the period space of Enriques surfaces. I, Math. Z. **207** (1991), 1–36, ibid. II. **220**(1995), 427–444.

[**S-T67**] Singer, I. and J. Thorpe: *Lecture notes on elementary topology and geometry,* Scott-Foresman and Co, Glenview (1967).

[**S-T69**] Singer, I. and J. Thorpe: The curvature of 4-dimensional Einstein spaces, in *Global analysis*, Princeton Univ. Press, Princeton (1969), 355–365.

[**Su**] Suwa, T.: On ruled surfaces of genus 1, J. Math. Soc. Jap. **21** (1969), 291–311.

[**Sup**] Supino, P.: A note on Campedelli surfaces, Geom. Dedicata **71** (1998), 19–31.

[**Sv**] Svarcman, O: Simple-connectedness of the factor spaces of the Hilbert-modular group, Funct. anal. appl. **8** (1974), 188–189.

[**Te**] Teleman, A.-D.: Projectively flat surfaces and Bogomolov's theorem on class VII_0-surfaces, Int. J. Math. **5** (1994), 253–264.

[Tj64] Tyurin, A. N.: On the classification of two-dimensional fibre bundles over an algebraic curve of arbitrary genus, Izv. Ak. Nauk. SSSR, Ser. Mat. **28** (1964), 21–52.

[Tj67] Tyurin, A. N.: Classification of vector bundles over an algebraic curve of arbitrary genus, Am. Math. Soc. Transl. **63** (1967), 245–279.

[To] Todorov, A.: Application of the Kaehler-Einstein-Calabi-metric to moduli of K3-surfaces, Invent. Math. **81** (1982), 251-304.

[T-Zh] Tsunoda, S. and De-Qi Zhang: Noether's inequality for noncomplete algebraic surfaces of general type, Publ. Res. Inst. Math. Sci. **28** (1992), 21–38.

[Ue75] Ueno, K.: *Classification theory of algebraic varieties and compact complex spaces,* Springer Lect. Notes in Math. **439** (1975).

[Ue76] Ueno, K.: A remark on automorphisms of Enriques surfaces, J. Fac. Sc. Univ. Tokyo, Sec. IA **23** (1976), 149–165.

[Ue77] Ueno, K.: Kodaira dimensions for certain fibre spaces, in *Complex analysis and algebraic geometry,* Cambr. Univ. Press (1977), 279–292.

[Ue80] Ueno, K.: Classification of algebraic manifolds, Int. Congr. Math. Helsinki 1978 (1980), 549–556.

[Ue86] Ue, M.: On the diffeomorphism types of elliptic surfaces with multiple fibers, Invent. Math. **84** (1986), 633–643.

[Ve66] Van de Ven, A.: On the Chern numbers of certain complex and almost complex manifolds, Proc. Natl. Acad. Sci. USA **55** (1966), 1624–1627.

[Ve76] Van de Ven, A.: On the Chern numbers of surfaces of general type, Invent. Math. **36** (1976), 285–293.

[Ve78a] Van de Ven, A.: On the Enriques classification of algebraic surfaces, in Sem. Bourbaki exp. 506 (1977), Springer Lect. Notes in Math. **677** (1978), 237–251.

[Ve78b] Van de Ven, A.: Some recent results on surfaces of general type, *Sem.Bourbaki 1976/77,* Springer Lect. Notes in Math **677** (1978), 155–165.

[Ve79] Van de Ven, A.: On the 2-connectedness of very ample divisors on a surface, Duke Math.J. **46** (1979), 403–407.

[Vie77] Viehweg, E.: Canonical divisors and the additivity of the Kodaira dimensions for a morphism of relative dimension one, Comp. Math. **135** (1977), 197–223.

[Vie82] Viehweg, E.: Vanishing theorems, J. f. reine u. angew. Math. **335** (1982), 1–8.

[Vie95] Viehweg, E.: *Quasi-projective moduli for polarized manifolds,* Springer Verlag, Berlin etc., Erg. Math. **30** (1995).

[Vin] Vinberg, E. B.: Discrete groups generated by reflections, Izv. Ak. Nauk. SSSR., Ser. Math. **35** (1971), 1083–1119.

[Wal] Walker, R.: *Algebraic curves,* Princeton Univ. Press, Princeton (1950).

[Wall] Wall, C. T. C: Diffeomorphisms of 4-manifolds, J. Lond. Math. Soc. **39** (1964), 131–140.

[**Wall86**] Wall, C. T. C.: Geometric structures on compact complex analytic surfaces, Topology **25** (1986), 119–153.

[**Wav**] Wavrik, J. J .: Obstructions to the existence of a space of moduli, in *Global analysis*, Princeton Univ.Press (1969), 403–413.

[**We94**] Werner, C.: A surface of general type with $p_g = q = 0$, $K^2 = 1$, Manuscr. Math. **84** (1994), 327–341.

[**We97**] Werner, C.: A four-dimensional deformation of a numerical Godeaux surface, Trans. Am. Math. Soc. **349** (1997), 1515–1525.

[**Weh**] Wehler, J.: Versal deformations of Hopf surfaces, J. f. reine u. angew. Math. **328** (1981), 22–32.

[**Wei71**] Weil, A.: *Variétés kählériennes,* Hermann, Paris (1971).

[**Wei80**] Weil, A.: *Œvres scientifiques (Collected papers)*, I–III, Springer Verlag, Berlin etc. (1980).

[**Wi78**] Wilson, P. M. H.: The behaviour of the plurigenera of surfaces under algebraic smooth deformations, Invent. Math. **47** (1978), 289–299.

[**Wi87**] Wilson, P. M. H.: Towards birational classification of algebraic varieties, Bull. London Math. Soc. **19** (1987), 1–48.

[**Wit**] Witten, E.: Monopoles and four-manifolds, Math. Res. Letters **1** (1994), 769–796.

[**W-R**] Wu Wen Tsun and G. Reeb: *Sur les espaces fibrés et les variétés feuilletés,* Hermann, Paris (1952).

[**Xi85a**] Xiao, Gang: *Surfaces fibrées en courbes de genre deux,* Lecture Notes in Mathematics **1137** (1985), Springer Verlag, Berlin etc.

[**Xi85b**] Xiao, Gang: L'irregularité des surfaces de type général dont le système canonique est composé d'un pinceau, Comp. Math. **56** (1985), 251–258.

[**Xi85c**] Xiao, Gang: Finitude de l'application bicanonique des surfaces de type general, Bull. Soc. Math. Fr. **113** (1985), 23–51.

[**Xi86a**] Xiao, Gang: Algebraic surfaces with high canonical degree, Math. Ann. **274** (1986), 473–483.

[**Xi86b**] Xiao, Gang: An example of hyperelliptic surfaces with positive index, Northeast. Math. J. **2** (1986), 255-257.

[**Xi87a**] Xiao, Gang: Fibered algebraic surfaces with low slope, Math. Ann. **276** (1987), 449–466.

[**Xi87b**] Xiao, Gang: Irregularity of surfaces with a linear pencil, Duke Math. J. **55** (1987), 597–602.

[**Xi87c**] Xiao, Gang: Hyperelliptic surfaces of general type with $K^2 < 4\chi$, Manuscr. Math. **57** (1987), 125–148.

[**Xi90**] Xiao, Gang: Degree of the bicanonical map of a surface of general type, Am. J. Math. **112** (1990), 713–736.

[**Y76**] Yau, S.-T.: Parallizable manifolds without complex structure, Topology **15** (1976), 51-54.

[**Y77**] Yau, S.-T.: Calabi's conjecture and some new results in algebraic geometry, Proc. Nat. Ac. Sc. USA **74** (1977), 1798–1799.

[Y78] Yau, S.-T.: On the Ricci curvature of a complete Kaehler manifold and the complex Monge-Ampère equation, Comm. Pure Appl. Math. **31** (1978), 339–411.

[Y-M] Yang, G. and M. Miyanishi: Surfaces of general type whose canonical map is composed of a pencil of genus 3 with small invariants, J. Math. Kyoto Univ. **38** (1998), 123–149.

[Y-Z] Yau, S.-T and E. Zaslow: BPS states, string duality, and nodal curves, Nuclear Phys. B **471** (1996), 503–512.

[Za58a] Zariski, O.: *Introduction to the problem of minimal models in the theory of algebraic surfaces,* Publ. Math. Soc. Japan **4** (1958).

[Za58b] Zariski, O.: On Castelnuovo's criterion $p_a = P_2 = 0$, Ill. J. Math. **2** (1958), 303–315.

[Za71] Zariski, O.: *Algebraic surfaces (second supplemented edition),* Springer Verlag, Heidelberg etc. (1971).

[Zh] Zhang, De-Qi: Noether's inequality for noncomplete algebraic surfaces of general type. II, Publ. Res. Inst. Math. Sci. **28** (1992), 679–707.

[Zu] Zucconi, F.: Numerical inequalities for surfaces with canonical map composed with a pencil, Indag. Math. **9** (1998), 459–476.

Notation

(see also p. 12)

Index

GPSR Compliance

The European Union's (EU) General Product Safety Regulation (GPSR) is a set of rules that requires consumer products to be safe and our obligations to ensure this.

If you have any concerns about our products, you can contact us on ProductSafety@springernature.com

In case Publisher is established outside the EU, the EU authorized representative is:

Springer Nature Customer Service Center GmbH
Europaplatz 3
69115 Heidelberg, Germany

Batch number: 10761599

Printed by Printforce, the Netherlands